Seventh Edition

HUMAN HEREDITY
Principles & Issues

Michael R. Cummings

Research Professor
Department of Biological Chemical and Physical Sciences
Illinois Institute of Technology
Chicago, Illinois

THOMSON
™
BROOKS/COLE

Australia • Canada • Mexico • Singapore • Spain
United Kingdom • United States

THOMSON

BROOKS/COLE

Executive Editor: Nedah Rose
Development Editor: Kari Hopperstead
Editorial Assistant: Sarah Lowe
Technology Project Manager: Travis Metz
Marketing Manager: Ann Caven
Marketing Assistant: Leyla Jowza
Advertising Project Manager: Kelley McAllister
Project Manager, Editorial Production: Belinda Krohmer
Art Director: Rob Hugel
Print/Media Buyer: Karen Hunt

Permissions Editor: Sarah Harkrader
Production Service: G & S Book Services
Text Designer: Gopa
Photo Researcher: Linda Sykes
Copy Editor: Kathleen Deselle
Cover Designer: Hiroko Chastain
Cover Image: Getty
Cover Printer: Phoenix Color Corp.
Compositor: G & S Book Services
Printer: Courier

For more information about our products, contact us at:
Thomson Learning Academic Resource Center
1-800-423-0563

For permission to use material from this text or product, submit a request online at **http://www.thomsonrights.com.** Any additional questions about permissions can be submitted by email to **thomsonrights@thomson.com.**

Thomson Brooks/Cole
10 Davis Drive
Belmont, CA 94002
USA

Asia
Thomson Learning
5 Shenton Way #01-01
UIC Building
Singapore 068808

Australia/New Zealand
Thomson Learning
102 Dodds Street
Southbank, Victoria 3006
Australia

Canada
Nelson
1120 Birchmount Road
Toronto, Ontario M1K 5G4
Canada

Europe/Middle East/Africa
Thomson Learning
High Holborn House
50/51 Bedford Row
London WC1R 4LR
United Kingdom

Latin America
Thomson Learning
Seneca, 53
Colonia Polanco
11560 Mexico D.F.
Mexico

Spain/Portugal
Paraninfo
Calle Magallanes, 25
28015 Madrid, Spain

To Maggie and Colin

About the Author

MICHAEL R. CUMMINGS received his Ph.D. in Biological Sciences from Northwestern University in 1968. His doctoral work, conducted in the laboratory of Dr. R. C. King, centered on ovarian development in *Drosophila melanogaster.* After a year on the faculty at Northwestern, he moved to a teaching and research position at the University of Illinois at Chicago. He is currently Research Professor in the Department of Biological Chemical and Physical Sciences at Illinois Institute of Technology.

At the undergraduate level, he has focused on teaching genetics, human genetics, and general biology for non-majors. About fifteen years ago, Dr. Cummings developed a strong interest in scientific literacy. He is now working to integrate the use of the Internet and World Wide Web into the undergraduate teaching of genetics and general biology and into textbooks. He has received awards given by the university faculty for outstanding teaching, has twice been voted by graduating seniors as the best teacher in their years on campus, and has received several teaching awards from student organizations.

His current research interests involve the molecular organization of the short-arm/centromere region of human chromosome 21. His laboratory is engaged in a collaborative effort to construct a physical map of this region of chromosome 21 to explore molecular mechanisms of chromosome interactions.

Dr. Cummings is the author and co-author of a number of widely used college textbooks, including *Biology: Science and Life, Concepts of Genetics, Genetics, A Molecular Perspective,* and *Essentials of Genetics.* He has also written articles on aspects of genetics for the *McGraw-Hill Encyclopedia of Science and Technology,* and has published a newsletter on advances in human genetics for instructors and students.

He and his wife, Lee Ann, are parents of two adult children, Brendan and Kerry, and have two grandchildren, Colin and Maggie. He is an avid sailor, enjoys reading and collecting books (biography, history), music (baroque, opera, and urban electric blues), eating the fine cuisine at Al's Beef, and is a long-suffering Cubs fan.

Contents

Chapter 5: Complex Patterns of Inheritance 94

Chapter 6: Cytogenetics: Karyotypes and Chromosome Aberrations 120

Chapter 7: Development and Sex Determination 150

Chapter 8: DNA Structure and Chromosomal Organization 178

Chapter 9: Gene Expression: From Genes to Proteins   198

Chapter 10: From Proteins to Phenotypes   218

Chapter 11: Mutation: The Source of Genetic Variation 244

Chapter 12: Genes and Cancer 268

Chapter 13: An Introduction to Cloning and Recombinant DNA   290

Chapter 14: Biotechnology and Society   314

Chapter 15: The Human Genome Project and Genomics   342

Chapter 16: Reproductive Technology, Gene Therapy, and Genetic Counseling   364

Chapter 17: Genes and the Immune System 382

Chapter 18: Genetics of Behavior 408

Chapter 19: Population Genetics and Human Evolution   434

Preface

WHEN THE FIRST EDITION of this book was published, recombinant DNA technology was just starting to have an impact on human genetics, and the beginnings of the Human Genome Project were several years in the future. As this edition goes to press, websites advertise the availability of genetic identification cards, television crime shows routinely solve cases using DNA forensic evidence, and genomic scans use microarrays carrying the human genome to screen for disorders that will not develop for decades. The supermarket and the pharmacy are filled with the products of biotechnology, and terms such as genetically modified organisms, DNA profiles, and prenatal diagnosis are parts of everyday speech.

Human Heredity has developed in parallel with advances in human genetics. In the second edition, recombinant DNA technology moved from being a section to a freestanding chapter. In the third edition, the impact of recombinant DNA technology gradually spread into other chapters. By the fourth edition, a chapter on biotechnology became necessary, and in the fifth edition, half a chapter was devoted to the Human Genome Project. The sixth edition was published shortly after the draft sequence of the human genome was reported, giving us a glimpse into the size, organization, and evolutionary history of our genome. The complete sequence of the coding portion of the genome was reported in 2003. What is now apparent, even in the short time following completion of the human genome sequence, is that this information is transforming the diagnosis and treatment of disease and accelerating the pace of research by an order of magnitude. Genome sequence information, coupled with technology such as DNA microarrays, is causing a fundamental shift in the way human genetics research is done. These developments allow researchers to use automated processes to gather data on a scale far beyond what was imagined only a few years ago, rapidly advancing our knowledge of genetics, creating new applications and products. Reflecting these changes, a chapter devoted to genomics has been added to this edition.

I wrote this book for one-term non-major human genetics courses to help undergraduates in the humanities, social sciences, business, engineering, and other fields understand these fast-moving developments. It is written for students with little or no background in biology, chemistry, or mathematics, but who want to learn something about human genetics. Some descriptive chemistry is used after an appropriate introduction and definition of terms. In the same vein, no advanced math skills are required to calculate elementary probabilities or to calculate genotype and allele frequencies.

Knowledge from genetics is rapidly being transferred to many other fields. The spread of technology and the issues it raises make it clear that difficult but informed decisions are required at many levels, from the personal to the political. The public, elected officials, and policy makers outside the scientific community need a working knowledge of genetics to help shape the course of research and its applications in our society. *Human Heredity* is written to transmit the principles of genetics in a straightforward and accessible way, without unnecessary jargon, detail, or the use of anecdotal stories in place of content.

GOALS OF THE TEXT

Being in the lab doing research in human genetics has offered me opportunities to train and nurture the development of graduate students and to present papers at meetings, symposia, and seminars where I meet and interact with other researchers and students to exchange ideas and help shape the course of genetics research and its applications. In addition, working with clinician colleagues and training medical students in my lab has given me valuable insight into the changes genetics, genomics, and technology have brought to medicine. I also work with undergraduates on a daily basis in my laboratory and through teaching courses in a biology curriculum to both majors and non-majors.

Using my research background in human genetics, my experience in training clinicians, and my classroom skills developed in teaching genetics to undergraduate and graduate students, I have written and continuously refined *Human Heredity*, combining the latest advances in research with the newest methods of teaching. In all editions, however, I have not worked in isolation. The book has been shaped by the contributions of others. Students in the course have suggested how to best introduce topics, identified the most effective examples and analogies in explaining concepts, and more important, been forthright in clarifying what does not work in the classroom. I have incorporated ideas refined by feedback from students and from faculty who have used the book at other institutions, and by comments and suggestions from reviewers. From the first edition, this book has been written to achieve several well-defined goals. This edition continues that tradition, incorporating the following goals:

1. Present the principles of human genetics in clear, concise language giving students a working knowledge of genetics. The premise in this approach is that a limited number of clearly presented, linked concepts is the best way to learn a complex subject such as genetics.
2. As in the classroom, begin the discussion of concepts at a level that students can understand and provide relevant examples that students can apply to themselves, their families, and their work environment.
3. Examine the social, cultural, and ethical implications associated with the use of genetic technology.
4. Communicate an understanding of the origin and amount of genetic diversity present in the human population, and how this diversity has been shaped by natural selection.

To achieve these goals, emphasis has been placed on clear writing with the use of accompanying photographs and artwork that teach rather than merely illustrate the ideas under discussion.

In general, the text consists of four sections: Chapters 1 through 7 cover cell division, transmission of traits from generation to generation, and development. Chapters 8 through 12 emphasize molecular genetics, mutation, and cancer. Chapters 13 through 16 include recombinant DNA, genomics, and biotechnology. These chapters cover gene action, mutation, cloning, the applications of genetic technology, the Human Genome Project, and the social, legal, and ethical issues related to genetics, as well as genetic screening, genetic testing, and genetic counseling. Chapters 17 through 19 consider specialized topics, including the immune system, population genetics, as well as the social aspects of genetics, including behavior.

Courses in human genetics use a wide range of formats. To accommodate these, the book is organized so it will be easy to use, no matter what order of topics an instructor chooses. After the section on transmission genetics, the chapters can be used in any order. Within each chapter, the outline lets the instructor and students easily identify central ideas.

NEW FEATURES OF THE SEVENTH EDITION

New Chapter

Chapter 15, titled "The Human Genome Project and Genomics," summarizes what we have learned about the human genome and introduces students to the new and dynamic field of genomics. The chapter begins by reviewing genetic mapping using linkage analysis to construct genetic maps, and the slow progress made with this method. The next section introduces the use of positional cloning, a method using recombinant DNA techniques to clone, map, and sequence disease-causing genes with no previous knowledge of the gene product or its function.

To identify, catalog, and elucidate the function of all human genes, the chapter examines the history, scope, methods, and results of the Human Genome Project, and the scientific intrigue surrounding its down-to-the wire race for the sequence. The steps in genome analysis are summarized, and the major features of the human genome are outlined. The challenges of the future derived from the impact of genomics on research, disease diagnosis, and treatment are reviewed in the last section of the chapter, using a keystone article written by genome scientists as a basis for the discussion.

Organizing the Book to Meet the Challenge of Genomics

The order of chapters has been revised, creating a middle section devoted to recombinant DNA techniques, biotechnology, and genomics as a central theme in human genetics. The reorganization and reordering of chapters has created a first section that outlines cell structure, mitosis, and meiosis (Chapter 2). Chapter 3 uses peas as a model system to cover Mendelian inheritance. Subsequent chapters in this section cover the basic methods of human genetics (Chapter 4), quantitative inheritance (Chapter 5), cytogenetics and karyotypes (Chapter 6), and human development (Chapter 7). While this first section precedes a discussion of genomics, the impact of recombinant DNA and genomics has been scattered throughout these chapters. In all cases, emphasis has been placed on ensuring that each chapter focuses on a small number of basic concepts.

The second section builds on the concepts of the first and moves the discussion to the molecular level, outlining the steps in the replication, storage, and expression of genetic information in the nucleotides of DNA molecules (Chapters 8 and 9), and the relationship between proteins and phenotype (Chapter 10). Chapter 11 explores how mutation alters phenotypes at the molecular and phenotypic level, leading into the final chapter of this section (Chapter 12), which discusses cancer, one of the consequences of mutation.

The core of the book contains chapters dealing with cloning, the impact of biotechnology, and the development of genomics and the Human Genome Project. Chapter 13 begins with a discussion of cloning, using organisms as the first set of examples. The chapter then considers the tools and methods used in cloning DNA and the methods available for analyzing clones, including DNA sequencing. Chapter 14 outlines the use of recombinant DNA techniques in biotechnology, using examples from agriculture, the development of model organisms for research, genetic testing, including the use of microarrays in genomic scanning, DNA profiles, and the ethical and social issues raised by the use of biotechnology. Chapter 15 begins by reviewing gene mapping in the pre–recombinant DNA era and the impact of positional cloning on mapping. The chapter moves on to explain the methods used in genomics and then discusses the results of the Human Genome Project and examines how genomics is used in investigating a genetic disorder. A section on proteomics as the next step in understanding our genome is followed by consideration of the ethical issues related to genomics and the future of genomics in research and medicine. Chapter 16 continues this theme by showing how biotechnology and genomics are used in reproductive technology, gene therapy, and genetic counseling. The ethical concerns about the use of biotechnology in assisted reproductive technology (ART) and gene therapy are outlined.

The final section of the book contains chapters on the immune system (Chapter 17) and behavior genetics (Chapter 18). Chapter 19 covers the essentials of population genetics and the use of molecular methods to study human evolution.

A Revised Art Program and Animations

The art program for this edition has been greatly revised and updated. Many figures have been replaced and new photos can be found throughout the book. The most significant change in the art program is the addition of dozens of **Active Figures**, art linked via the Human GeneticsNow™ website to animations that lead students step by step through the concepts. These animations are valuable assets for teaching and learning processes, including mitosis, meiosis, DNA replication, and gene expression.

Personalized Learning Resources and New Ways to Assess Learning

Recognizing that many students have difficulty solving genetics problems, the end-of-chapter questions and problems are now supplemented by Human GeneticsNow™, a password-protected website integrated with each chapter. All of the Active Figures from the text are located on this site, along with dozens more animations, interactive media, and tutorials. On Human GeneticsNow™, students can take diagnostic pre-tests that guide them to text, art, and animations that help them learn what they haven't yet mastered. After going over this personalized course of study, students finish with post-learning quizzes to assess their grasp of this new knowledge. Both the pre- and post-tests results can be mailed to instructors, who can also keep track of students' progress through their own access of the site. Access to Human GeneticsNow™ comes free with every new copy of *Human Heredity,* seventh edition.

Genetics in Practice: Relevant Case Studies

To make issues in human genetics relevant to situations that students may encounter outside the classroom, case studies are included at the end of each chapter, demonstrating the effects of "Genetics in Practice" in our society. This section contains scenarios and examples of genetic issues related to health, reproduction, personal decision making, public health, and ethics. Many of the case studies and the accompanying questions can be used for classroom discussions, student papers and presentations, and role playing. The cases and their questions are also located on the book's companion website along with links to resources for further research and exploration.

Genetic Databases as Resources

To foster awareness of the vast array of databases dealing with genetics and to integrate electronic resources into the text, genetic disorders mentioned in the book are referenced using their assigned indexing number from the comprehensive catalog assembled by Victor McKusick and his colleagues. This catalog is available online as *Online Mendelian Inheritance in Man™ (OMIM).* OMIM™ (updated daily) contains text, pictures, and videos along with literature references. OMIM™ is cross-linked to other databases containing DNA sequences, protein sequences, chromosome maps, and other resources. Students and an informed public need to be aware of the existence and relevance of such databases, and to be up to date, textbooks must incorporate these resources.

Students can use OMIM™ to obtain detailed information about a genetic disorder, its mode of inheritance, phenotype and clinical symptoms, mapping information, biochemical properties, the molecular nature of the disorder, and a bibliography of relevant papers. In the classroom, OMIM™ and its links are valuable resources for student projects and presentations.

For further reading about genetics, students can log on to InfoTrac® College Edition, an online library of articles from nearly 5,000 periodicals. This resource can be used in conjunction with electronic databases as resources for papers, class discussions, and presentations. Access to InfoTrac® is provided free with each new copy of *Human Heredity,* seventh edition.

New Internet Activities

The World Wide Web (WWW) is an important and valuable resource in teaching human genetics, and the *Human Heredity* companion website hosts quizzes, glossary, learning objectives, and a number of activities and links that can be used to expand on concepts and topics covered in the text. The website content can also be used to introduce the social, legal, and ethical aspects of human genetics into the classroom and serve as a point of contact with support groups and testing services. All of the website features, exercises, and activities described below can be easily completed online and e-mailed to instructors, making them ideal for assignments.

A new online feature is tied to the "How would you vote?" questions that follow every chapter's opening vignette. These are targeted questions about an issue related to the story and the chapter content. On the website, each "How would you vote?" question is accompanied by background information, links to helpful sites and materials, thought questions, and a chance to cast an online vote on the topic and view the resulting tallies.

Another new online feature is the inclusion of the "Genetics in Practice" case studies on the website. The cases, found at the end of each chapter, are repeated in their entirety on the website and accompanied by helpful links to resources for further exploration. This extra information makes them ideal as starting points for research projects and presentations.

In addition to the new features, this edition of *Human Heredity* contains updated end-of-chapter Internet Activities for students. These activities use WWW resources to enhance the topics covered in the chapter and are designed to develop critical thinking skills and to generate interaction and thought rather than passive observation. As with the other features, these activities are repeated on the book's companion website, along with links to websites and resources.

PEDAGOGICAL FEATURES

In this edition, the book has been substantially rewritten, and the order of chapters has been re-organized to reflect current findings and research directions, to engage student interest, and to improve the pedagogy. The basic organization within chapters, which has been successful as a teaching resource, has been continued in this edition.

(New!) Numbered Chapter Outlines

At the beginning of each chapter, an outline of the primary chapter headings provides an overview of the main concepts, secondary ideas, and examples. To help students grasp the central points, many of the headings are written as narratives or summaries of the ideas that follow. These outlines also serve as convenient starting points for students to review the material in the chapter. To make the outlines more useful, they have been numbered and used to organize both the summary and the questions and problems at the end of each chapter. In this way, students can more easily and clearly relate examples and questions to specific topics in the chapter.

Opening Case Study

Each chapter begins with a short prologue directly related to the main ideas of the chapter, often drawn from real life. Topics include the use of DNA fingerprinting in court cases, the cloning of milk cows, the creation of a DNA vaccine for SARS, and the development of *in vitro* fertilization (IVF) and the birth of Louise Brown, the first IVF baby. These vignettes are designed to promote student interest in the topics covered in the chapter and to demonstrate that laboratory research often has a direct impact on everyday life. In this edition, many of the opening stories are brand new or rewritten, and all are tied to a new feature, "How would you vote?"

(New!) How Would You Vote?

To stimulate thought and discussion, each chapter has a section, **How Would You Vote?** which presents an issue directly related to the opening story. It asks students to think about the topic and then visit the book's website where they can explore related links and cast a vote pro or con on the question posed. At the end of the chapter, the question is posed again, against the information presented in the chapter and after students have had a chance to learn more about the concepts related to the issue. These questions are intended to encourage students to think seriously about the genetic issues and concerns, provoking individual reflection and group discussion, which can be applied in a variety of ways both in and out of the classroom.

(New!) Keep in Mind Points

To keep students focused on the basic concepts in the chapter, a **Keep in Mind** box in the margin of the chapter opener contains a bulleted list of the main topics and key concepts presented in the chapter. Each item on the list is repeated in a highlighted text box at the conclusion of the section related to the concept, reinforcing the importance of the concept and providing students an aid in focusing their studies on fundamental points.

(New!) Active Figures

The most significant pedagogical addition to this edition is the use of **Active Figures**, linking art in the text to animations of important concepts, processes, and technologies discussed throughout the book. These animations convey an immediate appreciation of how a process works in a way that cannot be effectively shown in a static series of illustrations. These Active Figure animations can be found on the password-protected Human GeneticsNow™ site.

Genetic Journeys

Genetic Journeys feature boxes present ideas and applications that are related to and extend the central concepts in a chapter. The interesting but tangential examples presented provide context and connection of real examples to the ideas in the chapter.

Genetics in Society

Genetics in Society feature boxes provide a wider context to the ideas presented in the text. These essays elaborate and examine controversies that arise as genetic knowledge is transferred into technology and services.

Spotlight On . . .

Located in margins throughout the book, **Spotlight On** sidebars highlight applications of concepts, present the latest findings, and point out controversial ideas without interrupting the flow of the text.

Margin Glossary

A glossary in the page margins gives students immediate access to definitions of terms as they are introduced in the text. This format also allows definitions to be identified when students are studying or preparing for examinations. The definitions have been gathered into an alphabetical glossary at the back of the book. Because an understanding of the concepts of genetics depends on understanding the relevant terms, more than 350 terms are included in the glossary.

END-OF-CHAPTER FEATURES

The end-of-chapter features have been revised and updated for this edition. The questions are organized to reflect the order of topics in the chapter. Questions have been added to the case studies to enhance their use in the classroom, and new Internet activities have been added.

Genetics in Practice: Case Studies

Genetics in Practice case studies present specific examples of individuals and families using various genetic services, large-scale issues such as radioactive pollution, and the impact of the Human Genome Project. Many of these can be used as the basis for classroom discussions, student presentations, and role playing. The cases and their accompanying questions are also available on the book's companion website, where they are supplemented by links to other relevant sites. Students can e-mail instructors their responses to the case study questions through the website.

Summary

Each chapter ends with a summary that restates the major ideas covered in the chapter. The beginning outline and ending summary for each chapter use the same content and order to emphasize major concepts and their applications. Each point of the summary outline is followed by a brief restatement of the chapter material covered under the same heading. This helps students recall the concepts, topics, and examples presented in the chapter and, it is hoped, minimizes the chance they will attempt to learn by rote memorization.

Questions and Problems

The summary's focus on the chapter's main points is continued in the **Questions and Problems** at the end of each chapter. The questions and problems are presented under the headings from the chapter outline. This allows students to relate the problems and questions to specific topics presented in the chapter, to focus on concepts they find difficult, and to work the problems that illustrate these topics. The questions and problems are designed to test students' knowledge of the facts and their ability to reason from the facts to conclusions. To this end, they use an objective question format and a problem-solving format. Because some quantitative skills are necessary in human genetics, almost all chapters include some problems that require students to organize the concepts in the chapter and use these concepts in reasoning to a conclusion. Answers to selected problems are provided in an appendix. Answers to all questions and problems are available in the Instructor's Manual with Test Items.

Internet Activities

Internet Activities at the end of each chapter use websites to engage the student in activities related to the concepts discussed in the text. Internet resources are now an essential part of teaching genetics, and this section introduces students to the many databases, instructional sites, and support groups available to them. The activities are repeated and expanded on the book's companion website.

ANCILLARY MATERIALS

Instructor's Resources

The ancillary materials that accompany this edition are designed to assist the instructor in preparing lectures and examinations and to help keep instructors abreast of the latest developments in the field. Instructor materials are available to qualified adopters. Please consult your local Thomson Brooks/Cole sales representative for details. You may also visit the Brooks/Cole biology site at **http://biology.brookscole.com** to see samples of these materials, request a desk copy, locate your sales representative, or purchase a copy online.

Multimedia Manager
This easy-to-use, dual-platform digital library and presentation tool provides *all* of the art, photos, and tables from the text in PowerPoint® and JPG formats, along with a pre-created PowerPoint® lecture outline for each chapter, which you can modify and adapt to your own needs. A unique feature allows you to manipulate and resize figures and remove labels to customize your presentations.

Instructor's Manual with Test Items
An expanded and updated instructor's manual is available to help instructors in preparing class materials. This manual, prepared by Carl Frankel of Pennsylvania State University, Hazleton campus, contains chapter outlines, chapter summaries, teaching/learning objectives, key terms, additional test questions, and discussion questions. It also contains answers to all the end-of-chapter questions and problems found in the book.

ExamView®
This computerized test bank, available on CD-ROM, helps you create and deliver customized tests both online and in print. ExamView® guides you through the process while its "what you see is what you get" capability allows you to see the test you are creating on screen exactly as it will print or display online.

Human Heredity Companion Website for Instructors
The password-protected site for instructors at **http://biology.brookscole.com/cummings7** contains all the features of the student site listed below, plus chapter summaries and outlines, answers to end-of-chapter questions, and other helpful instructor resources.

Human GeneticsNow™
Instructor access to Human GeneticsNow™ at **http://now.ilrn.com/cummings7** includes the ability to monitor students' progress through the various chapter tests and media assets.

CNN® Today Videos
These short clips compiled from high-interest news stories related to genetics are a great way to launch your lectures. There are now two volumes of CNN® Today Videos for Genetics, published in 2003 and 2005. Both are available on videocassette and also on CD-ROM (in Quicktime® format).

Transparency Acetates
A set of 100 color transparencies featuring key figures—including drawings, charts, and diagrams from the text—is available to adopters.

Student Resources

Human GeneticsNow™
Located online at **http://now.ilrn.com/cummings7**, Human GeneticsNow™ is an exciting new assessment-centered learning tool developed in concert with the text. The site offers diagnostic pre-tests, personalized learning plans using media and animations located on the site, and

confirming post-tests. PIN code access to Human GeneticsNow™ is packaged *free* with every new copy of the text.

Human Heredity Companion Website for Students
A valuable partner to this text, the companion website at **http://biology.brookscole.com/ cummings7** features focused quizzing for each text chapter, glossary flashcards, learning objectives, "Internet Activities" with questions, "Genetics in Practice" cases with questions and links, "How would you vote?" exercises with voting tallies, annotated web links, and InfoTrac® keywords.

Study Guide
A student study guide has been prepared by Nancy Shontz of Grand Valley State University. The study guide includes chapter objectives and summaries, key terms, case worksheets (based on the "Genetics in Practice" case studies found in the text), discussion problems and questions, and other practice test items in multiple-choice, fill-in-the-blank, and modified true/false formats.

A Problem-Based Guide to Basic Genetics
Written and illustrated by Donald Cronkite of Hope College, this useful little manual provides students with a thorough and systematic approach to solving transmission genetics problems, along with numerous solved problems and practice problems.

Virtual Biology Laboratories: Genetics and Genetics 2 (Pedigree Analysis) Modules
Available Fall Term 2005. These "virtual" online experiments expose students to the tools used in modern biology, support and illustrate lecture material, and allow students to actually "do" science by performing experiments, acquiring data, and using the data to explain biological phenomena.

Gene Discovery Lab
This is a CD-ROM lab manual that provides a virtual laboratory experience for the student in doing experiments in molecular biology. It includes experiments that use nine of the most common molecular techniques in biology, an overview of the scientific method and experimental techniques, and Web links to provide access to data and other resources.

Contacting the Author

I welcome questions and comments from faculty and students about the book or about human genetics. Please contact me at: cummings@iit.edu

Acknowledgments

When Jerry Westby of West Publishing approached me almost exactly 10 years ago to ask if I would be interested in writing a text for my undergraduate nonmajors human genetics course, I was somewhat reluctant to consider a project of that dimension. In the end, Jerry's arguments were persuasive, and over the last decade, this book has grown to be a labor of love. As human genetics becomes more entwined with social and legal issues, it is essential that nonspecialists become familiar with the concepts of genetics. I am indebted to him for his insight, his creative contributions, and his commitment to bridging the gap between scientist and nonscientist.

Over the years, many reviewers, including those who helped with this edition, have given their time to improve the pedagogy, presentation of concepts, and nuances of language. Three past reviewers have gone to extraordinary lengths to help me learn and in some cases, re-learn details of genetics and have generously given me access to their collective wisdom: George Hudock of Indiana University, H. Eldon Sutton of the University of Texas, and Werner Heim of Colorado College. I am grateful for their efforts to help make this book an effective teaching tool.

To all the reviewers who helped in the preparation of this edition, I offer my thanks and gratitude for their efforts.

Kelly Bidle, *Rider University*
Jay Brewster, *Pepperdine University*
Anna G. Brownell, *Chapman University*

Drew Cressman, *Sarah Lawrence College*
Larry R. Eckroat, *Pennsylvania State University–Erie*
Cheryl L. Emmons, *Alfred University*
Edison R. Fowlks, *Hampton University*
Beth Gaydos, *San Jose City College*
L. Michael Hill, *Bridgewater College*
Robert Hinrichsen, *Indiana University of Pennsylvania*
Jody E. Johnson, *Arapahoe Community College*
Ron Kaltreider, *York College of Pennsylvania*
Mary King Kananen, *Pennsylvania State University–Altoona*
John Lennox, *Pennsylvania State University–Altoona*
Albert MacKrell, *Bradley University*
Patricia Matthews, *Grand Valley State University*
James B. Olesen, *Ball State University*
Dennis F. Rio, *Lewis University*
Jeanine Seguin, *Keuka College*
Jean M. Shingle, *Immaculata University*
Jan Trybula, *State University of New York–Potsdam*

In past editions, Michelle Murphy Whaley of the University of Notre Dame and Peter Follette took on the daunting task of revising and adding to the end-of-chapter questions and problems as well as writing questions for the Genetics in Practice case studies. For this edition, Peter once again took on the task of writing insightful questions that challenge students to think and apply their knowledge.

At Brooks/Cole, the book has had creative input from many talented individuals. I am especially indebted to my editor Nedah Rose and to Mary Arbogast, senior developmental editor, who developed a revision plan that integrates text with web-based resources including animated figures, web-linked learning assessments, and a multimedia manager for instructors. Ann Caven, the marketing manager, provided many insights during the planning stages, and her suggestions have helped make this a better book.

As the project got under way, Sarah Lowe assembled and analyzed the reviews and helped establish priorities for the revision. I was especially fortunate to have Kari Hopperstead oversee this revision. Her organizational skills kept me moving and provided me with vast amounts of analyses, information, and resources, all of which allowed me to focus on revising, updating, and writing. She was also a sounding board for my ideas, and contributed many important ideas of her own that improved the order of chapters, pedagogy, and student-oriented learning features. I am thankful for her creative input, ideas, and suggestions, which are reflected in the many new features added to this edition. Travis Metz coordinated all the web-based features of the book and helped integrate them with the text. The new artwork in this edition was done by Raychel Ciemma, Lisa Starr, and Gary Head, along with contributions from other sources. Gopa designed the layout, Hiroko Chastain designed the cover, and Sarah Harkrader oversaw the process of obtaining permissions to use figures and photos from other publications. It was a pleasure to once again work with Linda Sykes who did the photo research for this edition. Even as the deadlines tightened, she was willing to start anew looking for the best photos.

As the book moved into production, Belinda Krohmer at Brooks/Cole coordinated the project, tracked the loose ends, and helped me develop priority lists to keep things moving. At G&S Book Services, Jamie Armstrong was always unruffled as the schedule became a moving target and managed to get it all done and make it look easy. Kathleen Deselle not only copyedited my mangled syntax, but also researched many things for accuracy. Fran Andersen, the project editor, helped clarify and standardize definitions in the text and glossary, easing the student's task of learning the language of genetics.

Once again, I offer special thanks and appreciation to my colleague Suzanne McCutcheon for guiding my students during my many absences from the lab. Finally, I want to congratulate my graduate student Holly Dimitropoulos who launched her scientific career by winning a Fullbright fellowship for postdoctoral work on the molecular biology of aging.

Michael R. Cummings

1

A Perspective on Human Genetics

IN December 1998, after much debate and over a determined opposition, the Icelandic Parliament (Althingi) passed a controversial bill allowing de-CODE, a biotech company, to establish and operate a database called HSD (Icelandic Health Sector Database) containing the medical records of all residents of Iceland. In addition, deCODE is compiling the genealogies of the approximately 800,000 Icelanders who have lived there since the colonization of the island in the ninth and tenth centuries. In combination with blood and tissue samples (for DNA extraction) provided by patients, these databases are powerful tools in the hunt for disease-causing genes. The new law grants the company the right to sell this information (and the DNA samples) to third parties, including the research labs of pharmaceutical companies, with the hope that once disease genes are identified, diagnostic tests and therapies will soon follow.

Why establish such a database in Iceland? The most important reasons are that Iceland has a small, genetically isolated population, with little genetic variation. The first humans came to Iceland in the ninth and tenth centuries as a small founding population and until 50 years ago were almost completely isolated from outside immigration. Plague (in the 1400s) and volcanoes (in the 1700s) decimated the population, further reducing genetic variation. The 270,000 inhabitants of Iceland share a remarkably similar set of genes, providing fertile ground for gene hunters seeking to identify disease genes.

Why the controversy? Opponents point out that the privacy provisions of the law are inadequate and may violate the ethical principle that health records must be kept confidential. Abuses and misunderstandings may affect employment, insurance, and even marriage. In addition, critics question

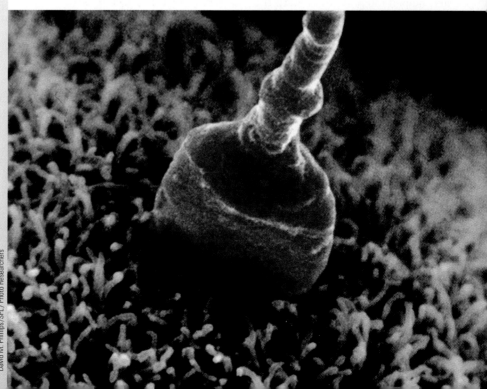

David M. Phillips /SPL / Photo Researchers

whether a single company should have exclusive rights to medical information and whether the Icelandic population will derive health benefits from this arrangement. The argument is far from settled and may involve court action and censure by international medical societies.

Aside from the controversies, deCODE has been successful in identifying genes associated with more than 25 common diseases with genetic components, including heart disease, asthma, stroke, and osteoporosis.

deCODE's success has fostered the development of similar projects elsewhere. The UK Biobank, launched in Great Britain in 2003, is screening 1.2 million individuals to establish a database of medical records and DNA samples from 500,000 Britons. The Biobank plans to use this information to search for genes that control complex traits such as hypertension and heart disease. Other programs are being developed in many other countries, including Estonia, Latvia, Singapore, and the Kingdom of Tonga. In the United States, programs using medical records and DNA samples from tens of thousands of individuals are underway at the Marshfield Clinic in Marshfield, Wisconsin, Northwestern University in Chicago, and Howard University in Washington, D.C.

How would you vote?

Several different countries, organizations, and corporations are compiling databases of genetic information for individuals within a population. Generally, these databases are intended as aids for medical research; however, the extremely private nature of the information being gathered makes many people concerned about its misuse. If a major medical center asked you to donate a DNA sample and allow them access to your medical records, how would you respond? What if they explained the information would be used in a project to search for genes that control complex traits such as hypertension, cardiovascular disease, or mental retardation? Visit the Human Heredity Companion website to find out more on the issue, then cast your vote online at **http://biology.brookscole.com /cummings7**

1.1 Genetics Is the Key to Biology

With gene-based programs like these becoming common, as we begin this book, we might pause and remember that genetics is more than a laboratory science; unlike some other areas of science, genetics and biotechnology have a direct impact on society.

As a first step in studying human genetics, we should ask what *is* genetics? As a

Keep in mind as you read

- Genes control cellular function and link generations together.

- Gregor Mendel discovered many properties of genes and founded genetics.

- Wrong ideas about genetics have influenced past laws and court decisions.

- Recombinant DNA and biotechnology affect many aspects of our daily lives.

Genetics   The scientific study of heredity.

working definition, we can say that **genetics** is the science of heredity. Like all definitions, this leaves a lot unsaid. To be more specific, what geneticists do is study how traits (like eye color and hair color) and diseases (like cystic fibrosis and sickle cell anemia) are passed from generation to generation. They also study the molecules that make up genes and gene products, and how genes are turned on and off. Some geneticists study why some genes are more frequent in some populations than in others. Other geneticists work in industry to develop products for agricultural and pharmaceutical firms. This work is called biotechnology and is a multi-billion-dollar component of the U.S. economy.

In a sense, genetics is the key to all of biology, because genes control what cells look like and what they do, as well as how babies develop and how we reproduce. An understanding of what genes are, how they are passed from generation to generation, and how they work is essential to our understanding of all life on Earth, including our species, *Homo sapiens*.

In the chapters that follow, we will ask and answer questions about genes and genetics: How are genes passed from parents to their children? What are genes made of? Where are they located? How do they make products called proteins, and how do proteins help create the differences among individuals that we can see and study? Because this book is about human genetics, we will use human genetic disorders as examples of inherited traits (see Genetics in Society: Genetic Disorders in Culture and Art). In addition, we will examine how genetic knowledge and genetic technology interact with and shape many of our social, political, legal, and ethical institutions and policies.

Almost every day, the media contains a story about human genetics. These stories may involve the discovery of a gene responsible for a genetic disorder, a controversy

Genetics in Society

Genetic Disorders in Culture and Art

It is difficult to pinpoint when the inheritance of specific traits in humans was first recognized. Descriptions of heritable disorders often appear in myths and legends of many different cultures. In some ancient cultures, assigned social roles—from prophets and priests to kings and queens—were hereditary. The belief that certain traits were heritable helped shape the development of many cultures and social customs.

In some ancient societies, the birth of a deformed child was regarded as a sign of impending war or famine. Clay tablets excavated from Babylonian ruins record more than 60 types of birth defects, along with the dire consequences thought to accompany such births. Later societies, ranging from the Romans to eighteenth-century Europe, regarded malformed individuals (such as dwarfs) as curiosities rather than figures of impending doom, and they were highly prized by royalty as courtiers and entertainers.

Whether motivated by fear, curiosity, or an urge to record the many variations of the human form, artists

have portrayed both famous and anonymous individuals with genetic disorders in paintings, sculptures, and other forms of the visual arts. These portrayals are often detailed, highly accurate, and easily recognizable today. In fact, across time, culture, and artistic medium, affected individuals in these portraits often resemble each other more closely than they do their siblings, peers, or family members. In some cases, the representations allow the disorder to be clearly diagnosed at a distance of several thousand years.

Throughout the book, you will find fine-art representations of individuals with genetic disorders. These portraits represent the long-standing link between science and the arts in many cultures. They are not intended as a gallery of freaks or monsters, but as a reminder that being human encompasses a wide range of conditions. A more thorough discussion of genetic disorders in art is in *Genetics and Malformations in Art*, by J. Kunze and I. Nippert, published by Grosse Verläg, Berlin, 1986.

about genetic testing, or a debate on the wisdom of genetically modifying our children. In many cases, as we will see, technology is far ahead of public policy and laws. To make informed decisions about genetics and biotechnology in your personal and professional life, you will need to know what genes are and how they work. In the rest of this chapter, we will preview some of the basic concepts of human genetics and introduce some of the social issues and controversies generated by genetic research. These concepts and issues will be explored in more detail in the chapters that follow.

1.2 What Are Genes and How Do They Work?

Simply put, a gene is the basic structural and functional unit of genetics. In molecular terms, a gene is a string of chemical subunits (nucleotides) in a **DNA** molecule (▶ Figure 1.1). (DNA is shorthand for deoxyribonucleic acid.) There are four different nucleotide subunits in DNA, and the sequence of these subunits stores information in the form of a **genetic code.** The sequence of "letters" encoded in the gene (each nucleotide is a letter in the code), in turn, defines the chemical subunits (amino acids) that make up gene products (proteins). When a gene is turned on, its stored information is decoded and used to make a protein molecule that folds into a three-dimensional shape and becomes functional (▶ Figure 1.2). The action of proteins produces characteristics we can see (like eye color or hair color) or measure (blood proteins or height). Understanding how different proteins are produced and how they work in the cell are an important part of genetics. We will cover these topics in Chapters 9 and 10.

■ **DNA** A helical molecule consisting of two strands of nucleotides that is the primary carrier of genetic information.

■ **Genetic code** The sequence of nucleotides that encodes the information for amino acids in a polypeptide chain.

Keep in mind

■ Genes control cellular function and link generations together.

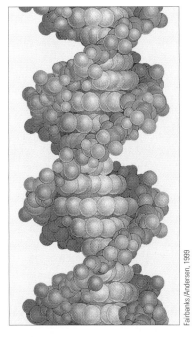

▲ FIGURE 1.1 The structure of DNA, discovered by James Watson and Francis Crick in 1952, marked the beginning of molecular genetics.

Fairbanks/Andersen, 1999

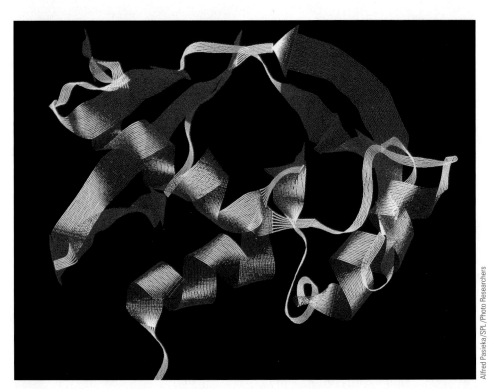

▲ FIGURE 1.2 A model of how a polypeptide chain folds into the three-dimensional structure forming a protein.

Alfred Pasieka/SPL/Photo Researchers

○ We can also define genes by their properties. Genes are copied (replicated), they mutate (undergo change), they are expressed (they can be turned on and off), and they can recombine (they can move from one chromosome to another). In later chapters, we will explore these properties and see how they are involved in genetic diseases.

1.3 How Are Genes Transmitted from Parents to Offspring?

Thanks to the work of Gregor Mendel (▶ Figure 1.3), a European monk, we understand how genes are passed from parents to offspring in all plants and animals, including humans. When Mendel began his experiments in the mid-nineteenth century, many people thought that traits carried by parents were blended together in their offspring. According to this idea, crossing a plant with red flowers to one with white flowers would produce plants with pink flowers (the pink color is a blend of red and white). In reality, when such plants are crossed, new plants with red and white flowers are produced in every generation. Mendel's experiments on pea plants provided the key to understanding how genes are passed from one generation to the next. As we will see, however, things are not always simple. There *are* cases when crossing plants with red flowers to plants with white flowers *does* produce plants with pink flowers. We will discuss these cases in Chapter 3 and show that plants with pink flowers do not overturn the principles of inheritance discovered by Mendel.

Working at a monastery in what is now the Czech Republic, Mendel began ten years of work on pea plants. In his work, the parental plants were chosen so that each had a different, distinguishing characteristic called a **trait**. For example, Mendel bred tall pea plants with short plants. Plant height is the trait in this case, and has two variations, tall and short. He also bred plants carrying green seeds with plants having yellow seeds. In this work, seed color is the trait; green and yellow are the

▲ FIGURE 1.3 Gregor Mendel, the Augustinian monk whose work on pea plants provided the foundation for genetics as a scientific discipline.

Portrait by Marcus Alan Vincent

▦ **Trait** Any observable property of an organism.

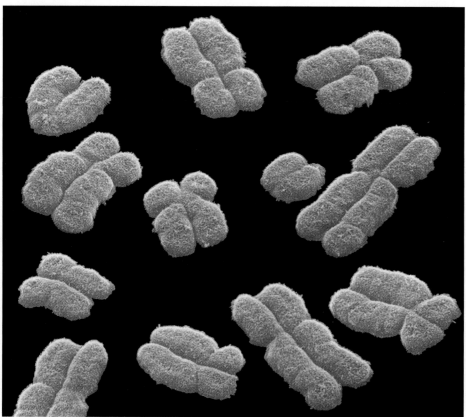

◀ FIGURE 1.4 Replicated human chromosomes as seen by scanning electron microscopy.

Andrew Syred/Photo Researchers

variations he studied. In these breeding experiments, he wanted to see how seed color was passed from generation to generation. Mendel kept careful records of the number and type of traits present in each generation. He also recorded the number of individual plants that carried each trait. He discovered patterns in the way traits were passed from parent to offspring through several generations. Based on these patterns, Mendel developed clear ideas about how traits are inherited. He concluded that traits such as plant height or flower color are passed from generation to generation by "factors" that are passed from parent to offspring. What he called "factors" we now call genes. Mendel reasoned that each parent carries two genes (a gene pair) for a given trait (flower color, plant height, etc.) but that each parent contributes only one of these genes to its offspring; otherwise the number of genes for a trait would double in each generation and soon reach astronomical numbers.

Mendel proposed that the two copies of each gene separate from each other during the formation of egg and sperm. As a result, only one copy of each gene is present in the sperm or egg. When an egg and sperm fuse together at fertilization, the genes from the mother and father become members of a new gene pair in the offspring. In the mid-twentieth century, researchers discovered that genes are made up of DNA molecules combined with proteins to form structures known as chromosomes. Chromosomes are found in the nucleus of human cells and other higher organisms (▶ Figure 1.4). As we will see in Chapter 2, the separation of genes during formation of the sperm and egg and the reunion of genes at fertilization is explained by the behavior of chromosomes in a form of cell division called meiosis.

When Mendel published his work on the inheritance of traits in pea plants (discussed in Chapter 3), there was no well-accepted idea of how traits were transmitted from parents to offspring; his evidence changed all this. To many, Mendel was the first geneticist and the founder of genetics, a field that has expanded in numerous directions in the past 125 years.

Keep in mind

- Gregor Mendel discovered many properties of genes and founded genetics.

1.4 How Do Scientists Study Genes?

Ideas that form the foundation of genetics were discovered by studying many different organisms including bacteria, yeast, insects, and plants, as well as humans. Because these principles are universal, discoveries made in one organism (like yeast) can be applied to other species, including humans. Although geneticists study many different species, they use a small number of basic approaches in their work, some of which are outlined in the following section.

There are different approaches to the study of genetics.

The most basic approach, called **transmission genetics** (Chapters 3 and 4), studies the pattern of inheritance that results when traits are passed from generation to generation. Using experimental organisms, geneticists carry out mating experiments to study the transmission of traits (such as height, eye color, flower color, and so on) from parents to offspring. These experimental results are analyzed to establish how a trait is inherited. As we discussed in an earlier section, Gregor Mendel did the first significant work in transmission genetics, using pea plants as his experimental organism. His methods form the foundation of transmission genetics.

Mating experiments in humans are not possible, thus a more indirect method, called pedigree analysis, is used. **Pedigree analysis** begins with a detailed family history. This history is used to reconstruct the pattern followed by a trait as it passes through several generations of a family. The results are used to determine how a trait

- **Transmission genetics** The branch of genetics concerned with the mechanisms by which genes are transferred from parent to offspring.

- **Pedigree analysis** The construction of family trees and their use to follow the transmission of genetic traits in families. It is the basic method of studying the inheritance of traits in humans.

Cytogenetics The branch of genetics that studies the organization and arrangement of genes and chromosomes using the techniques of microscopy.

Karyotype A complete set of chromosomes from a cell that has been photographed during cell division and arranged in a standard sequence.

Molecular genetics The study of genetic events at the biochemical level.

Recombinant DNA technology A series of techniques in which DNA fragments are linked to self-replicating vectors to create recombinant DNA molecules, which are replicated in a host cell.

Population genetics The branch of genetics that studies inherited variation in populations of individuals and the forces that alter gene frequency.

is inherited and to establish the risk of having affected children (▶ Figure 1.5). Pedigrees are constructed from interviews, letters, diaries, photographs, and family records.

Cytogenetics is a branch of genetics that studies chromosome number and structure (discussed in Chapter 6). At the beginning of the twentieth century, observations on chromosome behavior were used to correctly propose that genes are located on chromosomes. Cytogenetics is one of the most important investigative approaches in human genetics and is used, among other things, to map genes and to study chromosome structure and abnormalities. In clinical settings, cytogeneticists prepare **karyotypes**. These are standardized arrangements of chromosomes used to diagnose or to rule out genetic disorders (▶ Figure 1.6). In a karyotype, chromosomes are arranged by size, shape, and other characteristics that we will describe in Chapter 6.

A third approach, **molecular genetics,** has had the greatest impact on human genetics over the last four or five decades. Molecular genetics uses **recombinant DNA technology** to identify, isolate, clone (produce multiple copies), and analyze genes. Cloned genes can be used to study how genes are organized and how they work. Cloned genes can also be transferred between organisms and between species. Gene therapy to treat human genetic disorders uses cloned genes.

Recombinant DNA technology is also used for prenatal diagnosis of genetic disorders and to sequence the DNA carried by an individual. Advances in molecular genetics, especially those using recombinant DNA technology have generated much of the debate about the social, legal, and ethical aspects of genetics, including the genetic modification of plants and animals, the use of genetic testing for employment and insurance, and the genetic modification of humans by gene therapy.

A fourth approach studies the distribution of genes in populations. Population geneticists are interested in the forces that change the frequency of a given gene over many generations in a population and how these changes are involved in evolution. **Population genetics** has defined how much genetic variation exists in populations and how forces such as migration, population size, and natural selection change this variation. The coupling of population genetics with recombinant DNA technology has helped us understand the evolutionary history of our species and the migrations that distributed humans across Earth. It has also been used to develop methods of DNA fingerprinting and DNA identification, techniques widely used in paternity testing and criminal cases.

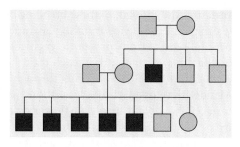

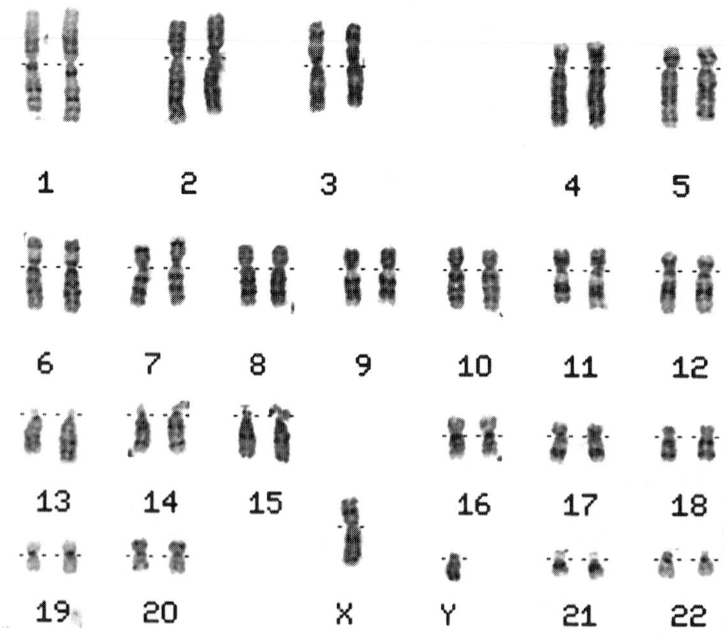

▲ FIGURE 1.5 A pedigree represents the inheritance of a trait through several generations of a family. In this pedigree, males are symbolized by squares, females by circles. Filled symbols indicate those carrying the trait being studied.

Genetics is used in basic and applied research.

Because the principles of genetics have many different uses, genetics is a discipline that crosses and recrosses the line between basic and applied research, often blurring distinctions between the two fields. In general, scientists do basic research

▲ FIGURE 1.6 A karyotype arranges the chromosomes in a standard format so they can be analyzed for abnormalities. This karyotype is that of a normal male.

Courtesy of Ifti Ahmed

in laboratory and field settings to understand how something works or why it works the way it does. In basic research, there is no immediate goal of solving a practical problem or making a commercial product; knowledge itself is the goal. In turn, results of basic research generate new ideas and more basic research. In this way, we gain detailed information about how things work inside cells, why animals behave in certain ways, and how plants turn carbon dioxide into sugar. Among other things, basic research in genetics has provided us with details about genes, how they work, and, more importantly, what happens when they don't work properly.

Applied research is usually done to solve a practical problem or turn a discovery into a commercial product. Applied research uses basic methods such as transmission genetics to study how a trait is inherited but also uses biotechnology to make products such as vaccines. In agriculture, applied genetic research has increased crop yields, lowered the fat content of pork, and created new forms of corn and soybeans that are disease resistant. In medicine, new diagnostic tests, the synthesis of customized proteins for treating disease, and the production of vaccines are just a few examples of applied genetic research.

Some uses of applied research are controversial and have generated debate about the merits and risks of biotechnology. Current controversies include environmental release of genetically modified organisms, the sale and consumption of food that has been modified by recombinant DNA technology (▶ Figure 1.7), the use of recombinant DNA-derived growth hormone in milk production, and the irradiation of food.

An understanding of the basic concepts of genetics will help all of us make informed decisions about the use of biotechnology in our lives, including the food we eat, the diagnostic tests we elect to have performed, and even the breeding of our pets. This course will provide you with the basic concepts of genetics and human genetics that can be used to make these informed decisions.

▲ FIGURE 1.7 Transgenic tomatoes have been genetically modified by recombinant DNA techniques to slow softening.

1.5 Has Genetics Affected Social Policy and Law?

Genetics and biotechnology not only affect our personal lives but also raise larger questions about ethics, social policy, and law. We will consider these current controversies surrounding genetics and biotechnology in several chapters, but you may be surprised to learn that controversies involving genetics are nothing new. In fact, genetics has had a significant impact on law and social policy for most of the last century. As we face decisions about how to use new forms of genetic technology, it is important to know and understand the history and outcomes of past controversies so we can avoid repeating mistakes and pitfalls.

Genetics has directly affected social policy.

After the publication of *The Origin of Species* by Charles Darwin, his cousin Francis Galton proposed that natural selection should be used to improve the human species. Galton started the field of **eugenics**, which he claimed could improve the intellectual, economic, and social level of humankind by selective breeding. Bypassing legal and ethical considerations, Galton's proposals were simple: people with desirable traits such as leadership or musical ability should be encouraged to have large families, whereas those with undesirable traits such as mental retardation or physical deformities should be discouraged from reproducing. Galton's reasoning was flawed because he believed that human traits were handed down without any

■ Eugenics   The attempt to improve the human species by selective breeding.

► **FIGURE 1.8** In the early part of the twentieth century, eugenics exhibits were a common feature at fairs and similar events. Such exhibits served to educate the public about genetics and the benefits of eugenics as public policy. These exhibits often included contests to find the eugenically perfect family.

American Philosophical Society

■ **Hereditarianism** The idea that human traits are determined solely by genetic inheritance, ignoring the contribution of the environment.

environmental influence. The idea that all human traits are determined only by genes is known as **hereditarianism.** His proposals contained another flaw: who defines what is a desirable or undesirable trait?

In spite of these flaws, eugenics took hold in the United States, and eugenicists worked to promote selective breeding in the human population (► Figure 1.8) and to prevent reproduction by those defined as genetically defective. Although almost unknown today, eugenics was a powerful and influential force in many aspects of American life from about 1905 through 1933.

Keep in mind

■ Wrong ideas about genetics have influenced past laws and court decisions.

Eugenics helped change immigration laws.

In the early decades of the twentieth century, European immigrants flooded into the United States as a result of World War I. Eugenicists argued that the high levels of unemployment, poverty, and crime among immigrants from southern and eastern Europe proved that people from these regions were genetically inferior and were polluting the genes of Americans. Based on testimony by eugenics experts, Congress passed the Immigration Restriction Act of 1924. As he signed the new law, President Coolidge commented that "America must remain American." This law, based on faulty and unproven eugenic assumptions, effectively closed the door to America for millions of people in southern and eastern Europe by reducing entry quotas from countries such as Italy and Russia by two-thirds, while allowing large numbers of immigrants from western European countries such as France, Germany, and Great Britain. Europeans were not alone in facing restrictions. The Chinese Exclusion Acts of 1882 and 1902 had already restricted immigration from Asia. In addition, a 1907 agreement between the U.S. and Japanese governments restricted the immigration of Japanese citizens. In the early decades of the twentieth century, there was little immigration from Africa, and lawmakers saw no need to specifically regulate entry from this continent.

Immigration laws based on faulty eugenics were in effect for just over 40 years. These laws were finally changed by the Immigration and Nationality Act of 1965, which was sponsored by Representative Emanuel Cellar of New York, a grandson of immigrants. Under this law, national quotas were abolished, and immigrants from all parts of the world were welcomed.

Eugenics helped restrict reproductive rights.

In addition to setting immigration policy, the eugenics movement in the United States helped pass laws that required sterilization of what were defined as genetically, intellectually, and morally inferior individuals. A committee of eugenicists concluded that up to 10% of the U.S. population should be prevented from reproducing by being institutionalized or sterilized. Other eugenicists testified before committees of state legislatures urging states to regulate reproductive rights. State laws allowing sterilization for those with certain genetic disorders and those convicted of certain crimes were passed beginning in 1907. In 1927, the U.S. Supreme Court (*Buck v. Bell*) upheld the right of states to use eugenic sterilization in an 8–1 decision. Oliver Wendell Holmes, one of the most respected justices of the Supreme Court, wrote the opinion. This ruling, which has never been overturned, includes the following statement:

> It is better for all the world, if instead of waiting to execute degenerate offspring for crime, or to let them starve for their imbecility, society can prevent those who are manifestly unfit from continuing their kind. The principle that sustains compulsory vaccination is broad enough to cover cutting the fallopian tubes.... Three generations of imbeciles are enough.

The three generations referred to by Holmes represent Carrie Buck, her mother, Emma, and Carrie's daughter, Vivian (▶ Figure 1.9). The case came to the U.S. Supreme Court to appeal the decision by a Virginia court that Carrie should be steril-

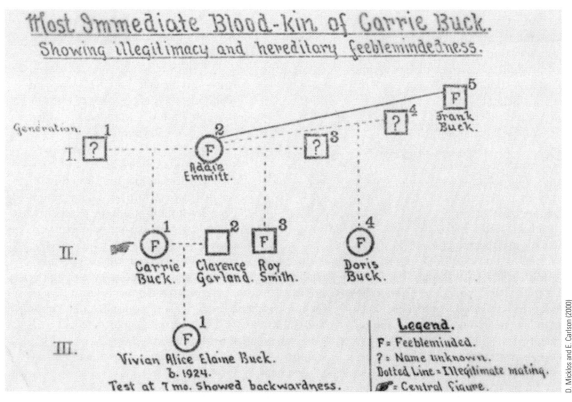

▲ FIGURE 1.9 A pedigree of the family of Carrie Buck, made at the Virginia Colony for the Epileptic and Feebleminded.

Spotlight on...

Eugenic Sterilization

Thirty states passed laws providing for sterilization of feebleminded individuals, a catchall term that covered both real and imagined disabilities. Behavior was used as a way of diagnosing someone as feebleminded, including alcoholism, criminal convictions, and sexual promiscuity. More than 60,000 people were sterilized before the practice was ended in 1979. Of these, five states, California (20,108), Virginia (7,450), North Carolina (6,297), Michigan (3,786), and Georgia (3,284) accounted for almost 70% of this total.

ized because she was feebleminded and promiscuous. Evidence presented at trial showed that Carrie, her mother, and Carrie's daughter were all mentally unfit. Soon after the Supreme Court ruling, Carrie Buck was sterilized. At the time, she was an unmarried teenager living in a foster home with her daughter. Later investigation showed that her child, Vivian, was not "feebleminded" as claimed, and that Carrie was not promiscuous, but had been raped by a relative of her foster parents.

Within a few years after the U.S. Supreme Court decision, sterilization laws were passed in many states, and at least 60,000 individuals were sterilized over the next decades (see Spotlight on Eugenic Sterilization). In recent years, states including Virginia, North and South Carolina, and Oregon have officially and publicly apologized for their involvement in eugenic sterilization.

Eugenics became associated with the Nazi movement.

In Germany, eugenics (known as *Rassenhygiene*) fused with genetics and the political philosophy of the Nazi movement (see Genetics in Society: Genetics, Eugenics, and Nazi Germany). Sterilization laws in the United States served as models for the 1933 "Law for the Protection Against Genetically Defective Offspring" passed in Germany. This law was gradually expanded to allow the systematic killing of people defined as socially defective, physically deformed, mentally retarded, and/or mentally ill. Later eugenics was used as a justification for the eradication of entire ethnic

Genetics in Society

Genetics, Eugenics, and Nazi Germany

In the first decades of the twentieth century, eugenics advocates in Germany were concerned with preservation of racial "purity," as were their colleagues in other countries, including the United States and England. By 1927, many states in the United States had enacted laws that prohibited marriage of "social misfits" and made sterilization compulsory for the "genetically unfit" and for those found guilty of certain crimes. In Germany, the laws of the Weimar government prohibited sterilization, and there were no laws restricting marriage on eugenic grounds. As a result, several leading eugenicists became associated with the National Socialist Party (Nazis), which advocated forced sterilization and other eugenic measures to preserve the purity of the Aryan "race."

Adolf Hitler and the Nazi Party came to power in January 1933. By July of that year, a sterilization law was in effect. Under the law, those regarded as having lives not worth living, including the feebleminded, epileptics, the deformed, those having hereditary forms of blindness or deafness, and alcoholics, were to be sterilized.

By the end of 1933, the law was amended to include

mercy killing (*Gnadentod*) of newborns who were incurably ill with hereditary disorders or birth defects. This program was gradually expanded to include children up to 3 or 4 years of age, then adolescents, and finally, all institutionalized children, including juvenile delinquents and Jewish children. More than two dozen institutions in Germany, Austria, and Poland were assigned to carry out this program. Children were usually killed by poison or starvation.

In 1939, the program was extended to include mentally retarded and mentally defective adults and adults with certain genetic disorders. This program began by killing adults in psychiatric hospitals. As increasing numbers were marked for death, gas chambers were installed at several institutions to kill people more efficiently, and crematoria were used to dispose of the bodies. This practice spread from mental hospitals to include defective individuals in concentration camps and then to whole groups of people in concentration camps, most of whom were Jews, Gypsies, Communists, homosexuals, or political opponents of the government.

groups such as the Gypsies and Jews. The close association between eugenics and the government of Nazi Germany quickly led to the decline of the eugenics movement in the United States by the late 1930s.

1.6 What Impact Is Genetics Having Now?

In the 1970s, recombinant DNA technology began with the discovery that bacteria protect themselves from viral infections by making proteins that cut the DNA of invading viruses into pieces. These proteins, called **restriction enzymes,** cut DNA from an organism at specific sites, producing a predictable pattern of fragments (▶ Figure 1.10). Soon after this discovery, scientists learned how to make recombinant DNA molecules by inserting these fragments into carrier DNA molecules. Placed inside bacterial cells, the recombinant molecules were copied, or **cloned.** DNA made by cloning was used for research and is the foundation for many applications, including genetic testing, gene therapy, and the biotechnology industry.

Newer methods made it possible to clone larger and larger DNA fragments, establishing collections of clones that included all the genes carried by an organism. The set of genes carried by an organism is called its **genome,** and the collection of clones that contain a whole genome is called a genomic library.

■ **Restriction enzyme** A bacterial enzyme that cuts DNA at specific sites.

■ **Clone** Genetically identical molecules, cells, or organisms all derived from a single ancestor.

■ **Genome** The set of genes carried by an individual.

The Human Genome Project has been completed.

With genomic libraries available, geneticists began planning ways to sequence all the clones in a genomic library and organize this information to identify all the genes in

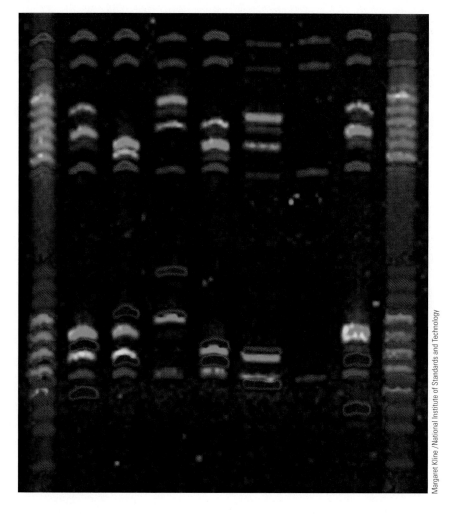

◀ FIGURE 1.10 A gel showing DNA fragments from different Y chromosomes produced by restriction enzymes.

Margaret Kline / National Institute of Standards and Technology a genome. The Human Genome Project (HGP) began as a federal program in 1990. In 2001, the HGP and a project undertaken by private industry reported the first draft of the Human Genome sequence, and in 2003 finished the rest of the gene-coding portion of the sequence. We now have a catalog of the 3 billion nucleotides and the 25,000 to 30,000 genes carried in human cells. The information gathered from genome projects gave rise to **genomics,** a new field of study that focuses on the organization, function, and evolution of genomes.

Genomics The study of the organization, function, and evolution of genomes.

Information from the HGP along with other advances in research and technology make it possible to diagnose genetic disorders before birth, test children and adults to reveal carriers of genetic disease, and now to test anyone's entire genome to detect genetic disorders and predispositions to cardiovascular disease, diabetes, and cancer.

Health care uses genetic testing and genome scanning.

Genetic technology is now an important part of medicine, and this impact will continue to grow as the information from the HGP is analyzed and applied to the diagnosis and treatment of human diseases. More than 10 million children and adults in the United States suffer from a genetic disorder, and every newborn has a 3% chance of having a genetic disorder, underscoring the need for tests that accurately diagnose heritable diseases at all stages of life, from prenatal to adult. The genes associated with hundreds of genetic diseases, including cystic fibrosis, sickle cell anemia, and muscular dystrophy, have been cloned and used to develop genetic tests. All 50 states and the District of Columbia test newborns for a range of genetic disorders. In addition, adults can be tested to determine if they are at risk of having a child with a genetic disorder. Couples can now obtain information they can use to make informed decisions about family planning when genetic testing is combined with genetic counseling.

New technology now makes it possible to screen an individual's entire genome, instead of testing for one genetic disorder at a time. This technology uses DNA microarrays, also called DNA chips, which carry the entire human genome (▶ Figure 1.11) and are being used to determine which genetic disorders someone has, will develop,

▲ FIGURE 1.11 A gene chip carrying the human gene set. This chip can be used to diagnose genetic disorders.

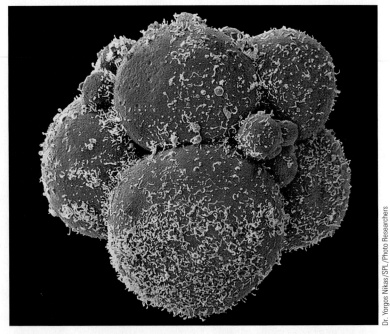

▲ FIGURE 1.12 Human embryo, shortly after fertilization in the laboratory. Embryos at this stage of development can be analyzed for genetic disorders before implantation into the uterus of the egg donor or that of another, surrogate mother.

or is predisposed to. DNA micro-arrays are also used in diagnosing infectious diseases and cancer.

In addition to the diagnosis of inherited diseases, genetic technology makes it possible to produce human embryos by fusion of sperm and eggs in a laboratory dish (▶ Figure 1.12) and to transfer the developing embryo to the womb of a surrogate mother. Embryos can also be frozen for transfer to a womb at a later time. We are beginning to treat genetic diseases by transplanting normal genes that act in place of defective copies, a procedure called **gene therapy**. We can even insert human genes into animals, creating new types of organisms to produce human proteins used in treating diseases such as emphysema.

PA Photo Library

▲ FIGURE 1.13 Dolly the sheep (right) with her offspring. Dolly was the first mammal cloned by nuclear transfer from a somatic cell.

■ **Gene therapy** Procedure in which normal genes are transplanted into humans carrying defective copies as a means for treating genetic diseases.

Biotechnology is impacting everyday life.

Recombinant DNA technology moved quickly from research laboratories into the business world; products and services using this technology are now commonplace. The commercial use of genetically modified organisms or their products is called biotechnology. They are found in hospitals, clinics, doctor's offices, drug stores, supermarkets, department stores, in law enforcement and the courts, and even in the production of industrial chemicals and the cleanup of waste sites.

The genetic modification of food is one of the most rapidly expanding and controversial areas of biotechnology. More than 60% of the corn and 80% of the soybeans grown in the United States is genetically modified. It is estimated that more than 60% of the processed food in supermarkets contains ingredients from transgenic plants.

Critics of this use of biotechnology have raised concerns that the use of herbicide-resistant corn and soybeans will speed the development of herbicide-resistant weeds and increase our use and dependence on chemical herbicides. Others point to the possibility that genetically engineered traits may be transferred to other organisms, leading to irreversible and deleterious changes in ecosystems.

Animals are also being cloned and genetically modified. The cloning of Dolly the sheep (▶ Figure 1.13) represented a breakthrough in cloning methods that along with other techniques makes it possible to produce dozens or hundreds of offspring with desirable traits such as high levels of milk production, meat with low fat content, or even speed in racehorses.

Recombinant DNA technology has been used for 20 years to produce human insulin in bacteria and other host cells for the treatment of diabetes. Now, genetically modified sheep and rabbits are being used to produce human proteins in their milk. These proteins are being used in clinical trials to treat human disease such as emphysema and Pompe disease. If successful, these proteins will soon be on the market.

Keep in mind

■ Recombinant DNA and biotechnology affect many aspects of our daily lives.

1.7 What Choices Do We Make in the Era of Genomics and Biotechnology?

In the span of about 35 years, we have learned how to predict the sex of unborn children, diagnose genetic disorders prenatally, and manufacture gene products to treat genetic diseases. We are now at a transition point where we are not only learning more about human genetics, we are starting to *apply* genetic knowledge in ways that were unforeseen just a few years ago. These applications are colliding with social standards, public policy, and laws, forcing us to rethink what is acceptable and unacceptable in our personal and public lives.

Should we buy and eat food from genetically modified plants and animals? Is milk from cloned cows safe to drink? Should we test ourselves or our children for genetic diseases, even if there is no treatment available? Is medicine produced from genetically modified animals safe? Should we vaccinate our children with edible vaccines produced from genetically altered bananas? We are faced with an increasing number of seemingly bewildering choices. Sorting through the rhetoric and hype to find the facts that allow us to make intelligent and informed choices is a problem in modern-day life. Beyond these immediate personal choices is the fact that the development of biotechnology is raising new ethical questions that we must face and answer in the near future.

We can make informed personal decisions and formulate relevant laws and public policy only if we have a working knowledge of the principles of genetics as they apply to humans and if we understand how genetics is used in biotechnology. As a student of human genetics, you have elected to become involved in the search for answers to these important questions.

Genetics in Practice

 Genetics in Practice case studies are critical thinking exercises that allow you to apply your new knowledge of human genetics to real-life problems. You can find these case studies and links to relevant websites at **http://biology.brookscole.com/cummings7**

CASE 1

In 1936, Fred Aslin and his eight brothers and sisters were sent to the Lapeer State School, a psychiatric institution in Michigan, after his father died and his mother was unable to care for her children. Neither he nor any of his sibs was mentally retarded. While he was institutionalized there, he and most of his siblings were labeled as feebleminded, and in 1944 at the age of 18, Fred was sterilized, as were three of his brothers and one of his sisters. After release from the institution, Fred became a farmer, served in the Korean War, and in 1996, he filed a request under the Freedom of Information Act to obtain copies of his records from the Lapeer School. What he found in the files infuriated him, and he filed suit against the state of Michigan, seeking compensation for the forced sterilization he underwent. In a March 2000 decision, the court ruled that the statute of limitations had expired, and he was denied compensation.

Fred's case is similar to those of many of the 60,000 U.S. citizens forcibly sterilized between 1907 and 1979. Michigan was one of the leading states in the number of sterilizations performed. Four states have issued formal apologies for the use of forced sterilization, but none have offered to compensate those who were sterilized.

1. Do you think states should apologize to individuals who were sterilized in the name of eugenics?

2. Do you think that states should compensate those who were sterilized? Why or why not?

1.1 Genetics Is the Key to Biology

■ Genetics is the science of heredity. In a sense, genetics is the key to all of biology, because genes control what cells look like and what they do. Understanding how genes work is essential to our understanding of how life works.

1.2 What Are Genes and How Do They Work?

■ The gene is the basic structural and functional unit of genetics. It is a string of chemical building blocks (nucleotides) in a DNA molecule. When a gene is turned on, the information stored in the gene is decoded and used to make a molecule that folds into a three-dimensional shape. This molecule is known as a protein (▷ Figure 1.2). The actions of proteins produce the traits we see (like eye color or hair color).

1.3 How Are Genes Transmitted from Parents to Offspring?

■ From his experiments on pea plants, Mendel concluded that pairs of genes separate from each other during the formation of egg and sperm. When the egg and sperm fuse during fertilization to form a zygote, the genes from the mother and father become members of a new gene pair in the offspring. The separation of genes during formation of the sperm and egg and the reunion of genes at fertilization is explained by the behavior of chromosomes in a form of cell division called meiosis.

1.4 How Do Scientists Study Genes?

■ Genes are studied using several different methods. Transmission genetics studies how traits are passed from generation to generation. Cytogenetics studies chromosome structure and the location of genes on chromosomes.

Molecular geneticists study the molecular makeup of genes and gene products and the function of genes. Population genetics focuses on the dynamics of populations and their interaction with the environment that results in changing gene frequencies over several generations.

1.5 Has Genetics Affected Social Policy and Law?

■ Eugenics was an attempt to improve the human race by using the principles of genetics. In the early years of the twentieth century, eugenics was a powerful force in shaping laws and public policy in the United States. This use of genetics was based on the mistaken assumption that genes alone determined human behavior and disorders and neglected the role of the environment. Eugenics fell into disfavor when it became part of the social programs of the Nazis in Germany.

1.6 What Impact Is Genetics Having Now?

■ The development of recombinant DNA technology is the foundation for DNA cloning, genome projects, and biotechnology. These developments are causing large-scale changes in many aspects of life and are impacting medicine, agriculture, and the legal system.

1.7 What Choices Do We Make in the Era of Genomics and Biotechnology?

■ With the completion of the Human Genome Project, the ability to manipulate human reproduction, and the ability to transfer genes, we are faced with many personal and social decisions. The ethical use of genetic information and biotechnology will require participation by a broad cross section of society.

Questions and Problems

 HUMAN Genetics Now™ Preparing for an exam? Assess your understanding of this chapter's topics with a pre-test, a personalized learning plan, and a post-test at **http://now.ilrn.com/cummings7**

1. Summarize Mendel's conclusions about traits and how they are passed from generation to generation.
2. What is population genetics?
3. What is hereditarianism, and what is the invalid assumption it makes?
4. Why are restriction enzymes important tools in recombinant DNA technology?
5. What are genomes? What is genomics?
6. In what way has biotechnology had an impact on agriculture in the United States?

7. We each carry 25,000 to 30,000 genes in our genome. Genes can be patented, and over 6,000 human genes have been patented. Do you think that companies or individuals should be able to patent human genes? Why or why not?
8. If your father were diagnosed with an inherited disease that develops around the age of 50, would you want to be tested to know if you will develop this disease? If so, when would you want to be tested? As a teenager, or sometime in your 40s? If not, would you have children?

Internet Activities

Internet Activities are critical thinking exercises using the resources of the World Wide Web to enhance the principles and issues covered in this chapter. For a full set of links and questions investigating the topics described below, log on to **http://biology.brookscole.com/cummings7**

1. *Learning Styles.* You can learn more from your studies in any subject if you know something about your personal learning preferences. At the *Active Learning Site* you may take a simple, informal assessment of your learning style. After completing the VARK learning style inventory, explore the tips for using your preferred style(s) to enhance learning.
2. *How to Study Biology.* The University of Texas maintains a website that provides suggestions on how to approach the study of biology, including genetics. Check out the general study suggestions for biology courses. Try developing a concept map, as outlined at this website, for some of the topics being covered in your genetics course.

3. *Genetics as a Contemporary Field of Research.* Genetics is one of the most active research fields in biology today. Go to the website for the Genetics Society of America and browse the information on the journal, meetings, and awards. Using the link to the "Careers Brochure," read what a number of prominent geneticists have to say about their careers.
4. *The Ongoing Eugenics Debate.* For a history of the eugenics movement in the United States, take a look at the "Eugenics Slide Show." Although the eugenics movement in the United States declined by the mid-1930s, there are those who argue that eugenicists are alive and active among us. Check out the "Eugenics" page for links to several points of view on this issue.

How would you vote now?

The understanding of genetics, as well as the applications of this understanding and their impact on society are rapidly growing. Not all applications of genetic knowledge are for the good, and individuals in our society need to be aware of the principles and issues involved so they can make informed decisions about their own genetic matters. At the beginning of this chapter, you were asked how you would respond if a major medical center asked you to donate a sample of DNA and allow access to your medical records for a project searching for genes that control complex traits such as hypertension, cardiovascular disease, or mental retardation. Now that you know more about how genetics and genetic information have been used and abused, what do you think? Would you consent to participate in a project to identify human genes for complex traits? Visit the Human Heredity Companion website to find out more on the issue, then cast your vote online at **http://biology.brookscole.com /cummings7**

For further reading and inquiry, log on to **InfoTrac College Edition,** your world-class online library, including articles from nearly 5,000 periodicals, at **http://infotrac.thomsonlearning.com**

2

Cells and Cell Division

MACROPHAGES (the word literally means *big eaters*) are white blood cells that begin life in the bone marrow and when they mature are released into the bloodstream. These cells prowl through the body's nooks and crannies seeking out dead and dying cells. If a macrophage encounters an old or injured red blood cell, for example, it forms a pocket around it and pulls the cell into itself. In healthy people, the red blood cell is slowly digested within the macrophage when small packets in the macrophage, called lysosomes, surround the red blood cell, releasing molecules that break down the red blood cell. The breakdown products are saved, recycled, and used to make new cells. After digesting the red blood cell, the macrophage expels any remaining debris and continues the hunt for other aging cells.

Lysosomes are an important part of the body's recycling program, and genetic disorders that cause lysosomal defects can have serious consequences. In one of these disorders, Gaucher disease, lysosomes lack an enzyme needed to break down membranes. As a result, cell parts remain undigested, and harmful amounts of molecules called glucosylceramides accumulate in macrophages, causing them to swell to several times their normal size and become nonfunctional. These enlarged cells, called Gaucher cells (see inset), accumulate in liver, spleen, and bone marrow. The liver and spleen enlarge and become damaged as Gaucher cells collect in these organs. Gaucher cells in bone marrow cause bone and joint pain and make the bones fragile and easily fractured.

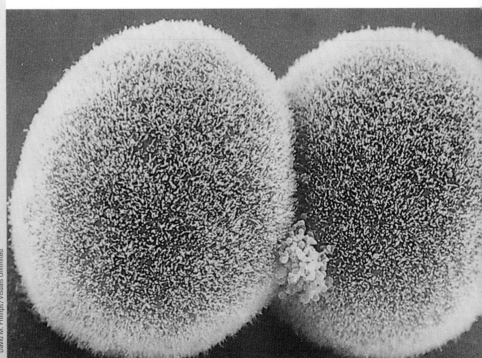

David M. Phillips/Visuals Unlimited

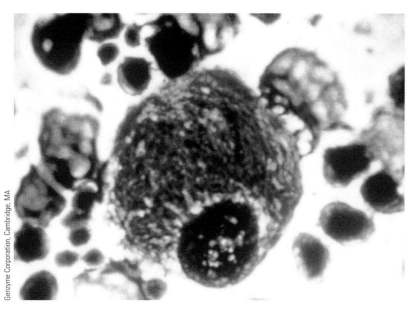

Genzyme Corporation, Cambridge, MA

An enlarged Gaucher cell.

Keep in mind as you read

■ Many genetic disorders alter cellular structure or function.

■ Gaucher disease affects lysosomal function.

■ Cancer is a disease of the cell cycle.

■ Meiosis maintains a constant chromosome number from generation to generation.

Gaucher disease is prevalent in populations with Eastern European Jewish ancestry. This disorder can be diagnosed by genetic testing and treated with a recombinant DNA–produced form of the enzyme given intravenously. Each treatment is done on an outpatient basis, takes 1 to 2 hours, and is usually done for life. While effective, treatment is expensive, costing $125,000 to $150,000 a year.

How would you vote?

Bone marrow transplantation is an alternative treatment for Gaucher disease and offers a permanent cure in place of costly, twice-weekly enzyme infusions. Some have argued that bone marrow donors are in short supply and because Gaucher disease is not life-threatening and can be treated by other means, these patients should have a lower priority as candidates for transplantation than those with high-risk diseases such as leukemia. Do you think candidates for transplants should be prioritized according to their illness? Visit the Human Heredity Companion website to find out more on the issue, then cast your vote online at **http://biology.brookscole.com/cummings7**

A Fatal Membrane Flaw

Cystic fibrosis is a genetic disorder that leads to early death. Affected individuals have thick, sticky secretions of the pancreas and lungs. Diagnosis is often made by finding elevated levels of chloride ions in sweat. According to folklore, mid-wives would lick the forehead of newborns. If the sweat was salty, they predicted that the infant would die a premature death. Despite intensive therapy and drug treatments, the average survival of persons with this disorder is only about 25 years. Cystic fibrosis is caused by a functional defect in a membrane protein that controls the movement of chloride ions across the plasma membrane. In normal cells, this protein functions as a channel controlling the flow of chloride, but in cystic fibrosis, the channel is unable to open. This causes chloride ions to accumulate inside the cell. To balance the chloride ions, the cells absorb excess sodium. In secretory glands, this leads to decreases in fluid production, resulting in blockage of flow from the pancreas and the accumulation of thick mucus in the lungs. The symptoms and premature death associated with this disorder point out the important role of membranes in controlling cell function.

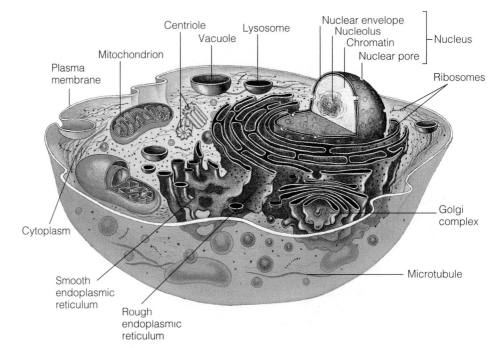

▲ FIGURE 2.1 A generalized human cell showing the organization and distribution of organelles as they would appear in the transmission electron microscope. The type, number, and distribution of organelles found in cells are related to cell function.

2.1 Cell Structure Reflects Function

We will review some of the basic aspects of human cell structure and then discuss the functions of cell components. Although cells differ widely in their size, shape, functions, and life cycle, they are fundamentally similar to one another—they all have a plasma membrane, cytoplasm, membranous organelles, and a membrane-bound nucleus. An idealized human cell is shown in ▶ Figure 2.1. A cell's structure and function are under genetic control, and many genetic disorders cause changes in cellular structure and/or function.

Keep in mind

■ Many genetic disorders alter cellular structure or function.

There are two cellular domains: the plasma membrane and the cytoplasm.

A double-layered plasma membrane separates the cell from the external environment. This membrane is a dynamic and active component of cell function and controls the exchange of materials with the environment outside the cell (▶ Figure 2.2). Gases, water, and some small molecules pass through the membrane easily, but large molecules are transported by energy-requiring systems. Molecules in and on the plasma membrane give the cells a form of molecular identity. The type and number of these molecules are genetically controlled and responsible for many important properties of cells, including blood type and compatibility in organ transplants. Several genetic disorders, including cystic fibrosis (OMIM 219700; see Spotlight on a Fatal Membrane Flaw), are associated with the plasma membrane. (See Chapter 4 for an explanation of OMIM numbers and the catalog of human genetic disorders.) The plasma membrane encloses the cytoplasm, which is a complex mixture of **molecules** and structural components. The cytoplasm also contains a number of specialized structures known collectively as **organelles.**

■ **Molecules** Structures composed of two or more atoms held together by chemical bonds.

■ **Organelles** Cytoplasmic structures that have a specialized function.

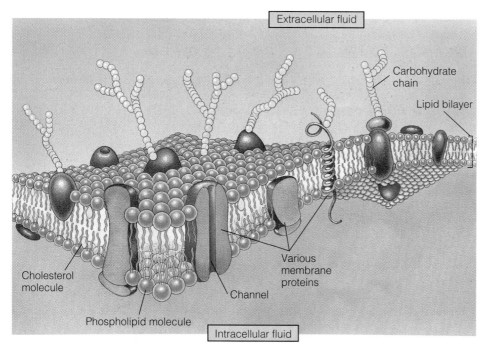

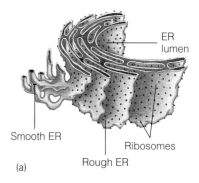

FIGURE 2.3 (a) Three-dimensional representation of the endoplasmic reticulum (ER), showing the relationship between the smooth and rough ER. (b) An electron micrograph of ribosome-studded rough ER.

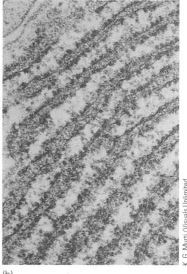

(b)

FIGURE 2.2 The plasma membrane. Proteins are embedded in a double layer of lipids. Short carbohydrate chains are attached to some proteins on the outer surface of the membrane.

Organelles are specialized structures in the cytoplasm.

The cytoplasm in a human cell has an organization related to its function, and this is reflected in its organelle content. In eukaryotes, cytoplasmic organelles divide the cell into a number of functional compartments. ▶ Table 2.1 summarizes the major organelles and their functions. We will review some of them here.

Endoplasmic Reticulum   The **endoplasmic reticulum (ER)** is a network of membranes that form channels in the cytoplasm (▶ Figure 2.3). The space inside the ER is a separate compartment called the lumen. In the lumen, proteins are folded, modified, and readied for transport to other locations in the cell or are tagged for export from the cell.

The outer surface of the ER is often covered with **ribosomes,** another cytoplasmic component (▶ Figure 2.3). Ribosomes are the most numerous cellular structures and can be found in the cytoplasm or attached to the outer surface of the ER. Ribosomes are the site of protein synthesis. (The process of protein synthesis is discussed in Chapter 9.)

Golgi Apparatus   Animal cells contain clusters of flattened membrane sacs called the **Golgi apparatus** (▶ Figure 2.4). The Golgi receive proteins from the ER and distribute them to their destinations inside and outside the cell. Abnormalities of the Golgi are responsible for Menkes disease (OMIM 309400) and other genetic disorders. The Golgi apparatus is also a source of membranes for other organelles, including lysosomes.

Lysosomes   The **lysosomes** are membrane-enclosed vesicles that contain digestive enzymes. The enzymes are made in the ER and transported to the Golgi where they are packaged into vesicles that bud off the Golgi to form the lysosomes (▶ Figure 2.4). Lysosomes are the processing centers of the cell. Materials brought into the cell including proteins, fats, carbohydrates, and viruses that are marked for destruction end up in lysosomes, where they are broken down and recycled or exported for

■ **Endoplasmic reticulum (ER)**   A system of cytoplasmic membranes arranged into sheets and channels that function in synthesizing and transporting gene products.

■ **Ribosomes**   Cytoplasmic particles composed of two subunits that are the site of protein synthesis.

■ **Golgi apparatus**   Membranous organelles composed of a series of flattened sacs. They sort, modify, and package proteins synthesized in the ER.

■ **Lysosomes**   Membrane-enclosed organelles that contain digestive enzymes.

Table 2.1   Overview of Cell Organelles

Organelle	Structure	Function
Nucleus	Round or oval body; surrounded by nuclear envelope.	Contains the genetic information necessary to control cell structure and function. DNA contains heredity information.
Nucleolus	Round or oval body in the nucleus consisting of DNA and RNA.	Produces ribosomal RNA.
Endoplasmic reticulum	Network of membranous tubules in the cytoplasm of the cell. Smooth endoplasmic reticulum contains no ribosomes. Rough endoplasmic reticulum is studded with ribosomes.	Smooth endoplasmic reticulum (SER) is involved in producing phospholipids and has many different functions in different cells. Rough endoplasmic reticulum (RER) is the site of the synthesis of lysosomal enzymes and proteins for extracellular use.
Ribosomes	Small particles found in the cytoplasm; made of RNA and protein.	Aid in the production of proteins on the RER and ribosome complexes (polysomes).
Golgi apparatus	Series of flattened sacs usually located near the nucleus.	Sorts, chemically modifies, and packages proteins produced on the RER.
Secretory vesicles	Membrane-bound vesicles containing proteins produced by the RER and repackaged by the Golgi apparatus; contain protein hormones or enzymes.	Store protein hormones or enzymes in the cytoplasm awaiting a signal for release.
Lysosome	Membrane-bound structure containing digestive enzymes.	Combines with food vacuoles and digests materials engulfed by cells.
Mitochondria	Round, oval, or elongated structures with a double membrane. The inner membrane is thrown into folds.	Complete the breakdown of glucose, producing NADH and ATP.

disposal. These organelles are important in cellular maintenance, and several genetic disorders, including Gaucher disease (OMIM 230800), described at the beginning of the chapter, disrupt or stop lysosome function. In most of these diseases, molecules transferred to lysosomes cannot be broken down and thus are stored there, causing the lysosome to enlarge and become distorted, eventually altering normal cell structure and function. Some lysosomal storage diseases are fatal. For example, Tay-Sachs disease (OMIM 272800) and Pompe disease (OMIM 232300) cause severe mental retardation, blindness, and death by the age of 3 or 4 years. These and other disorders that affect the structure or function of cell organelles reinforce the point made earlier that the functioning of the organism can be explained by events that occur within its cells.

Keep in mind

■ Gaucher disease affects lysosomal function.

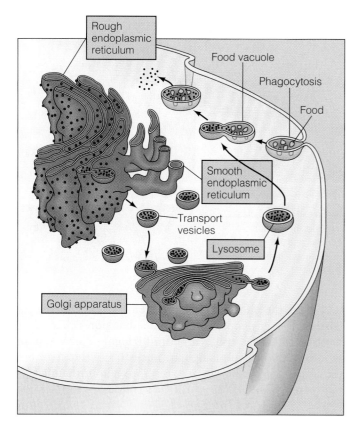

◀ FIGURE 2.4 The relationship between the Golgi complex and lysosomes. Digestive enzymes are synthesized in the ER and move to the Golgi in transport vesicles. In the Golgi, the enzymes are modified and packaged. Lysosomes pinch off the end of the Golgi membrane. In the cytoplasm, lysosomes fuse with and digest the contents of vesicles that are internalized from the plasma membrane.

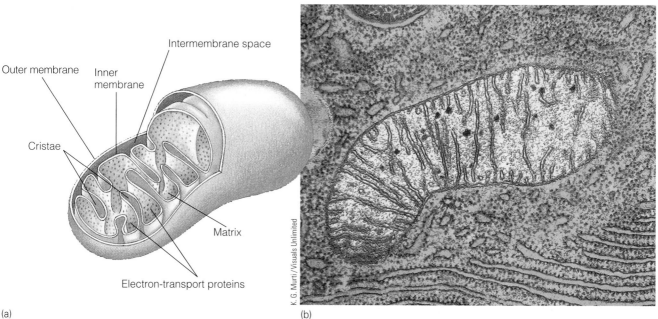

(a)

(b)

K. G. Murti/Visuals Unlimited

▲ FIGURE 2.5 The mitochondrion is a cell organelle involved in energy transformation. (a) The infolded inner membrane forms two compartments where chemical reactions transfer energy from one form to another. (b) A transmission electron micrograph of a mitochondrion.

Mitochondria   Energy transformation takes place in **mitochondria** (▶ Figure 2.5). Mitochondria carry their own genetic information in the form of circular DNA molecules. Mutations in mitochondrial DNA can cause a number of genetic disorders, including Kearns-Sayre syndrome (OMIM 530000) and MELAS syndrome (OMIM 535000). These and other genetic disorders affecting mitochondria are discussed in Chapter 4.

■ **Mitochondria** (singular: mitochondrion)   Membrane-bound organelles, present in the cytoplasm of all eukaryotic cells, that are the sites of energy production within the cells.

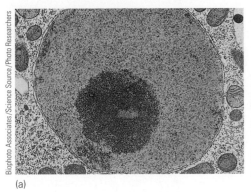
(a)

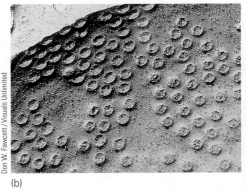
(b)

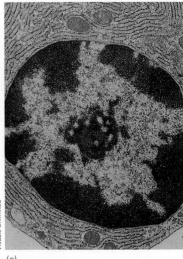
(c)

▲ FIGURE 2.6 (a) The nucleus is bounded by a double-layered plasma membrane called the nuclear membrane or nuclear envelope. The nucleolus is a prominent structure in the nucleus. (b) The nuclear membrane is studded with pores to allow exchange of materials between the nucleus and the cytoplasm. (c) During interphase, the chromosomes are uncoiled and dispersed throughout the nucleus as clumps of chromatin, clustered near the nuclear membrane.

■ **Nucleus**   The membrane-bounded organelle in eukaryotic cells that contains the chromosomes.

■ **Nucleolus (plural: nucleoli)**   A nuclear region that functions in the synthesis of ribosomes.

■ **Chromatin**   The component material of chromosomes, visible as clumps or threads in nuclei under a microscope.

■ **Chromosomes**   The threadlike structures in the nucleus that carry genetic information.

■ **Sex chromosomes**   In humans, the X and Y chromosomes that are involved in sex determination.

■ **Autosomes**   Chromosomes other than the sex chromosomes. In humans, chromosomes 1 to 22 are autosomes.

■ **Genes**   The fundamental units of heredity.

■ **Cell cycle**   The sequence of events that takes place between successive mitotic divisions.

■ **Interphase**   The period of time in the cell cycle between mitotic divisions.

■ **Mitosis**   Form of cell division that produces two cells, each of which has the same complement of chromosomes as the parent cell.

■ **Cytokinesis**   The process of cytoplasmic division that accompanies cell division.

Nucleus   The largest organelle is the **nucleus** (▶ Figure 2.6a). It is enclosed by a double membrane called the nuclear envelope. The envelope has pores that allow direct communication between the nucleus and cytoplasm (▶ Figure 2.6b). Within the nucleus, dense regions known as **nucleoli** (singular: **nucleolus;** ▶ Figure 2.6a) synthesize ribosomes. Dark strands of **chromatin** are seen throughout the nucleus (▶ Figure 2.6c). As a cell prepares to divide, the chromatin condenses to form the **chromosomes.**

In humans, chromosomes exist in pairs. Most human cells, called *somatic cells*, carry 23 pairs, or 46 chromosomes, but certain cells, such as sperm and eggs, carry only one copy of each chromosome and have 23 unpaired chromosomes. Human males have one pair of chromosomes that are not completely matched. Members of this pair are known as **sex chromosomes** and are involved in sex determination (see Chapter 7 for a discussion of this topic). There are two types of sex chromosomes, X and Y. Males carry an X and a Y chromosome, and females carry two X chromosomes. All other chromosomes are known as **autosomes.**

The chromosomes, carried in the nucleus, contain the genetic information that ultimately determines the structure and shape and function of the cell. The genetic information is composed of DNA and organized into units called **genes.** DNA and its associated proteins are organized into chromosomes.

2.2 The Cell Cycle Describes the Life History of a Cell

Cells in the body alternate between two states: division and nondivision. The time between cell divisions varies from minutes to months or even years. The sequence of events from division to division is called the **cell cycle.** A cycle consists of three parts: **interphase, mitosis,** and **cytokinesis** (▶ Figure 2.7). The time between divisions is the interphase, which is the first part of the cell cycle. The other two parts, mitosis (division of the chromosomes) and cytokinesis (division of the cytoplasm), define cell division.

Interphase has three stages.

Let's begin a discussion of the cell cycle with a cell that has just finished division. After division, the two daughter cells are about one-half the size of the parental cell. Before they can divide again, they must undergo a period of growth. These events take place during the three stages of interphase: G1, S, and G2.

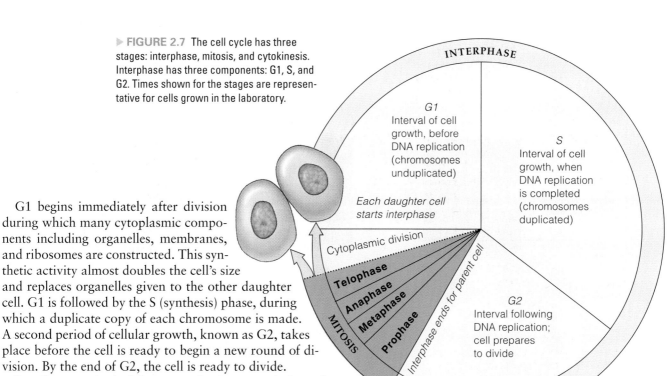

▶ **FIGURE 2.7** The cell cycle has three stages: interphase, mitosis, and cytokinesis. Interphase has three components: G1, S, and G2. Times shown for the stages are representative for cells grown in the laboratory.

G1 begins immediately after division during which many cytoplasmic components including organelles, membranes, and ribosomes are constructed. This synthetic activity almost doubles the cell's size and replaces organelles given to the other daughter cell. G1 is followed by the S (synthesis) phase, during which a duplicate copy of each chromosome is made. A second period of cellular growth, known as G2, takes place before the cell is ready to begin a new round of division. By the end of G2, the cell is ready to divide.

In cells grown in the laboratory, the time spent in interphase (G1, S, and G2) varies from 18 to 24 hours. Mitosis usually takes less than 1 hour, so cells spend most of their time in interphase. ▶ Table 2.2 summarizes the phases of the cell cycle.

The life history and cell cycles vary for different cell types. Some cells, like those in bone marrow, pass through the cell cycle continuously and divide regularly to form blood cells. At the other extreme, some cell types permanently enter an inactive state called G0 and never divide. In between are cell types, such as white blood cells, that are stopped in G1 but can divide in response to an infection.

When cells escape from the controls that are part of the cell cycle, they can become cancerous (see Genetic Journeys: Sea Urchins, Cyclins, and Cancer).

Table 2.2 Phases of the Cell Cycle

Phase	Characteristics
Interphase	
G1 (gap 1)	Stage begins immediately after mitosis. RNA, protein, and other molecules are synthesized.
S (synthesis)	DNA is replicated. Chromosomes become double stranded.
G2 (gap 2)	Mitochondria divide. Precursors of spindle fibers are synthesized.
Mitosis	
Prophase	Chromosomes condense. Nuclear envelope disappears. Centrioles divide and migrate to opposite poles of the dividing cell. Spindle fibers form and attach to chromosomes.
Metaphase	Chromosomes line up on the midline of the dividing cell.
Anaphase	Chromosomes begin to separate.
Telophase	Chromosomes migrate or are pulled to opposite poles. New nuclear envelope forms. Chromosomes uncoil.
Cytokinesis	Cleavage furrow forms and deepens. Cytoplasm divides.

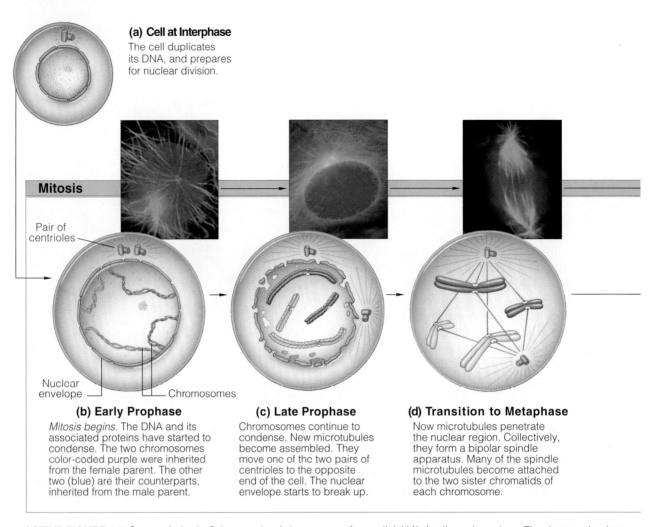

(a) Cell at Interphase
The cell duplicates its DNA, and prepares for nuclear division.

Mitosis

Pair of centrioles

Nuclear envelope — Chromosomes

(b) Early Prophase
Mitosis begins. The DNA and its associated proteins have started to condense. The two chromosomes color-coded purple were inherited from the female parent. The other two (blue) are their counterparts, inherited from the male parent.

(c) Late Prophase
Chromosomes continue to condense. New microtubules become assembled. They move one of the two pairs of centrioles to the opposite end of the cell. The nuclear envelope starts to break up.

(d) Transition to Metaphase
Now microtubules penetrate the nuclear region. Collectively, they form a bipolar spindle apparatus. Many of the spindle microtubules become attached to the two sister chromatids of each chromosome.

▲ ACTIVE FIGURE 2.8 Stages of mitosis. Only two pairs of chromosomes from a diploid (2*n*) cell are shown here. The photographs show mitosis in a mouse cell; the DNA is stained blue and the microtubules (spindle fibers) are stained green.

HUMAN
Genetics ⓔ Now™ Learn more about mitosis by viewing the animation on Human GeneticsNow at **http://now.ilrn.com/cummings7**

Cell division by mitosis occurs in four stages.

When a cell reaches the end of G2, it begins division, the second major part of the cell cycle. During division, two important steps are completed. A complete set of chromosomes is distributed to each daughter cell (mitosis), and the cytoplasm is distributed more or less equally to the two daughter cells (cytokinesis). Although cytoplasmic division can be somewhat imprecise and still be operational, the division and distribution of the chromosomes must be accurate and unfailing for the cell to function properly.

The net result of division is two daughter cells. In humans, each daughter cell receives a set of 46 chromosomes derived from a single parental cell with 46 replicated chromosomes. Although the distribution of chromosomes in cell division is usually precise, errors in this process do occur. These mistakes often have serious genetic consequences and are discussed in detail in Chapter 6.

Although mitosis is a continuous process, for the sake of discussion it is divided into four phases: prophase, metaphase, anaphase, and telophase (▶ Active Figure 2.8). These phases are accompanied by changes in chromosome organization as described in the following sections.

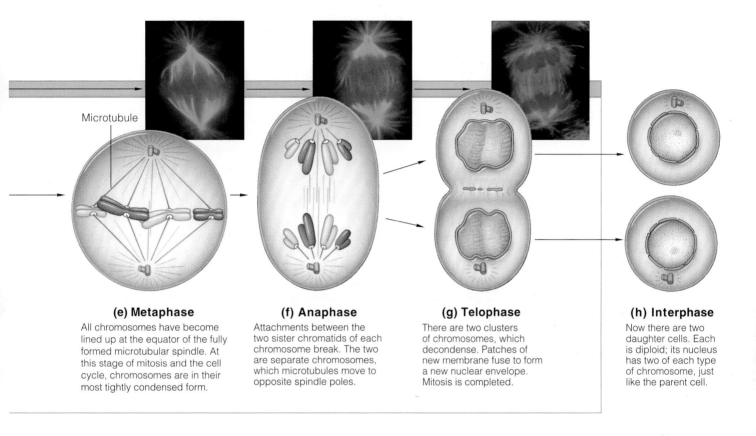

(e) Metaphase

All chromosomes have become lined up at the equator of the fully formed microtubular spindle. At this stage of mitosis and the cell cycle, chromosomes are in their most tightly condensed form.

(f) Anaphase

Attachments between the two sister chromatids of each chromosome break. The two are separate chromosomes, which microtubules move to opposite spindle poles.

(g) Telophase

There are two clusters of chromosomes, which decondense. Patches of new membrane fuse to form a new nuclear envelope. Mitosis is completed.

(h) Interphase

Now there are two daughter cells. Each is diploid; its nucleus has two of each type of chromosome, just like the parent cell.

Prophase   Prophase marks the beginning of mitosis. In the interphase just before prophase starts (▶ Active Figure 2.8a) the cell has replicated its DNA and chromosomes. Chromosomes are not usually visible in the nuclei of nondividing cells because they are uncoiled to form chromatin. At the beginning of **prophase,** the chromosomes coil and become recognizable as chromosomes (▶ Active Figure 2.8b). At first, chromosomes appear as long, thin, intertwined threads. As prophase continues, the chromosomes become shorter and thicker (▶ Active Figure 2.8c). In human cells, 46 chromosomes are present. Near the end of prophase, each chromosome consists of two longitudinal strands known as **chromatids.** The chromatids are held together by a structure called the **centromere.** Two chromatids joined by a centromere are known as **sister chromatids** (▶ Figure 2.9). Near the end of prophase, the nuclear membrane breaks down and a network of specialized fibers known as spindle fibers forms in the cytoplasm. When fully formed, the spindle fibers stretch across the cell (▶ Active Figure 2.8d).

Metaphase   **Metaphase** begins when the chromosomes are free in the cytoplasm. The chromosomes move to the middle, or equator, of the cell, where spindle fibers attach to the centromeres (▶ Active Figure 2.8d and e). At this stage there are 46 centromeres, each attached to two sister chromatids.

▣ **Prophase**   A stage in mitosis during which the chromosomes become visible and split longitudinally except at the centromere.

▣ **Chromatid**   One of the strands of a duplicated chromosome, joined by a single centromere to its sister chromatid.

▣ **Centromere**   A region of a chromosome to which microtubule fibers attach during cell division. The location of a centromere gives a chromosome its characteristic shape.

▣ **Sister chromatids**   Two chromatids joined by a common centromere. Each chromatid carries identical genetic information.

▣ **Metaphase**   A stage in mitosis during which the chromosomes move and become arranged near the middle of the cell.

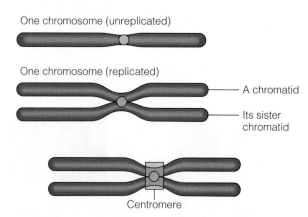

One chromosome (unreplicated)

One chromosome (replicated)

— A chromatid

— Its sister chromatid

Centromere

▲ FIGURE 2.9 Chromosomes replicate during the S phase. While attached to the centromere, the replicated chromosomes are called sister chromatids.

■ **Anaphase** A stage in mitosis during which the centromeres split and the daughter chromosomes begin to separate.

■ **Telophase** The last stage of mitosis, during which division of the cytoplasm occurs, the chromosomes of the daughter cells disperse, and the nucleus re-forms.

Anaphase In **anaphase,** the centromeres divide, converting each sister chromatid into a chromosome (▶ Active Figure 2.8f). A genetic disorder called Roberts syndrome is caused by a malfunction in centromere splitting during development (OMIM 268300; ▶ Figure 2.10). Late in anaphase, the chromosomes migrate toward opposite ends of the cell. By the end of anaphase, there is a complete set of chromosomes at each end of the cell. Although anaphase is the briefest stage of mitosis, it is essential for ensuring that each daughter cell receives a complete and identical set of 46 chromosomes.

Telophase and Cytokinesis At **telophase,** as the chromosomes reach opposite ends of the cell they begin to uncoil, the spindle fibers break down, and membranes from the ER begin to form a new nuclear membrane (▶ Active Figure 2.8g). At this point, mitosis is completed (▶ Active Figure 2.8h). The major features of mitosis are summarized in ▶ Table 2.3.

Cytokinesis, the division of the cytoplasm, begins with the formation of a constriction called a cleavage furrow at the equator of the cell (▶ Figure 2.11). The constriction gradually tightens by contraction of filaments just under the plasma membrane, which divides the cell in two.

▲ FIGURE 2.10 Roberts syndrome is a genetic disorder caused by malfunction of centromeres during mitosis. In this painting by Goya (1746–1828), the child on the mother's lap lacks limb development, which is characteristic of this syndrome.

Table 2.3	Summary of Mitosis
Stage	**Characteristics**
Interphase	Replication of chromosomes takes place.
Prophase	Chromosomes become visible as threadlike structures. As they continue to condense, they are seen as double structures, with sister chromatids joined at a single centromere.
Metaphase	Chromosomes become aligned at equator of cell.
Anaphase	Centromeres divide, and chromosomes move toward opposite poles.
Telophase	Chromosomes uncoil, nuclear membrane forms.
Cytokinesis	Cytoplasm divides.

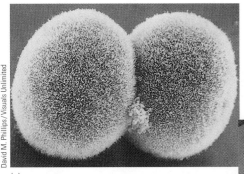

(a)

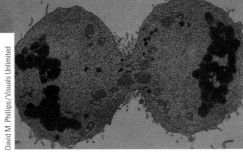

▲ **FIGURE 2.11** Cytokinesis. (a) A scanning electron micrograph of cleavage as seen from the outside of the cell. (b) A transmission electron micrograph of cytokinesis in a cross section of a dividing cell.

(b)

2.3 Mitosis Is Essential for Growth and Cell Replacement

Mitosis is an essential process in humans and all multicellular organisms. Some cells retain their capacity to divide throughout their life cycle, whereas others no longer divide after reaching adulthood. For example, cells in bone marrow continually move through the cell cycle and produce about 2 million red blood cells each second. Skin cells divide to replace dead cells that are continually sloughed off the surface of the body. By contrast, other cells, like many muscle cells, enter G0 and do not divide (see Spotlight on Cell Division and Spinal Cord Injuries).

Occasionally, cells escape from cell cycle regulation and grow uncontrollably, forming cancerous tumors. The mechanisms that regulate the cell cycle operate in G1. Much is known about how these systems work, and they will be described in Chapter 12, Genes and Cancer.

Cells grown in the laboratory undergo a characteristic number of divisions. Once this number, known as the Hayflick limit, is reached, the cells stop dividing. Cells from human embryos have a limit of about 50 divisions, enough to produce an adult and for cell replacement during a lifetime. Cells from adults can divide only about 10 to 30 times.

Keep in mind

■ Cancer is a disease of the cell cycle.

In human cells, the maximum number of divisions is under genetic control; several genetic disorders that affect cell division are associated with accelerated aging. One of these is progeria (OMIM 176670), in which 7- or 8-year-old affected children look like they are 70 or 80 years old (▶ Figure 2.12). Affected individuals usually die of coronary artery disease in their teens. Werner syndrome (OMIM 277700) is another genetic disorder associated with premature aging. In this case, the disease process begins between the ages of 15 and 20 years, and affected individuals die of age-related problems by 45 to 50 years.

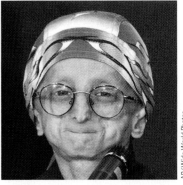

▲ **FIGURE 2.12** John Tacket, in spring of 2003 at age 15. He died in 2004, the oldest living person with progeria.

Meiosis I

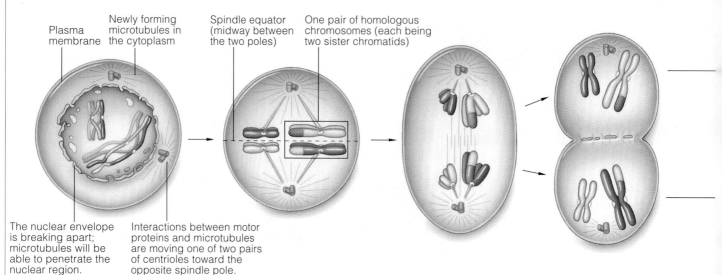

Plasma membrane

Newly forming microtubules in the cytoplasm

Spindle equator (midway between the two poles)

One pair of homologous chromosomes (each being two sister chromatids)

The nuclear envelope is breaking apart; microtubules will be able to penetrate the nuclear region.

Interactions between motor proteins and microtubules are moving one of two pairs of centrioles toward the opposite spindle pole.

(a) Prophase I

At the end of interphase, chromosomes are duplicated and in threadlike form. Now they start to condense. Each pairs with its homologue, and the two usually swap segments. The swapping, called crossing over, is indicated by the break in color on the pair of larger chromosomes. Newly forming spindle microtubules become attached to each chromosome.

(b) Metaphase I

Motor proteins projecting from the microtubules move the chromosomes and spindle poles apart. Chromosomes are tugged into position midway between the spindle poles. The spindle becomes fully formed by the dynamic interactions among motor proteins, microtubules, and chromosomes.

(c) Anaphase I

Some microtubules extend from the spindle poles and overlap at its equator. These *lengthen* and push the poles apart. Other microtubules extending from the poles *shorten* and pull each chromosome away from its homologous partner. These motions move the homologous partners to opposite poles.

(d) Telophase I

The cytoplasm of the cell divides at some point. There are now two haploid (*n*) cells with one of each type of chromosome that was present in the parent (2*n*) cell. *All chromosomes are still in the duplicated state.*

▲ **ACTIVE FIGURE 2.13** The stages of meiosis. In this form of cell division, homologous chromosomes physically associate to form a chromosome pair. Members of each pair separate from each other at meiosis I. In meiosis II, the centromeres of unpaired chromosomes divide, producing four cells, each with the haploid (*n*) number of chromosomes.

HUMAN
Genetics ℰNow™ Learn more about meiosis by viewing the animation on Human GeneticsNow at **http://now.ilrn.com/cummings7**

2.4 Cell Division by Meiosis: The Basis of Sex

The genetic information we inherit comes from two cells: a sperm and an egg. These cells are produced by a form of cell division called **meiosis** (▶ Active Figure 2.13). Recall that in mitosis, each daughter cell receives two copies of each chromosome. Cells with two copies of each chromosome are **diploid** or (**2n**) and have 46 chromosomes. In meiosis, members of a chromosome pair separate from each other, and so each cell receives a **haploid** (**n**) set of 23 chromosomes. These haploid cells form gametes (sperm and egg). Fusion of two gametes in fertilization restores the chromosome number to 46 and provides a full set of genetic information to the fertilized egg.

The distribution of chromosomes in meiosis is an exact process. Each gamete contains one member of each chromosome pair, not a random selection of 23 of the 46 chromosomes. The two rounds of division (meiosis I and meiosis II) accomplish the precise reduction in the chromosome number.

■ **Meiosis** The process of cell division during which one cycle of chromosomal replication is followed by two successive cell divisions to produce four haploid cells.

■ **Diploid (2n)** The condition in which each chromosome is represented twice as a member of a homologous pair.

■ **Haploid (n)** The condition in which each chromosome is represented once in an unpaired condition.

Keep in mind

■ Meiosis maintains a constant chromosome number from generation to generation.

Meiosis II

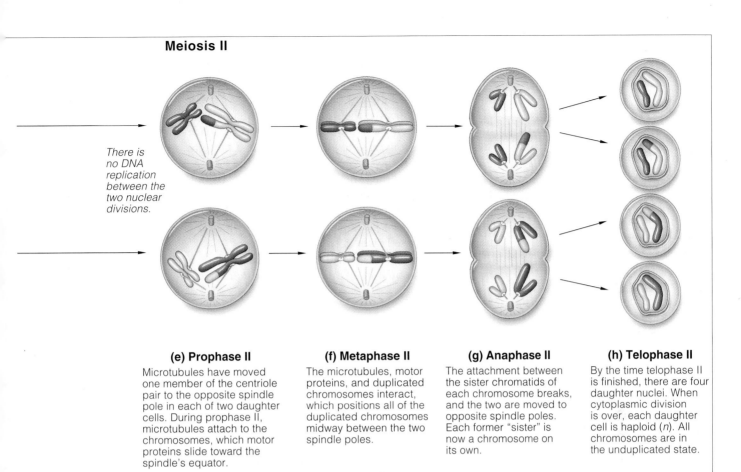

There is no DNA replication between the two nuclear divisions.

(e) Prophase II
Microtubules have moved one member of the centriole pair to the opposite spindle pole in each of two daughter cells. During prophase II, microtubules attach to the chromosomes, which motor proteins slide toward the spindle's equator.

(f) Metaphase II
The microtubules, motor proteins, and duplicated chromosomes interact, which positions all of the duplicated chromosomes midway between the two spindle poles.

(g) Anaphase II
The attachment between the sister chromatids of each chromosome breaks, and the two are moved to opposite spindle poles. Each former "sister" is now a chromosome on its own.

(h) Telophase II
By the time telophase II is finished, there are four daughter nuclei. When cytoplasmic division is over, each daughter cell is haploid (n). All chromosomes are in the unduplicated state.

Cells in the testis and ovary called germ cells undergo meiosis and produce gametes. In meiosis, diploid ($2n$) cells undergo one chromosomal replication followed by two divisions, resulting in four cells, each of which contains the haploid (n) number of chromosomes.

Meiosis I reduces the chromosome number.

Before cells enter meiosis, the chromosomes replicate during interphase. In prophase I, the chromosomes coil and become visible under a microscope (▷ Active Figure 2.13a). As the chromosomes coil, the nuclear membrane disappears, and the spindle becomes organized. Each chromosome physically associates with the other member of its pair. Members of a chromosome pair are **homologous chromosomes.** Once paired, the sister chromatids of each chromosome become visible so that each consists of two sister chromatids joined by a single centromere.

In metaphase I (▷ Active Figure 2.13b), members of a homologous pair line up across the middle of the cell. In anaphase I, members of each pair separate from each other and move toward opposite sides of the cell (▷ Active Figure 2.13c). Cytokinesis (division of the cytoplasm) occurs after telophase I, producing two haploid cells (▷ Active Figure 2.13d).

■ **Homologous chromosomes**
Chromosomes that physically associate (pair) during meiosis. Homologous chromosomes have identical gene loci.

Meiosis II begins with haploid cells.

In prophase II, the chromosomes coil and a spindle forms (▶ Active Figure 2.13e). Each unpaired chromosome consists of two sister chromatids joined by a centromere. At metaphase II (▶ Active Figure 2.13f), the 23 unpaired chromosomes attach to spindle fibers at their centromeres. Anaphase II (▶ Active Figure 2.13g) begins when the centromere of each chromosome divides for the first time. The 46 chromatids are converted to chromosomes and move to opposite ends of the cell.

In telophase II, the chromosomes uncoil, the nuclear membrane forms (▶ Active Figure 2.13h), and the process of meiosis is now complete. Cytokinesis then divides the cytoplasm, producing haploid cells. In meiosis, one diploid cell with 46 chromosomes has undergone one round of chromosome replication and two rounds of division to produce four haploid cells, each of which contains one copy of each chromosome (▶ Active Figure 2.13h).

The movement of chromosomes is summarized in ▶ Figure 2.14, and the events of meiosis are presented in ▶ Table 2.4. ▶ Figure 2.15 compares the events of mitosis and meiosis.

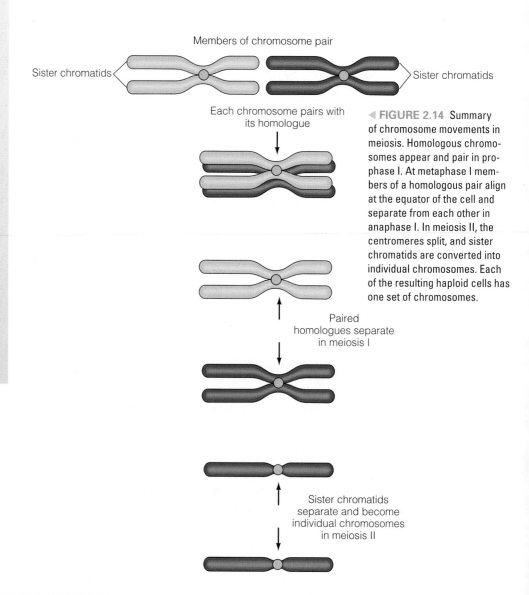

Members of chromosome pair

Sister chromatids — Sister chromatids

Each chromosome pairs with its homologue

Paired homologues separate in meiosis I

Sister chromatids separate and become individual chromosomes in meiosis II

◀ FIGURE 2.14 Summary of chromosome movements in meiosis. Homologous chromosomes appear and pair in prophase I. At metaphase I members of a homologous pair align at the equator of the cell and separate from each other in anaphase I. In meiosis II, the centromeres split, and sister chromatids are converted into individual chromosomes. Each of the resulting haploid cells has one set of chromosomes.

Genetic Journeys

Sea Urchins, Cyclins, and Cancer

Advances in human genetics and cancer research sometimes come from unexpected directions. One such story began at the Marine Biological Laboratories at Woods Hole, Massachusetts, in 1982. There, a group of young scientists led by Tim Hunt gathered for the summer to study biochemical changes that take place after fertilization in sea urchin eggs. They fertilized a batch of sea urchin eggs and, at 10-minute intervals, analyzed the newly made proteins during the first 2 to 3 hours of development. The fertilized egg first divides at about 1 hour and again about 2 hours after fertilization, resulting in a four-cell embryo.

Several new proteins appeared almost immediately after fertilization, including one that was continuously synthesized but then destroyed just before each round of cell division. Because of its cyclic behavior, this protein was called cyclin. Work with newly fertilized clam eggs revealed that this species also has cyclins that disappear just before mitosis. Because of their pattern of synthesis and destruction, Hunt and his colleagues concluded that cyclins might be involved in controlling cell division.

Subsequent work showed that cyclins are present in the cells of many organisms and act as important switches in controlling cell division. Sea urchins have only one cyclin, but humans and other mammals have as many as 8 to 12 different cyclins, each of which con-

trols one or more steps in cell division. What does all this have to do with cancer? It turns out that some non-dividing cells are arrested in the G1 phase. The mechanism that determines whether cells move through the cycle operates in G1. A critical switch point commits a cell to enter the S phase, G2, and mitosis or causes the cell to leave the cycle and become nondividing. The nature of this switch point, one of the central regulatory mechanisms in all of biology, is slowly being revealed by research in genetics and cell biology. The synthesis and action of cyclins generate the chemical signals that are part of this switch point. At the G1 control point, a cyclin combines with another protein, causing a cascade of events that moves the cell from G1 into S.

Cancer cells have disabled this signal and can divide continuously. Mutations in genes that control the synthesis or action of cyclins are important in the transition of a normal cell into a cancer cell. This important discovery is built on a foundation of work done on sea urchin embryos. Because eukaryotic cells share many properties, work done on yeast, sea urchin eggs, or clam embryos can be used to understand and predict events in normal human cells and in cells that have undergone mutations and become cancerous. For his work on cyclins, Tim Hunt shared the 2001 Nobel Prize for Physiology or Medicine with two other scientists who also worked on cell division.

Table 2.4 Summary of Meiosis

Stage	Characteristics
Interphase I	Chromosome replication takes place.
Prophase I	Chromosomes become visible, homologous chromosomes pair, and sister chromatids become visible. Recombination takes place.
Metaphase I	Paired chromosomes align at equator of cell.
Anaphase I	Homologous chromosomes separate. Members of each chromosome pair move to opposite poles.
Telophase I	Cytoplasm divides, producing two cells.
Prophase II	Chromosomes re-coil.
Metaphase II	Unpaired chromosomes become aligned at equator of cell.
Anaphase II	Centromeres split. Daughter chromosomes pull apart.
Telophase II	Chromosomes uncoil, nuclear membrane re-forms. Meiosis ends.
Cytokinesis	Cytoplasm divides after meiosis.

Meiosis produces new combinations of genes in two ways.

Meiosis produces new combinations of parental genes by random **assortment** of maternal and paternal chromosomes. Each pair of chromosomes we carry contains one from our mother (M) and one from our father (P). When chromosome pairs line up in metaphase I, the maternal and paternal members of each pair line up at random with respect to all other pairs (▶ Active Figure 2.16). In other words, the arrangement of any chromosomal pair can be maternal:paternal or paternal:maternal. As a result, cells produced in meiosis I are much more likely to receive a *combination* of maternal and paternal chromosomes than they are to receive a complete set of maternal chromosomes or a complete set of paternal chromosomes.

The number of chromosome combinations produced by meiosis is equal to 2^n, where 2 represents the chromosomes in each pair, and n represents the number of chromosomes in the haploid set. Humans have 23 chromosomes in the haploid set, so 2^{23}, or 8,388,608, different combinations of maternal and paternal chromosomes are possible in cells produced in meiosis I. Because each parent can produce 2^{23} combinations of chromosomes, more than 7×10^{13} combinations are possible in their children, each of which would carry a different assortment of parental chromosomes.

In meiosis, another process, called crossing-over, involves the physical exchange of parts between chromosome pairs (▶ Active Figure 2.17). This can produce many more combinations of paternal and maternal genes. When the variability generated by recombination is added to that produced by the random combination of maternal and paternal chromosomes, the number of different chromosome combinations that a couple can produce in their offspring has been estimated at 80^{23}. Obviously, the offspring of a couple represents only a very small fraction of all these possible gamete combinations. For this reason, it is almost impossible for any two children (aside from identical twins) to be genetically identical.

▼**ACTIVE FIGURE 2.16** The orientation of members of a chromosome pair at meiosis is random. Here three chromosomes (1, 2, and 3) have four possible alignments (maternal members of each chromosome pair are light blue; paternal members are dark blue). There are eight possible combinations of maternal and paternal chromosomes in the resulting haploid cells.

HUMAN Genetics ⑧ Now™ Learn more about assortment of chromosomes by viewing the animation on Human GeneticsNow at **http://now.ilrn.com/cummings7**

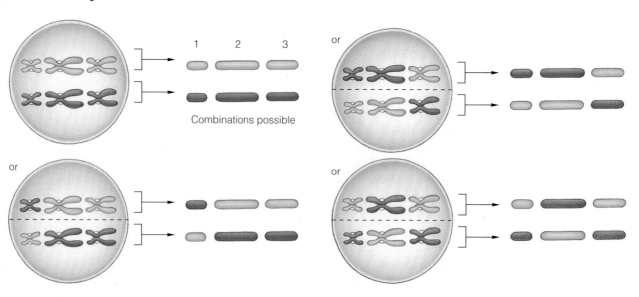

A comparison of the events in mitosis and meiosis. In mitosis (*left*), a diploid parental cell undergoes chromosomal replication and then enters prophase. The chromosomes appear doubled during late prophase, and unpaired chromosomes align at the middle (equator) of the cell during metaphase. In anaphase, the centromeres split, converting the sister chromatids into chromosomes. The result is two daughter cells, each of which is genetically identical to the parental cell. In meiosis I (*right*), the parental diploid cell undergoes chromosome replication and then enters prophase. Homologous chromosomes pair, and each chromosome appears doubled, except at the centromeres. Paired homologues align at the equator of the cell during metaphase, and members of a chromosome pair separate during anaphase. In meiosis II, the unpaired chromosomes in each cell align at the equator of the cell. During anaphase II, the centromeres split, and one copy of each chromosome is distributed to daughter cells. The result is four haploid daughter cells, which are not genetically equivalent to the parental cell.

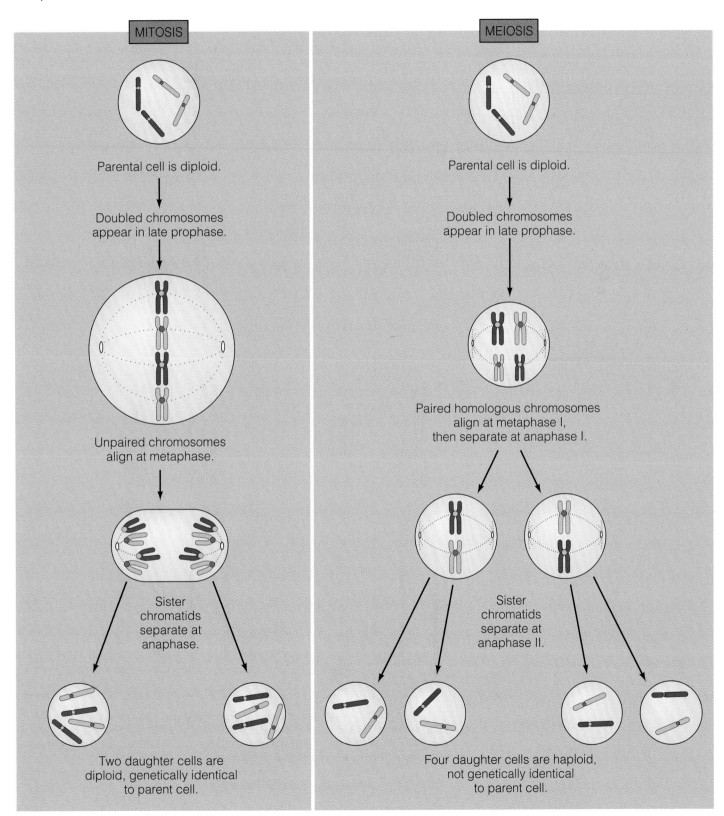

MITOSIS

MEIOSIS

Parental cell is diploid.

Parental cell is diploid.

Doubled chromosomes appear in late prophase.

Doubled chromosomes appear in late prophase.

Unpaired chromosomes align at metaphase.

Paired homologous chromosomes align at metaphase I, then separate at anaphase I.

Sister chromatids separate at anaphase.

Sister chromatids separate at anaphase II.

Two daughter cells are diploid, genetically identical to parent cell.

Four daughter cells are haploid, not genetically identical to parent cell.

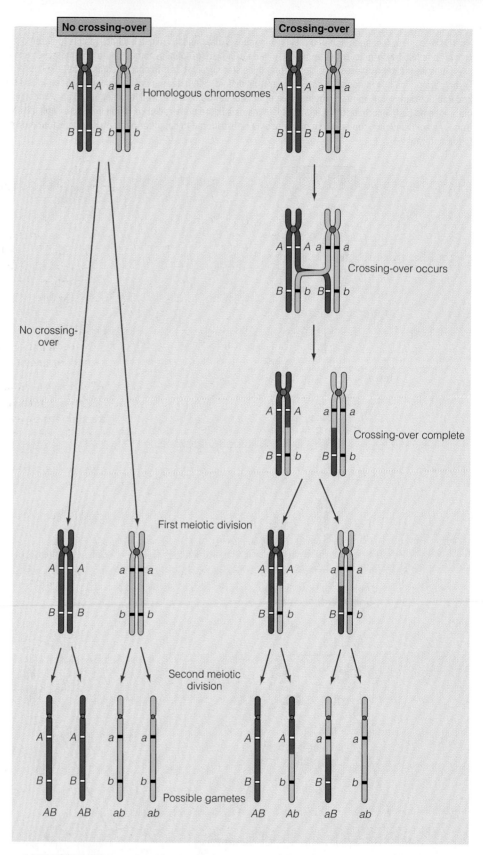

▲ **ACTIVE FIGURE 2.17** Crossing-over increases genetic variation by combining genes from both parents on the same chromatid. At left is shown the combination of maternal and paternal genes when no crossing-over occurs. At right, new combinations (*Ab, aB*) are produced by crossing-over, increasing genetic variability in the haploid cells that will form gametes.

HUMAN Genetics Ⓔ Now™   Learn more about crossing-over by viewing the animation on Human GeneticsNow at **http://now.ilrn.com/cummings7**

Genetics in Practice

 Genetics in Practice case studies are critical thinking exercises that allow you to apply your new knowledge of human genetics to real-life problems. You can find these case studies and links to relevant websites at **http://biology.brookscole.com/cummings7**

CASE 1

It is May 1989 and the scene is a crowded research laboratory with beakers, flasks, and pipettes covering the lab bench. People and equipment take up every possible space. One researcher, Joe, passes a friend staring into a microscope. Another student wears gloves while she puts precisely measured portions of various liquids into tiny test tubes. Joe glances at the DNA sequence results he is carrying. Something is wrong. There it is, a unique type of genetic mutation in a DNA sequence. The genetic information required to make a complete protein is missing, as if one bead had fallen from a precious necklace. Instead of returning to his station, Joe rushes to tell his supervisor, Dr. Tsui (pronounced "Choy"), that he has found a specific mutation in a person with cystic fibrosis (CF), but he does not see this same mutation in a normal person's genes. CF is a fatal disease that kills about 1 out of every 2,000 Caucasians (mostly children). Dr. Tsui examines the findings and is impressed but wants more evidence to prove that the result is real. He has had false hopes before, so he is not going to celebrate until they check this out carefully. Maybe the difference between the two gene sequences is just a normal variation among individuals. Five months later, Dr. Tsui and his team identify a "signature" pattern of DNA on either side of the mutation, and using that as a marker, they compare the genes of 100 normal people with the DNA sequence from 100 CF patients.

By September 1989, they are sure they have identified the CF gene. After several more years, Tsui and his team discover that the DNA sequence with the mutation encodes the information for a protein called CFTR (cystic fibrosis transmembrane conductance regulator), a part of the plasma membrane in cells that make mucus. This protein regulates a channel for chloride ions. Proteins are made of long chains of amino acids. The CFTR protein has 1,480 amino acids. Most children with CF are missing one single amino acid in their CFTR. Because of this, their mucus becomes too thick, causing all the other symptoms of CF. Thanks to Tsui's research, scientists now have a much better idea of how the disease works. We can now easily predict when a couple is at risk for having a child with CF. With increasing understanding, scientists may also be able to devise improved treatments for children born with this disease.

CF is the most common genetic disease among those of European ancestry. Children who have CF are born with it. Half of them will die before they are 25 and few make it past 30. It affects all parts of the body that secrete mucus: the lungs, stomach, nose, and mouth. The mucus of children with CF is so thick sometimes they cannot breathe. Why do 1 in 25 Caucasians carry the mutation for CF? Tsui and others think that people who carry it may also have resistance to diarrhea-like diseases.

1. Dr. Tsui's research team discovered the gene for cystic fibrosis. What medical advances can be made after a gene is cloned?

2. Why do you think a change in one amino acid in the CF gene can cause such severe effects in CF patients? Relate your answer to the CFTR protein function and the cell membrane.

CASE 2

Jim, a 37-year-old construction worker, and Sally, a 42-year-old business executive, were eagerly preparing for the birth of their first child. They, like more and more couples, chose to wait to have children until they were older and more financially stable. Sally had an uneventful pregnancy with prenatal blood tests and an ultrasound indicating that the baby looked great and everything seemed "normal." Then, a few hours after Ashley was born, they were told she was born with Down syndrome. In shock and disbelief, the couple questioned how this could happen to them. It has long been recognized that the risk of having a child with Down syndrome increases with maternal age. For example, the risk of having a child with Down syndrome when the mother is 30 years old is 1 in 1,000 and at age 40 is 9 in 1,000.

Well-defined and distinctive physical features characterize Down syndrome, which is the most frequent form of mental retardation caused by a chromosomal aberration. Most individuals (95%) with Down syndrome, or trisomy 21, have three copies of chromosome 21. Errors in meiosis that lead to trisomy 21 are almost always of maternal origin; only about 5% occur during spermatogenesis. It has been estimated that meiosis I errors account for 76% to 80% of maternal meiotic errors. In about 5% of patients, one copy is translocated to another chromosome, *(continued)* most often chromosome 14 or 21. No one is at fault when a child is born with Down syndrome, but the chances of it occurring increase with advanced maternal age.

Children with Down syndrome often have specific major congenital malformations such as those of the heart (30% to 40% in some studies), and have an increased incidence (10 to 20 times higher) of leukemia than the normal population. Ninety percent of all Down syndrome patients have significant hearing loss. The frequency of trisomy 21 in the population is 1 in 650 to 1,000 live births.

1. What prenatal tests could have been done to detect Down syndrome before birth? Should they have been done?

2. Down syndrome is characterized by mental retardation. Can individuals with Down syndrome go to school or hold a job?

3. Should people with mental disabilities be integrated into the community? Why or why not?

Summary

2.1 Cell Structure Reflects Function

■ The cell is the basic unit of structure and function in all organisms, including humans. Because genes control the number, size, shape, and function of cells, the study of cell structure helps us understand how genetic disorders disrupt cellular processes.

■ The nucleus contains the genetic information that controls the structure and function of the cell. This genetic information is carried in chromosomes. The number, size, and shape of chromosomes are species-specific traits. In humans, 46 chromosomes—the $2n$, or diploid, number—is present in most cells, whereas specialized cells known as gametes contain half of that number—the n, or haploid number—of chromosomes.

2.2 The Cell Cycle Describes the Life History of a Cell

■ At some point in their life, cells pass through the cell cycle, a period of nondivision (interphase) that alternates with division of the nucleus (mitosis) and division of the cytoplasm (cytokinesis). Cells must contain a complete set of genetic information. This is ensured by replication of each chromosome and by the distribution of a complete chromosomal set in the process of mitosis.

■ Mitosis (division) is one part of the cell cycle. During interphase (nondivision), a duplicate copy of each chromosome is made. The process of mitosis is divided into four stages: prophase, metaphase, anaphase, and telophase. In mitosis, one diploid cell divides to form two diploid cells. Each cell has an exact copy of the genetic information contained in the parental cell.

2.3 Mitosis Is Essential for Growth and Cell Replacement

■ Human cells are genetically programmed to divide about 50 times. This limit allows growth to adulthood and repairs such as wound healing. Alterations in this program can lead to genetic disorders of premature aging or to cancer.

2.4 Cell Division by Meiosis: The Basis of Sex

■ Meiosis is a form of cell division that produces haploid cells containing only the paternal or maternal copy of each chromosome. In meiosis, members of a chromosome pair physically associate. At this time, each chromosome consists of two sister chromatids joined by a common centromere. In metaphase I, pairs of homologous chromosomes align at the equator of the cell. In anaphase I, members of a chromosome pair separate from each other. Meiosis I produces cells that contain one member of each chromosome pair. In meiosis II, the unpaired chromosomes align at the middle of the cell. In anaphase II, the centromeres divide, and the daughter chromosomes move to opposite poles. The four cells produced in meiosis contain the haploid number (23 in humans) of chromosomes.

HUMAN
Genetics ⒺNow™   Preparing for an exam? Assess your understanding of this chapter's topics with a pre-test, a personalized learning plan, and a post-test at **http://now.ilrn.com/cummings7**

Cell Structure Reflects Function

1. What advantages are there in having the interior of the cell divided into a number of compartments such as the nucleus, the ER, lysosomes, and so forth?
2. Assign a function(s) to the following cellular structures:
   a. plasma membrane
   b. mitochondrion
   c. nucleus
   d. ribosome
3. How many autosomes are present in a body cell of a human being? In a gamete?
4. Define the following terms:
   a. chromosome   b. chromatin
5. Human haploid gametes (sperm and eggs) contain:
   a. 46 chromosomes, 46 chromatids
   b. 46 chromosomes, 23 chromatids
   c. 23 chromosomes, 46 chromatids
   d. 23 chromosomes, 23 chromatids

The Cell Cycle Describes the Life History of a Cell

6. What are sister chromatids?
7. Draw the cell cycle. What is meant by the term "cycle" in the cell cycle? What is happening at the S phase and the M phase?
8. In the cell cycle, at which stages do *two* chromatids make up *one* chromosome?
   a. beginning of mitosis
   b. end of G1
   c. beginning of S
   d. end of mitosis
   e. beginning of G2
9. Does the cell cycle refer to mitosis as well as meiosis?
10. It is possible that an alternative mechanism for generating germ cells could have evolved. Consider meiosis in a germ cell precursor (a cell that is diploid but will go on to make gametes). If the S phase were skipped, which meiotic division (meiosis I or meiosis II) would no longer be required?

11. Identify the stages of mitosis, and describe the important events that occur during each stage.
12. Why is cell furrowing important in cell division? If cytokinesis did not occur, what would be the end result?
13. A cell from a human female has just undergone mitosis. For unknown reasons, the centromere of chromosome 7 failed to divide. Describe the chromosomal contents of the daughter cells.
14. During which phases of the mitotic cycle would the terms *chromosome* and *chromatid* refer to identical structures?
15. Describe the critical events of mitosis that are responsible for ensuring that each daughter cell receives a full set of chromosomes from the parent cell.

Mitosis Is Essential for Growth and Cell Replacement

16. Mitosis occurs daily in a human being. What type of cells do humans need to produce in large quantities on a daily basis?
17. Speculate on how the Hayflick limit may lead to genetic disorders such as progeria or Werner syndrome. How is this related to cell division?
18. How can errors in the cell cycle lead to cancer in humans?

Cell Division by Meiosis: The Basis of Sex

19. List the differences between mitosis and meiosis in the following chart:

Attribute	Mitosis	Meiosis
Number of daughter cells produced		
Number of chromosomes per daughter cell		
Do chromosomes pair? (Y/N)		
Does crossing-over occur? (Y/N)		
Can the daughter cells divide again? (Y/N)		
Do the chromosomes replicate before division? (Y/N)		
Type of cell produced		

20. In the following diagram, designate each daughter cell as diploid (2n) or haploid (n).

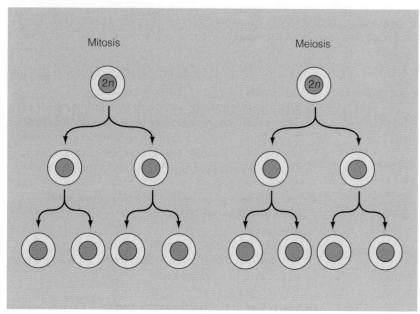

21. Which of the following statements is *not* true when comparing mitosis and meiosis?
  a. Twice the number of cells are produced in meiosis versus mitosis.
  b. Meiosis is involved in the production of gametes, unlike mitosis.
  c. Crossing-over occurs in meiosis I, not in meiosis II or mitosis.
  d. Meiosis and mitosis both produce cells that are genetically identical.
  e. In both mitosis and meiosis, the parental cell is diploid.

22. Match the phase of cell division with the following diagrams. In these cells, 2n = 4.
  a. anaphase of meiosis I
  b. interphase of mitosis
  c. metaphase of mitosis
  d. metaphase of meiosis I
  e. metaphase of meiosis II

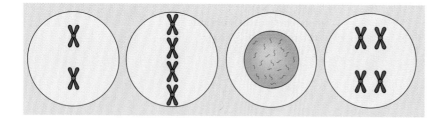

23. A cell has a diploid number of 6 (2n = 6).
  a. Draw the cell in metaphase of meiosis I.
  b. Draw the cell in metaphase of mitosis.
  c. How many chromosomes are present in a daughter cell after meiosis I?

  d. How many chromatids are present in a daughter cell after meiosis II?
  e. How many chromosomes are present in a daughter cell after mitosis?
  f. How many tetrads are visible in the cell in metaphase of meiosis I?

24. A cell (2n = 4) has undergone cell division. Daughter cells have the following chromosome content. Has this cell undergone mitosis, meiosis I, or meiosis II?

  a.

  b.

  c.

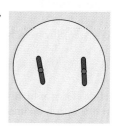

25. We are following the progress of human chromosome 1 during meiosis. At the end of prophase I, how many chromosomes, chromatids, and centromeres are present to ensure that chromosome 1 faithfully traverses meiosis?

26. What is physically exchanged during crossing-over?

27. Compare meiotic anaphase I with meiotic anaphase II. Which meiotic anaphase is more similar to the mitotic anaphase?

28. Provide two reasons why meiosis leads to genetic variation in diploid organisms.

Internet Activities are critical thinking exercises using the resources of the World Wide Web to enhance the principles and issues covered in this chapter. For a full set of links and questions investigating the topics described below, log on to **http://biology.brookscole.com/cummings7**

1. *Structure and Function of the Nucleus.* The *Cell Biology Topics* website maintained by the University of Texas presents basic information about cell biology arranged by organelle system.
   a. Choose the "Nucleus" link and explore the numerous structures within the nucleus.
   b. Within the "Nucleus" topic, choose the "chromosome" link to compare heterochromatin and euchromatin, the two different forms of DNA in the nucleus.

2. *Diversity of Cell Types.* The cellular world is almost unimaginably diverse, and modern technology has not only permitted new ways of viewing this diversity, it has made it possible to share this information worldwide. At the *Molecular Expressions Photo Gallery*, check out any of the "Galleries" on the contents page to view a variety of cells, organisms, cellular structures, and (occasionally) everyday objects photographed using a variety of different photomicrographic techniques. For an overview of different types of cellular structure (with colorful line drawings but no photomicrographs), follow the "Cell and Virus Structure" link.

3. *Mitosis Overview.* The "Mitosis" link at the *Molecular Expressions Photo Gallery* has both photomicrographs and an interactive tutorial for reviewing the phases of mitosis.

4. *Cell Size—and More Mitosis.* At the *Cells Alive!* website, follow the "Cell Biology" link and compare the sizes of different cells at the "How Big Is a . . . ?" page.
   Further Exploration. There is also a mitosis tutorial at the *Cells Alive!* website. Compare the mitosis tutorial at this site to the mitosis overview at the *Molecular Expressions Photo Gallery.*

How would you vote now?

It is possible to treat Gaucher disease, a genetic disorder resulting from a missing enzyme, with bone marrow transplantation. Transplanted bone marrow allows a Gaucher patient to produce the missing enzyme and inhibits the formation of the abnormal Gaucher cells. But bone marrow donors are in short supply, and there are other life-threatening diseases that can only be treated through such a transplant. Now that you know more about cells, what do you think? Should candidates for transplants be prioritized according to their illness? Visit the Human Heredity Companion website to find out more on the issue, then cast your vote online at **http://biology.brookscole.com/cummings7**

For further reading and inquiry, log on to **InfoTrac College Edition,** your world-class online library, including articles from nearly 5,000 periodicals, at **http://infotrac.thomsonlearning.com**

3

Transmission of Genes from Generation to Generation

O NE Friday evening in July, Patricia Stallings fed her 3-month-old son, Ryan, his bottle and put him to bed. Ryan became ill and threw up, but the next day he seemed better. By Sunday, however, Ryan was vomiting and had trouble breathing. Patricia drove him to a hospital in St. Louis. Tests there uncovered high levels of ethylene glycol, a component of antifreeze, in his blood. A pediatrician at the hospital believed that Ryan had been poisoned and had the infant placed in foster care. Patricia and David, her husband, could see him only during supervised visits. On one of those visits early in September, Patricia was left alone with Ryan briefly, fed him a bottle, and went home. After she left, Ryan became ill and died. The next day, Patricia Stallings was arrested and charged with murder. Authorities found large quantities of ethylene glycol in Ryan's blood and traces on a bottle Patricia used to feed Ryan on her visit. At trial, Patricia was found guilty of first-degree murder and sentenced to life in prison.

While in jail, she gave birth to another son, David Jr., who was immediately placed in foster care. Within 2 weeks, the baby developed similar symptoms, but since Patricia had no contact with her baby, she could not have poisoned him. Hearing of the case, two scientists at St. Louis University performed additional tests on blood taken from Ryan when he was hospitalized. They found no ethylene glycol in his blood and consulted with a human geneticist from Yale University, who conducted additional tests on Ryan's blood. His results also showed no traces of ethylene glycol, but he did find other compounds present, which helped solve the mystery. Based on these and further tests done on David Jr., the scientists presented evidence that previous testing was done improperly and that both Ryan and his brother suffered from a rare genetic disorder called methylmalonic acidemia (MMA). Biochemical evidence from blood samples supported this conclusion. Symp-

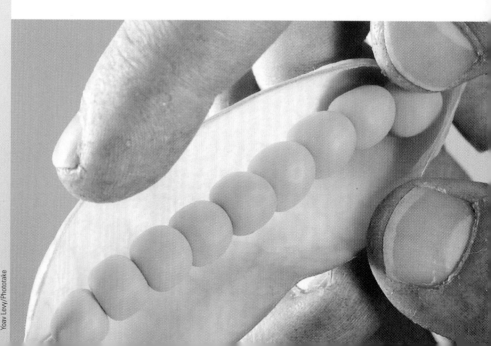

Yoav Levy/Phototake

toms of MMA are similar to those seen in ethylene glycol poisoning, but the cause is an inability to break down proteins in food. Faced with this evidence, Patricia Stallings's conviction was overturned, and she was released from jail after serving 14 months for a crime she did not commit.

How would you vote?

Laws in all 50 states and the District of Columbia require newborns to be screened for genetic disorders (from 4 disorders to more than 35, depending on the state). Many states screen for MMA, the disease that killed two of Patricia Stallings's children. Although some states allow exemptions for religious reasons, screening is mandatory in all states. Public health officials in favor of mandatory screening point out that for every $1 spent in screening, almost $9 is saved in health care costs. Others feel that mandatory screening violates patients' rights and express concern about the risk of having personal genetic information stored in newborn screening databases maintained by the state. These opponents also feel that screening may be used as the basis for future eugenics programs that restrict the reproductive rights of those diagnosed with a genetic disorder. Do you think that such screening should be mandatory, or should parents be able to refuse to have their child tested? Should schools, insurance companies, or employers have access to the results of such genetic testing without parental consent? Visit the Human Heredity Companion website to find out more on the issue, then cast your vote online at **http://biology.brookscole.com /cummings7**

Keep in mind as you read

■ Some traits can appear in offspring even when the parents don't have the trait.

■ We can identify genetic traits because they have a predictable pattern of inheritance worked out by Gregor Mendel.

■ Pedigrees are constructed to follow the inheritance of human traits.

3.1 Heredity: How Are Traits Inherited?

Before we get to a discussion of how traits in humans such as eye color and hair color are passed from generation to generation, let's ask the obvious question: Why are we starting with Gregor Mendel and pea plants if we are going to discuss human genetics? The answer won't be fully evident for a chapter or two, but there are two main reasons we start with pea plants. First, Mendel used experimental genetics to uncover the fundamental principles of genetics, principles that apply to pea plants as well as humans, and for ethical reasons, humans can't be used in experimental genetics; you can't make crosses. Second, humans have very few offspring compared to pea plants, and it takes a long time for one generation (20 or so years in humans compared to about 100 days in peas). As you will see in Chapters 4 and 5, studying how traits are

Mendel and Test Anxiety

Mendel entered the Augustinian monastery in 1843 and took the name Gregor. While studying at the monastery, he served as a teacher at the local technical high school. In the summer of 1850 he decided to take the examinations that would allow him to have a permanent appointment as a teacher. The exam was in three parts. Mendel passed the first two parts but failed one of the sections in the third part. In the fall of 1851, he enrolled at the University of Vienna to study natural science (the section of the exam he flunked). He finished his studies in the fall of 1853, returned to the monastery, and again taught at a local high school.

In 1855 he applied to take the teacher's examination again. The test was held in May of 1856, and Mendel became ill while answering the first question on the first essay examination. He left and never took another examination. As a schoolboy and again as a student at the monastery, Mendel had also suffered bouts of illness, all associated with times of stress.

In an analysis of Mendel's illnesses made in the early 1960s, a physician concluded that Mendel had a psychological condition that today would probably be called "test anxiety." If you are feeling stressed at exam time, take some small measure of comfort in knowing that it was probably worse for Mendel.

inherited in humans can be somewhat ambiguous. Thus, we begin with a model system in which the mechanisms of inheritance are clearly defined.

Johann Gregor Mendel was born in 1822 in Hynice, Moravia, a region now part of the Czech Republic. At the age of 21, he entered the Augustinian monastery at Brno as a way of continuing his interests in natural history (see Spotlight on Mendel and Test Anxiety). After completing his monastic studies, Mendel enrolled at the University of Vienna in the fall of 1851. In his course work, Mendel encountered the new ideas that cells are the fundamental unit of all living things. This new theory raised several questions about inheritance. Do both parents contribute equally to the traits of the offspring? In plants and most animals, the female gametes are much larger than those of the male, so this was a logical and much debated question. Related to this was the question of whether the traits in the offspring result from blending the traits of the parents. In 1854 Mendel returned to Brno to teach physics and began a series of experiments that were to resolve these questions.

3.2 Mendel's Experimental Design Resolved Many Unanswered Questions

Mendel's success in discovering the fundamental principles of inheritance was the result of carefully planned experiments. First, he set about choosing an organism for

▲ FIGURE 3.1 The study of the way traits such as flower color in pea plants and pod shape are passed from generation to generation provided the material for Mendel's work on heredity.

his experiments. Near the beginning of his landmark paper on inheritance, Mendel wrote:

> The value and validity of any experiment are determined by the suitability of the means as well as by the way they are applied. In the present case as well, it cannot be unimportant which plant species were chosen for the experiments and how these were carried out. Selection of the plant group for experiments of this kind must be made with the greatest possible care if one does not want to jeopardize all possibility of success from the very outset.

He then listed the properties that an experimental organism should have. First, it should have a number of different traits that can be studied. Second, the plant should be self-fertilizing and have a flower structure that minimizes accidental pollination. Third, the offspring of self-fertilized plants should be fully fertile so that further crosses can be made.

He paid particular attention to pea plants because more than 30 varieties of pea plants with different traits were available from seed dealers. The plant had a relatively short growth period, could be grown in the ground or in pots in the greenhouse, and could be self-fertilized or artificially fertilized by hand (▶ Figure 3.1).

Mendel then tested all available varieties of peas for 2 years to ensure that the traits they carried were true-breeding, that is, that self-fertilization gave rise to the same traits in all offspring, generation after generation. From these, he selected 22 varieties to plant in the monastery garden for his work (▶ Figure 3.2). Mendel studied seven characters that affected the seeds, pods, flowers, and stems of the plant (▶ Table 3.1). Each character was represented by two distinct forms: plant height by tall and short plants, seed shape by wrinkled and smooth peas, and so forth.

To avoid errors caused by small sample sizes, he planned experiments on a large scale, using some 28,000 pea plants in his experiments. He began by studying one pair of traits at a time and repeated his experiments for each trait to confirm his results. Using his training in physics and mathematics, Mendel analyzed his data according to the principles of probability and statistics. His methodical and thorough approach to his work and his lack of preconceived notions were the secrets of his success.

▲ FIGURE 3.2 The monastery garden where Mendel carried out his experiments in plant genetics.

Malcolm Gutter/Visuals Unlimited

Table 3.1	Traits Selected for Study by Mendel	
Structure Studied	**Dominant**	**Recessive**
SEEDS		
Shape	Smooth	Wrinkled
Color	Yellow	Green
Seed coat color	Gray	White
PODS		
Shape	Full	Constricted
Color	Green	Yellow
FLOWERS		
Placement	Axial (along stems)	Terminal (top of stems)
STEMS		
Length	Long	Short

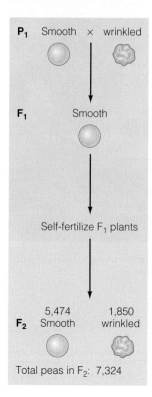

P₁ Smooth × wrinkled

F₁ Smooth

Self-fertilize F₁ plants

F₂ 5,474 Smooth   1,850 wrinkled

Total peas in F₂: 7,324

▲ FIGURE 3.3 One of Mendel's crosses. True-breeding varieties of peas (smooth and wrinkled) were used as the P₁ generation. All the offspring in the F₁ generation had smooth seeds. Self-fertilization of F₁ plants gave rise to both smooth and wrinkled progeny in the F₂ generation. About three-fourths of the offspring were smooth and about one-fourth were wrinkled.

■ **Genes**   The fundamental units of heredity.

■ **Recessive trait**   The trait unexpressed in the F₁ but reexpressed in some members of the F₂ generation.

■ **Dominant trait**   The trait expressed in the F₁ (or heterozygous) condition.

■ **Phenotype**   The observable properties of an organism.

■ **Genotype**   The specific genetic constitution of an organism.

■ **Segregation**   The separation of members of a gene pair from each other during gamete formation.

3.3 Crossing Pea Plants: Mendel's Study of Single Traits

To show how Mendel developed his ideas about how traits are inherited, we will first describe his experiments and his results. Then we will follow his reasoning in drawing conclusions and outline some of the further experiments that confirmed his ideas.

In his first set of experiments, Mendel studied the inheritance of seed shape. He took plants with smooth seeds and crossed them to plants with wrinkled seeds. In making this cross, flowers from one variety were fertilized using pollen from the other variety. The seeds that formed as a result of these crosses were all smooth. This was true whether the pollen used for fertilization came from a plant with smooth peas or a plant with wrinkled peas. Mendel planted the smooth seeds from this cross, and when the plants matured, the flowers were self-fertilized, and 7,324 seeds were collected. Of these, 5,474 were smooth and 1,850 were wrinkled. This set of experiments can be diagrammed as follows:

P₁: Smooth × wrinkled

F₁: All Smooth

F₂: 5,474 Smooth and 1,850 wrinkled

Mendel called the parental generation P₁; the offspring were called the F₁ (first filial) generation. The second generation, produced by self-fertilizing the F₁ plants, was called the F₂ (or second filial) generation. His experiments with seed shape are summarized in ▶ Figure 3.3.

What were the results and conclusions from Mendel's first series of crosses?

The results from experiments with all seven characters were the same as those Mendel observed with smooth and wrinkled seeds (▶ Figure 3.4). In all crosses, the following results were obtained:

■ The F₁ offspring showed only one of the two parental traits.
■ In all crosses, it did not matter which plant the pollen came from. The results were always the same.
■ The trait not shown in the F₁ offspring reappeared in about 25% of the F₂ offspring.

These were Mendel's first discoveries. His work showed that traits were not blended as they passed from parent to offspring; they remained unchanged, even though they might not be expressed in a given generation.

In all his experiments, it did not matter whether the male or female plant in the P₁ generation had smooth or wrinkled seeds; the results were the same. From these experiments he concluded that each parent makes an equal contribution to the genetic makeup of the offspring.

Keep in mind

■ Some traits can appear in offspring even when the parents don't have the trait.

Based on the results of his crosses with each of the seven characters, Mendel came to several conclusions:

■ **Genes** (Mendel called them factors) determine traits and can be hidden or unexpressed. For example, if you cross plants with smooth seeds to plants with wrinkled seeds, all the F₁ seeds will be smooth. When these seeds are grown and

Trait Studied
Results in F_2
Seed shape
5,474 smooth
1,850 wrinkled
Seed color
6,022 yellow
2,001 green
Seed coat color
705 gray
224 white
Pod shape
882 inflated
299 constricted
Pod color
428 green
152 yellow
Flower position
651 along stem
207 at tip
Stem length
787 tall
277 dwarf

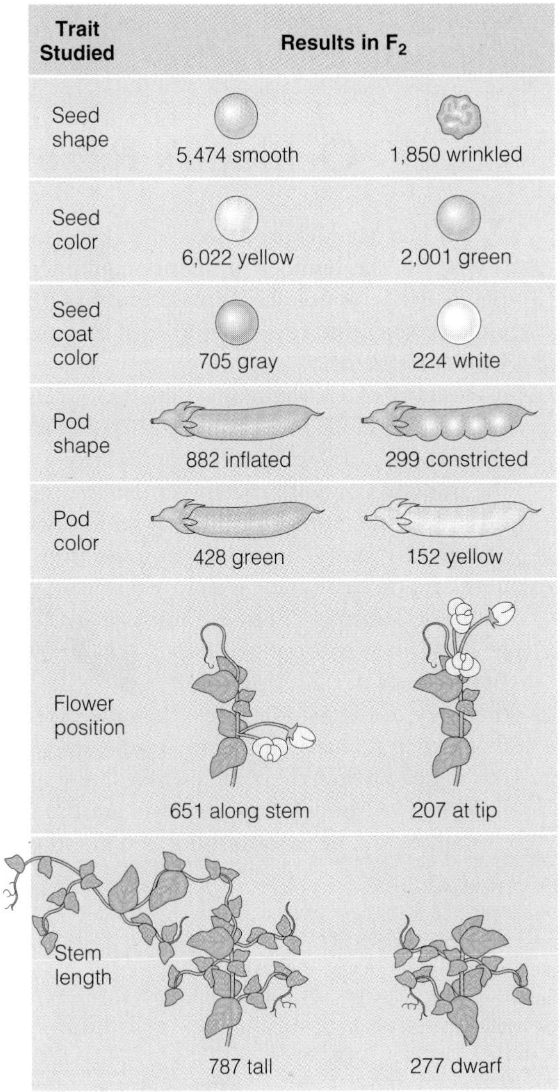

▶ FIGURE 3.4 Results of Mendel's crosses in peas. The numbers represent the F_2 plants showing a given trait. On average, three-fourths of the offspring showed one trait, and one-fourth showed the other (a 3:1 ratio). Crosses that involve a single trait are called monohybrid crosses.

self-fertilized, the next generation of plants (the F_2) will have some wrinkled seeds. This means that the F_1 seeds contain a gene for wrinkled that was present but not expressed. He called the trait not expressed in the F_1 but expressed in the F_2 plants a **recessive trait.** The trait present in F_1 plants he called the **dominant trait.** Mendel called this phenomenon dominance.

- Mendel concluded that despite their identical appearances, the P_1 and F_1 plants must be genetically different. When P_1 plants with smooth seeds are self-fertilized, all the plants in the next generation have only smooth seeds. But when F_1 plants with smooth seeds are self-fertilized, the F_2 plants have both smooth and wrinkled seeds. Mendel realized that it was important to make a distinction between the appearance of an organism and its genetic constitution. We now use the term **phenotype** to describe the appearance of an organism and the term **genotype** to describe the genetic makeup of an organism. In our example, the P_1 and F_1 plants with smooth seeds have identical phenotypes, but different genotypes.
- The results of self-fertilization experiments show that the F_1 plants must carry genes for smooth and wrinkled traits because both types of seeds are present in the F_2 generation. The question is, how many genes for seed shape are carried in the F_1 plants? Mendel had already reasoned that the male and female parent contributed equally to the traits of the offspring. The simplest explanation is that each F_1 plant carried two genes for seed shape, one for smooth that was expressed, and one for wrinkled that was unexpressed (see Genetic Journeys: Ockham's Razor). By extension, each P_1 and F_2 plant must also contain two genes for seed shape. To symbolize genes, uppercase letters are used to represent forms of a gene with a dominant pattern of inheritance, and lowercase letters are used to represent those with a recessive pattern of inheritance (S = smooth, and s = wrinkled). Using this shorthand, we can reconstruct the genotypes and phenotypes of the P_1 and F_1 as shown in ▶ Figure 3.5.

P1
Phenotypes
Genotypes
Smooth
SS
×
wrinkled
ss
Meiosis
Gametes
S S
s s
Fertilization
F1
Phenotype
Genotype
Smooth
Ss

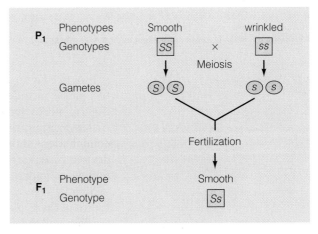

▲ FIGURE 3.5 The phenotypes and genotypes of the parents (P_1) and offspring (F_1), a cross involving seed shape.

The principle of segregation describes how a single trait is inherited.

If genes exist in pairs, there must be some way to prevent their number from doubling in each succeeding generation. (If each parent has two genes for a given trait, why doesn't the offspring have four?) Mendel reasoned that members of a gene pair must separate or segregate from each other during gamete formation. As a result, each gamete receives only one of the two genes that control a given trait. The separation of members of a gene pair during gamete formation is called the principle of **segregation,** or Mendel's First Law.

Genetic Journeys

Ockham's Razor

When Mendel proposed the simplest explanation for the number of factors contained in the F_1 plants in his monohybrid crosses, he was using a principle of scientific reasoning known as parsimony, or Ockham's razor.

William of Ockham (also spelled Occam) was a Franciscan monk and scholastic philosopher who lived from about 1300 to 1349. He had a strong interest in the study of thought processes and in logical methods. He is the author of the maxim known as Ockham's razor: "Pluralites non est pondera sine necessitate," which translates from the Latin as "Entities must not be multiplied without necessity." In the study of philosophy and theology of the Middle Ages, this was taken to mean that when constructing an argument, you should never go beyond the simplest argument unless it is necessary. Although Ockham was not the first to use this approach, he used this tool of logic so well and so often to dissect the arguments of his opponents that it became known as Ockham's razor.

The principle was transferred to scientific hypotheses in the fifteenth century. Galileo used the principle of parsimony to argue that because his model of the solar system was the simplest, it was probably correct (he was right). In modern terms, the phrase is taken to mean that in proposing a mechanism or hypothesis, use the least number of steps possible. The simplest mechanism is not necessarily correct, but it is usually the easiest to prove or disprove by doing experiments and the most likely to produce scientific progress.

For a given trait, Mendel concluded that both parents contribute an equal number of factors to the offspring. In this case, the simplest assumption is that each parent contributed one such factor and that the F_1 offspring contained two such factors. Further experiments proved this conclusion correct.

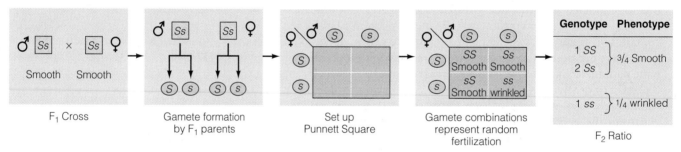

▲ **ACTIVE FIGURE 3.6** How a Punnett square can be used to generate the F_2 ratio in a cross from the F_1 generation.

HUMAN Genetics⊗Now™ Learn more about monohybrid crosses by viewing the animation on Human GeneticsNow at **http://now.ilrn.com/cummings7**

▶ Active Figure 3.6 diagrams the separation of a gene pair so that only one member of a pair is included in each gamete. In our example, each member of the F_1 generation can make two kinds of gametes in equal proportions (S gametes and s gametes). At fertilization, the random combination of these gametes produces the genotypic combinations shown in the Punnett square (a method for analyzing genetic crosses devised by R. C. Punnett). The F_2 has a genotypic ratio of $1\ SS : 2\ Ss : 1\ ss$ and a phenotypic ratio of 3 smooth : 1 wrinkled (dominant to recessive).

Mendel's reasoning allows us to predict the genotypes of the F_2 generation. One-fourth of the F_2 plants should carry only genes for smooth seeds (SS), and, when self-fertilized, all the offspring will have smooth seeds. Half (two-fourths) of the F_2 plants should carry genes for both smooth and wrinkled (Ss) and give rise to plants with smooth and wrinkled seeds in a 3:1 ratio when self-fertilized (▶ Figure 3.7). Finally, one-fourth of the F_2 plants should carry only genes for wrinkled (ss) and have all wrinkled progeny if self-fertilized. In fact, Mendel fertilized a number of plants from the F₂ generation and five succeeding generations to confirm these predictions.

Mendel carried out his experiments before the discovery of mitosis and meiosis and before the discovery of chromosomes. As we discuss in a later section, his conclusions about how traits are inherited are, in fact, descriptions of the way chromosomes behave in meiosis. Seen in this light, his discoveries are all the more remarkable.

Today, we call Mendel's factors genes and refer to the alternative forms of a gene as **alleles**. In the example we have been discussing, the gene for seed shape (S) has two alleles, smooth (S) and wrinkled (s). Individuals that carry identical alleles of a given gene (SS or ss) are **homozygous** for the gene in question. Similarly, when two different alleles are present in a gene pair (Ss), the individual has a **heterozygous** genotype. The SS homozygotes and the Ss heterozygotes show the dominant smooth phenotype (because S is dominant to s), and ss homozygotes show the recessive, wrinkled phenotype.

■ **Allele** One of the possible alternative forms of a gene, usually distinguished from other alleles by its phenotypic effects.

■ **Homozygous** Having identical alleles for one or more genes.

■ **Heterozygous** Carrying two different alleles for one or more genes.

3.4 More Crosses with Pea Plants: The Principle of Independent Assortment

Mendel realized the need to extend his studies on the inheritance from monohybrid crosses to more complex situations. He wrote:

> In the experiments discussed above, plants were used which differed in only one essential trait. The next task consisted in investigating whether the law of development thus found would also apply to a pair of differing traits.

For this work, he selected seed shape and seed color as traits to be studied, because, as he put it, "Experiments with seed traits lead most easily and assuredly to success."

Mendel performed crosses involving two traits.

As before, we will analyze the actual experiments of Mendel, outline his results, and summarize the conclusions he drew from them. From previous crosses, Mendel knew that for seeds, smooth is dominant to wrinkled and yellow is dominant to green. In our reconstruction of these experiments, we will use the following symbols: smooth (S), wrinkled (s), yellow (Y), and green (y). Mendel selected true-breeding plants with smooth, yellow seeds and crossed them with true-breeding plants with wrinkled, green seeds (▶ Figure 3.8).

Analyzing the results and drawing conclusions

The F₁ plants were crossed, producing an F₂ generation with four phenotypic combinations. Mendel counted a total of 556 seeds with these phenotypes:

315 smooth and yellow
108 smooth and green
101 wrinkled and yellow
32 wrinkled and green

The F₂ included the parental phenotypes (smooth yellow, wrinkled green) and two new phenotypes (smooth green and wrinkled yellow). These phenotypic classes occurred in a 9:3:3:1 ratio.

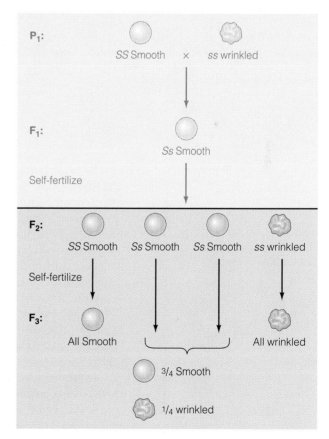

▲ FIGURE 3.7 Self-crossing F₂ plants demonstrate that there are two different genotypes among the plants with smooth peas in the F₂ generation.

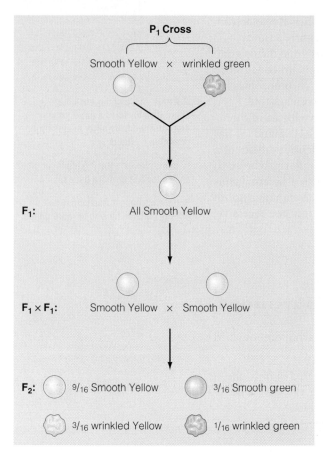

P₁ Cross

Smooth Yellow × wrinkled green

F₁: All Smooth Yellow

F₁ × F₁: Smooth Yellow × Smooth Yellow

F₂: 9/16 Smooth Yellow   3/16 Smooth green

3/16 wrinkled Yellow   1/16 wrinkled green

▲ FIGURE 3.8 The phenotypic distribution in a cross with two traits. Plants in the F₂ generation show the parental phenotypes and two new phenotypic combinations. Crosses involving two traits are called dihybrid crosses.

▪ **Independent assortment** The random distribution of genes into gametes during meiosis.

The principle of independent assortment explains the inheritance of two traits.

Before we discuss what is meant by independent assortment, let's see how the phenotypes and genotypes of the F₁ and F₂ were generated. The F₁ plants with smooth yellow seeds were heterozygous for both seed shape and seed color. The genotype of the F₁ plants was *SsYy*, with *S* and *Y* alleles dominant to *s* and *y*. Mendel had already concluded that members of a gene pair separate or segregate from each other during gamete formation.

During meiosis in the F₁ plants, the *S* and *s* alleles went into gametes independently of the *Y* and *y* alleles (▷ Active Figure 3.9). Because each gene pair segregated independently, the F₁ plants produced gametes with all combinations of these alleles in equal proportions: *SY*, *Sy*, *sY*, and *sy*. If fertilizations occur at random (as expected), 16 combinations result (▷ Active Figure 3.9).

The Punnett square in ▷ Active Figure 3.9 shows the following:

▪ Nine combinations have at least one copy of *S* and *Y*.
▪ Three combinations have at least one copy of *S* and are homozygous for *yy*.
▪ Three combinations have at least one copy of *Y* and are homozygous for *ss*.
▪ One combination is homozygous for *ss* and *yy*.

In other words, the 16 combinations of genotypes fall into four phenotypic classes in a 9:3:3:1 ratio:

9 smooth and yellow
3 smooth and green
3 wrinkled and yellow
1 wrinkled and green

The results of Mendel's cross involving two traits can be explained by assuming (as Mendel did) that during gamete formation, alleles of one gene pair segregate into gametes independently of the alleles belonging to other gene pairs, resulting in the production of gametes containing all combinations of alleles. This second fundamental principle of genetics outlined by Mendel is called the principle of **independent assortment,** or Mendel's Second Law.

The results of this cross raise an interesting question: how can we be sure that the number of offspring in each phenotypic class is close enough to what we expect? For example, if we do a cross and expect a 3:1 phenotypic ratio in the offspring, finding 75 plants with the dominant phenotype and 25 with the recessive phenotype in every 100 offspring would be ideal. What happens if 80 offspring have the dominant phenotype and 20 have the recessive phenotype, or what if the results are 65 dominant and 35 recessive? Is this close enough to a 3:1 ratio, or is our expectation wrong? To determine whether the observed results of an experiment meet expectations, geneticists use a statistical test. We are not going to pursue this analysis, but scientists use a variety of statistical tools to analyze data; in this case something called the chi square test would be used to evaluate how closely the results of the cross fit our expectations.

After 10 years of work, Mendel presented his results in 1865 at the meeting of the local Natural Science Society and published his paper the following year in the *Proceedings* of the society. Although copies of the journal were widely circulated,

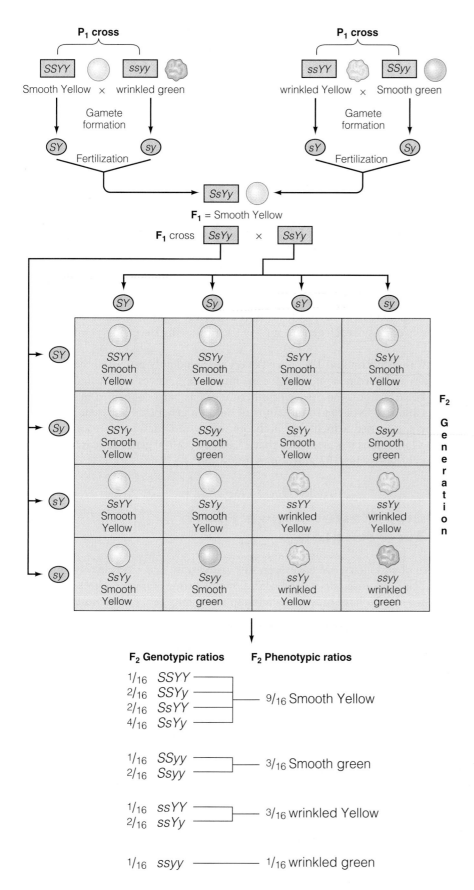

P₁ cross

| SSYY | Smooth Yellow | × | ssyy | wrinkled green |

Gamete formation

SY   sy

Fertilization

P₁ cross

| ssYY | wrinkled Yellow | × | SSyy | Smooth green |

Gamete formation

sY   Sy

Fertilization

SsYy

F₁ = Smooth Yellow

F₁ cross SsYy × SsYy

SY   Sy   sY   sy

	SY	Sy	sY	sy
SY	SSYY Smooth Yellow	SSYy Smooth Yellow	SsYY Smooth Yellow	SsYy Smooth Yellow
Sy	SSYy Smooth Yellow	SSyy Smooth green	SsYy Smooth Yellow	Ssyy Smooth green
sY	SsYY Smooth Yellow	SsYy Smooth Yellow	ssYY wrinkled Yellow	ssYy wrinkled Yellow
sy	SsYy Smooth Yellow	Ssyy Smooth green	ssYy wrinkled Yellow	ssyy wrinkled green

F₂ Generation

F₂ Genotypic ratios   **F₂ Phenotypic ratios**

1/16 *SSYY*
2/16 *SSYy*
2/16 *SsYY*   9/16 Smooth Yellow
4/16 *SsYy*

1/16 *SSyy*
2/16 *Ssyy*   3/16 Smooth green

1/16 *ssYY*
2/16 *ssYy*   3/16 wrinkled Yellow

1/16 *ssyy* ——— 1/16 wrinkled green

◄ **ACTIVE FIGURE 3.9** Punnett square of the dihybrid cross shown in Figure 3.8. There are two combinations of dominant and recessive traits that can result in doubly heterozygous F₁ plants. One (*left*) is a cross between Smooth, Yellow and wrinkled, green. The other (*right*) is a cross between wrinkled, Yellow and Smooth, green.

HUMAN Genetics ⑥ Now™ Learn more about dihybrid crosses by viewing the animation on Human GeneticsNow at **http://now.ilrn.com/cummings7**

■ **Genetics** The scientific study of heredity.

■ **Locus** The position occupied by a gene on a chromosome.

■ **Multiple alleles** Genes that have more than two alleles.

the significance of Mendel's findings was unappreciated. Finally, in 1900 three scientists independently confirmed Mendel's work and brought his paper to widespread attention. These events stimulated a great interest in what is now called **genetics.** Unfortunately, Mendel died in 1884—unaware he had founded an entire scientific discipline.

3.5 Meiosis Explains Mendel's Results: Genes Are on Chromosomes

When Mendel was working with pea plants, the behavior of chromosomes in mitosis and meiosis was unknown. By 1900, however, the details of mitosis and meiosis had been described. As scientists confirmed that Mendelian inheritance operated in many organisms, it became obvious that genes and chromosomes had much in common (▶ Table 3.2). Both chromosomes and genes occur in pairs. In meiosis, members of a chromosome pair separate from each other, and members of a gene pair separate from each other during gamete formation (▶ Active Figure 3.10). Finally, the fusion of gametes during fertilization restores the diploid number of chromosomes and two copies of each gene to the zygote, producing the genotypes of the next generation.

In 1903 Walter Sutton and Theodore Boveri independently proposed the idea that because genes and chromosomes behave in similar ways, genes are located on chromosomes. This chromosome theory of inheritance has been confirmed in many different ways and is one of the foundations of modern genetics. Each gene is located at a specific site (called a **locus**) on a chromosome and each chromosome carries many genes. In humans it is estimated that 25,000 to 30,000 genes are carried on the 24 different chromosomes (22 autosomes, the X and Y).

3.6 Many Genes Have More Than Two Alleles

For the sake of simplicity, we have been discussing only situations involving genes with two alleles. Because alleles represent different forms of a gene, there is no reason why a gene has to have only two alleles. In fact, many genes have more than two alleles. Any *individual* can carry only two alleles of a gene, but a group of individuals can carry many different alleles of a gene. In humans, the gene for ABO blood types has three alleles. Such genes are said to have **multiple alleles.** Your ABO blood type is determined by genetically encoded molecules (called antigens) present on the surface of your red blood cells (▶ Figure 3.11). These molecules are an identity tag recognized by the body's immune system.

Table 3.2 Genes, Chromosomes, and Meiosis	
Genes	**Chromosomes**
Occur in pairs (alleles)	Occur in pairs (homologues)
Members of a gene pair separate from each other during meiosis	Members of a pair of homologues separate from each other during meiosis
Members of one gene pair independently assort from other gene pairs during meiosis	Members of one chromosome pair independently assort from other chromosome pairs during meiosis

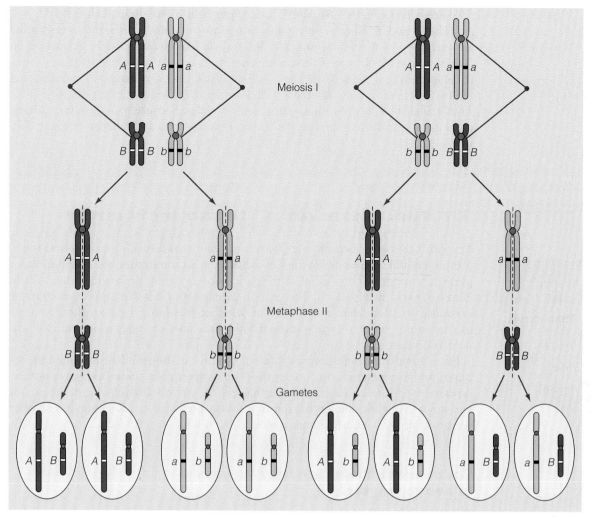

▲ ACTIVE FIGURE 3.10 Mendel's observations about segregation and independent assortment are explained by the behavior of chromosomes during meiosis. The arrangement of chromosomes at metaphase I is random. As a result, four combinations of the two genes are produced in the gametes.

HUMAN Genetics ⊗ Now™ Learn more about independent assortment by viewing the animation on Human GeneticsNow at **http://now.ilrn.com/cummings7**

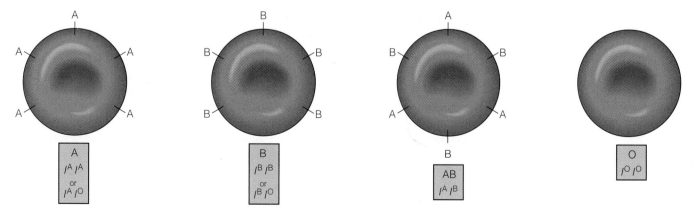

▲ FIGURE 3.11 Each allele of codominant genes is fully expressed in the heterozygote. Type A blood has A antigens on the cell surface, and type B has B antigens on the surface. In type AB, both the A and the B antigen are present on the cell surface. Thus, the *A* and *B* alleles of the *I* gene are codominant. In type O blood, no antigen is present. The *O* allele is recessive to both the *A* and the *B* allele.

There is one gene (I) for ABO blood types, and it has three alleles, I^A, I^B, and I^O. The A and B alleles control the formation of slightly different forms of the antigen. If you are homozygous for the A allele (AA), you carry the A antigen on cells and have blood type A. If you are homozygous for the B allele (BB), you carry the B antigen and are type B. The third allele (O) does not make any antigen, and individuals homozygous for the I^O allele carry no encoded antigen on their cells. The O allele is recessive to both the A and B alleles. Because there are three alleles, there are six possible genotypes, including homozygotes and heterozygotes (▶ Figure 3.11).

3.7 Variations on a Theme by Mendel

After Mendel's work became widely known, geneticists turned up cases where the F_1 phenotypes did not resemble either of the parents. In some cases, the offspring had a phenotype intermediate to that of the parents or a phenotype in which the traits of both parents were expressed. This led to a debate about whether such cases could be explained by Mendelian inheritance or whether there might be another, separate mechanism of inheritance that did not follow the laws of segregation and independent assortment.

Eventually, work with several different organisms showed that although phenotypes can be somewhat complex, at the level of genotypes, these cases were not exceptions to Mendelian inheritance. In this section, we will discuss some of these cases and show that while phenotypes may not follow predicted ratios, genotypes do obey the principles of Mendelian inheritance.

Codominant alleles are fully expressed in heterozygotes.

The first variation we consider is called **codominance.** In this situation, neither allele is dominant, and heterozygotes fully express both alleles. In the ABO system, *AB* heterozygotes carry *both* the A and B antigens on their red cells and are blood type AB. In this case, neither allele is dominant to the other. Because each is fully expressed, they are codominant alleles (▶ Figure 3.11). As a result, the three alleles of the ABO system have six genotypic combinations that contribute to four phenotypes (▶ Table 3.3).

Incomplete dominance has a distinctive phenotype in heterozygotes.

Another variation involves **incomplete dominance.** In this case, the phenotype of heterozygotes is intermediate to those of the homozygous parents. In snapdragons (▶ Figure 3.12), crossing a true-breeding variety with red flowers to one with white flowers produces an F_1 with pink flowers. The F_1 phenotype is intermediate to the parental phenotypes. When the F_1 plants are self-fertilized, they produce an F_2 with red, pink, and white flowers in a 1:2:1 ratio.

These results can be explained by looking at the genotypes involved. Each genotype in this cross has a distinct phenotype (RR is red, Rr is pink, and rr is white), and the phenotypic ratio of 1 red:2 pink:1 white is the same as the genotypic ratio of 1 RR:2 Rr:1 rr. In this case, two copies of the R allele (RR) make enough pigment to produce red flowers. The single copy of the R allele carried by heterozygotes (Rr) produces less red pigment, and the result is pink flowers. In other words, the R allele is incompletely dominant over the r allele. Because the r allele produces no color, homozygous rr plants have white flowers.

Table 3.3
ABO Blood Types

Genotypes	Phenotypes
$I^A I^A$, $I^A I^O$	Type A
$I^B I^B$, $I^B I^O$	Type B
$I^A I^B$	Type AB
$I^O I^O$	Type O

■ **Codominance** Full phenotypic expression of both members of a gene pair in the heterozygous condition.

■ **Incomplete dominance** Expression of a phenotype that is intermediate between those of the parents.

FIGURE 3.12 Incomplete dominance in snapdragon flower color. Red-flowered snapdragons crossed with white-flowered snapdragons produce offspring that have pink flowers in the F₁. In heterozygotes, the allele for red flowers is incompletely dominant over the allele for white.

3.8 Mendelian Inheritance in Humans

Now that we know how segregation and independent assortment work in pea plants, let's turn our attention to humans. After Mendel's work was rediscovered, some scientists believed that inheritance of traits in humans might not work the same way as in plants and other animals. However, the first trait (a hand deformity called brachydactyly; OMIM 112500) analyzed in humans (in 1905) was found to follow the rules of Mendelian inheritance and so have all the 5,000 plus traits described since then.

Segregation and independent assortment occur with human traits.

To illustrate that segregation and independent assortment apply to human traits, let's follow the inheritance of a recessive trait called albinism (OMIM 203100). Homozygotes (*aa*) cannot make skin pigment and have very pale, white skin, white hair, and colorless eyes (▶ Figure 3.13). Anyone carrying at least one dominant allele (*A*) can make enough pigment to have colored skin, hair, and eyes.

To apply Mendelian inheritance to humans, we'll start with parents who are heterozygotes (*Aa*) with normal pigmentation (▶ Figure 3.14). During meiosis, the dominant and recessive alleles separate from each other and end up in different gam-

FIGURE 3.13 Individuals with albinism lack pigment in the skin, hair, and eyes.

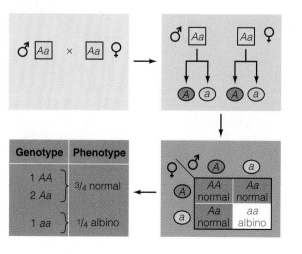

◀ **FIGURE 3.14** The segregation of albinism, a recessive trait in humans. As in pea plants, alleles of a gene pair separate from each other during gamete formation.

Genotype	Phenotype
1 *AA* 2 *Aa*	3/4 normal
1 *aa*	1/4 albino

etes. Because each parent can produce two different types of gametes (one with *A* and another with *a*), there are four possible combinations of gametes at fertilization. If they have enough children (say 20 or 30), we will see something close to the predicted phenotypic ratio of 3 pigmented : 1 albino offspring and a genotypic ratio of 1*AA* : 2*Aa* : 1*aa* (▶ Figure 3.14). In other words, segregation of alleles during gamete formation produces the same outcome in both pea plants and humans. Important: This does not mean that there will be one albino child and three normally pigmented children in every such family with four children. It does mean that if the parents are heterozygotes, each child has a 25% chance of being albino and a 75% chance of having normal pigmentation.

The inheritance of two traits in humans also follows the Mendelian principle of independent assortment (▶ Figure 3.15). To illustrate, let's examine a family in which each parent is heterozygous for albinism (*Aa*) and heterozygous for another recessive trait, hereditary deafness (OMIM 220290). Homozygous dominant (*DD*) or heterozygous individuals (*Dd*) can hear, but homozygous recessive (*dd*) individuals are deaf. During meiosis, alleles for skin color and alleles for hearing assort into gametes independently. As a result, each parent produces equal proportions of four different gametes (*AD*, *Ad*, *aD*, and *ad*). There are 16 possible combinations of gametes at fertilization (four types of gametes in all possible combinations), resulting in four different phenotypic classes (▶ Figure 3.15). An examination of the possible genotypes shows that there is a 1 in 16 chance that a child would be both deaf and an albino.

In pea plants and other organisms such as *Drosophila*, genetic analysis is done using experimental crosses with predetermined genotypes. In humans, experimental crosses are not possible, and geneticists must often infer genotypes based on the pat-

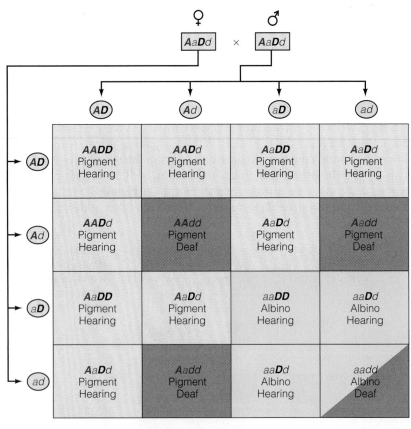

▲ FIGURE 3.15 Independent assortment for two traits in humans follows the same pattern of inheritance as traits in pea plants.

tern of inheritance observed in a family. In human genetics, the study of a trait begins with a family history, as outlined in the following section.

Pedigree construction is important to human genetics.

The fundamental method of genetic analysis in humans is collecting a family history to follow the inheritance of a trait. This method is called **pedigree construction**. A **pedigree** is the orderly presentation of family information in the form of an easily readable chart. Using a pedigree, the inheritance of a trait can be followed through several generations. Analysis of the pedigree using the principles of Mendelian inheritance can determine whether a trait has a dominant or recessive pattern of inheritance.

■ **Pedigree construction** Use of family history to determine how a trait is inherited and to determine risk factors for family members.

■ **Pedigree** A diagram listing the members and ancestral relationships in a family; used in the study of human heredity.

Pedigrees use a standardized set of symbols, many of which are borrowed from genealogy. ▶ Figure 3.16 shows many of these symbols. In pedigrees, squares represent males, and circles represent females. Someone with the phenotype in question is represented by a filled-in symbol. Heterozygotes, when known, are indicated by a shaded dot inside a symbol. Relationships between individuals in a pedigree are shown as a series of lines. Parents are connected by a horizontal line, and a vertical line leads to their offspring. If the parents are closely related (such as first cousins), they are connected by a double line. The offspring are connected by a horizontal sibship line and listed in birth order from left to right along the sibship line:

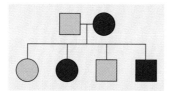

A numbering system is used in pedigree construction. Each generation is identified by a Roman numeral (I, II, III, and so on), and within a generation, each individual is identified by an Arabic number (1, 2, 3, and so on) as shown:

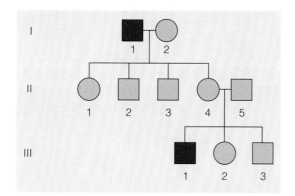

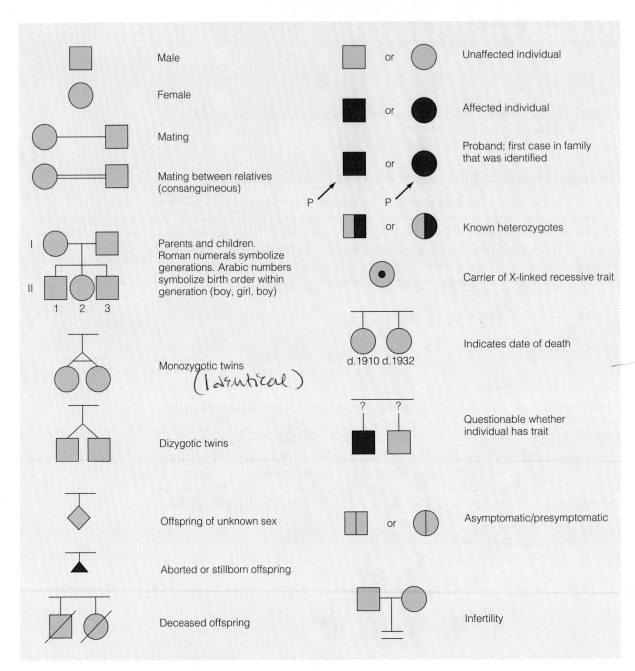

Male

Female

Mating

Mating between relatives (consanguineous)

I

II
1 2 3

Parents and children. Roman numerals symbolize generations. Arabic numbers symbolize birth order within generation (boy, girl, boy)

Monozygotic twins

(Identical)

Dizygotic twins

Offspring of unknown sex

Aborted or stillborn offspring

Deceased offspring

☐ or ○ Unaffected individual

■ or ● Affected individual

■ or ● Proband; first case in family that was identified

P   P

▨ or ◑ Known heterozygotes

⊙ Carrier of X-linked recessive trait

d.1910 d.1932 Indicates date of death

? ? Questionable whether individual has trait

▯▯ or ◑ Asymptomatic/presymptomatic

Infertility

▲ FIGURE 3.16 Symbols used in pedigree analysis.

Pedigrees are often constructed after a family member afflicted with a genetic disorder has been identified. This individual, known as the **proband,** is indicated on the pedigree by an arrow and the letter P:

■ Proband   First affected family member who seeks medical attention for a genetic disorder.

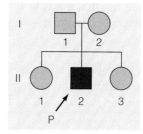

Because pedigree construction is a family history, details about earlier generations may be uncertain as memories fade. If the sex of someone is unknown, a diamond is used. If there is doubt that a family member had the trait in question, this is indicated by a question mark above the symbol.

A pedigree is a form of symbolic communication used by clinicians and researchers in human genetics (▶ Active Figure 3.17). It contains information that can help establish how a trait is inherited, identify those at risk of developing or transmitting the trait, and is a resource for establishing biological relationships within a family. In Chapter 4, we will see how pedigree analysis is used to establish the genotypes of individuals and to predict the chances of having children affected with a genetic disorder.

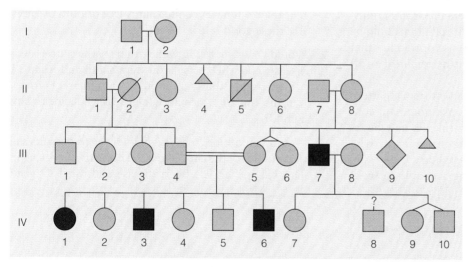

▲ **ACTIVE FIGURE 3.17**   A pedigree showing the inheritance of a trait through several generations in a family. This pedigree and all those in this book use the standardized set of symbols adopted in 1995 by the American Society of Human Genetics.

HUMAN Genetics ⒺNow™   Learn more about pedigree analysis by viewing the animation on Human GeneticsNow at **http://now.ilrn.com/cummings7**

Genetics in Practice

Genetics in Practice case studies are critical thinking exercises that allow you to apply your new knowledge of human genetics to real-life problems. You can find these case studies and links to relevant websites at **http://biology.brookscole.com/cummings7**

CASE 1

Pedigree analysis is a fundamental tool for investigating whether a trait is following a traditional Mendelian pattern of inheritance. It can also be used to help identify individuals within a family that may be at risk for the trait.

Adam and Sarah, a young couple of Eastern European Jewish ancestry, went to a genetic counselor because they were planning a family and wanted to know what their chances were for having a child with a genetic condition. The genetic counselor took a detailed family history from both of them and discovered several traits in their respective families.

Sarah's maternal family history is suggestive of an inherited form of breast and ovarian cancer with an autosomal dominant pattern of cancer from her grandmother to mother because of the young ages at which they were diagnosed with their cancers. If an altered gene that predisposed to breast and ovarian cancer were in Sarah's family, she, her sister, and any of her own future children could be at risk for inheriting this gene. The counselor told her that genetic testing is available that may help determine if an altered gene is in her family.

Adam's paternal family history has a very strong pattern of early-onset heart disease. An autosomal dominant condition known as familial hypercholesterolemia may be responsible for the number of individuals in the family who have died from heart attacks. Like hereditary breast and ovarian cancer, there is genetic testing available to see if

Adam carries this altered gene. Testing may give the couple more information on the chances that their children could inherit the gene. Adam had a first cousin who died from Tay-Sachs disease (TSD), a fatal autosomal recessive condition more commonly found in people of Eastern European Jewish descent. For his cousin to have TSD, both parents must have been carriers for the disease-causing gene. If that is the case, Adam's father could be a carrier as well. If Adam's father and mother were both carriers, each of Adam and Sarah's children have a 25% chance of being afflicted with TSD. If Adam's father has the TSD gene, it is possible that Adam inherited the gene. Because Sarah is also of Eastern European ancestry, she could be a carrier of this gene, although no one in her family has been affected with TSD. A simple blood test performed on both Sarah and Adam could determine if they are carriers of this gene.

1. If Sarah carries the mutant cancer gene and Adam carries the mutant heart disease gene, what is the chance that they will have a child that is free of both diseases? Are these good odds?

2. Would you want to know the results of the cancer, heart disease, and TSD tests if you were Sarah and Adam? Is it their responsibility as good potential parents to find out this kind of information before they decide to have a child?

3. Would you decide to have a child if the test results said that you carry the cancer gene? The heart disease gene? The TSD gene? The heart disease and the TSD gene?

Summary

3.1 Heredity: How Are Traits Inherited?

- In the centuries before Gregor Mendel experimented with the inheritance of traits in the garden pea, several competing theories attempted to explain how traits were passed from generation to generation, but none were completely successful.

3.2 Mendel's Experimental Design Resolved Many Unanswered Questions

- Mendel carefully selected an organism to study, kept careful records, and studied the inheritance of traits for several generations. In his decade-long series of experiments, Mendel established the foundation for the science of genetics.

3.3 Crossing Pea Plants: Mendel's Study of Single Traits

- Mendel studied crosses in the garden pea that involved one pair of alleles and demonstrated that the phenotypes associated with these traits are controlled by pairs of factors, now known as genes. These factors separate or segregate from each other during gamete formation and exhibit dominant/recessive relationships. This is known as the principle of segregation.

3.4 More Crosses with Pea Plants: The Principle of Independent Assortment

- In later experiments, Mendel discovered that members of one gene pair separate or segregate independently of other gene pairs. This principle of independent assortment leads to the formation of all possible combinations of gametes with equal probability in a cross between two individuals.

3.5 Meiosis Explains Mendel's Results: Genes Are on Chromosomes

- Segregation and independent assortment of genes results from the behavior of chromosomes in meiosis. At the turn of the twentieth century, it became apparent that genes are located on chromosomes.

3.6 Many Genes Have More Than Two Alleles

- Although any individual can carry only two alleles of a gene, many genes have multiple alleles, carried by members of a population.

3.7 Variations on a Theme by Mendel

- Variations in phenotypes can change the expected outcome of crosses, but the genotypic ratios do not change. Codominant alleles are both expressed in the phenotype, whereas in incomplete dominance, the heterozygote has a phenotype intermediate to that of the parents.

3.8 Mendelian Inheritance in Humans

- Because genes for human genetic disorders exhibit segregation and independent assortment, the inheritance of certain human traits is predictable, making it possible to provide genetic counseling to those at risk of having children affected with genetic disorders.

- Instead of direct experimental crosses, traits in humans are traced by constructing pedigrees that follow a trait through several generations.

HUMAN
Genetics ⊜ Now™ Preparing for an exam? Assess your understanding of this chapter's topics with a pre-test, a personalized learning plan, and a post-test at **http://now.ilrn.com/cummings7**

Crossing Pea Plants: Mendel's Study of Single Traits

1. Explain the difference between the following terms:
   a. Gene versus allele versus locus
   b. Genotype versus phenotype
   c. Dominant versus recessive
   d. Complete dominance versus incomplete dominance versus codominance

2. Of the following, which are phenotypes, and which are genotypes?
   a. *Aa*
   b. Tall plants
   c. *BB*
   d. Abnormal cell shape
   e. *AaBb*

3. Define Mendel's Law of Segregation.

4. Define Mendel's Law of Independent Assortment.

5. Suppose that organisms have the following genotypes. What types of gametes will these organisms produce, and in what proportions?
   a. *Aa*
   b. *AA*
   c. *aa*

6. Given the following matings, what are the predicted genotypic ratios of the offspring?
   a. *Aa* × *aa*
   b. *Aa* × *Aa*
   c. *AA* × *Aa*

7. Brown eyes (*B*) are fully dominant over blue eyes (*b*).
   a. A 3:1 phenotypic ratio of F_1 progeny indicates that the parents are of what genotype?
   b. A 1:1 phenotypic ratio of F_1 progeny indicates that the parents are of what genotype?

8. An unspecified character controlled by a single gene is examined in pea plants. Only two phenotypic states exist for this trait. One phenotypic state is completely dominant to the other. A heterozygous plant is self-crossed. What proportion of the progeny of plants exhibiting the dominant phenotype is homozygous?

9. Sickle cell anemia (SCA) is a human genetic disorder caused by a recessive allele. A couple plans to marry and wants to know the probability that they will have an affected child. With your knowledge of Mendelian inheritance, what can you tell them if (1) both are normal, but each has one affected parent and the other parent has no family history of SCA; or (2) the man is affected by the disorder, but the woman has no family history of SCA?

10. If you are informed that being right- or left-handed is heritable and that a right-handed couple is expecting a child, can you conclude that the child will be right-handed?

11. Stem length in pea plants is controlled by a single gene. Consider the cross of a true-breeding, long-stemmed variety to a true-breeding, short-stemmed variety in which long stems are completely dominant.
    a. If 120 F_1 plants are examined, how many plants are expected to be long-stemmed? Short-stemmed?
    b. Assign genotypes to both P_1 varieties and to all phenotypes listed in (a).
    c. A long-stemmed F_1 plant is self-crossed. Of 300 F_2 plants, how many should be long-stemmed? Short-stemmed?
    d. For the F_2 plants mentioned in (c), what is the expected genotypic ratio?

More Crosses with Pea Plants: The Principle of Independent Assortment

12. Organisms have the following genotypes. What types of gametes will these organisms produce and in what proportions?
    a. *Aabb*
    b. *AABb*
    c. *AaBb*

13. Given the following matings, what are the predicted phenotypic ratios of the offspring?
    a. *AABb* × *Aabb*
    b. *AaBb* × *aabb*
    c. *AaBb* × *AaBb*

14. A woman is heterozygous for two genes. How many different types of gametes can she produce, and in what proportions?

15. Two traits are simultaneously examined in a cross of two pure-breeding pea-plant varieties. Pod shape can be either swollen or pinched. Seed color can be either green or yellow. A plant with the traits swollen and green is crossed with a plant with the traits pinched and yellow, and a resulting F_1 plant is self-crossed. A total of 640 F_2 progeny are phenotypically categorized as follows:

    swollen, yellow 360
    120 swollen and green
    120 pinched and yellow
    40 pinched and green

    a. What is the phenotypic ratio observed for pod shape? Seed color?
    b. What is the phenotypic ratio observed for both traits considered together?
    c. What is the dominance relationship for pod shape? Seed color?
    d. Deduce the genotypes of the P_1 and F_1 generations.

16. Consider the following cross in pea plants, in which smooth seed shape is dominant to wrinkled, and yellow

seed color is dominant to green. A plant with smooth, yellow seeds is crossed to a plant with wrinkled, green seeds. The peas produced by the offspring are all smooth and yellow. What are the genotypes of the parents? What are the genotypes of the offspring?

17. Consider another cross in pea plants involving the genes for seed color and shape. As before, yellow is dominant to green, and smooth is dominant to wrinkled. A plant with smooth, yellow seeds is crossed to a plant with wrinkled, green seeds. The peas produced by the offspring are as follows: one-fourth are smooth, yellow; one-fourth are smooth, green; one-fourth are wrinkled, yellow; and one-fourth are wrinkled, green.
    a. What is the genotype of the smooth, yellow parent?
    b. What are the genotypes of the four classes of offspring?

18. Determine the possible genotypes of the following parents by analyzing the phenotypes of their children. In this case, we will assume that brown eyes (*B*) is dominant to blue (*b*) and that right-handedness (*R*) is dominant to left-handedness (*r*).
    a. Parents: brown eyes, right-handed × brown eyes, right-handed
       Offspring: 3/4 brown eyes, right-handed
                  1/4 blue eyes, right-handed
    b. Parents: brown eyes, right-handed × blue eyes, right-handed
       Offspring: 6/16 blue eyes, right-handed
                  2/16 blue eyes, left-handed
                  6/16 brown eyes, right-handed
                  2/16 brown eyes, left-handed
    c. Parents: brown eyes, right-handed × blue eyes, left-handed
       Offspring: 1/4 brown eyes, right-handed
                  1/4 brown eyes, left-handed
                  1/4 blue eyes, right-handed
                  1/4 blue eyes, left-handed

19. Think about this one carefully. Albinism and hair color are governed by different genes. A recessively inherited form of albinism causes affected individuals to lack pigment in their skin, hair, and eyes. In hair color, red hair is inherited as a recessive trait, and brown hair is inherited as a dominant trait. An albino woman whose parents both have red hair has two children with a man who is normally pigmented and has brown hair. The brown-haired partner has one parent who has red hair. The first child is normally pigmented and has brown hair. The second child is albino.
    a. What is the hair color (phenotype) of the albino parent?
    b. What is the genotype of the albino parent for hair color?
    c. What is the genotype of the brown-haired parent with respect to hair color? Skin pigmentation?
    d. What is the genotype of the first child with respect to hair color and skin pigmentation?
    e. What are the possible genotypes of the second child for hair color? What is the phenotype of the second child for hair color? Can you explain this?

20. Consider the following cross:

    P$_1$: *AABBCCDDEE* × *aabbccddee*
    F$_1$: *AaBbCcDdEe* (self-cross to get F$_2$)

    What is the chance of getting an *AaBBccDdee* individual in the F$_2$ generation?

21. In the following trihybrid cross, determine the chance that an individual could be phenotypically *A*, *b*, *C* in the F$_1$ generation.

    P$_1$: *AaBbCc* × *AabbCC*

22. In pea plants, long stems are dominant to short stems, purple flowers are dominant to white, and round seeds are dominant to wrinkled. Each trait is determined by a single, different gene. A plant that is heterozygous at all three loci is self-crossed, and 2,048 progeny are examined. How many of these plants would you expect to be long-stemmed with purple flowers, producing wrinkled seeds?

Variations on a Theme by Mendel

23. A character of snapdragons amenable to genetic analysis is flower color. Imagine that a true-breeding red-flowered variety is crossed to a pure line having white flowers. The progeny are exclusively pink-flowered. Diagram this cross, including genotypes for all P$_1$ and F$_1$ phenotypes. What is the mode of inheritance? Let *F* = red, and *f* = white.

24. In peas, straight stems (*S*) are dominant to gnarled (*s*), and round seeds (*R*) are dominant to wrinkled (*r*). The following cross (a test cross) is performed: *SsRr* × *ssrr*. Determine the expected phenotypes of the progeny and what fraction of the progeny should exhibit each phenotype.

25. Pea plants usually have white or red flowers. A strange pea-plant variant is found that has pink flowers. A self-cross of this plant yields the following phenotypes:

    30 red flowers
    62 pink flowers
    33 white flowers

    What are the genotypes of the parents? What is the genotype of the progeny with red flowers?

26. A plant geneticist is examining the mode of inheritance of flower color in two closely related species of exotic plants. Analysis of one species has resulted in the identification of two pure-breeding lines—one produces a distinct red flower and the other produces no color at all; however, she cannot be sure. A cross of these varieties produces all pink-flowered progeny. The second species exhibits similar pure-breeding varieties; that is, one variety produces red flowers, and the other produces an albino flower. A cross of these two varieties, however, produces orange-flowered progeny exclusively. Analyze the mode of inheritance of flower color in these two plant species.

27. What are the possible genotypes for the following blood types?

a. type A
b. type B
c. type O
d. type AB

28. A man with blood type A and a woman with blood type B have three children: a daughter with type AB, and two sons, one with type B and one with type O blood. What are the genotypes of the parents?

29. What is the chance that a man with type AB blood and a woman with type A blood whose mother is type O can produce a child that is
    a. type A
    b. type AB
    c. type O
    d. type B

30. A hypothetical human trait is controlled by a single gene. Four alleles of this gene have been identified: *a*, *b*, *c*, and *d*. Alleles *a*, *b*, and *c* are all codominant; allele *d* is recessive to all other alleles.
    a. How many phenotypes are possible?
    b. How many genotypes are possible?

Meiosis Explains Mendel's Results: Genes Are on Chromosomes

31. Discuss the pertinent features of meiosis that provide a physical correlate to Mendel's abstract genetic laws of random segregation and independent assortment.

32. The following diagram shows a hypothetical diploid cell. The recessive allele for albinism is represented by *a*, and *d* represents the recessive allele for deafness. The normal alleles for these conditions are represented by *A* and *D*, respectively.

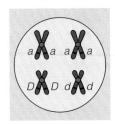

a. According to the principle of segregation, what is segregating in this cell?

b. According to Mendel's principle of independent assortment, what is independently assorting in this cell?
c. How many chromatids are in this cell?
d. How many tetrads are in this cell?
e. Write the genotype of the individual from whom this cell was taken.
f. What is the phenotype of this individual?
g. What stage of cell division is represented by this cell (prophase, metaphase, anaphase, or telophase of meiosis I, meiosis II, or mitosis)?
h. After meiosis is complete how many chromatids and chromosomes will be present in one of the four progeny cells?

Mendelian Inheritance in Humans

33. Define the following pedigree symbols:

a.
b.
c.
d.
e.

34. Draw the following simple pedigree. A man and a woman have three children: a daughter, then two sons. The daughter marries and has monozygotic (identical) twin girls. The youngest son in generation II is affected with albinism.

35. Construct a pedigree given the following information. Mary is 16 weeks' pregnant and was referred for genetic counseling because of advanced maternal age. Mary has one daughter, Sarah, who is 5 years old. Mary has three older sisters and four younger brothers. The two oldest sisters are married and each has one son. All her brothers are married, but none has any children. Mary's parents are both alive, and she has two maternal uncles and three paternal aunts. Mary's husband, John, has two brothers, one older and one younger, neither of whom is married. John's mother is alive, but his father is deceased.

Internet Activities are critical thinking exercises using the resources of the World Wide Web to enhance the principles and issues covered in this chapter. For a full set of links and questions investigating the topics described below, log on to **http://biology.brookscole.com/cummings7**

1. *Mendelian Genetics and Plant Genetics.* Gregor Mendel crossed pea plants to investigate the results of hybridization experiments. Now you give it a try! At the CUNY Brooklyn Mendelian Genetics site, read the Introduction carefully (some of the steps are a little tricky) and then click on the "Plant Hybridization" link at the site to choose and perform some crosses of your own.

2. *Mendel's Discoveries in His Own Words.* You can read Mendel's original paper in English and German at the *MendelWeb* website. In addition to Mendel's original text, this site also has links to essays and commentary on his works and writings, and on the state of knowledge about heredity before Mendel.

3. *Meet Gregor Mendel?* Check out Professor John Blamire's fictionalized account of Mendel's life.

How would you vote now?

Using the principles Mendel discovered and modern pedigree analysis, it is possible for couples planning families to determine the approximate risk their children have of inheriting certain genetic disorders. To know for certain whether a child has inherited a genetic disorder, genetic testing can be performed. However, in the United States, some genetic testing is required by law and is performed on all newborns regardless of their individual risk. Some states only test for a few genetic disorders, but others test for nearly 3 dozen diseases. Not everyone is comfortable with mandatory testing, feeling it is an invasion of privacy, and fearing that the results could be misused to restrict reproductive rights. Now that you know more about inheritance, what do you think? Should all states be required to test for as many genetic conditions as possible, or should this be left up to the parents? If genetic testing is mandatory, who should have access to the results? Visit the Human Heredity Companion website to find out more on the issue, then cast your vote online at **http://biology.brookscole.com/cummings7**

For further reading and inquiry, log on to **InfoTrac College Edition,** your world-class online library, including articles from nearly 5,000 periodicals, at **http://infotrac.thomsonlearning.com**

4

Pedigree Analysis in Human Genetics

WAS Abraham Lincoln, the sixteenth president of the United States, affected with a genetic disorder? Evidence in support of this idea is based on two observations: Lincoln's physical appearance and the report of an inherited disorder in a distant relative.

Photographs, written descriptions, and medical reports give us detailed information about Lincoln's physical appearance. He was 6 ft. 4 in. tall and thin, weighing between 160 and 180 lbs. for most of his adult life. He had long arms and legs with large, narrow hands and feet. Contemporary descriptions of his appearance indicate that he was stoop-shouldered, loose jointed, and walked with a shuffling gait. In addition, he wore eyeglasses to correct a visual problem.

Lincoln's physical appearance and eye problems are suggestive of an inherited disorder called Marfan syndrome. This genetic disease affects the connective tissue of the body, causing visual problems, blood vessel defects, and loose joints. In addition to the physical evidence, a child diagnosed with Marfan syndrome in the 1960s was found by pedigree construction and analysis to have ancestors in common with Lincoln (the common ancestor was Lincoln's great-great-grandfather). In the mid-1960s, these observations led to widespread speculation that Lincoln had Marfan syndrome.

Others disagree with this idea, arguing that Lincoln's long arms and legs and body proportions were well within the normal limits for tall, thin individuals. In addition, although Lincoln wore eyeglasses, he was farsighted,

Sinclair Stammers/SPL/Photo Researchers

whereas those with the usual form of Marfan syndrome are nearsighted. Lastly, Lincoln showed no outward signs of problems with major blood vessels, such as the aorta.

The gene for Marfan syndrome was identified and cloned in 1991. Using DNA testing, it is possible to determine whether Lincoln or anyone else carries the gene for Marfan syndrome. Soon after the gene was isolated, a group of scientists proposed extracting DNA from fragments of Lincoln's skull (preserved in the National Museum of Health and Medicine in Washington, D.C.) for DNA analysis to see if he had Marfan syndrome. As described later in this chapter, this test has not yet been done, but the proposal raises several important questions related to the emerging field of biohistory. Is there an overriding public interest in knowing if Lincoln had a genetic disorder that had no bearing on his performance in office? Is there any justifiable scientific or societal gain from such knowledge? Does genetic testing violate Lincoln's right to privacy, or that of his family from the disclosure of medical information?

How would you vote?

In 1991, a committee of scientists, historians, and Lincoln scholars recommended testing tissue samples from Abraham Lincoln to determine if he had Marfan syndrome. One bioethicist called the proposal a form of voyeurism, but others pointed out that public officials do not have the same expectation of privacy as the rest of us, and supported the idea of testing. Do you think there is a compelling reason to determine whether Lincoln, who died in 1865, had Marfan syndrome? Is there a scientific or social benefit to having such information or is it simply an invasion of privacy? Visit the Human Heredity Companion website to find out more on the issue, then cast your vote online at **http://biology.brookscole.com/cummings7**

4.1 Studying the Inheritance of Traits in Humans

Mendel used pea plants for two important reasons. First, they can be crossed in many combinations. Second, each cross is likely to produce large numbers of offspring, an important factor in understanding how a trait is inherited. If you were picking an organism for genetic studies, humans would not be a good choice. With pea plants, it is easy to carry out crosses between plants with purple flowers and plants with white flowers and to repeat this cross as often as necessary. For obvious reasons, experimental matings in humans are not possible. You can't ask albino humans to

Keep in mind as you read

- Pedigree construction and analysis are basic methods in human genetics.

- Genetic disorders can be inherited in a number of different ways. We will consider six patterns of inheritance.

- Patterns of gene expression are influenced by many different environmental factors.

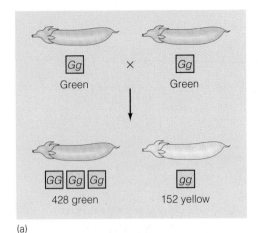

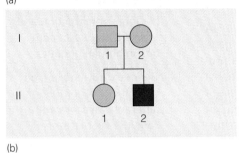

(a)

(b)

▲ FIGURE 4.1 Inheritance in pea plants and humans. (a) In pea plants, a cross between two heterozygotes provides enough offspring in each phenotypic class to allow the pattern of inheritance to be determined. (b) Humans have relatively few offspring, often making it difficult to interpret how a trait is inherited.

mate with homozygous normally pigmented individuals and have their progeny interbreed to produce an F_2. For the most part, human geneticists base their work on the offspring of matings that have already taken place, regardless of whether such matings are the most genetically informative.

Compared with the progeny that can be counted in a single cross with peas, humans produce very few offspring, and these usually represent only a small fraction of the possible genetic combinations. If two heterozygous pea plants are crossed ($Aa \times Aa$), about three-fourths of the offspring will express the dominant phenotype, and the recessive phenotype will be expressed in the remaining one-fourth of the progeny (▶ Figure 4.1). Mendel was able to count hundreds and sometimes thousands of offspring from such a cross to record progeny in all expected phenotypic classes and to clearly establish a phenotypic ratio of 3:1 for recessive traits. As a parallel, consider two humans, each of whom is phenotypically normal. Suppose this couple has two children, one an unaffected daughter, and the other a son affected with a genetic disorder. The ratio of phenotypes in this case is 1:1. This makes it difficult to decide whether the trait is carried on an autosome or a sex chromosome (see Chapter 2 to review autosomes and sex chromosomes), whether it is a dominant or recessive trait, and whether it is controlled by a single gene or by two or more genes.

This example reminds us that the basic method of genetic analysis in humans is observational and indirect rather than experimental and requires reconstructing events that have already taken place rather than designing experiments to test a hypothesis directly. As outlined in the last chapter, one of the first steps in studying a human trait is pedigree construction. Then, the information in the pedigree is used to determine how a trait is inherited. This chapter focuses on the analysis of pedigrees and their use in human genetics.

Keep in mind

■ Pedigree construction and analysis are basic methods in human genetics.

4.2 Pedigree Analysis Is a Basic Method in Human Genetics

A pedigree is an orderly presentation of family information, using standardized symbols. Analysis of the pedigree using knowledge of Mendelian principles can determine whether the trait has a dominant or recessive pattern of inheritance and whether the gene in question is located on an X or Y chromosome, or one of the other 22 chromosomes (the autosomes). In addition, the information in the pedigree can be used in other ways, and we will discuss some of those applications later in this chapter.

Collection of pedigree information is not always straightforward. Knowledge about distant relatives is often incomplete, and recollections about medical conditions can be blurred by the passage of time. Older family members are sometimes reluctant to discuss relatives who had abnormalities or who were placed in institutions. As a result, gathering information for pedigree construction can be a challenge for the geneticist. In addition, organizing and storing the pedigree information for several generations in a large family can be a difficult task. The collection, storage, and analysis of pedigree information can be done using software such as Cyrillic (▶ Figure 4.2). These programs give on-screen displays of pedigrees and genetic information that can be used to analyze patterns of inheritance.

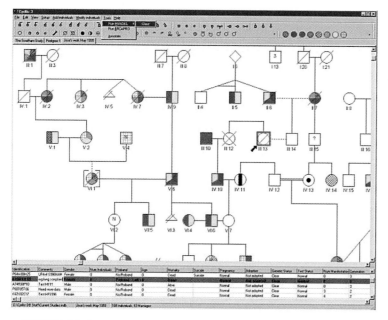
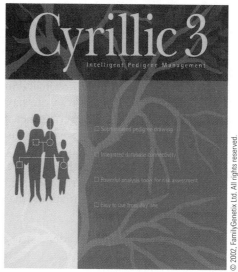

▲ **FIGURE 4.2** Software programs such as Cyrillic can be used to prepare pedigrees, store information, and analyze pedigrees.

Once a pedigree has been constructed, the principles of Mendelian inheritance are used to determine how the trait in question is inherited. The patterns of inheritance we consider in this chapter include

- autosomal recessive
- autosomal dominant
- X-linked dominant
- X-linked recessive
- Y-linked inheritance
- mitochondrial inheritance

Keep in mind

■ Genetic disorders can be inherited in a number of different ways. We will consider six patterns of inheritance.

Pedigree analysis proceeds in several steps. In analyzing a pedigree, a geneticist tries to rule out all patterns of inheritance that are inconsistent with the pedigree. For example, only males carry a Y chromosome. If a trait is controlled by a gene on the Y, then only males will be affected. If the pedigree shows affected females, then Y-linked inheritance can be ruled out. Analysis of the pedigree is complete only when all possible patterns of inheritance have been considered. If all other possible ways of inheritance have been ruled out and only one pattern of inheritance is supported by the information in the pedigree, it is accepted as the pattern of inheritance for the trait being examined.

However, it may turn out that there is not enough information to rule out all other possible patterns of inheritance. Analysis of a pedigree may indicate that a trait can be inherited in an autosomal dominant *or* an X-linked dominant fashion. If this is so, the pedigree is examined to determine if one manner of transmission is more likely than the other. If that is the case, then the most likely type of inheritance is used as the basis for further work. If one pattern is as likely as the other, the geneticist is

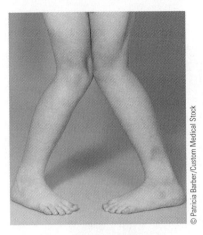

▲ FIGURE 4.3 Ehlers-Danlos syndrome. This disorder can be inherited as an autosomal dominant, autosomal recessive, or X-linked recessive trait. People who have the common autosomal dominant form have loose joints and highly elastic skin, which can be stretched by several inches but returns to its normal position when released.

forced to conclude that the trait can be explained by autosomal dominant or X-linked dominant inheritance, and that more work is necessary to identify the pattern of inheritance. This may require adding more family members to the pedigree or the analysis of pedigrees from other families with the same trait.

As a further complication, some genetic disorders have more than one pattern of inheritance. Ehlers-Danlos syndrome (▶ Figure 4.3; OMIM 130000 and other numbers), characterized by loose joints and easily stretched skin, can be inherited as an autosomal dominant, autosomal recessive, or X-linked recessive trait. In other cases, a trait can have a single pattern of inheritance but be caused by mutation in any of several genes. Porphyria (OMIM 176200 and other numbers), a metabolic disorder associated with abnormal behavior, is inherited as an autosomal dominant trait. However, it can be caused by mutation in genes on chromosomes 1, 9, 11, and 14.

For several reasons, it is important to establish how a trait is inherited. If the pattern of inheritance can be established, it can be used to predict genetic risk in several situations, including

- pregnancy outcome
- adult onset disorders
- recurrence risks in future offspring

4.3 There Is a Catalog of Human Genetic Traits

In this chapter we will limit our discussion to traits controlled by a single gene. Near the end of the chapter, we consider factors that can influence gene expression, and in the next chapter, we discuss traits that are controlled by two or more genes.

To keep track of genetic disorders and the genes that control them, Victor McKusick, a geneticist at Johns Hopkins University, and his colleagues have compiled a catalog of human genetic traits. The catalog is published on the World Wide Web as "Online Mendelian Inheritance in Man" (OMIM). The online version contains text, pictures, references, and links to other databases (▶ Figure 4.4). Each trait is assigned a catalog number (called the OMIM number). In this chapter and throughout the book, the OMIM number for each trait discussed is listed. You can obtain more information about any inherited trait through an integrated series of databases called Entrez, one part of which is OMIM. Access to Entrez and OMIM is available through the book's home page or through search engines.

4.4 Analysis of Autosomal Recessive Traits

Although human families are relatively small, analysis of affected and unaffected members over several generations usually provides enough information to determine whether a trait has a recessive pattern of inheritance and is carried on an autosome or a sex chromosome. Recessive traits carried on autosomes have several distinguishing characteristics:

- For rare or relatively rare traits, most affected individuals have unaffected parents.
- All of the children of two affected (homozygous) individuals are affected.
- The risk of an affected child from a mating of two heterozygotes is 25%.
- Because the trait is autosomal, it is expressed in both males and females, who are affected in roughly equal numbers. Either the male or female parent can transmit the trait.
- In pedigrees involving rare traits, the unaffected (heterozygous) parents of an affected (homozygous) individual may be related to each other.

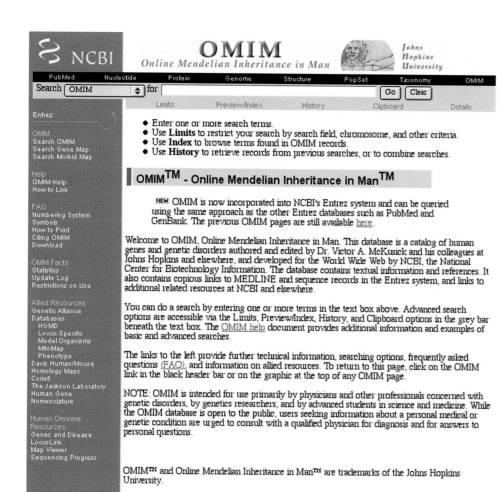

FIGURE 4.4 Online Mendelian Inheritance in Man (OMIM) is an online
database that contains information
about human genetic disorders.

A number of autosomal recessive genetic disorders are listed in ▶ Table 4.1. A pedigree illustrating a pattern of inheritance typical of autosomal recessive genes is shown
in ▶ Active Figure 4.5. Characteristic of a rare recessive trait, the trait appears in individuals (III-2, III-5, and III-6) with unaffected parents (II-1 and II-2). In addition,
two affected parents (III-2 and III-3) have affected children. Although the number of
children is small, the outcome fits the expectations for an autosomal recessive trait.

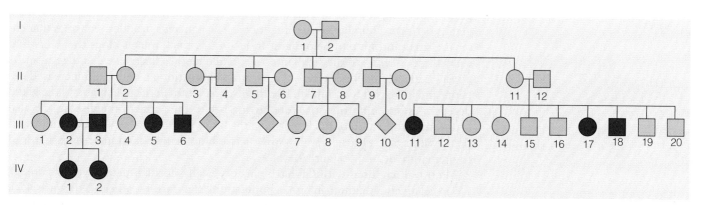

▲ ACTIVE FIGURE 4.5 A pedigree for an autosomal recessive trait. This pedigree has many of the characteristics associated with an autosomal recessive pattern of inheritance. Most affected individuals have normal parents, about one-fourth of the children in large affected families show the trait, both
sexes are affected in roughly equal numbers, and affected parents produce only affected children.

HUMAN Genetics Now™ Learn more about autosomal recessive inheritance by viewing the animation on Human GeneticsNow at
http://now.ilrn.com/cummings7

Table 4.1 Some Autosomal Recessive Traits

Trait	Phenotype	OMIM Number
Albinism	Absence of pigment in skin, eyes, hair	203100
Ataxia telangiectasia	Progressive degeneration of nervous system	208900
Bloom syndrome	Dwarfism; skin rash; increased cancer rate	210900
Cystic fibrosis	Mucus production that blocks ducts of certain glands, lung passages; often fatal by early adulthood	219700
Fanconi anemia	Slow growth; heart defects; high rate of leukemia	227650
Galactosemia	Accumulation of galactose in liver; mental retardation	230400
Phenylketonuria	Excess accumulation of phenylalanine in blood; mental retardation	261600
Sickle cell anemia	Abnormal hemoglobin, blood vessel blockage; early death	141900
Thalassemia	Improper hemoglobin production; symptoms range from mild to fatal	141900/141800
Xeroderma pigmentosum	Lack of DNA repair enzymes, sensitivity to UV light; skin cancer; early death	278700
Tay-Sachs disease	Improper metabolism of gangliosides in nerve cells; early death	272800

Some autosomal recessive traits represent minor variations in phenotype, such as hair color and eye color (see Genetic Journeys: Was Noah an Albino?). Others can be life-threatening or even fatal. Examples of these more severe phenotypes include cystic fibrosis and sickle cell anemia.

Cystic fibrosis is a recessive trait.

Cystic fibrosis (CF; OMIM 219700) is a disabling and fatal genetic disorder inherited as an autosomal recessive trait. CF affects the glands that produce mucus, digestive enzymes, and sweat. This disease has far-reaching effects because the affected glands perform a number of vital functions. The pancreas produces enzymes that enter the small intestine to help digest food. In CF, thick mucus clogs the ducts that carry these enzymes to the small intestine, reducing the effectiveness of digestion. As a result, affected children often suffer from malnutrition in spite of an increased appetite and increased food intake. Eventually, cysts form in the pancreas and the gland degenerates into a fibrous structure, giving rise to the name of the disease. CF also causes the production of thick mucus in the lungs that blocks airways, and most patients with cystic fibrosis develop obstructive lung diseases and infections that lead to premature death (▶ Figure 4.6).

Almost all children with CF have phenotypically normal, heterozygous parents. CF is relatively common in some populations but rare in others (▶ Figure 4.7). Among the U.S. white population, CF has a frequency of 1 in 2,000 births, and 1 in 22 members of this group are heterozygous carriers. In the U.S. black population, the disease is less common and has a frequency of about 1 in 18,000. Among U.S. citizens with origins in Asia, CF is a rare disease whose frequency is about 1 in 90,000. Heterozygous carriers are extremely rare in this population.

■ **Cystic fibrosis** A fatal recessive genetic disorder associated with abnormal secretions of the exocrine glands.

Genetic Journeys

Was Noah an Albino?

The biblical character Noah, along with the ark and its animals, is among the most recognizable figures in the Book of Genesis. His birth is recorded in a single sentence, and although the story of how the ark was built and survived a great flood is told later, there is no mention of Noah's physical appearance. But other sources contain references to Noah that are consistent with the idea that Noah was one of the first albinos mentioned in recorded history.

The birth of Noah is recorded in several sources, including the Book of Enoch the Prophet, written about 200 BC. This book, quoted several times in the New Testament, was regarded as lost until 1773, when an Ethiopian version of the text was discovered. In describing the birth of Noah, the text relates that his "flesh was white as snow, and red as a rose; the hair of whose head was white like wool, and long, and whose eyes were beautiful."

A reconstructed fragment of one of the Dead Sea Scrolls describes Noah as an abnormal child born to normal parents. This fragment of the scroll also provides some insight into the pedigree of Noah's family, as does the Book of Jubilees. According to these sources, Noah's father (Lamech) and his mother (Betenos) were first cousins. Lamech was the son of Methuselah, and Lamech's wife was a daughter of Methuselah's sister. This is important because marriage between close relatives is sometimes involved in pedigrees of autosomal recessive traits, such as albinism.

If this interpretation of ancient texts is correct, Noah's albinism is the result of a consanguineous marriage, and not only is he one of the earliest albinos on record but his grandfather Methuselah and Methuselah's sister are the first recorded heterozygous carriers of a recessive genetic trait.

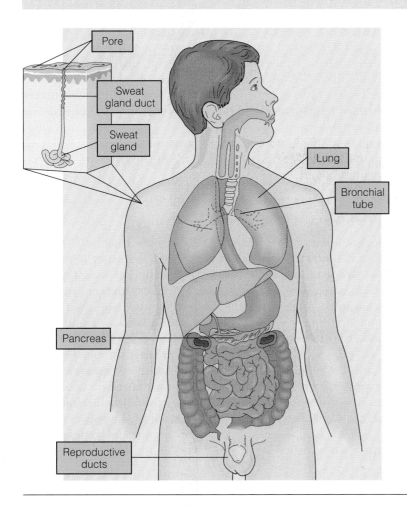

◀ FIGURE 4.6 Organ systems affected by cystic fibrosis. Sweat glands in affected individuals secrete excessive amounts of salt. Thick mucus blocks the transport of digestive enzymes in the pancreas. The trapped digestive enzymes gradually break down the pancreas. The lack of digestive enzymes results in poor nutrition and slow growth. Cystic fibrosis affects both the upper respiratory tract (the nose and sinuses) and the lungs. Thick, sticky mucus clogs the bronchial tubes and the lungs, making breathing difficult. It also slows the removal of viruses and bacteria from the respiratory system, resulting in lung infections. In males, mucus blocks the ducts that carry sperm, and only about 2% to 3% of affected males are fertile. In women who have cystic fibrosis, thick mucus plugs the entrance to the uterus, lowering fertility.

▶ FIGURE 4.7 About 1 in 25 Americans of European descent, 1 in 46 Hispanics, 1 in 60 to 65 African Americans, and 1 in 150 Asian Americans is a carrier for cystic fibrosis. A crowd such as this may contain a carrier.

© Jeff Greenberg/Visuals Unlimited

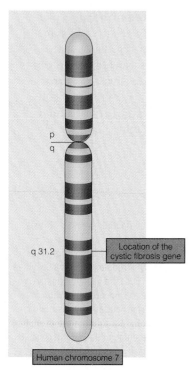

p
q

q 31.2

Location of the cystic fibrosis gene

Human chromosome 7

▲ FIGURE 4.8 Human chromosome 7. The gene for cystic fibrosis (CF) maps to region 7q31.2–31.3, about two-thirds of the way down the long arm of the chromosome.

■ Sickle cell anemia A recessive genetic disorder associated with an abnormal type of hemoglobin, a blood transport protein.

The molecular basis of CF was identified in 1989 by a team of researchers led by Lap-Chee Tsui and Francis Collins. Using recombinant DNA techniques, this team mapped the CF gene to a region of chromosome 7 (▶ Figure 4.8). They explored this region using molecular genetic mapping and identified the CF gene by comparing nucleotide sequence of genes in normal and CF individuals.

The CF gene encodes a protein called the *cystic fibrosis transmembrane conductance regulator* (CFTR), which is inserted in the plasma membrane of specific gland cells (▶ Figure 4.9). CFTR regulates the flow of chloride ions across the plasma membrane. In CF, the protein is either not present in the plasma membrane or, if present, is only partially functional. Because fluids move across plasma membranes in response to the movement of ions, an absent or defective CFTR protein reduces the amount of fluid added to glandular secretions, blocking ducts and obstructing airflow in the lungs.

Once the CF gene and the CFTR protein were isolated and studied in detail, new methods of treatment were developed, including the use of gene therapy, a process we will discuss in detail in Chapter 16.

Sickle cell anemia is a recessive trait.

Those with ancestors from parts of West Africa, the lowlands around the Mediterranean Sea, or parts of the Indian subcontinent have a high frequency of a genetic disorder called **sickle cell anemia** (SCA; OMIM 141900). This autosomal recessive disorder causes production of abnormal hemoglobin, a protein found in red blood cells. This protein transports oxygen from the lungs to the tissues of the body. In SCA, abnormal hemoglobin molecules polymerize to form rods (▶ Figure 4.10), which causes red blood cells to become crescent- or sickle-shaped (▶ Figure 4.11). The deformed cells are fragile and break open as they circulate through the body. New red blood cells are not produced fast enough to replace those that are lost, and the oxygen-carrying capacity of the blood is reduced, causing anemia.

Those with sickle cell anemia tire easily and can develop heart failure caused by an increased load on the circulatory system. The deformed blood cells clog small blood vessels and capillaries, further reducing oxygen transport and sometimes initiating a sickling crisis. As oxygen levels fall in the body, more and more red blood

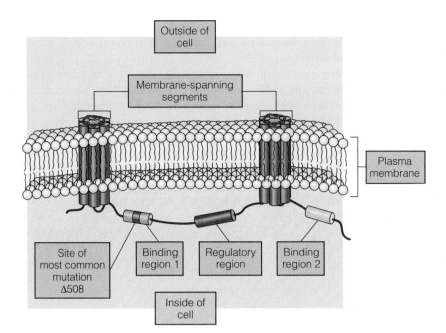

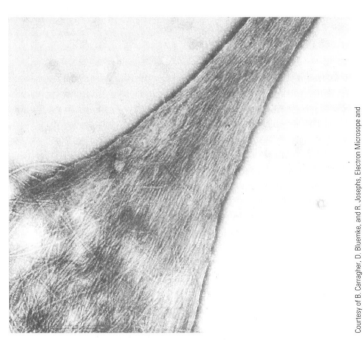

Courtesy of B. Carragher, D. Bluemke, and R. Josephs, Electron Microsope and Image Laboratory, University of Chicago

▶ FIGURE 4.9 The cystic fibrosis gene product. The CFTR protein is located in the plasma membrane of the cell and regulates the movement of chloride ions across the cell membrane. The regulatory region controls the activity of the CFTR molecule in response to signals from inside the cell. In most cases (about 70%), the protein is defective in binding region 1.

◀ FIGURE 4.10 Hemoglobin molecules aggregate in sickle cell anemia. The mutant hemoglobin molecules in red blood cells stack together to form rods. The formation of rods causes the red blood cells to deform and become elongated or sickle-shaped.

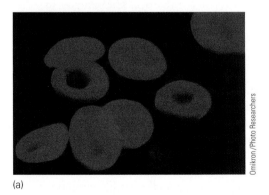

(a)   (b)

Omikron/Photo Researchers

▲ FIGURE 4.11 Red blood cells. (a) Normal red blood cells are flat, disk-shaped cells, indented in the middle on both sides. (b) In sickle cell anemia, the cells become elongated and fragile.

cells become sickled, causing intense pain as blood vessels are blocked. In some affected areas, ulcers and sores appear on the skin. Blockage of blood vessels in the brain can cause strokes and paralysis.

Because of the number of organ systems affected and the severity of the effects, untreated SCA can be lethal. Some affected individuals die in childhood or adolescence, but aggressive medical treatment allows survival into adulthood. As in CF, most affected individuals are children of phenotypically normal, heterozygous parents.

The high frequency of sickle cell anemia in certain populations is related to the frequency of malaria. Sickle cell heterozygotes are more resistant to malaria than homozygous normal individuals. The high frequency of this mutation in the U.S. black population is a genetic relic of West African origins, where malaria is present. In U.S. blacks, sickle cell anemia occurs with a frequency of 1 in every 500 births, and the frequency of heterozygotes is approximately 1 in 12. The same is true for those with ancestral origins in lowland regions of Italy, Sicily, Cyprus, Greece, and the Middle East. This abnormal gene has a double effect: it causes sickle cell anemia but also confers resistance to malaria. The molecular basis of SCA is well known and is discussed in later chapters.

4.5 Analysis of Autosomal Dominant Traits

In autosomal dominant disorders, heterozygotes and those with a homozygous dominant genotype have an abnormal phenotype. Unaffected individuals carry two recessive alleles and have a normal phenotype. Careful pedigree analysis is necessary to determine whether a trait is caused by a dominant allele. Dominant traits have a distinctive pattern of inheritance:

- Every affected individual should have at least one affected parent. Exceptions occur when the gene has a high mutation rate. (Mutation is a heritable change in a gene.)
- Because most affected individuals are heterozygotes with a homozygous recessive (unaffected) spouse, each child has a 50% chance of being affected.
- Because the trait is autosomal, the numbers of affected males and females are roughly equal.
- Two affected individuals may have unaffected children, again because most affected individuals are heterozygous. (In contrast, two individuals affected with an autosomal recessive trait have only affected children.)
- The phenotype in homozygous dominant individuals is often more severe than the heterozygous phenotype.

A number of autosomal dominant traits are listed in ▶ Table 4.2. The pedigree in ▶ Active Figure 4.12 is typical of the pattern found in autosomal dominant conditions.

Marfan syndrome is an autosomal dominant trait.

■ **Marfan syndrome** An autosomal dominant genetic disorder that affects the skeletal system, the cardiovascular system, and the eyes.

Marfan syndrome (OMIM 154700) is an autosomal dominant disorder that affects the skeletal system, the eyes, and the cardiovascular system. Those with Marfan syndrome are tall and thin with long arms and legs and long, thin fingers. Because of their height and long limbs, those with Marfan syndrome often excel in sports like basketball and volleyball, although nearsightedness and defects in the lens of the eye are also common (▶ Figure 4.13).

The most dangerous effects of Marfan syndrome are on the cardiovascular system, especially the aorta. The aorta is the main blood-carrying vessel in the body. As it leaves the heart, the aorta arches back and downward, feeding blood to all major or-

Table 4.2 Some Autosomal Dominant Traits

Trait	Phenotype	OMIM Number
Achondroplasia	Dwarfism associated with defects in growth regions of long bones	100800
Brachydactyly	Malformed hands with shortened fingers	112500
Camptodactyly	Stiff, permanently bent little fingers	114200
Crouzon syndrome	Defective development of mid-face region, protruding eyes, hook nose	123500
Ehlers-Danlos syndrome	Connective tissue disorder, elastic skin, loose joints	130000
Familial hypercholesterolemia	Elevated levels of cholesterol; predisposes to plaque formation, cardiac disease; may be most prevalent genetic disease	144010
Adult polycystic kidney disease	Formation of cysts in kidneys; leads to hypertension, kidney failure	173900
Huntington disease	Progressive degeneration of nervous system; dementia; early death	143100
Hypercalcemia	Elevated levels of calcium in blood serum	143880
Marfan syndrome	Connective tissue defect; death by aortic rupture	154700
Nail-patella syndrome	Absence of nails, kneecaps	161200
Porphyria	Inability to metabolize porphyrins; episodes of mental derangement	176200

▼ ACTIVE FIGURE 4.12 A pedigree for an autosomal dominant trait. This pedigree shows many of the characteristics of autosomal dominant inheritance. Affected individuals have at least one affected parent, about one-half of the children who have one affected parent are affected, both sexes are affected with roughly equal frequency, and affected parents can have unaffected children.

HUMAN Genetics⑧Now™ Learn more about autosomal dominant inheritance by viewing the animation on Human GeneticsNow at **http://now.ilrn.com/cummings7**

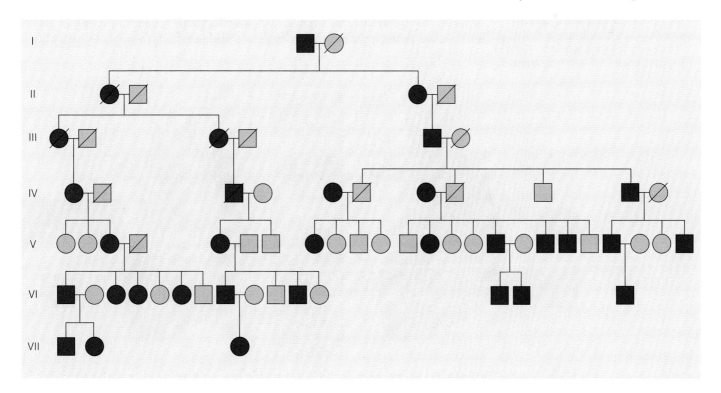

gan systems. Marfan syndrome weakens the connective tissue around the base of the aorta, causing it to enlarge and eventually to split open (▶ Figure 4.14). The enlargement can be repaired by surgery if it is detected in time.

The gene responsible for Marfan syndrome is located on chromosome 15. The normal product of the gene is a protein called fibrillin, which is part of connective tissue. The disorder affects males and females with equal frequency and is found in all ethnic groups with a frequency of about 1 in 10,000 individuals. About 25% of affected individuals appear in families with no previous history of Marfan syndrome, indicating that this gene has a high mutation rate.

As outlined at the beginning of the chapter, it has been suggested that Abraham Lincoln, the sixteenth president of the United States, had Marfan syndrome. To resolve this question, a group of research scientists met in 1991 to formulate a proposal to use bone fragments from Lincoln's body as a source of DNA to decide whether Lincoln did, in fact, have Marfan syndrome. The following year, it was decided that testing should be delayed until more information about the fibrillin gene was known. In 2001, scientists met again and concluded that enough information was known about the gene, and that testing should go forward, but as of this writing, it has not been done.

4.6 Sex-Linked Inheritance Involves Genes on the X and Y Chromosomes

Human females have two X chromosomes, and males have an X and a Y chromosome. These two chromosomes are very different in size and appearance. The X chromosome is medium-sized with a centromere near the middle, while the Y chromosome is much smaller (about 25% as large as the X) and has its centromere very close to one end (▶ Figure 4.15). At meiosis, the X and Y chromosomes pair only at a small region at the tip of the short arms, indicating that most genes on the X chromosome are not present on the Y.

Because the X and Y chromosomes carry different genes, these genes have a distinctive pattern of inheritance. Genes on the X chromosome are called **X-linked**, and genes on the Y chromosome are called **Y-linked**. Female humans have two copies of all X-linked genes and can be heterozygous or homozygous for any of them. Males,

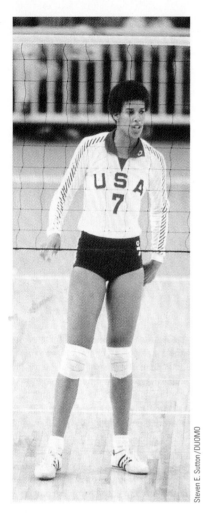

▲ FIGURE 4.13 Flo Hyman was a 6′5″ star on the U.S. women's volleyball team that won a silver medal in the 1984 Olympics. Two years later, at the age of 31, she died in a volleyball game from a ruptured aorta caused by Marfan syndrome.

Steven E. Sutton/DUOMO

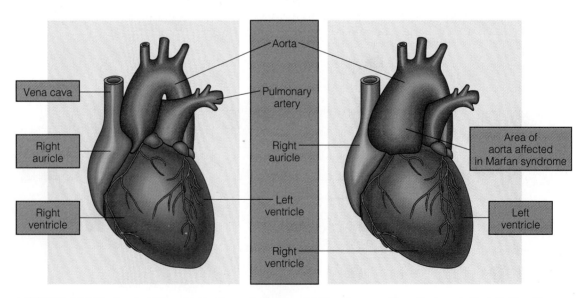

▲ FIGURE 4.14 The heart and its major blood vessels. Oxygen-rich blood is pumped from the lungs to the left side of the heart. From there, blood is pumped through the aorta to all parts of the body.

on the other hand, carry only one copy of the X chromosome. This means that males carrying a gene for a recessive disorder such as hemophilia or color blindness do not have a normal dominant allele of the gene to mask expression of the recessive allele. This explains why males are affected by X-linked recessive genetic disorders far more often than females.

Because a male cannot be homozygous or heterozygous for genes on the X chromosome, males are said to be **hemizygous** for all genes on the X chromosome. Traits controlled by genes on the X chromosome are defined as dominant or recessive by their phenotype in females.

Let's look at how the X and Y chromosomes are transmitted from parents to offspring. Males give an X chromosome to all daughters and a Y chromosome to all sons. Females give an X chromosome to all daughters and to all sons (▶ Figure 4.16). As a result, the X and Y chromosomes have a distinctive pattern of inheritance. Males pass X-linked traits to all their daughters (who may be heterozygous or homozygous for the condition). If a female is heterozygous for an X-linked trait, her sons have a 50% chance of receiving the recessive allele. In the following sections, we consider examples of sex-linked inheritance and explore the characteristic pedigrees in detail.

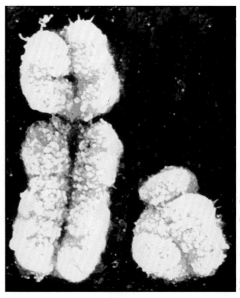

▲ FIGURE 4.15 The human X chromosome (*left*) and the Y chromosome (*right*). This false-color scanning electron micrograph shows the differences between these chromosomes.

4.7 Analysis of X-Linked Dominant Traits

Only a small number of dominant traits are carried on the X chromosome. Dominant X-linked traits have a distinctive pattern of transmission with three characteristics:

- Affected males produce all affected daughters and no affected sons.
- A heterozygous affected female will transmit the trait to half of her children, and sons and daughters are equally affected.
- On average, twice as many daughters as sons are affected.

As expected, a homozygous female will transmit the trait to all of her offspring. A pedigree for an X-linked dominant form of phosphate deficiency, **hypophosphatemia** (OMIM 307800) is shown in ▶ Figure 4.17. This disorder causes a type of rickets, or bowleggedness. To determine whether a trait is X-linked dominant or autosomal dominant, the children of affected males must be carefully analyzed. In X-linked dominant traits, affected males transmit the trait *only* to daughters, and never to sons. On the other hand, if the condition is inherited as an autosomal dominant trait,

■ **X-linked** The pattern of inheritance that results from genes located on the X chromosome.

■ **Y-linked** The pattern of inheritance that results from genes located only on the Y chromosome.

■ **Hemizygous** A gene present on the X chromosome that is expressed in males in both the recessive and dominant condition.

■ **Hypophosphatemia** An X-linked dominant disorder. Those affected have low phosphate levels in blood and skeletal deformities.

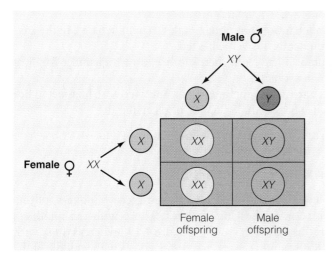

◀ FIGURE 4.16 Distribution of sex chromosomes by parents. All children receive an X chromosome from their mothers. Fathers pass their X chromosome to all daughters and a Y chromosome to all their sons. The sex chromosome content of the sperm determines the sex of the child.

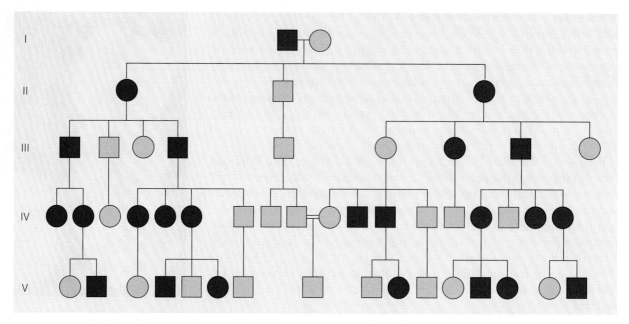

▲ **FIGURE 4.17** A pedigree for hypophosphatemia, an X-linked dominant trait. This pedigree shows characteristics of X-linked dominant traits. Affected males produce all affected daughters and no affected sons; affected females transmit the trait to roughly half their children, with males and females equally affected; and twice as many females as males are affected with the trait.

heterozygous affected males pass the trait to daughters *and* sons, so about half of all daughters and about half of all sons are affected. As seen in the pedigree (▶ Figure 4.17), males affected with X-linked dominant traits only transmit the trait to their daughters, but affected females have affected sons and affected daughters.

4.8 Analysis of X-Linked Recessive Traits

Males give an X chromosome to all daughters but do not give an X chromosome to their sons. Females give an X chromosome to all their children. In addition, males are hemizygous for all genes on the X chromosome and show phenotypes for all X-linked genes. These two factors produce a distinctive pattern of inheritance for X-linked recessive traits. This pattern can be summarized as follows:

- Hemizygous males and homozygous females are affected.
- Phenotypic expression is much more common in males than in females. In the case of rare alleles, males are almost exclusively affected.
- Affected males get the mutant allele from their mothers and transmit it to all of their daughters but not to any sons.
- Daughters of affected males are usually heterozygous and therefore unaffected but sons of heterozygous females have a 50% chance of receiving the recessive gene.

A pedigree for an X-linked recessive trait is shown in ▶ Active Figure 4.18.

Color blindness is an X-linked recessive trait.

The most common form of **color blindness,** known as red-green blindness, affects about 8% of the male population in the United States. Those with red blindness (OMIM 303900) do not see red as a distinct color (▶ Figure 4.19), whereas those with green blindness (OMIM 303800) cannot see green or other colors in the middle of the

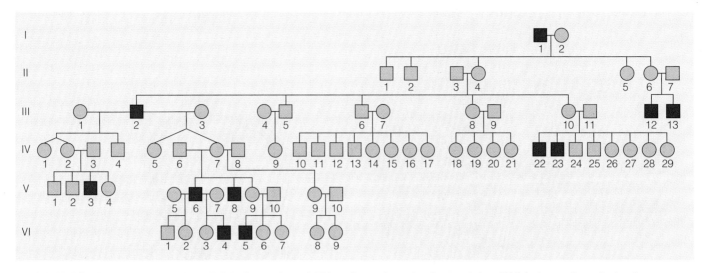

▲ **ACTIVE FIGURE 4.18** Pedigrees for an X-linked recessive trait. This pedigree shows the characteristics of X-linked recessive traits: hemizygous males are affected and transmit the trait to all daughters, who become heterozygous carriers, and phenotypic expression is much more common in males than in females.

HUMAN Genetics ⒺNow™ Learn more about X-linked recessive inheritance by viewing the animation on Human GeneticsNow at **http://now.ilrn.com/cummings7**

▲ **FIGURE 4.19** People who are color-blind see colors differently. (a) Those with normal vision see the red leaves. (b) Someone who is red-green color-blind sees the leaves as gray.

visual spectrum (▶ Figure 4.20). Both red blindness and green blindness are inherited as X-linked recessive traits. A rare form of blue color blindness (OMIM 190900) is inherited as an autosomal dominant condition that maps to chromosome 7.

These three genes encode different proteins found in color vision cells of the retina (▶ Active Figure 4.21). These proteins normally bind to visual pigments in retinal cells that are sensitive to red, green, or blue wavelengths of light. When light strikes

▪ **Color blindness**
Defective color vision caused by reduction or absence of visual pigments. There are three forms: red, green, and blue blindness.

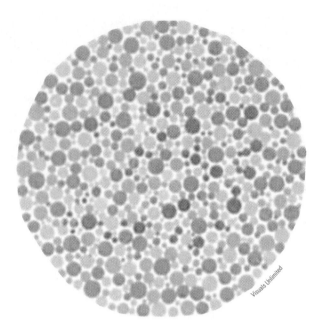

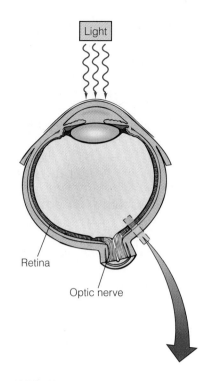

Light

Retina

Optic nerve

▲ FIGURE 4.20 People with normal color vision see the number 29 in the chart; however, those who are color-blind cannot see any number.

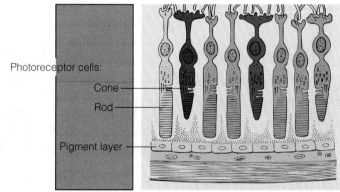

Photoreceptor cells:

Cone

Rod

Pigment layer

▶ ACTIVE FIGURE 4.21 In the retina, there are two types of light receptor cells: rods are sensitive to differences in light intensity, and cones are sensitive to differences in color. There are three types of cones: red-sensitive, green-sensitive, and blue-sensitive. Defects in the cones cause color blindness.

HUMAN Genetics ⊜ Now™ Learn more about eye structure and function by viewing the animation on Human GeneticsNow at **http://now.ilrn.com/cummings7**

these cells, they signal the brain, which processes the signals to produce color vision. If the protein for red color vision is defective or absent, retina cells that respond to red light are nonfunctional, resulting in red color blindness. Similarly, defects in the green or blue color vision proteins produce green and blue blindness.

Muscular dystrophy is an X-linked recessive trait.

Muscular dystrophy is a group of inherited diseases characterized by progressive weakness and loss of muscle tissue. There are autosomal and X-linked forms of muscular dystrophy. The most common form of muscular dystrophy is an X-linked disorder, Duchenne muscular dystrophy (DMD; OMIM 310200), which affects 1 in 3,500 males in the United States. DMD males appear healthy at birth and develop symptoms between 1 and 6 years of age. Progressive muscle weakness is one of the first signs of DMD, and affected individuals use a distinctive set of maneuvers to get up from a prone position (▶ Figure 4.22). The disease progresses rapidly, and by 12 years of age affected individuals are usually confined to wheelchairs because of muscle degeneration. Death usually occurs by the age of 20 due to respiratory infection or cardiac failure.

The DMD gene is located near one end of the X chromosome and encodes a pro-

■ **Muscular dystrophy** A group of genetic diseases associated with progressive degeneration of muscles. Two of these, Duchenne and Becker muscular dystrophy, are inherited as X-linked, allelic, recessive traits.

tein called *dystrophin*. Normal forms of dystrophin attach to the cytoplasmic side of the plasma membrane in muscle cells and stabilize the membrane during the mechanical stresses of muscle contraction (▶ Figure 4.23). When dystrophin is absent or defective, the plasma membranes are torn apart by the forces generated during muscle contraction, eventually causing the death of muscle tissue.

Most individuals with DMD have no detectable amounts of dystrophin in their muscle tissue. However, those with another form of muscular dystrophy, Becker muscular dystrophy (BMD; OMIM 310200), make a shortened form of dystrophin that is partially functional. As a result, those with BMD develop symptoms at a later age, have milder symptoms, and live longer than those with DMD. These two diseases are caused by different mutations in the same gene. The *DMD* gene has been isolated and cloned using recombinant DNA techniques. Future work on the structure and function of dystrophin will hopefully lead to the development of an effective treatment for muscular dystrophy.

There are over 850 X-linked recessive traits, including color blindness, muscular dystrophy, hemophilia (see Spotlight on Hemophilia, HIV, and AIDS; see also Genetics in Society: Hemophilia and History), and others (▶ Table 4.3).

4.9 Paternal Inheritance: Genes on the Y Chromosome

Genes carried on the Y chromosome are called Y-linked. Because only males have Y chromosomes, traits encoded by genes on the Y are passed directly from father to son. In addition, all Y-linked traits should be expressed because males are hemizygous for all genes on the Y chromosome. To date, only about three dozen Y-linked traits have been discovered. These include a gene for a protein found in the cell nucleus that may regulate gene expression. Another gene mapped to the Y chromosome, testis-determining factor (TDF/SRY; OMIM 480000), is involved in determining maleness in developing embryos. The *TDF/SRY* gene and early human development are discussed in Chapter 7. Some of the genes mapped to the Y chromosome are listed in ▶ Table 4.4. ▶ Figure 4.24 shows a pedigree for Y-linked inheritance.

▲ FIGURE 4.22 Children with Duchenne muscular dystrophy use a characteristic set of movements when rising from the prone position. Once the legs are pulled under the body, children use their arms to push the torso into an upright position.

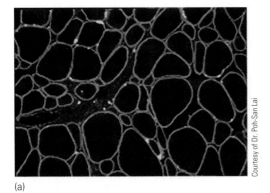

(a)

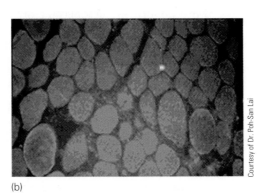

(b)

Courtesy of Dr. Poh-San Lai

▲ FIGURE 4.23 Photographs of cells stained for dystrophin. (a) In normal cells, dystrophin is localized to the membrane surrounding muscle cells. (b) In muscular dystrophy, no dystrophin is present in the membranes.

Hemophilia, HIV, and AIDS

People with hemophilia who used donated blood and blood components to control bleeding episodes in the early 1980s were exposed to HIV, the virus that causes AIDS. This occurred because some blood donors unknowingly had HIV infection and contaminated the blood supply. The result was that many people, including more than half of the hemophilia patients in the United States, developed HIV infection. Most of the blood contamination took place before the cause of AIDS was discovered and before a test to identify HIV-infected blood was instituted. It has been estimated that nearly 10,000 individuals who have hemophilia are infected with HIV.

Fortunately, blood donor screening and new clotting products have virtually eliminated the risk of HIV transmission through blood products. As of January 1991, there have been no reports that anyone who exclusively received heat-treated, donor-screened products is infected with HIV.

Table 4.3 Some X-Linked Recessive Traits

Trait	Phenotype	OMIM Number
Adrenoleukodystrophy	Atrophy of adrenal glands; mental deterioration; death 1 to 5 years after onset	300100
Color blindness		
Green blindness	Insensitivity to green light; 60 to 75% of color blindness cases	303800
Red blindness	Insensitivity to red light; 25 to 40% of color blindness cases	303900
Fabry disease	Metabolic defect caused by lack of enzyme alpha-galactosidase A; progressive cardiac and renal problems; early death	301500
Glucose-6-phosphate dehydrogenase deficiency	Benign condition that can produce severe, even fatal anemia in presence of certain foods, drugs	305900
Hemophilia A	Inability to form blood clots; caused by lack of clotting factor VIII	306700
Hemophilia B	"Christmas disease"; clotting defect caused by lack of factor IX	306900
Ichthyosis	Skin disorder causing large, dark scales on extremities, trunk	308100
Lesch-Nyhan syndrome	Metabolic defect caused by lack of enzyme hypoxanthine-guanine phosphoribosyl transferase (HGPRT); causes mental retardation, self-mutilation, early death	308000
Muscular dystrophy	Duchenne-type, progressive; fatal condition accompanied by muscle wasting	310200

4.10 Maternal Inheritance: Mitochondrial Genes

Mitochondria are cytoplasmic organelles that convert energy from food molecules into ATP, a molecule that powers many cellular functions. Billions of years ago, ancestors of mitochondria were free-living bacteria that adapted to live inside cells of primitive eukaryotes. Over time, most of the genes carried on the bacterial chromosome have been lost, but as an evolutionary relic of their free-living ancestry, mitochondria carry DNA molecules that encode information for 37 mitochondrial genes.

Mitochondria are transmitted from mothers to all their offspring through the cytoplasm of the egg. (Sperm lose all cytoplasm during maturation.) As a result, mitochondria and genetic disorders caused by mutations in mitochondrial genes are maternally inherited. Both males and females can be affected by these disorders, but only females transmit mitochondria and these disorders, producing a distinctive pattern of inheritance (▶ Figure 4.25).

Genetic disorders in mitochondrial DNA are associated with defects in energy conversion. Tissues with the highest energy requirements are most affected. These in-

Table 4.4 Some of the Genes Mapped to the Y Chromosome

Gene	Product	OMIM Number
ANT3 ADP/ATP translocase	Enzyme that moves ADP into, ATP out of mitochondria	403000
CSF2RA	Cell surface receptor for growth factor	425000
MIC2	Cell surface receptor	450000
TDF/SRY	Protein involved in early stages of testis differentiation	480000
H-Y antigen	Plasma membrane protein	426000
ZFY	DNA binding protein that may regulate gene expression	490000

clude the nervous system, muscle, liver, and kidneys. Some of the disorders associated with mutations in mitochondria genes are listed in ▶ Table 4.5.

4.11 Variations in Gene Expression

Many genes have regular and consistent patterns of expression, but others produce a wide range of phenotypes or have a delayed expression, any of which can cause problems in pedigree analysis. In some cases, a mutant genotype may not be expressed at all, producing a normal phenotype. Variation in phenotypic expression is caused by a number of factors, including age, interactions with other genes in the genotype, and interactions between genes and the environment.

Gene expression often changes with age.

Although many genes are expressed early in development or shortly after birth, some disorders do not develop until adulthood. One of the best known examples is **Huntington disease** (HD; OMIM 143100), an autosomal dominant disorder. The phenotype of HD is first expressed between the ages of 30 and 50 years. Affected individuals develop a progressive degeneration of the nervous system, causing mental deterioration and uncontrolled, jerky movements of the head and limbs. The disease progresses slowly, and death occurs some 5 to 15 years after onset. Because most affected individuals are heterozygotes, each child of an affected parent has a 50% chance of developing the disease. The gene for HD has been identified and cloned using recombinant DNA techniques. This makes it possible to test family members and

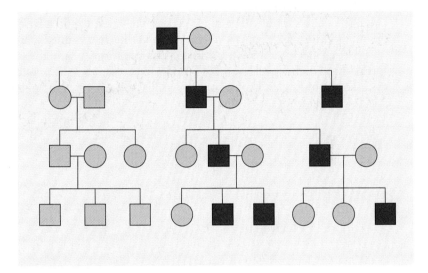

▲ FIGURE 4.24 A pedigree for a Y-linked trait.

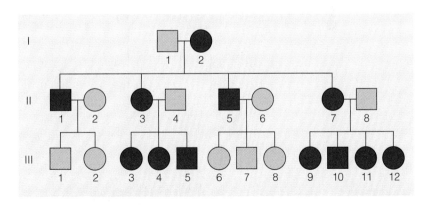

▲ FIGURE 4.25 A pedigree showing the pattern of inheritance associated with mitochondrial genes. Both males and females can be affected by mitochondrial disorders, but only females can transmit the traits to their children.

■ **Huntington disease** An autosomal dominant disorder associated with progressive neural degeneration and dementia. Adult onset is followed by death 10 to 15 years after symptoms appear.

Genetics in Society

Hemophilia and History

Hemophilia, an X-linked recessive disorder, is characterized by defects in the mechanism of blood clotting. This form of hemophilia, called hemophilia A, occurs with a frequency of 1 in 10,000 males. Because only homozygous recessive females can have hemophilia, the frequency in females is much lower, on the order of 1 in 100 million.

Pedigree analysis indicates that Queen Victoria of England carried this gene. Because she passed the mutant allele on to several of her children, it is likely that the mutation occurred in the X chromosome she received from one of her parents. Although this mutation spread through the royal houses of Europe, the present royal family of England is free of hemophilia because it is descended from Edward VII, an unaffected son of Victoria.

Perhaps the most important case of hemophilia among Victoria's offspring involved the royal family of Russia. Victoria's granddaughter Alix, a carrier, married Czar Nicholas II of Russia. She gave birth to four daughters and then a son, Alexis, who had hemophilia. Frustrated by the failure of the medical community to cure Alexis, the royal couple turned to a series of spiritualists, including the monk Rasputin. While under Rasputin's care, Alexis recovered from several episodes of bleeding, and Rasputin became a powerful adviser to the royal family. Some historians have argued that the czar's preoccupation with Alexis's health and the insidious influence of Rasputin contributed to the revolution that overthrew the throne. Other historians point out that Nicholas II was a weak czar and that revolution was inevitable, but it is interesting to speculate that much of twentieth-century Russian history turns on a mutation carried by an English queen.

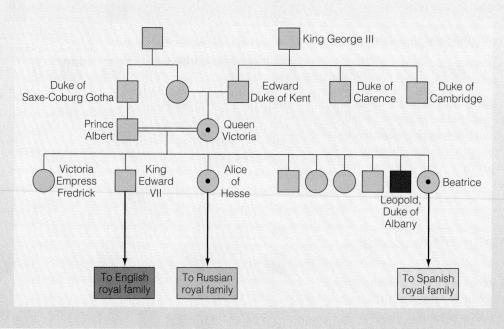

Table 4.5 Some Mitochondrial Traits		
Trait	**Phenotype**	**OMIM Number**
Kearns-Sayre syndrome	Short stature; retinal degeneration	530000
Leber optic atrophy (LHON)	Loss of vision in center of visual field; adult onset	535000
MELAS syndrome	Episodes of vomiting, seizures, and stroke-like episodes	540000
MERRF syndrome	Deficiencies in the enzyme complexes associated with energy transfer	545000
Oncocytoma	Benign tumors of the kidney	553000

identify those who will develop the disorder. This disorder is discussed in more detail in Chapter 16.

Porphyria (OMIM 176200), an autosomal dominant disorder, is also expressed later in life. This disease is caused by the inability to correctly metabolize porphyrin, a chemical component of hemoglobin. As blood levels of porphyrin increase, wine-colored urine is produced. Elevated levels also cause episodes of intense physical pain, seizures, dementia, and psychosis. These symptoms rarely appear before puberty and usually appear in middle age. King George III, the British king during the American Revolution, may have suffered from porphyria (▶ Figure 4.26). At the age of 50 (in 1788) he became delirious and had convulsions. He improved physically, but he remained irrational and confused for months until early in 1789, when his mental functions improved. Later, after two more episodes, his son, George IV, replaced him on the throne, and George III died years later, blind and senile.

Penetrance and expressivity cause variations in gene expression.

The terms *penetrance* and *expressivity* define two different aspects of phenotypic variation. **Penetrance** is the probability that a disease phenotype will be present when a disease genotype is present. For example, if all individuals carrying the gene for a dominant disorder have the mutant phenotype, the gene has 100% penetrance. If only 25% of those with the mutant gene show the mutant phenotype, penetrance is 25%. **Expressivity** refers to the degree of phenotype that is expressed. The following example shows the relationship between penetrance and expressivity.

An autosomal dominant trait **camptodactyly** (OMIM 114200) causes an unmovable, bent, little finger. In some cases, both little fingers are bent; in others, only one finger is affected; and in a small percentage of cases, neither finger is affected, even though the mutant genotype is present (▶ Figure 4.27).

Because the trait is dominant, all heterozygotes and all homozygotes should have a bent little finger on both hands. The pedigree in ▶ Figure 4.27 shows that nine people must carry the dominant allele for camptodactyly, but phenotypic expression is seen only in eight, giving 8/9, or 88%, penetrance. One individual (III-4) is not affected even though he passed the trait to his offspring and must carry the mutant gene. We can only estimate the degree of penetrance in this pedigree be-

▲ FIGURE 4.26 King George III of Great Britain (1738–1820) was probably afflicted with porphyria, a genetic disorder that appears in adulthood and affects behavior.

■ **Porphyria** A genetic disorder inherited as a dominant trait that leads to intermittent attacks of pain and dementia. Symptoms first appear in adulthood.

■ **Penetrance** The probability that a disease phenotype will appear when a disease-related genotype is present.

■ **Expressivity** The range of phenotypes resulting from a given genotype.

■ **Camptodactyly** A dominant human genetic trait that is expressed as immobile, bent, little fingers.

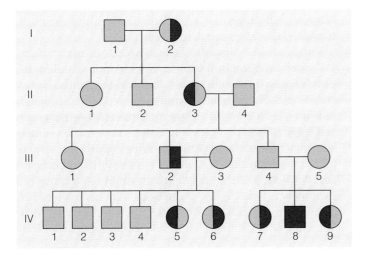

◀ FIGURE 4.27 Penetrance and expressivity. This pedigree shows the transmission of camptodactyly in a family. Fully shaded symbols indicate members with two affected hands. Those with affected left hands are indicated by shading the left half of the symbol, and those affected only in the right hand have the right half of the symbol shaded. Symbols with light shading indicate unaffected family members. There is no penetrance in individual III-4, even though he passed the gene for camptodactyly to all of his children. Expressivity includes no phenotypic expression, expression in one hand, and one individual (IV-8) with both hands affected.

cause II-1, II-2, and III-1 have normal phenotypes but no children, so we cannot be sure if they carry the gene for camptodactyly with no penetrance. More samples from other pedigrees would be necessary to establish a reliable figure for penetrance of this gene.

Keep in mind

■ Patterns of gene expression are influenced by many different environmental factors.

Expressivity defines the *degree* of expression for a given trait. In the camptodactyly pedigree, some members are affected on the left hand and others on the right hand; in one case both hands are affected; in another, neither hand is affected. This variable gene expression results from interactions with other genes and with nongenetic factors in the environment.

Variations in gene expression that we have discussed all result from the relationship between a gene and the mechanisms that produce the gene's phenotype. The genotypes for these genes follow the predictable pattern worked out by Mendel for traits in the pea plant, but expression can be complicated by factors that include temperature and age.

Genetics in Practice

 Genetics in Practice case studies are critical thinking exercises that allow you to apply your new knowledge of human genetics to real-life problems. You can find these case studies and links to relevant websites at **http://biology.brookscole.com/cummings7**

CASE 1

Florence is an active 44-year-old elementary school teacher who began experiencing severe headaches and nausea. She told her physician that her energy level had been dramatically reduced the last few months, and her arms and legs felt like they "weighed 100 pounds each," particularly after she worked out in the gym. Her doctor performed a complete physical and noticed that she did have reduced strength in her arms and legs and that her left eyelid was droopier than her right eyelid. He referred her to an ophthalmologist who discovered that she had an unusual pigment accumulation on her retina, which had not yet affected her vision. She then visited a clinical geneticist who examined the mitochondria in her muscles. She was diagnosed with a mitochondrial genetic disorder known as Kearns-Sayre syndrome.

Mitochondria are responsible for the conversion of food molecules into energy to meet the cell's energy needs. In mitochondrial disorders, these biochemical processes are abnormal, and energy production is reduced. Muscle tends to be particularly affected because it requires a lot of energy, but other tissues such as the brain may also be involved. Under the microscope, the mitochondria in muscle from people with mitochondrial disorders look abnormal, and they often accumulate around the edges of muscle fibers. This gives a particular staining pattern, known as a "ragged red" appearance, and this is usually how mitochondrial disorders are diagnosed.

Mitochondrial disorders affect people in many ways. The most common problem is a combination of mild muscle weakness in the arms and legs together with droopy eyelids and difficulty in moving the eyes. Some people do not have problems with their eye muscles, but have arm and leg weakness that gets worse after exertion. This weakness may be associated with nausea and headaches. Sometimes muscle weakness is obvious in small babies if the illness is severe, and they may have difficulties in feeding and swallowing. Other parts of the body may be involved, including the electrical conduction system of the heart. Most mitochondrial disorders are mildly disabling, particularly in people who have eye muscle weakness and limb weakness. The age at which the first symptoms develop is variable, ranging from early childhood to late adult life.

About 20% of those with mitochondrial disorders have similarly affected relatives. Because only mothers transmit this disorder, it was suspected that some of these conditions are caused by a mutation in the genetic information carried by mitochondria. Mitochondria have their own genes, separate from the genes in the chromosomes of the nucleus. Only mothers pass mitochondria and their genes on to children, whereas the nuclear genes come from both parents. In about one-third of people with mitochondrial disorders, substantial chunks of the mitochondrial genes are deleted. Most of these individuals do not have affected relatives, and it seems likely that the deletions arise either during development of the egg or during very early development of the embryo.

Deletions are particularly common in people with eye muscle weakness and the Kearns-Sayre syndrome.

1. Why would mitochondria have their own genomes?

2. How would mitochondria be passed from mother to offspring during egg formation? Why doesn't the father pass on mitochondria to offspring?

CASE 2

The Smiths had just given birth to their second child and were eagerly waiting to take the newborn home. At that moment, their obstetrician walked into the hospital room with some news about their daughter's newborn screening tests. The physician told them that the state's mandatory newborn screening test had detected an abnormally high level of phenylalanine in their daughter's blood. The Smiths asked if this was just a fleeting effect, like newborn jaundice, that would "go away" in a few days. When they were told that this was unlikely, they were even more confused. The pregnancy had progressed without any complications, and their daughter was born looking perfectly "normal." Mrs. Smith even had a normal amniocentesis early in the pregnancy. The physician asked a genetic counselor to come to their room to explain their daughter's newly diagnosed condition.

The counselor began her discussion with the Smiths by taking a family history from each of them. She explained that phenylketonuria (PKU) is a genetic condition that results when an individual inherits an altered gene from each parent. The counselor wanted to make this point early in the session in case either parent was casting blame for their daughter's condition. She explained that PKU is characterized by an increased concentration of phenylalanine in blood and urine and that mental retardation can be part of this condition if it is not treated at an early age.

To prevent the development of mental retardation, after early diagnosis, dietary therapy must begin before the child is 30 days old. The newborn needs to follow a special diet in which the bulk of protein in the infant's formula is substituted for an artificial amino acid mixture low in phenylalanine. The child must stay on this diet indefinitely for it to be maximally effective.

PKU is one of several diseases known as the hyperphenylalaninemias, which occur with a frequency of 1 in 10,000 births. Classic PKU accounts for two-thirds of these cases. PKU is an autosomal recessive disorder, widely distributed among whites and Asians, but rare in blacks. Heterozygous carriers do not show symptoms but may have slightly increased phenylalanine concentrations. If untreated, children with classic PKU can experience progressive mental retardation, seizures, and hyperactivity. EEG abnormalities; mousy odor of the skin, hair, and urine; tendency to hypopigmentation; and eczema complete the clinical picture.

1. Why did amniocentesis fail to detect PKU? What disorders can amniocentesis detect?

2. Assume you are the genetic counselor. How would you counsel the parents to help them cope with their situation if one or both were blaming themselves for their child's condition?

3. What foods contain phenylalanine? How disruptive do you think the diet therapy will be to everyday life?

4.1 Studying the Inheritance of Traits in Humans

- The inheritance of single gene traits in humans is often called Mendelian inheritance because of the pattern of segregation within families. These traits produce phenotypic ratios similar to those observed by Mendel in the pea plant. Although the results of studies in peas and humans may be similar, the methods are somewhat different.

4.2 Pedigree Analysis Is a Basic Method in Human Genetics

- Instead of direct experimental crosses, human traits are traced by constructing pedigrees that follow a trait through several generations of a family.

4.3 There Is a Catalog of Human Genetic Traits

- As genetic traits are identified, they are described, cataloged, and numbered in a database called "Online Mendelian Inheritance in Man" (OMIM). This online resource is updated on a daily basis and contains information about all known human genetic traits.

4.4 Analysis of Autosomal Recessive Traits

- Information in the pedigree is used to determine how a trait is inherited. These patterns include autosomal dominant, autosomal recessive, X-linked dominant, X-linked recessive, Y-linked, and mitochondrial. Autosomal recessive traits have several characteristics: for rare traits, most affected individuals have unaffected parents; all children of affected parents are affected; the risk of an affected child with heterozygous parents is 25%.

4.5 Analysis of Autosomal Dominant Traits

- Dominant traits have several characteristics: except in traits with high mutation rates, every affected individual has at least one affected parent; because most affected individuals are heterozygous, and have unaffected mates, each child has a 50% risk of being affected. Two affected individuals can have an unaffected child.

4.6 Sex-Linked Inheritance Involves Genes on the X and Y Chromosomes

- Males give an X chromosome to all daughters, but not to their sons. Females pass an X chromosome to all their children. Because of this and the fact that most genes on the X chromosome are not on the Y, genes on the sex chromosomes have a distinct pattern of inheritance.

4.7 Analysis of X-Linked Dominant Traits

- Affected males produce all affected daughters and no affected sons.

- A heterozygous affected female will transmit the trait to half of her children, and sons and daughters are equally affected.

- On average, twice as many daughters as sons are affected.

4.8 Analysis of X-Linked Recessive Traits

- X-linked recessive traits affect males more than females because males are hemizygous for genes on the X chromosome. In X-linked recessive inheritance: Affected males receive the mutant allele from their mother and transmit it to all daughters, but not to their sons; daughters of affected males are usually heterozygous; sons of heterozygous females have a 50% chance of being affected.

4.9 Paternal Inheritance: Genes on the Y Chromosome

- Because only males have Y chromosomes, genes on the Y chromosome are passed directly from father to son. All Y-linked genes are expressed because males are hemizygous for genes on the Y chromosome.

4.10 Maternal Inheritance: Mitochondrial Genes

- Mitochondria are cytoplasmic organelles that convert energy from food molecules into ATP, a molecule that powers many cellular functions. Mitochondria are transmitted from mothers to all their offspring through the cytoplasm of the egg. As a result, mitochondria and genetic disorders caused by mutations in mitochondrial genes are maternally inherited. Genetic disorders in mitochondrial DNA are associated with defects in energy conversion.

4.11 Variations in Gene Expression

- Several factors can affect the expression of a gene, including interactions with other genes in the genotype and interactions between genes and the environment. Some phenotypes are expressed only in adulthood, including Huntington disease. Penetrance affects the expression of a gene and is the probability that a disease phenotype will appear when the disease-producing genotype is present. Another variation in gene expression is expressivity, which is the range of phenotypic variation associated with a given genotype. These variations can affect pedigree analysis.

HUMAN
Genetics ⓔ Now™ Preparing for an exam? Assess your understanding of this chapter's topics with a pre-test, a personalized learning plan, and a post-test at **http://now.ilrn.com/cummings7**

Pedigree Analysis Is a Basic Method in Human Genetics

1. What are the reasons that pedigree charts are used?
2. Pedigree analysis permits all of the following, except:
   a. an orderly presentation of family information
   b. the determination of whether a trait is genetic
   c. the determination of whether a trait is dominant or recessive
   d. an understanding of which gene is involved in a heritable disorder
   e. the determination of whether a trait is sex-linked or autosomal
3. Using the pedigree provided, answer the following questions.
   a. Is the proband male or female?
   b. Is the grandfather of the proband affected?
   c. How many siblings does the proband have and where is he or she in the birth order?

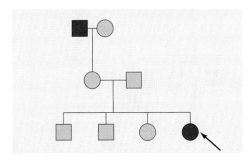

4. What does OMIM stand for? What kinds of information are in this database?

Analysis of Autosomal Recessive and Dominant Traits

5. What pattern of inheritance is suggested by the following pedigree?
6. Does the indicated individual (III-5) show the trait in question?

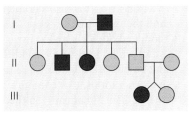

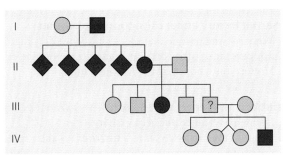

7. Use this information to respond to the following problems: (1) The proband (affected individual that led to the construction of the pedigree) exhibits the trait. (2) Neither her husband nor her only sibling, an older brother, exhibits the trait. (3) The proband has five children by her current husband. The oldest is a boy, followed by a girl, then another boy, and then identical twin girls. Only the second oldest fails to exhibit the trait. (4) Both parents of the proband show the trait.
   a. Construct a pedigree of the trait in this family.
   b. Determine how the trait is inherited (go step by step to examine each possible pattern of inheritance).
   c. Can you deduce the genotype of the proband's husband for this trait?
8. In the following pedigree, assume that the father of the proband is homozygous for a rare trait. What pattern of inheritance is consistent with this pedigree? In particular, explain the phenotype of the proband.

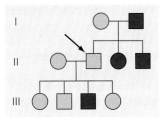

9. Using the following pedigree, deduce a compatible pattern of inheritance. Identify the genotype of the individual in question.

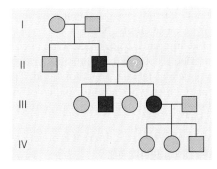

10. A proband female suffering from an unidentified disease seeks the advice of a genetic counselor prior to starting a family. Based on the following data, the counselor constructs a pedigree encompassing three generations: (1) The maternal grandfather of the proband suffers from the disease. (2) The mother of the proband is unaffected and is the youngest of five children, the three oldest being male. (3) The proband has an affected older sister, but the youngest siblings are unaffected twins (boy and girl). (4) All individuals suffering from the disease have been revealed. Duplicate the counselor's feat.

11. Describe the primary gene or protein defect and the resulting phenotype for the following diseases:
    a. cystic fibrosis   b. sickle cell anemia
    c. Marfan syndrome
12. List and describe two other diseases inherited in the following fashion:
    a. autosomal dominant   b. autosomal recessive
13. The father of 12 children begins to show symptoms of neurofibromatosis.
    a. What is the probability that Sam, the man's second oldest son (II-2), will suffer from the disease if he lives a normal life span? (Sam's mother and her ancestors do not have the disease.)
    b. Can you infer anything about the presence of the disease in Sam's paternal grandparents?
14. Huntington disease is a rare, fatal disease that usually develops in the fourth or fifth decade of life. It is caused by a single autosomal dominant allele. A phenotypically normal man in his twenties, who has a 2-year-old son of his own, learns that his father has developed Huntington disease. What is the probability that he himself will develop the disease? What is the chance that his young son might eventually develop the disease?

Analysis of X-Linked Dominant and Recessive Traits
15. The X and Y chromosomes are structurally and genetically distinct. However, they do pair during meiosis at a small region near the tips of their short arms, indicating that the chromosomes are homologous in this region. If a gene lies in this region, will its pattern of transmission be more like a sex-linked gene or an autosomal gene? Why?
16. What is the chance that a color-blind male and a carrier female will produce:
    a. a color-blind son?
    b. a color-blind daughter?
17. A young boy is color-blind. His one brother and five sisters are not. The boy has three maternal uncles and four maternal aunts. None of his uncles' children or grandchildren is color-blind. One of the maternal aunts married a color-blind man, and half of her children, both male and female, are color-blind. The other aunts married men who have normal color vision. All their daughters have normal vision, but half of their sons are color-blind.
    a. Which of the boy's four grandparents transmitted the gene for color blindness?
    b. Are any of the boy's aunts or uncles color-blind?
    c. Is either of the boy's parents color-blind?
18. Describe the phenotype and primary gene or protein defect of the X-linked recessive disease muscular dystrophy.
19. In the beginning of this chapter, we used an example of a couple, both phenotypically normal, with two children, one unaffected daughter and one son affected with a genetic disorder. The phenotype ratio is 1:1, making it difficult to determine whether the trait is autosomal or X-linked. With your knowledge of genetics, what are the genotypes of the parents and children in the autosomal case? In the X-linked case?

20. The following is a pedigree for a common genetic trait. Analyze the pedigree to determine whether the trait is inherited as:
    a. autosomal dominant   b. autosomal recessive
    c. X-linked dominant   d. X-linked recessive
    e. Y-linked

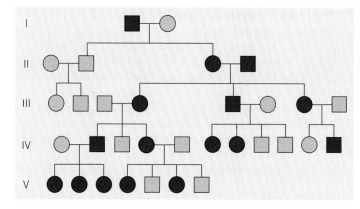

21. As a genetic counselor investigating a genetic disorder in a family, you are able to collect a four-generation pedigree that details the inheritance of the disorder in question. Analyze the information in the pedigree to determine whether the trait is inherited as:
    a. autosomal dominant   b. autosomal recessive
    c. X-linked dominant   d. X-linked recessive
    e. Y-linked

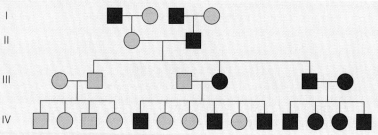

22. In the eighteenth century, a young boy suffered from a skin condition known as ichthyosis hystrix gravior. The phenotype of this disorder includes thickening of skin and the formation of loose spines that are periodically sloughed off. This man married and had six sons, all of whom had this same condition. He also had several daughters, all of whom were unaffected. In all succeeding generations, this condition was passed on from father to son. What can you theorize about the location of the gene that causes ichthyosis hystrix gravior?

Maternal Inheritance: Mitochondrial Genes
23. What are the unique features of mitochondria that are not present in other cellular organelles in human cells?
24. How is mitochondrial DNA transmitted?

Variations in Gene Expression
25. Define penetrance and expressivity.
26. Suppose that space explorers discover an alien species possessing the same genetic principles as apply to hu-

mans. Although all 19 aliens analyzed to date carry a gene for a third eye, only 15 display this phenotype. What is the penetrance of the third-eye gene in this population?

27. A genetic disorder characterized by falling asleep in genetics lectures is known to be 20% penetrant. All 90 students in a genetics class are homozygous for this gene. Theoretically, how many of the 90 students will fall asleep during the next lecture?

28. Explain how camptodactyly is an example of expressivity.

29. How can dominant lethal alleles survive in a population?

30. Why are humans difficult subjects for genetic analysis?

Internet Activities

Internet Activities are critical thinking exercises using the resources of the World Wide Web to enhance the principles and issues covered in this chapter. For a full set of links and questions investigating the topics described below, log on to **http://biology.brookscole.com/cummings7**

1. *A Database of Human Genetic Disorders and Traits.* The Internet site *Online Mendelian Inheritance in Man,* or OMIM, is an online database of human genetic disorders and genetically controlled traits that is updated daily. For any given disorder or trait, information including symptoms, mode of inheritance, molecular genetics, diagnosis, therapies, and more is given.

2. *Genetic Disorders and Support.* Information about many genetic disorders and support groups and organizations for persons with genetic disorders is available on the World Wide Web. You can find information about a particular disorder, its treatments, or parent groups through web search engines. In addition, this text's home page has a link to a list of genetic support groups from whom you can obtain more information about a specific genetic disorder.

3. *Would You Want to Know if You Carried the Gene for a Disorder?* Not all dominant genetic disorders are obvious in early life, and, of course, an individual may be a carrier for a recessive disorder without displaying the characteristics of the trait. At *Do You Really Want to Know If You Have a Disease Gene?* journalist and author Robin Henig explores this question, which we will return to in Chapter 16, Reproductive Technology, Gene Therapy, and Genetic Counseling.
Further Exploration. For a simple version of the genetics of left-handedness, in addition to a look at what life is like for a southpaw, check out *Lorin's Lefthandness Site.*

How would you vote now?

In the emerging field of biohistory, researchers use genetic testing to investigate the lives and deaths of historical figures. Based on a pedigree analysis and some contemporary accounts, some scientists and historians believe that Abraham Lincoln had the genetic disorder Marfan syndrome. Genetic testing could provide the final answer, but there has been debate about the value of such information and the ethics of researching it. Now that you know more about pedigree analysis and inheritance in humans, what do you think? Should scientists perform tests to determine whether Lincoln had Marfan syndrome? Visit the Human Heredity Companion website to find out more on the issue, then cast your vote online at **http://biology.brookscole.com/cummings7**

For further reading and inquiry, log on to **InfoTrac College Edition**, your world-class online library, including articles from nearly 5,000 periodicals, at **http://infotrac.thomsonlearning.com**

5

Complex Patterns of Inheritance

I N 1713, a new king was crowned in Prussia. He immediately began one of the largest military buildups of the eighteenth century. In the space of 20 years, King Frederick William I, ruler of fewer than 2 million citizens, enlarged his army from around 38,000 men to slightly less than 100,000 troops. Compare Frederick's army with the neighboring kingdom of Austria, with 20 million citizens and an army of just under 100,000 men, and you will understand why Frederick William was regarded as a military monomaniac. The crowning glory of this military machine was his personal troops, the Potsdam Grenadier Guards. This unit was composed of the tallest men obtainable. Frederick William was obsessed with having giants in this guard, and his recruiters used bribery, kidnapping, and smuggling to fill the ranks of this unit. It is said that members of the guard could lock arms while marching on either side of the king's carriage. Many members were close to 7 feet tall. Although someone 7 feet tall is not much of a novelty in today's NBA, any man taller than about 5 feet 4 inches in eighteenth-century Prussia was above average height.

King Frederick William was also rather miserly, and because this recruiting was costing him millions, he decided it would be more economical simply to breed giants to serve in his elite unit. To accomplish this, he ordered that every tall man in the kingdom was to marry a tall, robust woman, expecting that the offspring would all be giants. Unfortunately, this idea was a frustrat-

Mary Kay Denny/Photo Edit

ing failure. Not only was it slow, but most of the children were shorter than their parents. While continuing this breeding program, the king reverted to kidnapping and bounties, and he also let it be known that the best way for foreign governments to gain his favor was to send giants to be members of his guard. This human breeding experiment continued until shortly after King Frederick William's death in 1740, when his son, Frederick the Great, disbanded the Potsdam Guards.

How would you vote?

King Frederick William's program of selective breeding of tall humans was a failure, and today, such programs would be condemned as unethical. In our time, it is possible to fertilize eggs outside of the womb and to test the resulting embryos for their genetic characteristics before implanting them in a woman's uterus (discussed in Chapter 14). One possible application of this technology would be to test for genetic markers associated with higher IQ levels. Would you consider having such tests done, and implanting only those embryos carrying such markers? Visit the Human Heredity Companion website to find out more on the issue, then cast your vote online at **http://biology.brookscole.com/cummings7**

5.1 Some Traits Are Controlled by Two or More Genes

What exactly went wrong with King Frederick William's experiment in human genetics? After all, when Mendel intercrossed true-breeding tall pea plants, the offspring were all tall. Even when heterozygous tall pea plants are crossed, three-fourths of the offspring are tall. The situation in humans is more complex than King Frederick imagined, and as we will see, the chances that his breeding program would have worked were very small.

Phenotypes can be discontinuous or continuous.

The problem with comparing inheritance of height in pea plants with height in humans is that a single gene pair controls height in pea plants, but in humans, height is a complex trait determined by several gene pairs, nongenetic factors, and environmental interactions. The tall and short phenotypes in pea plants are two distinct phenotypes and show **discontinuous variation**. In measuring height in humans, it is difficult to set up only two phenotypes. Instead, height in humans is an example of **continuous variation**. Unlike Mendel's pea plants, people are not either 18 inches or 84 inches tall; they fall into a series of overlapping phenotypic classes (▶ Active Fig-

Keep in mind as you read

■ Many human diseases are controlled by the action of several genes.

■ Environmental factors interact with genes to produce variations in phenotype.

■ The genetic contribution to phenotypic variation can be estimated.

■ Twin studies provide an insight into the interaction of genotypes and environment.

■ Many multifactorial traits have social and cultural impacts.

■ **Discontinuous variation** Phenotypes that fall into two or more distinct, nonoverlapping classes.

■ **Continuous variation** A distribution of phenotypic characters that is distributed from one extreme to another in an overlapping, or continuous, fashion.

ure 5.1). Traits that show continuous variation in phenotype are often controlled by two or more separate gene pairs.

How are complex traits defined?

Understanding the distinction between discontinuous and continuous traits was an important advance in genetics and is based on accepting the idea that genes interact with each other and with the environment. **Polygenic traits** are determined by two or more gene pairs. **Multifactorial traits** are controlled by two or more genes *and* show significant interactions with the environment. Although each gene controlling multifactorial traits is inherited in Mendelian fashion, the interaction of genes with the environment produces variable phenotypes that often do not show clear-cut Mendelian ratios, producing a distribution of phenotypes. Height is a polygenic trait because it is determined by more than one gene and is multifactorial because environmental factors contribute to variations in expression. In humans, most multifactorial traits have a polygenic component because the phenotypes are the product of interaction between genes and the environment. The term **complex trait** is used to describe conditions such as hypertension, obesity, and cardiovascular disease, where the relative contributions of genes and environment are not yet established.

■ **Polygenic traits** Traits controlled by two or more genes.

■ **Multifactorial traits** Traits that result from the interaction of one or more environmental factors and two or more genes.

■ **Complex traits** Traits controlled by multiple genes and the interaction of environmental factors where the contributions of genes and environment are undefined.

Keep in mind

■ Many human diseases are controlled by the action of several genes.

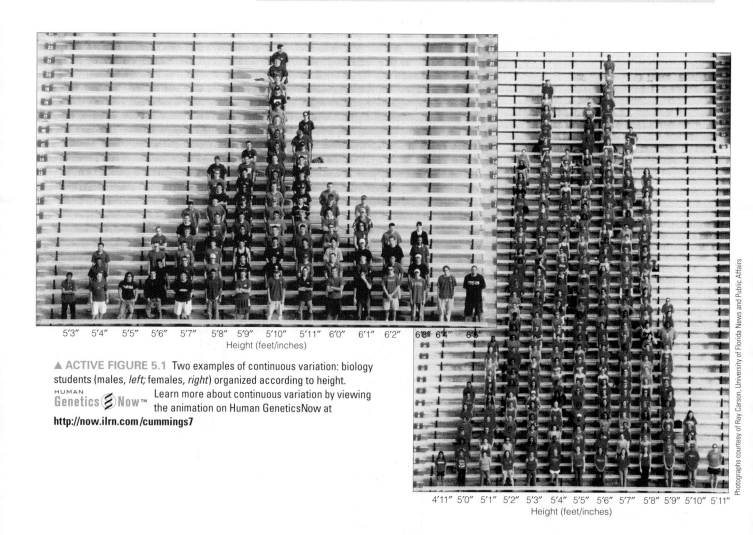

▲ ACTIVE FIGURE 5.1 Two examples of continuous variation: biology students (males, *left;* females, *right*) organized according to height.

HUMAN Genetics ⑤ Now™ Learn more about continuous variation by viewing the animation on Human GeneticsNow at
http://now.ilrn.com/cummings7

Height (feet/inches)

5′3″ 5′4″ 5′5″ 5′6″ 5′7″ 5′8″ 5′9″ 5′10″ 5′11″ 6′0″ 6′1″ 6′2″ 6′3″ 6′4″ 6′5″

4′11″ 5′0″ 5′1″ 5′2″ 5′3″ 5′4″ 5′5″ 5′6″ 5′7″ 5′8″ 5′9″ 5′10″ 5′11″

Height (feet/inches)

Assessing interaction of genes, environment, and phenotype can be difficult.

Many human diseases are complex traits controlled by several genes with environmental contributions. The complexity arises in part because each gene contributes only a small amount to the phenotype, and the environmental components can be hard to identify and measure. Complex traits can be fully understood only when all the genetic and environmental components are fully identified and their individual effects and their interactions have been measured.

Let's turn again to our example of human height. Recall that the average height of men in eighteenth-century Prussia was about 5 feet 4 inches. The average height of men in the United States is now 5 feet 9 inches. It's unlikely that that much genetic change has occurred over 300 years, so environmental factors are probably affecting the expression of a genotypically determined trait. However, identifying and measuring those environmental factors and assigning how much each factor has contributed to the increase in height is not an easy task.

In some cases, genes alone and environment alone produce no observable trait; only when a specific gene and a specific environmental factor interact will there be an effect. A good example is the role of the 5'-HTT gene and emotional stress in producing depression. The 5'-HTT gene (OMIM 182138) encodes a transporter for serotonin, a chemical involved in nerve signal transmission and a target for drugs used to treat depression. The 5'-HTT gene has two alleles, long and short. In one study, people with one or two copies of the short allele experienced more depression and suicidal thoughts when faced with stressful life events. People with two copies of the long allele had much better responses to stress. In other words, how people respond to emotional stress (an environmental factor) is affected by their genotype.

In this chapter, we examine traits controlled by genes at two or more loci (polygenic traits) and traits controlled by two or more genes with significant environmental influences (multifactorial traits). In multifactorial inheritance, the degree to which genetics controls a trait can be estimated by measuring heritability. We consider this concept and the use of twins as a means of measuring the heritability of a trait. In the last part of the chapter, we examine a number of human polygenic traits, some of which have been the subject of political and social controversy.

Keep in mind

■ Environmental factors interact with genes to produce variations in phenotype.

5.2 Polygenic Traits and Variation in Phenotype

In the early part of the twentieth century, it was found that many traits in plants and animals show continuous phenotypic variation. For example, crossing tall and short tobacco plants (▶ Figure 5.2a) produces an F$_2$ with most plants intermediate in height compared with the parents. Compare this result to the same experiment performed with pea plants (▶ Figure 5.2b). For a time, geneticists debated whether continuous variation was consistent with the principles of Mendelian inheritance or perhaps signaled the existence of another mechanism of inheritance. The outcome of this argument is important to human genetics because many human traits show continuous variation.

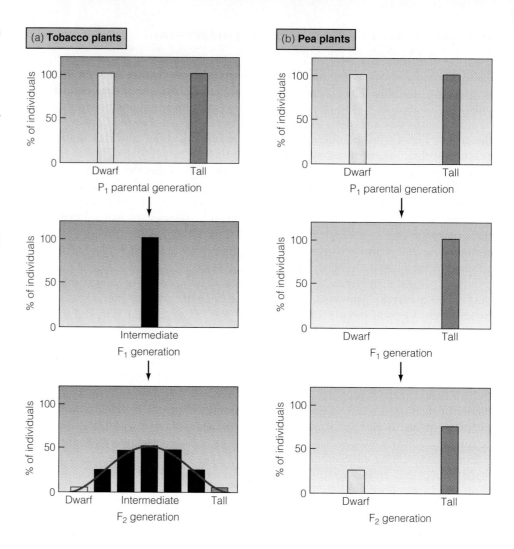

▶ FIGURE 5.2 A comparison of traits that have continuous and discontinuous phenotypes. (a) Histograms show the percentage of plants that have different heights in crosses between tall and dwarf strains of tobacco plants carried to the F_2 generation. The F_1 generation is intermediate to the parents in height, and the F_2 shows a range of phenotypes from dwarf to tall. Most plants have a height intermediate to those of the P_1 generation. (b) Histograms show the percentage of plants that have different heights in crosses between tall and dwarf strains of the pea plant. The F_1 generation has the tall phenotype, and the F_2 has two distinct phenotypic classes: 75% of the offspring are tall, and 25% are dwarf. The differences between tobacco plants and pea plants are explained by the fact that height in tobacco plants is controlled by two or more gene pairs, whereas height in peas is controlled by a single gene.

In the years immediately after the rediscovery of Mendel's work, interest in human genetics was largely centered on determining whether "social" traits, such as alcoholism, feeblemindedness, and criminal behavior, were inherited. Some geneticists constructed pedigrees and simply assumed that single genes controlled these traits. Other geneticists pointed out that these traits did not show the phenotypic ratios observed in experimental organisms and concluded that Mendelian inheritance might not apply to humans. In fact, the biomathematician Karl Pearson, who studied polygenic traits in humans, is reported to have said, "There is no truth in Mendelism at all."

The controversy over continuous variation was resolved by 1930. Experimental crosses with plants showed that traits determined by a number of different genes, each of which makes a small contribution to the phenotype, can explain the continuous distribution of phenotypes in the F_2 generation. This distribution of phenotypes follows a bell-shaped curve. A small number of individuals have phenotypes identical to the P_1 generation (very short or very tall, for example). Most F_2 offspring, however, have phenotypes between these extremes; their distribution follows a bell-shaped curve (▶ Figure 5.3a and b). Traits showing this pattern of polygenic inheritance are controlled by two or more genes, with each gene adding a small but equal amount to the phenotype.

Polygenic inheritance has several distinguishing characteristics:

- Traits are usually quantified by measurement rather than by counting.
- Two or more genes contribute to the phenotype. Each gene contributes in an additive way to the phenotype and the effect of individual genes may be small.

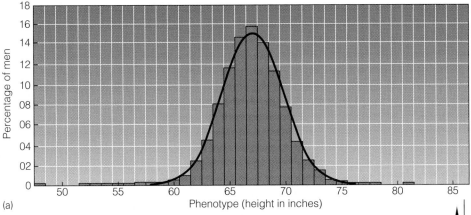

(a)

▲ FIGURE 5.3 (a) A bell-shaped, or "normal," curve shows the distribution of phenotypes for traits controlled by two or more genes. In a normal curve, few individuals are at the extremes of the phenotype, and most individuals are clustered around the average value. In this case, the phenotype is height measured in a population of human males. (b) A bell-shaped curve of the height distribution of the females shown in Figure 5.1 (*right*).

■ Phenotypic expression of polygenic inheritance varies across a wide range. This variation is best analyzed in populations rather than in individuals (▶ Figure 5.4).

Human eye color is a polygenic trait.

The distribution of phenotypes and F_2 ratios in traits controlled by two, three, or four genes is shown in ▶ Figure 5.5. If two genes control a trait such as eye color, there are five phenotypic classes in the F_2, each of which is controlled by four, three, two, one, or zero dominant alleles. The F_2 ratio of $1:4:6:4:1$ results from the genotypic combinations that produce each phenotype. At one extreme is the homozygous dominant (*AABB*) genotype with four dominant alleles; at the other extreme is the homozygous recessive (*aabb*) genotype with no dominant alleles. The largest phenotypic class (6/16) has six genotypic combinations of two dominant alleles (*AaBb*, *Aabb*, *aaBB*, etc.). The five basic human eye colors (▶ Figure 5.6) can be explained by a model using two genes (*A* and *B*), each of which has two alleles (*Aa* and *Bb*).

As the number of loci that controls a trait increases, the number of phenotypic classes increases. As the number of classes increases, there is less phenotypic difference between each class. This means that there is a greater chance for environmen-

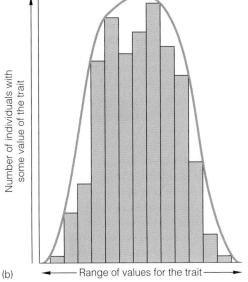

(b)

▲ FIGURE 5.4 Skin color is a polygenic trait controlled by three or four genes, producing a wide range of phenotypes. Environmental factors (exposure to the sun and weather) also contribute to the phenotypic variation, making this a multifactorial trait.

■ **Regression to the mean** In a polygenic system, the tendency of offspring of parents who have extreme differences in phenotype to exhibit a phenotype that is the average of the two parental phenotypes.

tal factors to override the small genotypic differences between classes, blending the phenotypes together. For example, exposure to sunlight can alter skin color and obscure genotypic differences.

Averaging out the phenotype is called regression to the mean.

Sir Francis Galton, a cousin of Charles Darwin, studied the inheritance of many traits in humans. He noticed that children of tall parents were usually shorter than their parents, and children of short parents were usually taller than their parents. Children with very tall parents or very short parents have heights closer to the average height of the population rather than the average height of their parents. This important concept is called **regression to the mean** and is caused by polygenic inheritance of height and the influence of environmental factors (such as diet and health) on expression of the genotype. Regression to the mean explains why King William Frederick's attempt to breed giants for his elite guard unit failed. Using very tall parents (say at least 5 ft. 9 in.) results in more children with average height (close to the population average of 5 ft. 4 in.) than tall children. When you take into account that many of the Potsdam Grenadier Guards were probably tall because of environmental factors (endocrine malfunctions) and did not have the genotypes to produce tall offspring under any circumstances, it is easy to see why Frederick's program didn't succeed.

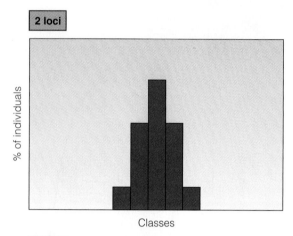

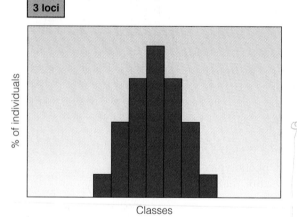

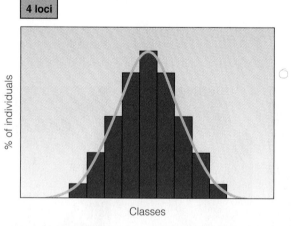

▲ **FIGURE 5.5** The number of phenotypic classes in the F_2 generation increases as the number of genes controlling the trait increases. This relationship allows geneticists to estimate the number of genes involved in expressing a polygenic trait. As the number of phenotypic classes increases, the distribution of phenotypes becomes a normal curve.

5.3 Multifactorial Traits: Polygenes and the Environment

In considering the interaction of polygenes and environmental factors, let's first review some basic concepts: (1) The genotype represents the genetic constitution of an individual. It is fixed at the moment of fertilization and, barring mutation, is unchanging. (2) The phenotype is the sum of the observable characteristics. It is variable and undergoes continuous change throughout the life of the organism. (3) The environment in which a gene exists and operates includes all other genes in the genotype and their effects and interactions; and all nongenetic factors, whether physical or social, that can interact with the genotype (see Genetic Journeys: Is Autism a Genetic Disorder?).

Multifactorial traits have several important characteristics:

■ Traits are polygenic (controlled by several genes).
■ Genes controlling the trait act additively, each contributing a small amount to the phenotype.
■ Environmental factors interact with the genotype to produce the phenotype.

In assessing interactions between the genotype and the environment, as in all science, you have to ask the right question. Suppose the question is, "How much of a given phenotype is caused by heredity and how much by environment?" Because each individual has a unique genotype and has been exposed to a unique set of environmental conditions, it is impossible to evaluate quantitatively the phenotype's genetic and environmental components. Thus, for a given individual, the question as posed cannot be answered. However, in a later section we see that if the question is changed, it is possible to estimate the genotypic contribution to a phenotype.

Frank Cezus/FPG/Getty Images

Frank Cezus/FPG/Getty Images

2001 PhotoDisc, Inc.

Ted Beaudin/FPG/Getty Images

Stan Sholik/FPG/Getty Images

▲ FIGURE 5.6 Samples from the range of continuous variation in human eye color. Different alleles of more than one gene interact to synthesize and deposit melanin in the iris. Combinations of alleles result in small differences in eye color, making the distribution for eye color appear to be continuous over the range from light blue to black.

Genetic Journeys

Is Autism a Genetic Disorder?

Autism is a developmental disorder characterized by impairment in social interactions and communication and by narrow and stereotypical patterns of abilities. As depicted in the movie *Rain Man,* symptoms can include aversion to human contact, language difficulties that show up as bizarre speech patterns, difficulty in understanding what others think, and repetitive body movements. These characteristics seem to be associated with malfunctions of the central nervous system. Autistic individuals have changes in brain anatomy and biochemistry. Symptoms usually appear before the age of 30 months in affected individuals. As outlined in Chapter 4, the information from pedigree construction is used to establish whether a trait is genetically determined and to ascertain its mode of inheritance. Although these steps seem simple and clearcut, in practice the decisions are often more difficult. To illustrate these difficulties, we will briefly consider two questions of current interest in human genetics: is autism a genetic disorder, and if so, how is this trait inherited?

Autism affects about 4 of every 10,000 individuals in the general population. There is a much higher frequency of autism in pairs of identical twins than in nonidentical twins, and siblings of an autistic child are 75 times more likely to be autistic than are members of the general population. These observations indicate that autism has a strong genetic component. Many teams of researchers are working to identify the chromosome regions and the genes involved in autism. In an early study, a team of researchers from UCLA and the University of Utah studied the incidence and inheritance of autism, using almost all the families in the state of Utah as a study group. In 187

families, there was a single autistic child, and in 20 families, there were multiple cases. In the multiple-case families, simple recessive or dominant Mendelian inheritance does not easily explain the pattern of transmission.

The accuracy of pedigree studies can be affected by several factors. Autism is a behavioral trait, and the phenotype is not always clearly defined. In addition, there may be a number of different diseases that all produce a similar set of symptoms. Some cases may have symptoms that are too mild to be diagnosed, whereas very severe cases may result in prenatal death, all of which can skew the pedigrees.

To resolve these problems, a group of 21 different institutions formed the International Molecular Genetic Study of Autism Consortium to use recombinant DNA technology to search for autism genes.

Using pedigree analysis and molecular markers, this team identified loci on several chromosomes that may contain such genes. A study of 153 families identified a region on the long arm of chromosome 7 that contains a susceptibility gene. Other studies have turned up genes on the long arm of chromosome 2 and the short arm of chromosome 16. Using a combination of DNA samples from affected families and sequence data from the Human Genome Project, researchers are working to identify genes on these chromosomes. These results, as well as those from twin studies, are most consistent with a polygenic mode of inheritance for autism or a susceptibility to autism. It is estimated that between 5 and 20 genes may contribute to autism. Once these genes have been identified, the role of environmental factors in triggering autism will need to be evaluated.

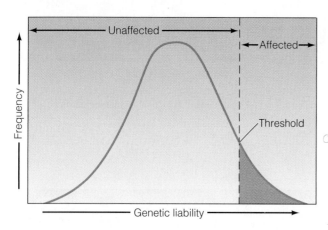

▲ **FIGURE 5.7** A model to explain the discontinuous distribution of some multifactorial traits. In this model, liability for a genetic disorder is distributed among individuals in a normal curve. This liability is caused by a number of genes, each acting additively. Only those individuals who have a genetic liability above a certain threshold are affected if exposed to certain environmental conditions.

Several methods are used to study multifactorial traits.

Although the degree of interaction between a genotype and the environment can be difficult to estimate, family studies indicate that such interactions do occur. We will briefly consider two ways of studying the genetic components of multifactorial traits: the use of a model (the threshold model) and estimating the risk that a multifactorial disease will recur in a family (recurrence risk).

Even though they are polygenic, many multifactorial diseases do not show a bell-shaped phenotypic distribution. These disorders have a discontinuous distribution; someone is affected or not affected. Congenital birth defects such as clubfoot or cleft palate are examples of traits that are distributed discontinuously but are, in fact, multifactorial.

Multifactorial diseases are best explained by the threshold model. In this model, liability is distributed among individuals in a bell-shaped curve (▶ Figure 5.7). Those with liability above a certain threshold develop the disease. The threshold can be reached by genotype (having more genes for the disease), environmental factors, or, in most cases, a combination of genetic and environmental factors. The threshold model is useful in explaining the frequency of certain disorders and congenital malformations. Evidence for a threshold in any given disorder is indirect and comes mainly from family studies.

To look for threshold effects, the frequency of the disorder among relatives of affected individuals is compared with the frequency of the disorder in the general population. In a family, first-degree relatives (parents-children) have one-half of their genes in common, second-degree relatives (grandparents-grandchildren) have one-fourth of their genes in common, and third-degree relatives (first cousins) have one-eighth of their genes in common. As the degree of relatedness declines, so does the probability that individuals will share the same combination of alleles for the genes that control the trait.

According to the threshold model, risk for a disorder should decrease as the degree of relatedness decreases. In fact, the distribution of risk for some congenital malformations, shown in ▶ Table 5.1, declines as the degree of relatedness declines. The multifactorial threshold model provides only indirect evidence for the effect of genotype on traits and for the degree of interaction between the genotype and the environment. The model is helpful, however, in genetic counseling for predicting recurrence risks in families that have certain congenital malformations and multifactorial disorders.

	Risk Relative to General Population			
Multifactorial Trait	**MZ Twins**	**First-Degree Relatives**	**Second-Degree Relatives**	**Third-Degree Relatives**
Clubfoot	300×	25×	5×	2.0×
Cleft lip	400×	40×	7×	3.0×
Congenital hip dislocation (females only)	200×	25×	3×	2.0×
Congenital pyloric stenosis (males only)	80×	10×	5×	1.5×

Table 5.1 Familial Risks for Multifactorial Threshold Traits

The interaction between genotype and environment can be estimated.

How can we measure the interaction between the genotype and the environment? To do this, we must first examine the total variation in phenotype of a population, rather than looking at individual members of the population. Phenotypic variation is derived from two sources: (1) different genotypes in the population and (2) different environments in which the genotypes are expressed. **Heritability** measures the roles of genotype and environment in producing phenotypic variability in a population.

Keep in mind

■ The genetic contribution to phenotypic variation can be estimated.

5.4 Heritability Measures the Genetic Contribution to Phenotypic Variation

Phenotypic variation caused by different genotypes is known as **genetic variance**. Phenotypic variation among individuals with the same genotype is known as **environmental variance**.

The heritability of a trait, symbolized by H, is the amount of phenotypic variation caused by *genetic* differences. Heritability is always a variable, and it is not possible to obtain an absolute value for any given trait. Heritability depends on several factors, including the population being measured and the amount of environmental variation present at the time of measurement. Remember, heritability is observed in populations, not in individuals. In other words, heritability is a statistical value (expressed as a percent) that defines the genetic contribution to the trait being analyzed in a population of related individuals (see later discussion).

In general, if heritability is high (it is 100% when $H = 1.0$), the phenotypic variation is largely genetic, and the environmental contribution is low. If the heritability value is low (it is zero when $H = 0.0$), there is little genetic contribution to the observed phenotypic variation, and the environmental contribution is high.

Heritability estimates are based on known levels of genetic relatedness.

Heritability is calculated using relatives because we know the fraction of genes shared by related individuals. As described in a previous section, parents and children share one-half their genes, grandparents and grandchildren share one-fourth their genes, and so forth. These relationships are expressed as **correlation coefficients**. The half-set of genes received by a child from its parent corresponds to a correlation coefficient of 0.5. The genetic relatedness of identical twins is 100% and the correlation coefficient is 1.0. Unless a mother and father are related by descent, they should be genetically unrelated, and the correlation coefficient for this relationship is 0.0.

Using the genetic relatedness among population members expressed as a correlation coefficient and using the measured phenotypic variation expressed in quantitative units (inches, pounds, etc.), a heritability value can be calculated for a specific phenotype in a population. If the heritability value for a trait is 0.72, this means that 72% of the phenotypic variability seen in the population is caused by genetic differences in the population.

Fingerprints can be used to estimate heritability.

Because of interactions between genes and the environment, it is difficult to find multifactorial traits that can be used to measure heritability. Fingerprints are one such trait that has been used to successfully measure heritability.

Heritability An expression of how much of the observed variation in a phenotype is due to differences in genotype.

Genetic variance The phenotypic variance of a trait in a population that is attributed to genotypic differences.

Environmental variance The phenotypic variance of a trait in a population that is attributed to differences in the environment.

Correlation coefficients Measures of the degree to which variables vary together.

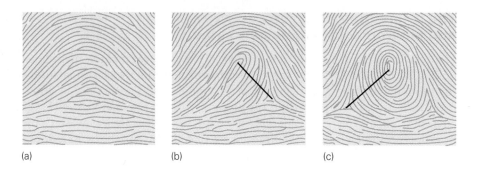

► FIGURE 5.8 The three basic patterns of fingerprints: (a) arch, (b) loop, and (c) whorl. The triangular areas in (b) and (c) where ridge patterns diverge are called triradii. Ridge counts are made from prints of loops and whorls by superimposing a line from the triradius to the center of the print and counting the number of ridges that cross the line.

(a) (b) (c)

Fingerprints are laid down in the first 3 months of embryonic development (weeks 6 to 13). They are a polygenic trait and can be influenced by the environment only during this short period. Everyone, including identical twins, has a unique set of fingerprints. Even though identical twins share the same set of genes and occupy the same uterus simultaneously, each lives in a slightly different prenatal environment. These subtle environmental factors are enough to create different fingerprint patterns (► Figure 5.8).

Using correlation coefficients, Sarah Holt studied fingerprint ridges (called total ridge counts, or TRCs) in males and females (► Table 5.2). The agreement between observed and expected values indicates that TRC is almost totally under genetic control and that environmental factors play only a minor role.

Ridge counts in mothers and their children estimate heritability of this trait as $H = 0.96$, meaning that 96% of the phenotypic variation seen in ridge counts is caused by differences in genotype. The small amount of nongenetic variation helps explain why identical twins have different fingerprint patterns.

5.5 Twin Studies and Multifactorial Traits

Using correlation coefficients to measure the amount of observed phenotypic variability provides an estimate of heritability. This method, however, has one main problem. The closer the genetic relationship, the more likely it is that the relatives share a common environment. In other words, how can we tell whether parents and children have similar phenotypes because they have one-half of their genes in common or because they share a similar environment? Is there a way we can separate the effects of genotype on phenotypic variation from the effects of the environment?

To solve this problem, human geneticists look for situations in which genetic and environmental influences are clearly separated. One way to do this is to study twins

Table 5.2 Correlations between Relatives for Total Ridge Count (TRC)

Relationship	Number of Pairs	Observed Correlation Coefficient	Expected Correlation Coefficient between Relatives	Heritability
Mother-child	405	0.48 ± 0.04	0.50	0.96
Father-child	405	0.49 ± 0.04	0.50	0.98
Husband-wife	200	0.05 ± 0.07	0.00	—
Sibling-sibling	642	0.50 ± 0.04	0.50	1.00
Monozygotic twins	80	0.95 ± 0.01	1.00	0.95
Dizygotic twins	92	0.49 ± 0.08	0.50	0.98

Note: From *Quantitative genetics of fingerprint patterns,* by S. B. Holt (1961). *Br. Med. Bull.,* 17, 247–250.

(▶ Figure 5.9). Identical twins share the same genotype. If identical twins are separated at birth and raised in different environments, the genotype is constant, and the environments are different. To reverse the situation, geneticists compare traits in unrelated adopted children with those of natural children in the same family. In this situation, there is a constant environment and maximum genotypic differences. As a result, twin studies and adoption studies are important tools in measuring heritability in humans.

The biology of twins includes monozygotic and dizygotic twins.

Before examining the results of twin studies, we will look briefly at the biology of twinning. There are two types of twins, **monozygotic (MZ)** (identical) and **dizygotic (DZ)** (fraternal). Monozygotic twins originate from a single egg fertilized by a single sperm (▶ Figure 5.10a). During an early stage of development, two separate embryos are formed. Additional splitting is also possible (see Genetic Journeys: Twins, Quintuplets, and Armadillos). Because they arise from a single fertilization event, MZ twins share the same genotype, have the same sex, and carry the

▲ **FIGURE 5.9** Identical twins (monozygotic twins) have the same sex and share a single genotype.

same genetic markers, such as blood types. Dizygotic twins originate from two separate fertilization events: two eggs, ovulated in the same ovarian cycle, are fertilized independently by two different sperm (▶ Figure 5.10b). DZ twins are no more related than other pairs of siblings, have half of their genes in common, can differ in sex, and may have different genetic markers, such as blood types.

For heritability studies, it is essential to know whether a pair of twins is MZ or DZ. Comparison of many traits such as blood groups, sex, eye color, hair color, fingerprints, palm and sole prints, DNA fingerprinting, and analysis of DNA molecular markers are used to identify twins.

Concordance can occur in twins.

To evaluate phenotypic differences between twins, traits are scored as present or absent rather than measured quantitatively. Twins show **concordance** if both have a

■ **Monozygotic (MZ)** Twins derived from a single fertilization involving one egg and one sperm; such twins are genetically identical.

■ **Dizygotic (DZ)** Twins derived from two separate and nearly simultaneous fertilizations, each involving one egg and one sperm. Such twins share, on average, 50% of their genes.

■ **Concordance** Agreement between traits exhibited by both twins.

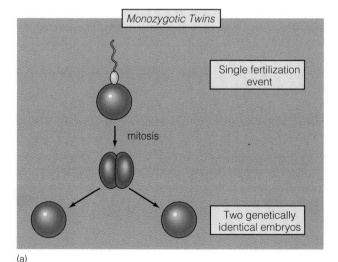

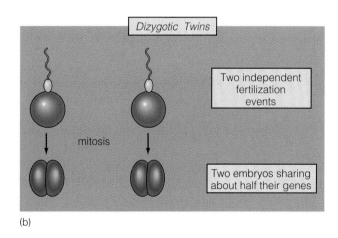

▲ **FIGURE 5.10** (a) Monozygotic (MZ) twins result from the fertilization of a single egg by a single sperm. After one or more mitotic divisions, the embryo splits in two and forms two genetically identical individuals. (b) Dizygotic (DZ) twins result from the independent fertilization of two eggs by two sperm during the same ovulatory cycle. Although these two embryos simultaneously share the same uterine environment, they share only about half of their genes.

Genetic Journeys

Twins, Quintuplets, and Armadillos

Monozygotic (MZ) twins are genetically identical because of the way in which they are formed. The process of embryo splitting that gives rise to MZ twins can be considered a form of human asexual reproduction. In fact, another mammal, the nine-banded armadillo, produces litters of genetically identical, same-sex offspring that arise by embryo splitting. In armadillo reproduction a single fertilized egg splits in two, and daughter embryos can split again, resulting in litters of two to six genetically identical offspring.

Multiple births in humans occur rarely. About 1 in 7,500 births are triplets, and 1 in 658,000 births are quadruplets. In many cases, both embryo splitting and multiple fertilizations are responsible for naturally occurring multiple births. Triplets may arise by fertilization of two eggs, in which one of them undergoes embryo splitting. The use of hormones to enhance fertility has slightly increased the frequency of multiple births. These drugs work by inducing the production of multiple eggs in a single menstrual cycle. The subsequent fertilizations have resulted in multiple births ranging from twins to septuplets.

Embryo splitting in naturally occurring births was documented in the Dionne quintuplets born in May 1934. This was the first case in which all five members of a set of quintuplets survived. Blood tests and physical similarities indicate that these quintuplets arose from a single fertilization followed by several embryo splits.

From this, it seems that MZ twins, armadillos, and the Dionne quintuplets have something in common—embryo splitting.

---

trait and are discordant if only one twin has the trait. As noted, MZ twins have 100% of their genes in common, whereas DZ twins, on average, have 50% of their genes in common. For a genetically determined trait, the correlation in MZ twins should be higher than that in DZ twins. If the trait is completely controlled by genes, concordance should be 1.0 in MZ twins and close to 0.5 in DZ twins.

The degree of difference in concordance between MZ and DZ twins is important; the greater the difference, the greater the heritability. Concordance values for several traits are listed in ▶ Table 5.3. The concordance value for cleft lip in MZ twins is higher than that for DZ twins (42% versus 5%). Although this difference suggests a genetic component to this trait, the value is so far below 100% that environmental factors are obviously important in the majority of cases. As this example shows, concordance values must be interpreted cautiously.

Concordance values can be converted to heritability values through a number of statistical methods. Some calculated heritability values for obesity are listed in the right column of ▶ Table 5.4. Obesity is measured by body mass index (BMI = weight in kilograms divided by height in meters squared). Remember that heritability is a relative value, valid only for the population measured and only under the environmental conditions in effect at the time of measurement. Heritability measurements made within one population cannot be compared with heritability measurements for the same trait in another population because the two groups differ in genotypes and environmental variables in unknown ways.

Keep in mind

■ Twin studies provide an insight into the interaction of genotypes and environment.

Table 5.3 Concordance Values in Monozygotic (MZ) and Dizygotic (DZ) Twins

	Concordance Values (%)	
Trait	MZ	DZ
Blood types	100	66
Eye color	99	28
Mental retardation	97	37
Hair color	89	22
Down syndrome	89	7
Handedness (left or right)	79	77
Epilepsy	72	15
Diabetes	65	18
Tuberculosis	56	22
Cleft lip	42	5

Table 5.4 Heritability Estimates for Obesity in Twins (from Several Studies)

Condition	Heritability
Obesity in children	0.77–0.88
Obesity in adults (weight at age 45)	0.64
Obesity in adults (body mass index at age 20)	0.80
Obesity in adults (weight at induction into armed forces)	0.77
Obesity in twins reared together or apart	
Men	0.70
Women	0.66

We can study obesity with twins and families.

Obesity is a trait that is said to "run" in families. Obesity is also a national health problem. A 1999 federal study has estimated that 61% of the adults in the United States are overweight and 26% are obese, an increase from the 1988–94 survey (▶ Figure 5.11). These individuals are at risk for diseases such as high blood pressure, elevated levels of cholesterol in the blood, coronary artery disease, and adult-onset diabetes. Increases in the incidence of obesity have taken place in the last 30–40 years and cannot be explained by changes in our genetic makeup. Instead, we must look to changing environmental factors interacting with our genes.

Twin studies have been used to estimate the heritability of obesity. The results show high values of heritability for obesity, suggesting that this condition has a strong genetic component, with heritability estimates that average close to 70% (▶ Table 5.4). However, heritability estimates from twin studies are indirect ways of studying multifactorial traits. Heritability estimates from adoption studies are much lower and cluster around a value of 30%. Family studies give heritability values between these two values. In spite of the wide range in heritability estimates, these studies all indicate that genes contribute to obesity.

What are some genetic clues to obesity?

Heritability estimates are performed at the phenotypic level and cannot tell us anything about how many genes control the trait being studied, whether such genes are inherited in a dominant, recessive, or sex-linked fashion, or how such genes act to produce the phenotype. Several methods are being used to identify genes that contribute to complex traits in humans (such as obesity).

Recent breakthroughs in understanding how genes regulate body weight have come from studies in mice. Several mouse genes that control body weight have been identified, isolated, cloned, and analyzed. Mice mutant for the genes obese (*ob*) and diabetes (*db*) are both obese (▶ Figure 5.12). The *ob* gene encodes the weight-controlling hormone **leptin** (from the Greek word for "thin"), which is produced in fat cells. In mice, the hormone is released from fat cells and travels through the blood to the brain, where cells of the hypothalamus have cell surface receptors for leptin. These receptors are encoded by the *db* gene. Binding of leptin activates the leptin receptor and initiates a response that involves changes in gene expression in the hypothalamus (see Spotlight on Leptin and Female Athletes). These two genes are part of a pathway in the central nervous system that regulates energy balance in the body (▶ Figure 5.13). Other genes in the pathway have been identified and cloned.

The human gene for leptin (OMIM 164160), equivalent to the mouse *ob* gene, maps to chromosome 7q31.1. The leptin receptor gene (OMIM 601007), equivalent to the mouse *db* gene, maps to chromosome 1p31. Mutations in genes of the energy regulation pathway and other single genes that result in obesity account for only a small percentage (~5%) of all cases of obesity in the human population and cannot explain the explosive increase in obesity in developed countries. Obesity is clearly a complex disorder involving the action and interaction of multiple genes and environmental factors.

To search for genes in the polygenic set that contribute to obesity in 95% of all cases, several research groups did genome-wide searches using large, multigenerational families and molecular markers to search for a link between the

■ **Leptin** A hormone produced by fat cells that signals the brain and ovary. As fat levels become depleted, secretion of leptin slows and eventually stops.

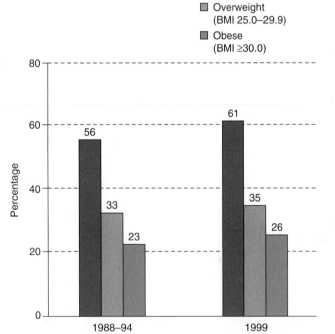

Legend:
- ■ Overweight or obese (BMI ≥25.0)
- ■ Overweight (BMI 25.0–29.9)
- ■ Obese (BMI ≥30.0)

▲ **FIGURE 5.11** Age-adjusted prevalence of overweight and obesity among U.S. adults age 20 years and older. The 1999 survey shows that 61% of the adult population is overweight or obese. Comparison with the 1988–94 survey shows that only 56% of the population was overweight or obese at that time. (BMI [body mass index] = weight in kilograms divided by height in meters squared.)

▲ **FIGURE 5.12** The obese (*ob*) mouse mutant, shown on the left (a normal mouse is on the right), has provided many clues about how weight is controlled in humans.

John Sholtis, The Rockefeller University, New York

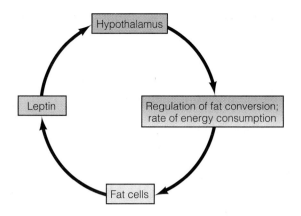

◄ FIGURE 5.13 The hormone leptin is produced in fat cells, moves through the blood, and binds to receptors in the hypothalamus. Binding presumably activates a control mechanism (still unknown) that controls weight by regulating the conversion of food energy into fat and the rate of energy consumption.

Leptin and Female Athletes

Leptin, like some other hormones, may have multiple effects. In addition to signaling the hypothalamus about body fat levels, leptin may also signal the ovary. The obese strain of mice in which leptin was discovered has a mutated leptin gene, and these mice do not make any leptin. Females of this strain are infertile and do not ovulate. Injection of leptin into these females leads to ovulation and normal levels of fertility. The discovery that leptin receptors are found in the ovary indicates that leptin may act directly to control ovulation. In mammalian females, including humans, ovulation stops when body fat falls below a certain level. Many female athletes, such as marathon runners, have low levels of body fat and stop menstruating. Because leptin is produced by fat cells, females who have low levels of body fat may have lowered their leptin levels to the point at which ovulation and menstruation cease.

molecular markers and obesity. The results of several of these genome scans indicates that important genes for obesity are located on chromosomes 2, 3, 5, 6, 7, 10, 11, 17, and 20 (▶ Figure 5.14). Further work may identify additional genes involved in obesity and provide a foundation for studying how these genes interact with environmental factors to cause obesity.

5.6 A Survey of Some Multifactorial Traits

Many important human diseases are multifactorial, such as cardiovascular disease (▶ Table 5.5). Some progress has been made in defining the genetic components of the multifactorial traits discussed in this section. New methods of screening for polygenes developed in the last few years and the results of the Human Genome Project are helping us understand how genetic and environmental factors contribute to these and other complex traits.

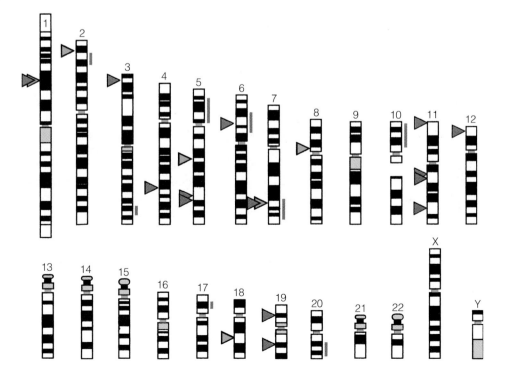

◄ FIGURE 5.14 Some obesity genes in the human genome. The green triangles represent single genes in which obesity is the main phenotype with little environmental influence. The red triangles represent genes that are candidates for obesity genes. Mutations in these genes have an association with obesity. Red bars represent chromosome regions that may contain other genes associated with obesity. More than 70 genes associated with obesity have been identified.

Table 5.5 Risk Factors for Cardiovascular Disease

Heredity (history of cardiovascular disease before age 55 in family members)	Smoking
	Obesity
Being male	Lack of exercise
Hypertension	Stress
High blood cholesterol (high LDL and/or low HDL)	

■ **Essential hypertension** Elevated blood pressure, consistently above 140/90 mm Hg.

■ **Familial hypercholesteremia** Autosomal dominant disorder with defective or absent LDL receptors. Affected individuals are at increased risk for cardiovascular disease.

■ **Lipoproteins** Particles that have protein and phospholipid coats that transport cholesterol and other lipids in the bloodstream.

Cardiovascular disease has genetic and environmental components.

Genetic disorders that lead to cardiovascular disease include **essential hypertension** (OMIM 145500) and **familial hypercholesterolemia** (OMIM 143890). Both traits have significant environmental contributions (▶ Table 5.5).

Hypertension occurs when blood pressure (▶ Active Figure 5.15) is consistently above 140/90 mm Hg (140 is the pressure generated when the heart ventricles contract, and 90 is the pressure when the ventricles are relaxed). At least 10 genes are involved in controlling blood pressure. Most work by controlling the amount of salt and water reabsorbed into the blood by the kidney. One of these is the gene for angiotensinogen (AGT). AGT (OMIM 106150) is a protein made in the liver that controls salt and water retention, which in turn controls blood pressure. Some variants of this protein have been linked to a predisposition to hypertension, which is a silent killer because no obvious symptoms appear in early stages of the disease. Some 10% to 20% of the adult population of the United States suffers from hypertension, making it a serious health problem.

Atherosclerosis results from an imbalance between dietary intake and the synthesis and breakdown of lipids, especially cholesterol. This imbalance can lead to blockage of blood vessels and the development of cardiovascular disease (▶ Figure 5.16). The most serious consequences come from blockages in the brain and heart, which cause strokes and heart attacks.

Cholesterol is not soluble in blood plasma and is wrapped in a coat of proteins and phospholipids for transport. The coat and its contents are known as **lipoproteins** (▶ Figure 5.17). Lipoproteins are classified by their size and density. Cholesterol is carried by low-density lipoprotein (LDL) and high-density lipoprotein (HDL). LDLs are about 45% cholesterol, and HDLs are about 20% cholesterol. The risk of atherosclerosis is related to the HDL/total cholesterol ratio. Higher HDL levels mean lower risk.

Genetic studies of cardiovascular disease are directed at finding genes that predispose to disease, identifying environmental factors (such as diet and exercise) that affect disease progression, and detecting individuals at risk for cardiovascular disease. Many genes, including those that encode apolipoproteins (the part of the lipoprotein that attaches to cell receptors) control cholesterol levels in the body. Other genes encode cell receptors that bind to and internalize lipoproteins and enzymes that degrade lipoproteins inside the cell.

The autosomal dominant disease familial hypercholesterolemia (FH; OMIM 143890) is caused by defects in cell surface receptors that control the uptake of LDLs. Affected heterozygotes have elevated cholesterol levels in their blood serum and usually develop coronary artery disease between the ages of 40 and 50. The frequency of FH heterozygotes is about 1 in 500 in European, Japanese, and U.S. populations, although in regions of Quebec the frequency is 1 in 122. The highest reported frequency is in a South African community, where the frequency is 1 in 71. The average frequency of 1 in 500 makes this disorder one of the most common mutations in our species, and one of the major causes of cardiovascular disease.

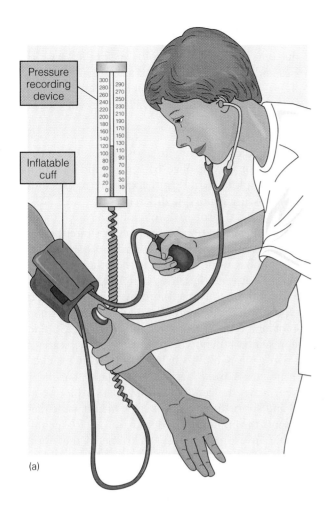

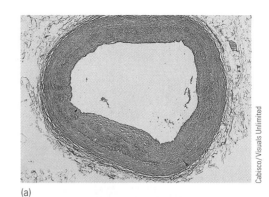

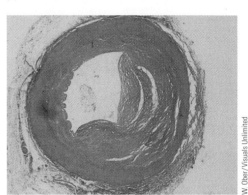

(a)

(b)

Cabisco/Visuals Unlimited

W. Ober/Visuals Unlimited

▲ FIGURE 5.16 (a) A cross section of a normal artery. (b) A cross section of an artery partially blocked by atherosclerotic plaque. As excess cholesterol accumulates in the body, it accumulates in plaques, leading to cardiovascular disease.

(a)

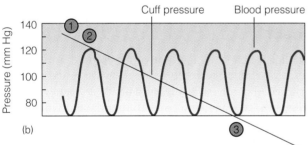

Cuff pressure   Blood pressure

①
②

③

(b)

▲ ACTIVE FIGURE 5.15 Blood pressure readings. (a) A blood pressure cuff is used to determine blood pressure. As shown in (b), blood pressure rises and falls as the left ventricle of the heart contracts and relaxes. To measure blood pressure, air is pumped into the cuff until it stops blood flow into the lower arm, and no sound can be heard (1). Air is gradually let out of the cuff, and when blood is heard flowing past the cuff, the pressure measured at this point is taken as the systolic pressure—the upper number in a blood pressure reading (2). As pressure in the cuff is gradually released, the artery becomes fully open, and no sound is heard again. This is the diastolic pressure—the lower number in a blood pressure reading (3).

HUMAN
Genetics ⒼNow™   Learn more about blood pressure readings by viewing the animation on Human GeneticsNow

at **http://now.ilrn.com/cummings7**

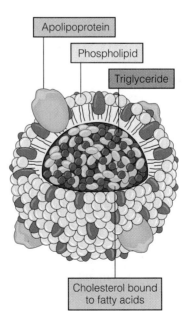

Apolipoprotein

Phospholipid

Triglyceride

Cholesterol bound
to fatty acids

◀ FIGURE 5.17 Lipoproteins carry cholesterol and other lipids through the blood enclosed in a protein and phospholipid coating. There are several types of lipoproteins that have different proportions of lipids. Low-density lipoproteins (LDLs) are composed of about 45% cholesterol, and high-density lipoproteins (HDLs) are made of about 20% cholesterol. High levels of LDLs and a low level of HDLs are risk factors for cardiovascular disease.

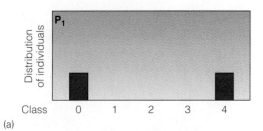

(a)

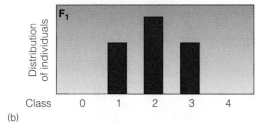

(b)

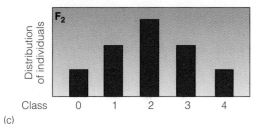

(c)

▲ **FIGURE 5.18** Frequency diagrams of skin colors. (a) Skin color distribution in the parents falls into two discontinuous classes. (b) Color values of seven children from the parents in (a) are intermediate to those of their parents. (c) Skin colors of 32 children of the parents in (b). Color values range from one phenotypic extreme to the other, and most are clustered around a mean value. This normal distribution of phenotypes is characteristic of a polygenic trait.

Skin color is a multifactorial trait.

Questions about the inheritance of skin color inevitably raise other questions about races and the genetic differences between various populations. We will discuss whether there is a genetic basis for separating human populations into races in Chapter 19. Here we will focus only on the multifactorial basis of skin color.

One of the first investigations on the genetics of skin color, done between 1910 and 1914, studied black-white marriages in Bermuda and in the Caribbean. Skin color is controlled by genetic and environmental factors, making it a multifactorial trait. Exposure to the sun can darken skin color and obscure genotypic differences.

The results illustrate several properties of polygenically controlled multifactorial traits. The F_1 generation had skin colors intermediate to their parents. In the F_2, a small number of children were as white as one grandparent, a small number were as black as the other grandparent, and most had skin color between these two extremes (▶ Figure 5.18). Because F_2 individuals could be grouped into five phenotypic classes, the investigators hypothesized that two gene pairs control skin color (see the distribution of genotypes for two loci in ▶ Figure 5.5). Each F_2 phenotypic class represented a genotype produced by the segregation and assortment of two gene pairs. To help explain their results, let's suppose that these genes are *A* and *B*, respectively. Class 0 has the lightest skin color and represents the genotype *aabb*, Class 1 has the genotype *Aabb* or *aaBb*, and so forth, up to Class 5, which has the darkest skin color and represents the homozygous dominant (*AABB*) genotype.

Later work by other investigators using instruments that measure light reflected from the skin surface showed that skin color is actually controlled by more than two gene pairs. The data are most consistent with a model that involves three or four genes (▶ Figure 5.19).

Intelligence and intelligence quotient (IQ): are they related?

The idea that intelligence can be quantitatively measured arose in the late eighteenth and early nineteenth centuries. Early on in the study of intelligence, phrenologists believed that physical measurements of regions of the skull revealed how much intelligence, courage, and so forth that the individual possessed (▶ Figure 5.20). Later, physical measurements gave way to overall brain size (craniometry). Large brains were associated with high intelligence, and small brains with lower intelligence.

At the turn of the twentieth century, psychological rather than physical methods were used to measure intelligence. Alfred Binet, a French psychologist, developed a graded series of tasks related to basic mental processes, such as comprehension, direction (sorting), and correction, and tested children for their ability to perform these tasks. Each child began by performing the simplest tasks and progressing until the tasks became too difficult. The age assigned for the last task performed be-

▶ **FIGURE 5.19** Distribution of skin color as measured by a reflectometer at a wavelength of 685 nm. The results are shown for an additive model of skin color, with environmental effects, for one to four gene pairs. Distributions observed in several populations indicate that three or four gene pairs control skin color.

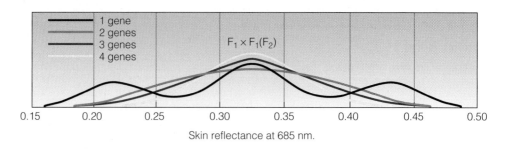

came the child's mental age, and the intellectual age was calculated by subtracting the mental age from the chronological age.

Wilhelm Stern, another psychologist, divided mental age by chronological age, and the number became known as the **intelligence quotient (IQ)**. If a child of 7 years (chronological age) was able to successfully perform tasks for a 7-year-old but could not do tasks for an 8-year-old, a mental age of 7 would be assigned. To determine the IQ for this child, divide mental age by chronological age: mental age (7) divided by chronological age (7) = 1.0. The quotient is multiplied by 100 to eliminate the decimal point (1.0 × 100 = 100), to obtain an IQ of 100.

The physical and psychological methods of measuring intelligence both assume that intelligence is a biological property that can be quantitatively expressed as a single number (see Spotlight on Building a Smarter Mouse). In fact, if anything, the use of IQ tests by governments and educational institutions has strengthened the assumption that IQ measures a fundamental, genetically determined physiological or biochemical property of the brain related to intelligence. The question is whether this assumption is correct, and whether psychological methods (IQ tests) measure intelligence any more accurately than the discredited physical methods of phrenology or craniometry.

The question can only be answered if intelligence is defined in such a way that it can be objectively measured, like we measure height, weight, or fingerprint ridges. Intelligence is often thought of as abilities in abstract reasoning, mathematical skills, verbal expression, problem solving, and creativity. There is no evidence that any of these properties are directly measured by an IQ test, and there is at present no objective way to quantify such components of intelligence.

IQ values are heritable.

The values obtained in IQ measurements, however, do have significant heritable components. The evidence that IQ has genetic components comes from two areas: studies that estimate IQ heritability and comparison of IQs in individuals raised together (unrelated individuals, parents and children, siblings, and MZ and DZ twins) and individuals raised separately (unrelated individuals, siblings, and MZ twins). Heritability estimates for IQ range from 0.6 to 0.8. The high correlation observed for MZ twins raised together indicates that genetics plays a significant role in determining IQ (▶ Figure 5.21). But rearing MZ twins apart or raising siblings in different environments significantly reduces the correlation and provides evidence that the environment has a substantial role in determining IQ.

Keep in mind

- Many multifactorial traits have social and cultural impacts.

What is the controversy about IQ and race?

The assumption that intelligence is solely determined by biological factors coupled with the misuse or misunderstanding of the limits of heritability estimates have misled people to conclude that differences in IQ among different racial groups are genetically determined. On standardized IQ tests, blacks score an average of 15 points lower than the average score of 100 by whites, and Asians score significantly above the average of 100. These differences are consistent across different tests, and the scores themselves do not seem to be a serious issue. The controversy is over what causes these differences. Are such differences genetic in origin, do they reflect environmental differences, or are both factors at work? If both, to what degree does inheritance contribute to the differences? The debate about these questions has been renewed by the 1994 publication of *The Bell Curve* by Herrnstein and Murray.

Spotlight on...

Building a Smarter Mouse

During the learning process, chemical changes occur at synapses, the gaps between nerve cells where signals are transferred between nerves. If both nerve cells are active at the same time, learning and memory are enhanced. By creating mice that overexpress or underexpress genes encoding proteins that transmit signals across the gap in the synapse, researchers can test the effects of overexpression or underexpression of these genes on learning and memory. One protein, called the NMDA receptor, has two subunits, NR1 and NR2. Young mice have higher levels of NR2 and learn better than adult mice. Researchers genetically engineered mice to overexpress the NR2 subunits and found that when these mice were adults, they learned faster and had better memories that other adult mice. This work shows that genetic enhancement of intelligence and memory is possible and has identified a key gene in learning and memory.

■ **Intelligence quotient (IQ)** A score derived from standardized tests that is calculated by dividing the individual's mental age (determined by the test) by his or her chronological age and multiplying the quotient by 100.

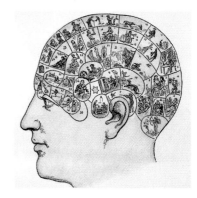

▲ **FIGURE 5.20** Phrenology model showing areas of the head overlying brain regions that control different traits. Intelligence was estimated by measuring the area of the skull overlying the region of the brain thought to control this trait.

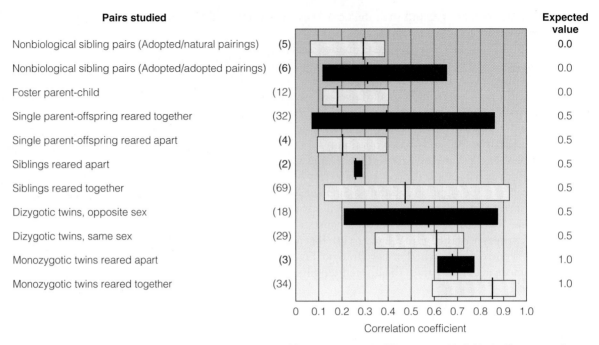

▲ FIGURE 5.21 A graphical representation of correlations in IQ measurements in different sets of individuals. The expected correlation coefficients are determined by the degree of genetic relatedness in each set of individuals. The vertical line represents the median correlation coefficient in each case.

The results of heritability studies have been used to support the argument that intelligence is mainly innate and inherited, citing heritability values of 0.8 for intelligence. In most cases, however, the reasoning used to support this argument misuses the concept of heritability. Recall that a measured heritability of 0.8 (for example) means that 80% of the phenotypic variation observed is due to genetic differences *within that* population. Heritability differences *between* two populations cannot be compared because heritability measures only variation within a population at the time of measurement. By definition, it cannot be used to estimate genetic variation between populations. In other words, we cannot use heritability differences between groups to conclude that there are genetic differences between those groups. As we will see in Chapter 19, genetic variation within any population is much greater than the amount of genetic variation between any two populations. In fact, the amount of genetic variation within a population is so great that it swamps out the genetic differences between populations, invalidating the idea that human populations can be sorted into racial groups.

It is quite evident that both genetic and environmental factors make important contributions to intelligence. Clearly, the relative amount each contributes cannot be measured accurately at this time. Several points about this debate should be kept in mind. First, it is clear that IQ test scores cannot be equated with intelligence. Second, IQ scores are not fixed and can be changed significantly by training in problem solving and, in fact, like many other phenotypes, change somewhat throughout the life of an individual. Variation in IQ scores is quite wide, and values measured in one racial or ethnic group greatly overlap those of other groups, making comparisons more difficult. In addition, worldwide IQ scores have been rising significantly for over 50 years. An IQ score regarded as average in 1944 would be well below average in 2004, emphasizing once again the role of environment in IQ scores.

The problem in discussing the differences in IQ scores between groups arises when quantitative differences in scores are converted into qualitative judgments used to rank groups as superior or inferior. Genetics, like all sciences, progresses by formulating hypotheses that can be rigorously and objectively tested but is often misused for ideological ends.

Scientists are searching for genes that control intelligence.

As discussed in the section on obesity, heritability studies cannot provide information about the number, location, or identity of genes involved in intelligence. To learn about the genes themselves, scientists are using an expanded definition of intelligence that goes beyond IQ and recognizes that many genes are involved in normal cognitive and intellectual function, defined as **general cognitive ability**.

Several approaches are being used to identify genes associated with intelligence. Animal models such as the fruit fly *Drosophila* and the mouse are being studied to identify single genes that control aspects of learning, memory, and spatial perception. *Drosophila* mutants such as *dunce, rutabaga, amnesiac* and others have been used to identify biochemical pathways in the nervous system that play important roles in learning and memory. Human brains have these same pathways, and they may play similar roles in our nervous system.

Another approach uses recombinant DNA techniques and information from the Human Genome Project to identify genes that affect specific polygenic traits, such as reading ability and IQ. These polygenes are also known as **quantitative trait loci (QTLs)**. The search has uncovered genes associated with reading disability (developmental dyslexia) on chromosome 6 (OMIM 600202) and chromosome 15. More recently, a gene associated with cognitive ability has been identified on chromosome 4 (OMIM 603783). Using approaches that combine behavioral genetics, twin studies, and molecular genetics, more genes controlling aspects of cognitive ability are likely to be identified in the near future. As results from the Human Genome Project are analyzed, it will become easier to define the number and actions of genes involved in higher mental processes and provide insight into the genetics of intelligence.

■ **General cognitive ability**
Characteristics that include verbal and spatial abilities, memory, speed of perception, and reasoning.

■ **Quantitative trait loci (QTLs)** Two or more genes that act on a single polygenic trait.

Genetics in Practice

 Genetics in Practice case studies are critical thinking exercises that allow you to apply your new knowledge of human genetics to real-life problems. You can find these case studies and links to relevant websites at **http://biology.brookscole.com/cummings7**

CASE 1

Sue and Tim were referred for genetic counseling after they inquired about the risk of having a child with a cleft lip. Tim was born with a mild cleft lip that was surgically repaired. He expressed concern that his future children could be at risk for a more severe form of clefting. Sue was in her twelfth week of pregnancy, and both were anxious about this pregnancy because Sue had a difficult time conceiving. The couple stated that they would not consider terminating the pregnancy for any reason but wanted to be prepared for the possibility of having a child with a birth defect. The genetic counselor took a three-generation family history from both Sue and Tim and found that Tim was the only person to have had a cleft lip. Sue's family history showed no cases of cleft lip. Tim and Sue had several mis-

conceptions about how clefting occurs, and the genetic counselor spent time explaining how cleft lips occur and some of the known causes of this birth defect. The following list summarizes the counselor's discussion with this couple.

- Fathers, as well as mothers, can pass on genes that cause clefting.
- Some clefts are caused by environmental factors, meaning the condition didn't come from the father or the mother.
- One child in 33 is born with some sort of birth defect. One in 700 is born with a cleft-related birth defect.
- Most clefts occur in boys; however, a girl can be born with a cleft.

(continued)

- If a person (male or female) is born with a cleft, the chances of that person having a child with a cleft, given no other obvious factor, is 7 in 100.
- Some clefts are related to identifiable syndromes. Of those, some are autosomal dominant. A person with an autosomal dominant gene has a 50% probability of passing the gene to an offspring.
- Many clefts run in families even when there does not seem to be any identifiable syndrome present.
- Clefting seems to be related to ethnicity, occurring most often among Asians, Latinos, and Native Americans (1:500); next most often among persons of European ethnicity (1:700); and least often among persons of African origin (1:1,000).
- A cleft condition develops during the fourth to the eighth week of pregnancy. After that critical period, nothing the mother does can cause a cleft. Sometimes a cleft develops even before the mother is aware that she is pregnant.
- Women who smoke are twice as likely to give birth to a child with a cleft.
- Women who ingest large quantities of vitamin A or low quantities of folic acid are more likely to have children with a cleft.
- In about 70% of cases, the fetal face is clearly visible using ultrasound. Facial disorders have been detected at the fifteenth gestational week of pregnancy. Ultrasound can be precise and reliable when diagnosing fetal craniofacial conditions.

1. After hearing this information, should Sue and Tim feel that their chances of having a child with a cleft lip are increased over that of the general population?

2. Can cleft lip be surgically corrected?

3. If the child showed a cleft lip through ultrasound analysis, and the parents then started blaming each other (because Sue is a smoker, and Tim was born with the defect), how would you counsel them?

CASE 2

Louise was an active 27-year-old gymnastic instructor at a local YMCA. She was recently accepted into law school and worked part time in the evenings at the local Gap clothing store. She was very busy and therefore thought nothing of her increasing fatigue until it started to affect her daily activities. She also complained of occasional dizziness, difficulty hearing, constipation, and problems with controlling her bladder. Her general practitioner conducted a series of blood tests to determine what was wrong. All tests showed she was in perfect health, but her symptoms were becoming progressively worse. Over a period of months, she was forced to quit teaching gymnastics and was soon confined to a wheelchair. Her doctor finally referred her for a genetic evaluation. The clinical geneticist immediately recognized Louise's symptoms as signs of multiple sclerosis (MS). The geneticist explained that fatigue, almost to the point of being disabling, is the most common symptom of MS. Some medications could help her fatigue but their effect may not last. The best cure for fatigue is to listen to her body and just rest whenever possible.

The geneticist warned her of other possible symptoms of MS, including: (1) numbness, which is most likely a direct result of the destruction of myelin in the nerves; (2) tingling or pins-and-needles sensation in her feet (the same sensation you get when your foot "falls asleep"), for which not much can be done other than to massage the affected area and rest; (3) tremors, usually of the hands, also due to myelin destruction; (4) muscle spasms (sustained or temporary muscle contractions) in the arms, legs, abdomen, back—just about anywhere; (5) depression and mood swings; (6) memory problems; (7) loss of balance with or without dizziness; and (8) bowel and bladder problems. MS is the most common autoimmune disease involving the nervous system, affecting approximately 250,000 individuals in the United States. The cause of MS is unknown; however, some studies have suggested that the risk to a first-degree relative (sibling, parent, or child) of a patient with MS is at least 15 times that for a member of the general population. Unfortunately, no definite genetic pattern is discernible.

1. What could explain why a first-degree relative of an MS patient is 15 times more likely to have MS than the general population? Does this mean that MS has a genetic component?

2. What is an autoimmune disease? What kinds of genes could be involved in this process?

3. What kind of questions should medical researchers ask in the study of MS and possible treatments?

5.1 Some Traits Are Controlled by Two or More Genes

- In this chapter we considered genes that affect traits additively or quantitatively. The pattern of inheritance that controls metric characters (those that can be measured quantitatively) is called polygenic inheritance because two or more genes are usually involved.

5.2 Polygenic Traits and Variation in Phenotype

- The distribution of polygenic traits through the population follows a bell-shaped, or normal, curve. Parents whose phenotypes are near the extremes of this curve usually have children whose phenotypes are less extreme than the parents' and are closer to the population mean. This phenomenon, known as regression to the mean, is characteristic of systems in which the phenotype is produced by the additive action of many genes.

5.3 Multifactorial Traits: Polygenes and the Environment

- Variations in the expression of polygenic traits are often due to the action of environmental factors. Polygenic traits with a strong environmental component are called multifactorial traits or complex traits. The impact of environment on genotype can cause genetically susceptible individuals to exhibit a trait discontinuously even though there is an underlying continuous distribution of genotypes for the trait.

5.4 Heritability Measures the Genetic Contribution to Phenotypic Variation

- The degree of phenotypic variation produced by a genotype in a given population can be estimated by cal-
culating the heritability of a trait. Heritability is estimated by observing the amount of variation among relatives having a known fraction of genes in common. MZ twins have 100% of their genes in common and, when raised in separate environments, provide an estimate of the degree of environmental influence on gene expression. Heritability is a variable that is validly calculated only for the population under study and the environmental condition in effect at the time of the study.

5.5 Twin Studies and Multifactorial Traits

- In twin studies, the degree of concordance for a trait is compared in MZ and DZ twins reared together or apart. MZ twins result from the splitting of an embryo produced by a single fertilization, whereas DZ twins are the products of multiple fertilizations. Although twin studies can be useful in determining whether a trait is inherited, they cannot provide any information about the mode of inheritance or the number of genes involved.

5.6 A Survey of Some Multifactorial Traits

- Many human traits are multifactorial, combining polygenes and action by environmental factors. Some of these include skin color, intelligence, and aspects of behavior. The genetics of these traits has often been misused and misrepresented for ideological or political ends. New genetic approaches using recombinant DNA methods are helping identify genes involved in these traits.

Questions and Problems

HUMAN
Genetics ⑤ Now™ Preparing for an exam? Assess your understanding of this chapter's topics with a pre-test, a personalized learning plan, and a post-test at http://now.ilrn.com/cummings7

Some Traits Are Controlled by Two or More Genes

1. Describe why continuous variation is common in humans, and provide examples of such traits.
2. The text outlines some of the problems Frederick William I encountered in his attempt to breed tall Potsdam Guards.
   a. Why were the results he obtained so different from those obtained by Mendel with short and tall pea plants?

   b. Why were most of the children shorter than their tall parents?
3. What role might environment have played in causing Frederick William's problems, especially at a time when nutrition varied greatly from town to town and from family to family?
4. Do you think Frederick William's experiment would have worked better if he had ordered brother-sister marriages within tall families instead of just choos-

ing the tallest individuals from throughout the country?

5. As it turned out, one of the tallest Potsdam Guards had an unquenchable attraction to short women. During his tenure as guard he had numerous clandestine affairs. In each case children resulted. Subsequently, some of the children, who had no way of knowing that they were related, married and had children of their own. Assume that two pairs of genes determine height. The genotype of the 7-foot-tall Potsdam Guard was $A'A'B'B'$, and the genotype of all of his 5-foot clandestine lovers was $AABB$, where an A' or B' allele adds 6 inches to the base height of 5 feet conferred by the $AABB$ genotype.
   a. What were the genotypes and phenotypes of all of the F_1 children?
   b. Diagram the cross between the F_1 offspring, and give all possible genotypes and phenotypes of the F_2 progeny.

Polygenic Traits and Variation in Phenotype

6. Describe why there is a fundamental difference between the expression of a trait that is determined by polygenes and the expression of a trait that is determined monogenetically.

Multifactorial Traits: Polygenes and the Environment

7. A clubfoot is a common congenital birth defect in the Smith family. This defect is caused by a number of genes but appears to be phenotypically distributed in a noncontinuous fashion. Geneticists use the multifactorial threshold model to explain the occurrence of this defect. Explain this model. Explain predisposition to the defect in an individual who has a genotypic liability above the threshold versus an individual who has a liability below the threshold.

Heritability Measures the Genetic Contribution to Phenotypic Variation

8. Define genetic variance.
9. Define environmental variance.
10. How is heritability related to genetic and environmental variance?
11. Why are relatives used in the calculation of heritability?
12. If there is no genetic variation within a population for a given trait, what is the heritability for the trait in the population?

Twin Studies and Multifactorial Traits

13. Can conjoined (Siamese) twins be dizygotic twins given the theory that conjoined twins arise from incomplete division of the embryo?
14. Dizygotic twins:
   a. are as closely related as monozygotic twins
   b. are as closely related as nontwin siblings
   c. share 100% of their genetic material
   d. share 25% of their genetic material
   e. none of the above
15. Why are monozygotic twins who are reared apart so useful in the calculation of heritability?
16. Monozygotic (MZ) twins have a concordance value of 44% for a specific trait, whereas dizygotic twins have a concordance value of less than 5% for the same trait. What could explain why the value for MZ twins is significantly less than 100%?
17. If monozygotic twins show complete concordance for a trait whether they are reared together or apart, what does this suggest about the heritability of the trait?
18. Researchers set up an obesity study in which MZ and DZ twins who served in the armed forces were studied at induction into the military and 25 years later. Results indicated that obesity has a strong genetic component.
   a. What are some of the problems with this study?
   b. Design a better study to test whether obesity has a genetic component.
19. What does the *ob* gene code for? How does it work? Is this just a gene found in animals, or do humans have it also?
20. What is the importance of the comparison of traits between adopted and natural children in determining heritability?

A Survey of Some Multifactorial Traits

21. Is cardiovascular disease (hypertension, atherosclerosis, and familial hypercholesterolemia) genetic?
22. Discuss the difficulties in attempting to determine whether intelligence is genetically based.
23. At the age of 9 years, your genetics instructor was able to perform the mental tasks of an 11-year-old. According to Wilhelm Stern's method, calculate his or her IQ.
24. Suppose that a team of researchers analyzes the heritability of high SAT scores and assigns a heritability of 0.75 for this ability. This team also determines that a certain ethnic group has a heritability value that is 0.12 lower when compared with that of other ethnic groups. The group concludes that there must be a genetic explanation for the differences in scores. Why is this an invalid conclusion?

Internet Activities are critical thinking exercises using the resources of the World Wide Web to enhance the principles and issues covered in this chapter. For a full set of links and questions investigating the topics described below, log on to **http://biology.brookscole.com/cummings7**

1. *Twin Studies*. Access the website of the International Society for Twin Studies. At this site, click on the link to the Genetic Epidemiology Group of the Queensland Institute of Medical Research. Once there, click on "Studies" to view information on several long-term twin study projects. The project overview mentions several different ongoing studies. This website also has a link to the International Society for Twin Studies. At the *ITST* website, open and read the link to the "Declaration of Rights and Statement of Needs of Twins and Higher Order Multiples."

2. *A Multifaceted Look at a Multifactorial Trait*. Body weight is an example of a multifactorial trait that results from the interaction between multiple genes and the environment. At the same time, the issue of body weight has taken on different connotations in different societies. The PBS series *Frontline* has a website that addresses both the question of what makes people "fat" and how fatness is perceived by our society.

   a. At the website, click on "What the Experts Said in This Report" regarding body weight, especially those comments regarding the various causes—cultural, societal, and genetic—of fatness in people.

   b. Follow the link to the chapter of Richard Klein's book *Eat Fat*. Klein discusses how fatness and even the word *fat* once had very positive connotations; however, this is generally untrue today, especially in the industrialized Western world.

How would you vote now?

The idea of selectively breeding humans, as King Frederick William of Prussia attempted to do with the Potsdam Guards, is generally considered highly unethical. However, with increased knowledge of genetics and advances in reproductive technology, it is possible in our time to test embryos for their genetic characteristics, such as genetic markers associated with higher IQ levels, before they are even implanted in a mother's womb. Now that you know more about multifactorial traits that involve polygenic inheritance and the effects of environmental factors, what do you think? Would you consider having such tests done, and implanting only those embryos carrying the desired markers? Visit the Human Heredity Companion website to find out more on the issue, then cast your vote online at **http://biology.brookscole.com/cummings7**

For further reading and inquiry, log on to **InfoTrac College Edition**, your world-class online library, including articles from nearly 5,000 periodicals, at **http://infotrac.thomsonlearning.com**

6

Cytogenetics: Karyotypes and Chromosome Aberrations

Chapter Outline

TIERNEY'S home pregnancy test confirmed the good news: she and her husband, Greg, were having a baby. As the pregnancy progressed, they went to their obstetrician for routine exams. After an ultrasound test, they were told the fetus had a heart defect. Worse was the news that the nature of the defect made the doctor suspect it was caused by Down syndrome. Further testing confirmed the diagnosis.

Now Greg and Tierney Fairchild were struggling with an array of emotional conflicts and difficult decisions. First, they faced the guilt and fear that accompanies the news that the fetus is not normal. Next, they had to decide whether to terminate the pregnancy or have a child that would face major surgery, a lifetime of retardation, and medical risks such as hypothyroidism and leukemia. The Fairchilds had made tough decisions before. As an interracial couple, the Fairchilds had faced the specter of discrimination and intolerance in their decision to marry. This decision would not only involve another person, but one burdened by racial discrimination and mental retardation. Before making their decision about the pregnancy, they sought information, advice, and help.

Unfortunately, there is no way to predict how retarded a Down syndrome child will be. Some are severely retarded—unable to dress, eat, or use the toilet by themselves. Others attend school, graduate, have jobs, and live in group homes or supervised settings as adults. After much anguish and conflicting advice, they decided to continue the pregnancy and had a daughter,

Naia. She underwent successful heart surgery, and both parents are now deeply involved in her education at home and at school. During this ordeal, they opened their lives to Mitchell Zukoff, a reporter from the *Boston Globe,* who wrote a series of articles and later a book, *Choosing Naia,* about the choices and decisions the Fairchilds faced during and after this pregnancy.

How would you vote?

About 1 in every 800 children is born with Down syndrome. It is the most common chromosomal disorder in humans. More than 90% of couples faced with the Fairchild's dilemma elect to terminate the pregnancy. The symptoms of Down syndrome are variable and cannot be accurately predicted before birth. Naia is now able to care for herself; attends school; and is developing her language, motor, and cognition skills. She is a loving child and an integral part of a family. Put into the same situation, would you elect to terminate or continue a pregnancy after a diagnosis of Down syndrome? Would you consider adopting a Down syndrome child? Visit the Human Heredity Companion website to find out more on the issue, then cast your vote online at **http://biology.brookscole.com/cummings7**

6.1 The Human Chromosome Set

The number of chromosomes in the nucleus of an organism is characteristic for a given species: cells of the fruit fly *Drosophila melanogaster* have 8 chromosomes, corn plants have 20 chromosomes, and humans have 46 chromosomes. The chromosome numbers for several species of plants and animals are given in ▶ Table 6.1. As discussed in Chapter 2, human chromosomes exist in pairs, with most cells having 23 pairs, or 46 chromosomes. This is the diploid, or *2n,* number of chromosomes. Certain cells, such as eggs and sperm (gametes), contain only one copy of each chromosome. These cells have 23 chromosomes, which is the haploid, or *n,* number of chromosomes.

As chromosomes become visible in the early stages of cell division, certain structural features can be recognized. Each chromosome contains a specialized region known as the **centromere,** which divides the chromosome into two arms. The location of the centromere is characteristic for a given chromosome (▶ Figure 6.1). Chromosomes with centromeres at or near the middle have arms of equal length and are known as **metacentric** chromosomes (▶ Figure 6.2). If the centromere is located away from the center, the arms are unequal in length, and the chromosome is called a **submetacentric** chromosome. If the centromere is located very close to one end, the chromosome is called an **acrocentric** chromosome.

Keep in mind as you read

■ Karyotype construction and analysis are used to identify chromosome abnormalities.

■ Polyploidy results when there are more than two haploid sets of chromosomes.

■ Monosomy and trisomy involve the loss and gain of a single chromosome to a diploid genome.

■ Age of the mother is the best known risk factor for trisomy.

■ Changes in the number of sex chromosomes have less impact than changes in autosomes.

■ Chromosomes can lose, gain, or rearrange genes.

■ Some fragile sites are associated with mental retardation.

■ **Centromere**   A region of a chromosome to which microtubule fibers attach during cell division. The location of a centromere gives a chromosome its characteristic shape.

■ **Metacentric**   Describes a chromosome that has a centrally placed centromere.

■ **Submetacentric**   Describes a chromosome whose centromere is placed closer to one end than the other.

■ **Acrocentric**   Describes a chromosome whose centromere is placed very close to, but not at, one end.

Table 6.1	Chromosome Number in Selected Organisms	
Organism	Diploid Number (2n)	Haploid Number (n)
Human (*Homo sapiens*)	46	23
Chimpanzee (*Pan troglodytes*)	48	24
Gorilla (*Gorilla gorilla*)	48	24
Dog (*Canis familiaris*)	78	39
Chicken (*Gallus domesticus*)	78	39
Frog (*Rana pipiens*)	26	13
Housefly (*Musca domestica*)	12	6
Onion (*Allium cepa*)	16	8
Corn (*Zea mays*)	20	10
Tobacco (*Nicotiana tabacum*)	48	24
House mouse (*Mus musculus*)	40	20
Fruit fly (*Drosophila melanogaster*)	8	4
Nematode (*Caenorhabditis elegans*)	12	6

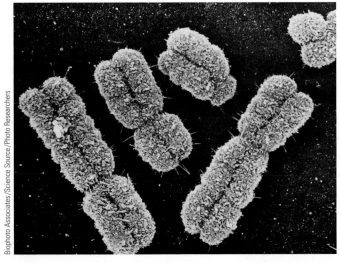

▲ FIGURE 6.1 Human chromosomes as seen at metaphase of mitosis in the scanning electron microscope. The replicated chromosomes appear as double structures, consisting of sister chromatids joined by a single centromere.

■ **Sex chromosomes** In humans, the X and Y chromosomes that are involved in sex determination.

■ **Autosomes** Chromosomes other than the sex chromosomes. In humans, chromosomes 1 to 22 are autosomes.

■ **Karyotype** A complete set of chromosomes from a cell that has been photographed during cell division and arranged in a standard sequence.

Human males and females (and other animal species) have one pair of **sex chromosomes** that are not completely homologous. Females have two homologous X chromosomes, and males have a nonhomologous pair, consisting of one X and one Y chromosome. Chromosomes other than sex chromosomes are called **autosomes.**

Human chromosomes are usually studied and are photographed while in metaphase of mitosis. For convenience, the chromosomes are cut out from a photograph and are arranged in pairs according to size and centromere location to form a **karyotype** (▶ Figure 6.3).

By convention, the chromosomes in the human karyotype are numbered and arranged into seven groups, A through G. Each group is defined by size and centromere location. Staining produces unique band patterns for each chromosome, allowing individual chromosomes to be clearly identified. The standardized G-banding pattern for the human chromosome set is shown in ▶ Figure 6.4. Chromosome banding patterns are used to identify specific regions on each chromosome (▶ Figure 6.5). For identification of regions, the short arm of each chromosome is designated the p arm, and the long arm the q arm. Each arm is subdivided into numbered regions beginning at the centromere. Within each region, the bands are identified by number. Thus, any region in the human karyotype can be identified by a descriptive address, such as 1q2.4. This address consists of the chromosome number (1), the arm (q), the region (2), and the band (4) (▶ Figure 6.5).

Karyotypic analysis of banded chromosomes is a powerful tool for chromosomal studies and is one of the basic techniques in human genetics.

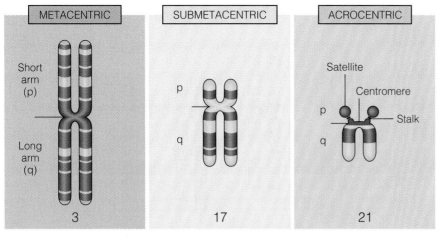

METACENTRIC

Short arm (p)

Long arm (q)

3

SUBMETACENTRIC

p

q

17

ACROCENTRIC

Satellite

Centromere

p

Stalk

q

21

◀ FIGURE 6.2 Human metaphase chromosomes are identified by size, centromere location, and banding pattern. The relative size, centromere locations, and banding patterns for three representative human chromosomes are shown. Chromosome 3 is one of the largest human chromosomes and, because the centromere is centrally located, is a metacentric chromosome. Chromosome 17 is a submetacentric chromosome because the centromere divides the chromosome into two arms of unequal size. Chromosome 21 has a centromere placed very close to one end and is called an acrocentric chromosome. In humans, the short arm of each chromosome is called the p arm, and the long arm is called the q arm.

◀ FIGURE 6.3 A human karyotype showing replicated chromosomes from a cell in metaphase of mitosis. This female has 46 chromosomes, including two X chromosomes.

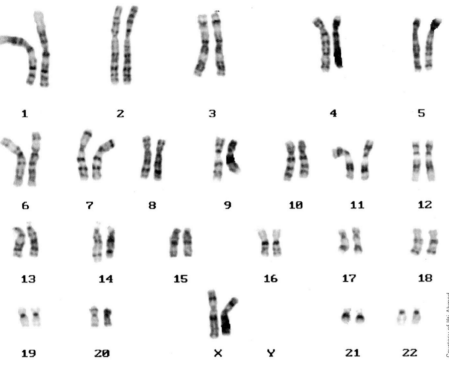

1 2 3 4 5

6 7 8 9 10 11 12

13 14 15 16 17 18

19 20 X Y 21 22

Courtesy of Ifti Ahmed

6.2 Making a Karyotype

Karyotypes are constructed using cells from a number of sources, including white blood cells (lymphocytes), skin cells (fibroblasts), amniotic fluid cells (amniocytes), and chorionic villus cells (placental cells). One of the most common methods of preparing cells begins with a blood sample and is shown in ▶ Figure 6.6. A few drops of the blood are added to a flask containing a nutrient growth medium. Because lymphocytes in the blood sample do not normally divide, a mitosis-inducing chemical, such as phytohemagglutinin, is added to the flask, and the cells are grown for 2 or 3 days at body temperature (37°C) in an incubator. Then a drug such as Colcemid is added to stop dividing cells at metaphase. Over a period of about 2 hours of treatment, all cells entering mitosis are arrested in metaphase.

The blood cells are concentrated by centrifugation; adding a salt solution breaks open and destroys the red blood cells (which are nondividing) and swells the lymphocytes. After fixation in a mixture of methanol and acetic acid, the swollen lymphocytes are dropped onto a microscope slide. The impact causes the fragile cells to

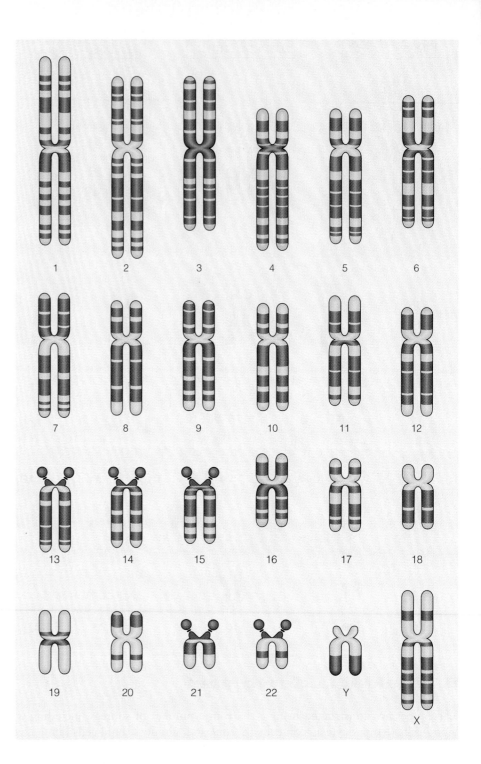

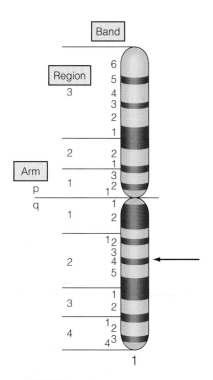

▲ FIGURE 6.5 The system of naming chromosome bands. Each autosome is numbered from 1 to 22. The sex chromosomes are X and Y. Within a chromosome, the short arm is the p arm, and the long arm is the q arm. Each arm is divided into numbered regions. Within each region, numbers designate the bands. The area marked by the arrow is designated as 1q2.4 (chromosome 1, long arm q, region 2, band 4).

break open, spreading metaphase chromosomes onto the slide. The chromosome preparation is partially digested with trypsin, an enzyme that enhances the banding pattern. After staining, the preparation is examined with a microscope and the cluster of metaphase chromosomes is photographed (▶ Figure 6.7a). Chromosomal images are cut from the photograph and arranged according to size, centromere location, and banding pattern to form a karyotype (▶ Figure 6.7b).

In most cytogenetics laboratories, computer-generated karyotypes have replaced cut-and-paste karyotypes. A video camera attached to a microscope transmits images of metaphase chromosomes to a computer, where they are recorded, digitized, and processed to make a karyotype. The metaphase chromosomes in ▶ Figure 6.7a were recorded by this method. ▶ Figure 6.7b shows the computer-derived karyotype.

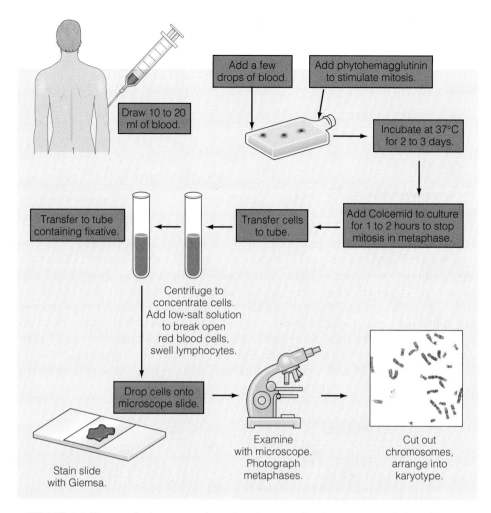

▲ FIGURE 6.6 The steps in the process of creating a karyotype for chromosome analysis.

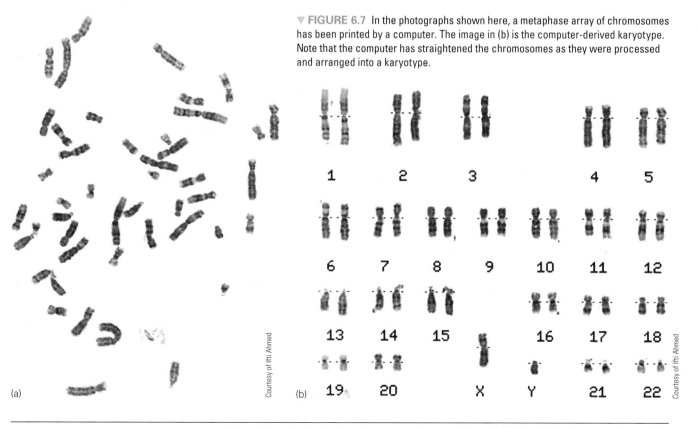

▼ FIGURE 6.7 In the photographs shown here, a metaphase array of chromosomes has been printed by a computer. The image in (b) is the computer-derived karyotype. Note that the computer has straightened the chromosomes as they were processed and arranged into a karyotype.

Courtesy of Ifti Ahmed

(a)

(b)

Courtesy of Ifti Ahmed

Note that the software has straightened the chromosomes as they were processed and arranged into a karyotype.

Keep in mind

■ Karyotype construction and analysis are used to identify chromosome abnormalities.

6.3 Constructing and Analyzing Karyotypes

Stains and dyes are used to produce a pattern of bands that is specific to each chromosome (although homologous chromosomes have the same pattern). One of the most common methods is G-banding, in which chromosomes are first treated with an enzyme (trypsin) that partially digests chromosomal proteins and then are stained with Giemsa stain (a mixture of dyes). The resulting pattern of bands is used to identify individual chromosomes in cytogenetic analysis. Metaphase chromosomes have a total of about 550 bands. More bands can be produced using cells in early metaphase or late prophase. In these stages, chromosomes are less condensed, and up to 2,000 bands can be identified in the normal human karyotype. Some other banding methods are reviewed in ▶ Figure 6.8.

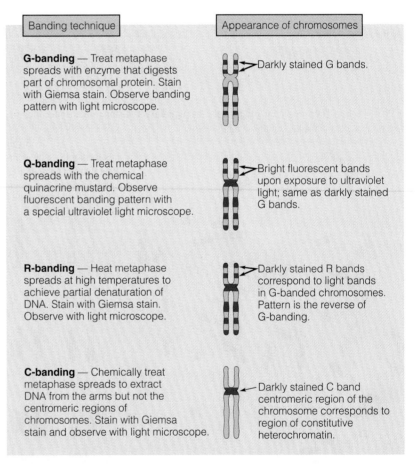

Banding technique	Appearance of chromosomes
G-banding — Treat metaphase spreads with enzyme that digests part of chromosomal protein. Stain with Giemsa stain. Observe banding pattern with light microscope.	Darkly stained G bands.
Q-banding — Treat metaphase spreads with the chemical quinacrine mustard. Observe fluorescent banding pattern with a special ultraviolet light microscope.	Bright fluorescent bands upon exposure to ultraviolet light; same as darkly stained G bands.
R-banding — Heat metaphase spreads at high temperatures to achieve partial denaturation of DNA. Stain with Giemsa stain. Observe with light microscope.	Darkly stained R bands correspond to light bands in G-banded chromosomes. Pattern is the reverse of G-banding.
C-banding — Chemically treat metaphase spreads to extract DNA from the arms but not the centromeric regions of chromosomes. Stain with Giemsa stain and observe with light microscope.	Darkly stained C band centromeric region of the chromosome corresponds to region of constitutive heterochromatin.

▲ FIGURE 6.8 Four common staining procedures used in chromosomal analysis. Most karyotypes are prepared using G-banding. Q-banding and R-banding produce a pattern of bands that is the reverse of those in G-banded chromosomes.

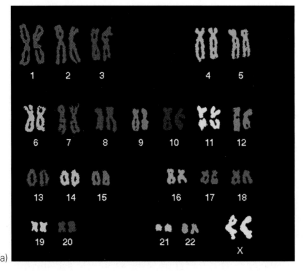
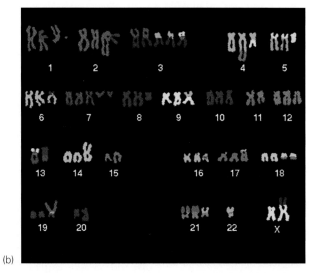

(a)

(b)

▲ FIGURE 6.9 Chromosome painting with five different-colored markers in a normal cell (a) produces a pattern that highlights the translocations and deletions found in cancer cells (b).

A karyotype is described by (1) the number of chromosomes, (2) the sex chromosome content, (3) the presence or absence of individual chromosomes, and (4) the nature and extent of any structural abnormality. The symbols for structural alterations include t for a translocation, dup for duplication, and del for deletion. These structural aberrations are discussed later in the chapter. If a male has a deletion in the short arm of chromosome 5, but otherwise is chromosomally normal, this is represented as 46,XY,del(5p). ▶ Table 6.2 lists the descriptions of some chromosomal aberrations using this system.

Chromosome analysis is a painstaking procedure; to make it easier to spot abnormalities, cytogeneticists are now using a technique called chromosome painting. This method uses DNA sequences attached to fluorescent dyes. The sequences attach to chromosome-specific regions, painting the chromosome a distinctive color. Using several different DNA sequences and fluorescent dyes produces a unique pattern for each of the 24 human chromosomes (22 autosomes and the X and Y chromosomes) (▶ Figure 6.9).

Table 6.2	Chromosomal Aberrations	
Chromosomal Abnormality		**Syndrome Phenotype**
46,del(4p)	Wolf-Hirschhorn syndrome	Mental retardation; midline facial defects consisting of broad nose, wide-set eyes, small lower jaw, and cleft palate; heart, lung, and skeletal abnormalities common; severely reduced survival
46,del(11)(p13)	WAGR syndrome	Tumors of the kidney (Wilms tumor) and of the gonad (gonadoblastoma); aniridia (absence of the iris); ambiguous genitalia; mental retardation
46,t(9;22)(q34a11)	CML (chronic myelogenous leukemia)	Enlargement of liver and spleen; anemia; excessive, unrestrained growth of white cells (granulocytes) in the bone marrow
46,t(8;14)	Burkitt's lymphoma	Malignancy of B lymphocytes that mature into the antibody-producing plasma cells; solid tumors, typically in the bones of the jaw and organs of the abdomen

What cells are obtained for chromosome studies?

Almost any cell with a nucleus (mature red blood cells have no nuclei) can be used to make a karyotype. In adults, white blood cells (lymphocytes), skin cells (fibroblasts), and cells from biopsies or surgically removed tumor cells are routinely used for chromosome studies.

Chromosomal abnormalities can be detected before birth using amniocentesis and chorionic villus sampling to collect cells from embryos and fetuses. A less invasive method now under development collects fetal cells that have crossed into the mother's circulatory system (see Genetic Journeys: Using Fetal Cells from the Mother's Blood).

Amniocentesis collects cells from the fluid surrounding the fetus.

Amniocentesis A method of sampling the fluid surrounding the developing fetus by inserting a hollow needle and withdrawing suspended fetal cells and fluid; used in diagnosing fetal genetic and developmental disorders; usually performed in the sixteenth week of pregnancy.

Chorionic villus sampling (CVS) A method of sampling fetal chorionic cells by inserting a catheter through the vagina or abdominal wall into the uterus. Used in diagnosing biochemical and cytogenetic defects in the embryo. Usually performed in the eighth or ninth week of pregnancy.

Amniocentesis is routinely used to collect fetal cells for analysis. First, the fetus and placenta are located by ultrasound, and a needle is inserted through the abdominal and uterine walls (avoiding the placenta and fetus) into the amniotic sac surrounding the fetus. Approximately 10 to 30 ml of fluid is withdrawn by syringe. Amniotic fluid is mostly fetal urine containing cells shed from the skin, respiratory tract, and urinary tract of the fetus. Cells are isolated from the fluid by centrifugation (▷ Active Figure 6.10). The fluid can be tested if a biochemical genetic disorder is suspected. More than 100 such disorders can now be diagnosed by amniocentesis.

Amniotic cells can be analyzed to detect biochemical disorders or chromosome abnormalities. Karyotype preparation makes it possible to diagnose the sex of the fetus and identify any chromosomal abnormalities.

Amniocentesis is usually not performed until the sixteenth week of pregnancy. Before this time, there is very little amniotic fluid, and contamination of the sample with maternal cells is often a problem. There is a small risk of maternal infection and a slight increase (less than 1%) in the probability of a spontaneous abortion. To offset these risks, amniocentesis is normally used only under certain conditions:

- Advanced maternal age. Because the risk for children with chromosome abnormalities increases dramatically after the age of 35, amniocentesis is recommended for pregnant women who are 35 years or older. The majority of all amniocentesis cases are performed because of advanced maternal age.
- A previous child with a chromosomal aberration. The recurrence risk in such cases is 1% to 2%.
- A parent with a chromosome rearrangement. If either parent carries a chromosomal translocation or other rearrangement that can cause an abnormal karyotype in the child, amniocentesis should be considered.
- X-linked disorder. If the mother is a carrier of an X-linked biochemical disorder that cannot be diagnosed prenatally, and if she is willing to abort if the fetus is male, amniocentesis is recommended.

Chorionic villus sampling retrieves fetal tissue from the placenta.

Chorionic villus sampling (CVS) has several advantages over amniocentesis. CVS can be performed earlier in the pregnancy (8 to 10 weeks, compared with 16 weeks for amniocentesis). Because placental cells are dividing, karyotypes are available within a few hours or a few days. With amniocentesis, fetal cells must be grown in the laboratory for several days before karyotyping. In CVS, a flexible catheter is inserted through the vagina or abdomen into the uterus, guided by ultrasound images. Some chorionic villi (fetal tissue that forms part of the placenta) are removed by suction (▷ Figure 6.11). Enough material is usually obtained by CVS to allow biochem-

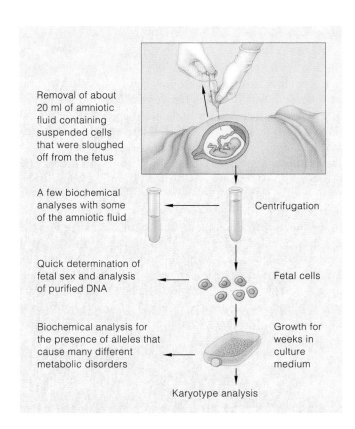

◄ ACTIVE FIGURE 6.10 In amniocentesis, a syringe needle is inserted through the abdominal wall and uterine wall to collect a small sample of the amniotic fluid. The fluid contains fetal cells that can be collected and used for prenatal chromosomal or biochemical analysis.

HUMAN
Genetics ⊜ Now™ Learn more about amniocentesis by viewing the animation on Human GeneticsNow at
http://now.ilrn.com/cummings7

Removal of about 20 ml of amniotic fluid containing suspended cells that were sloughed off from the fetus

A few biochemical analyses with some of the amniotic fluid

Centrifugation

Quick determination of fetal sex and analysis of purified DNA

Fetal cells

Biochemical analysis for the presence of alleles that cause many different metabolic disorders

Growth for weeks in culture medium

Karyotype analysis

ical testing or extraction of DNA for molecular analysis. The use of recombinant DNA techniques for prenatal diagnosis is discussed in Chapter 14.

CVS is used less often than amniocentesis. Although early studies indicated that CVS posed a higher risk to mother and fetus than amniocentesis, improvements in instrumentation and technique have lowered the risk somewhat. CVS offers early diagnosis of genetic diseases, and if termination of pregnancy is elected, maternal risks are lower at 9 to 12 weeks than at 16 weeks.

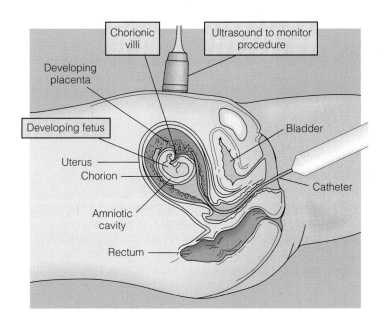

Chorionic villi

Ultrasound to monitor procedure

Developing placenta

Developing fetus

Bladder

Uterus
Chorion

Catheter

Amniotic cavity

Rectum

◄ FIGURE 6.11 The chorionic villus sampling technique. A catheter is inserted into the uterus through the vagina to remove a sample of fetal tissue from the chorion. Cells in the tissue can be used for chromosomal or biochemical analysis.

At the birth of a child, anxious parents have two questions: Is it a boy or a girl, and is the baby normal? The term *normal* usually means free from all birth defects. The causes of such defects, of course, are both environmental and genetic. Among the genetic causes, we have considered disorders such as sickle cell anemia or Marfan syndrome, caused by the mutation of single genes, and multifactorial disorders such as cleft palate. Changes in chromosome number or changes in chromosome structure can cause other genetic disorders. These changes may involve entire chromosome sets, individual chromosomes, or alterations within individual chromosomes.

The diploid, or $2n$, number of chromosomes in somatic cells and the haploid, or n, number in gametes are called the normal, or euploid, condition. An increase in the number of chromosome sets in a cell is called **polyploidy**. A cell with three sets of chromosomes is triploid, one with four sets is tetraploid, and so forth. **Aneuploidy** is a change in chromosome number that involves less than a whole chromosome set. The simplest form of aneuploidy involves the gain or loss of a single chromosome. Loss of a single chromosome is known as **monosomy** $(2n - 1)$ and the addition of one chromosome to the diploid set is known as trisomy $(2n + 1)$.

■ **Polyploidy** A chromosomal number that is a multiple of the normal haploid chromosomal set.

■ *Aneuploidy* A chromosomal number that is not an exact multiple of the haploid set.

■ **Monosomy** A condition in which one member of a chromosomal pair is missing; having one less than the diploid number $(2n - 1)$.

Genetic Journeys

Using Fetal Cells from the Mother's Blood

More than a century has passed since placental cells from the fetus were first discovered in the circulatory systems of pregnant women. In 1969, cytogeneticists observed cells with a Y chromosome in the blood of women who later gave birth to male infants. Since then, research has been directed at finding ways to recover and use fetal cells from the maternal circulation for prenatal diagnosis. The goal is to carry out genetic analysis on fetal cells recovered from the maternal circulation without using the invasive procedures of amniocentesis and chorionic villus sampling (CVS).

Use of fetal cells for prenatal diagnosis would lower the risk of injury to the mother and fetus. Several types of fetal cells enter the maternal circulation, including placental cells; white blood cells; and immature, nucleated red blood cells. These cells probably enter the bloodstream in detectable amounts between the sixth and twelfth week of pregnancy. But because less than 1 in every 100,000 cells in the mother's blood is from the fetus, collecting enough fetal cells from a blood sample is one of the challenges facing those working to develop this technique.

Several methods are being tried to isolate fetal cells from maternal blood, but no single technique has emerged as the best. Even though these methods are still under development, several studies have found it possible to diagnose chromosomal abnormalities in fetal cells collected from maternal blood. In addition, fetal cells have been used to diagnose some genetic disorders, including sickle cell anemia and thalassemia, and to determine fetal blood types.

The separation of fetal cells may be most useful for detecting trisomies of chromosomes 13, 18, 21, X, and Y, which together account for more than 95% of all fetal aneuploidies. The wider use of this technique will depend on further refinements in separating fetal cells, and in growing them in the laboratory. If these barriers can be overcome, fetal blood sampling may gradually replace more invasive procedures for prenatal chromosome analysis.

Changes in chromosome number are fairly common in humans and are a major cause of reproductive failure. It is now estimated that as many as one in every two conceptions may be aneuploid and that 35% to 70% of early embryonic deaths and spontaneous abortions are caused by aneuploidy. About 1 in every 170 live births is at least partially aneuploid, and from 5% to 7% of all early childhood deaths are related to aneuploidy. Humans have a rate of aneuploidy that is up to 10 times higher than that of other mammals, including other primates. Understanding the causes of aneuploidy in humans remains one of the great challenges in human genetics.

Polyploidy changes the number of chromosomal sets.

Abnormalities in the number of chromosomal sets can arise in several ways: (1) errors in meiosis during gamete formation, (2) events at fertilization, or (3) errors in mitosis following fertilization. Polyploidy can result through errors in mitosis or meiosis. If homologous chromosomes fail to separate during meiosis I, meiosis II will produce diploid gametes. Fusion of this diploid gamete with a normal haploid gamete will produce a triploid zygote (▶ Figure 6.12).

Polyploidy can also be produced at fertilization by the simultaneous penetration of a haploid egg by two haploid sperm (dispermy). The resulting zygote contains three haploid chromosome sets and is triploid. Some common polyploid conditions are discussed in the following sections.

Keep in mind

■ Polyploidy results when there are more than two haploid sets of chromosomes.

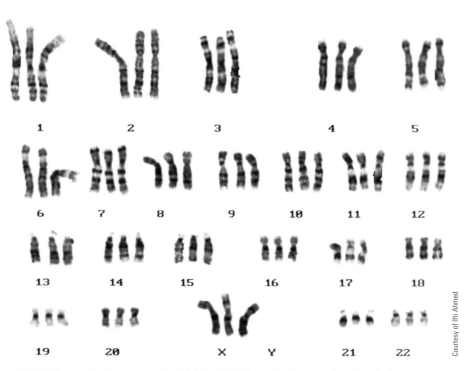

Courtesy of Ifti Ahmed

▲ FIGURE 6.12 The karyotype of a triploid individual contains three copies of each chromosome.

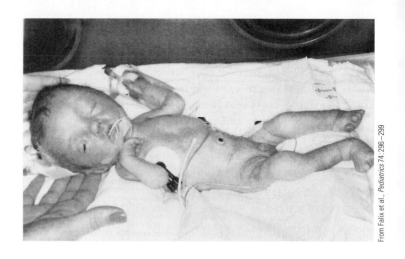

From Falix et al., *Pediatrics* 74:296–299

■ **Triploidy** A chromosomal number that is three times the haploid number, having three copies of all autosomes and three sex chromosomes.

■ **Tetraploidy** A chromosomal number that is four times the haploid number, having four copies of all autosomes and four sex chromosomes.

■ **Nondisjunction** The failure of homologous chromosomes to separate properly during meiosis or mitosis.

Triploidy The most common form of polyploidy in humans is **triploidy,** found in 15% to 18% of all spontaneous abortions. Three types of triploid chromosome sets are observed: 69,XXY; 69,XXX; and 69,XYY. Approximately 75% of all cases of triploidy have two sets of paternal chromosomes. Accidents in male gamete formation do not occur this often, so most triploid zygotes probably arise as the result of dispermy. Although biochemical changes that accompany fertilization normally prevent such fertilizations, this system is not fail-safe.

Almost 1% of all conceptions are triploid, but over 99% of these die before birth, and only 1 in 10,000 live births is triploid. Triploid infants do not survive, and most die within a month. Triploid newborns have multiple abnormalities, including an enlarged head; fusion of fingers and toes (syndactyly); and malformations of the mouth, eyes, and genitals (▶ Figure 6.13). The high rate of embryonic death and failure to survive indicates that triploidy is a lethal condition.

Tetraploidy Tetraploidy is found in about 5% of all spontaneous abortions and is extremely uncommon in live births. **Tetraploidy** can result from a failure of cytokinesis in the first mitotic division after fertilization. If tetraploidy arises sometime after the first mitotic division, two different cell types are present in the embryo, normal diploid cells and tetraploid cells. These mosaic individuals survive somewhat longer than full tetraploids, but the condition is still life-threatening.

In summary, polyploidy in humans can arise by at least two different mechanisms, errors in cell division or errors at fertilization. In either case, it is inevitably lethal. Polyploidy does not involve the mutation of any genes, but only changes in the number of gene copies. How this quantitative change in gene number is related to lethality in development is unknown.

Aneuploidy changes the number of individual chromosomes.

As defined earlier, aneuploidy is the addition or deletion of individual chromosomes from the normal diploid set of 46. The most common cause of aneuploidy is **nondisjunction,** the failure of chromosomes to separate at anaphase (▶ Active Figure 6.14). Although this failure can occur in either meiosis or mitosis, nondisjunction in meiosis is the leading cause of aneuploidy in humans. There are two cell divisions in meiosis, and nondisjunction can occur in either the first or second division with different genetic consequences.

Nondisjunction in meiosis I produces abnormal gametes. All gametes produced from this event will carry both members of a chromosomal pair or neither member of the pair. Nondisjunction in meiosis II produces two normal gametes and two abnormal gametes (▶ Active Figure 6.14). Gametes missing a copy of a specific chromosome will produce a monosomic zygote. Those with an extra copy of a chromosome will produce a trisomic zygote.

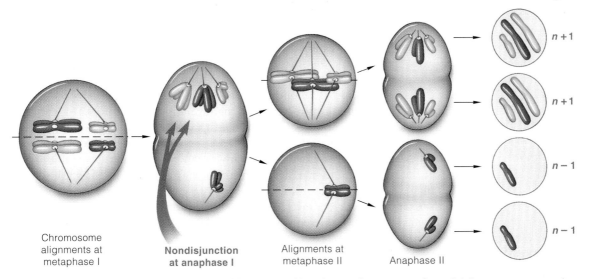

Chromosome number in gametes:

Chromosome
alignments at
metaphase I

**Nondisjunction
at anaphase I**

Alignments at
metaphase II

Anaphase II

▲ **ACTIVE FIGURE 6.14** Nondisjunction in a cell with two pairs of homologous chromosomes. One pair fails to separate properly at anaphase I of meiosis. Two gametes carry both members of a chromosome pair (*n* + 1), and two are missing one chromosome (*n* − 1).

**HUMAN
Genetics ⓔ Now™** Learn more about nondisjunction by viewing the animation on Human GeneticsNow at **http://now.ilrn.com/cummings7**

The phenotypic effects of aneuploidy range from minor physical symptoms to devastating and lethal deficiencies in major organ systems. Among survivors, phenotypic effects often include behavioral deficits and mental retardation. In the following section, we look at some of the important features of autosomal aneuploid phenotypes. Then we consider the phenotypic effects of sex chromosome aneuploidy.

Autosomal monosomy is a lethal condition.

Aneuploidy during gamete formation produces equal numbers of monosomic and trisomic gametes and embryos. However, autosomal monosomies are only rarely observed among spontaneous abortions and live births. The likely explanation is that the majority of autosomal monosomic embryos are lost very early, even before pregnancy is recognized.

Autosomal trisomy is relatively common.

Most autosomal trisomies are lethal. Autosomal **trisomy** is found in about 50% of all cases of chromosomal abnormalities in fetal death. Karyotypic analysis of miscarriages indicates that autosomes are differentially involved in trisomy (▶ Figure 6.15). Trisomies for chromosomes 1, 3, 12, and 19 are rarely observed, whereas trisomy for chromosome 16 accounts for almost one-third of all cases. As a group, the acrocentric chromosomes (13–15, 21, and 22) are represented in 40% of all miscarriages. Reasons include differences in the rate of nondisjunction in different chromosomes, differences in the rate of fetal death before recognition of pregnancy, or a combination of factors. Only a few autosomal trisomies result in live births (trisomy 8, 13, and 18). Trisomy 21 (Down syndrome) is the only au-

■ **Trisomy** A condition in which one chromosome is present in three copies, whereas all others are diploid; having one more than the diploid number (2*n* + 1).

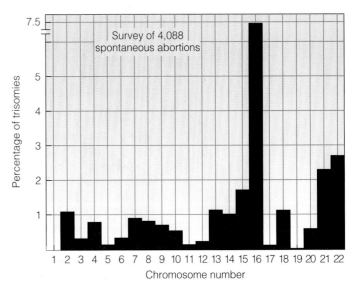

▲ **FIGURE 6.15** The results of a cytogenetic survey of over 4,000 spontaneous abortions show a wide variation in the presence of individual chromosomes in trisomic embryos.

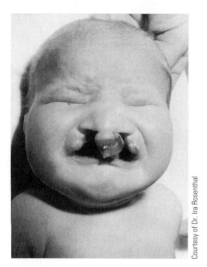

▲ FIGURE 6.16 An infant with trisomy 13, showing a cleft lip and palate (the roof of the mouth).

tosomal trisomy that allows survival into adulthood. In the following section, we will review some autosomal trisomies.

Trisomy 13: Patau Syndrome (47,+13)   Trisomy 13 was discovered by cytogenetic analysis of a malformed child. The karyotype revealed 47 chromosomes, and the extra chromosome was identified as chromosome 13 (47,+13). Only 1 in 15,000 live births involves trisomy 13, and the condition is lethal. Half of all affected individuals die in the first month, and the mean survival time is 6 months. The phenotype of trisomy 13 involves facial malformations (▶ Figure 6.16), eye defects, extra fingers or toes, and feet with large protruding heels. Internally, there are usually severe malformations of the brain and nervous system, as well as congenital heart defects. The involvement of so many organ systems indicates that abnormalities form early in embryonic development, perhaps as early as the sixth week. Parental age is the only factor known to be related to trisomy 13. Parents of children with trisomy 13 are older (averaging about 32 years) than parents who have normal children. The relationship between parental age and aneuploidy is discussed later in this chapter.

Keep in mind

■ Monosomy and trisomy involve the loss and gain of a single chromosome to a diploid genome.

Trisomy 18: Edwards Syndrome (47,+18)   Infants with Edwards syndrome are small at birth, grow very slowly, and are mentally retarded. For reasons still unknown, 80% of all trisomy 18 births are female. Clenched fists, with the second and fifth finger overlapping the third and fourth fingers, and malformed feet are also characteristic (▶ Figure 6.17). Heart malformations are almost always present, and heart failure or pneumonia usually causes death. Trisomy 18 occurs with a frequency of 1 in 11,000 live births, and the average survival time is 2 to 4 months. As in trisomy 13, advanced maternal age is a factor predisposing to trisomy 18.

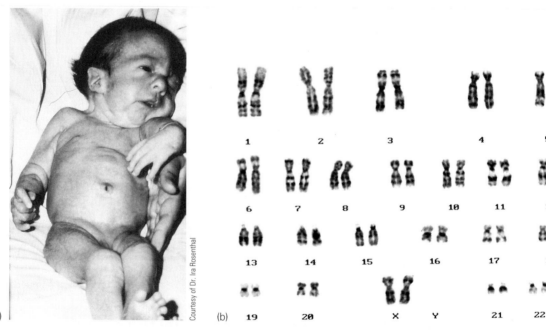

(a)   (b)

▲ FIGURE 6.17 (a) An infant with trisomy 18. (b) A trisomy 18 karyotype.

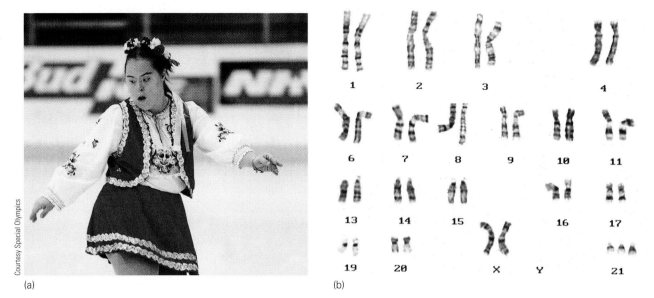

▲ FIGURE 6.18 (a) A child with trisomy 21. (b) A karyotype shows this child has three copies of chromosome 21.

Trisomy 21: Down Syndrome (47,+21) The features of **trisomy 21** (OMIM 190685) were first described by John Langdon Down in 1866. He called the condition "mongolism" because of the distinctive skin fold, known as an epicanthic fold, in the corner of the eye (▶ Figure 6.18). To remove the racist implications inherent in the term, Lionel Penrose and others changed the name to Down syndrome. Trisomy 21 was the first chromosomal abnormality discovered in humans. In 1959, Jerome Lejeune and his colleagues discovered that the presence of an extra copy of chromosome 21 is the underlying cause of Down syndrome.

Down syndrome occurs in about 0.5% of all conceptions and in 1 in 900 live births. It is a leading cause of childhood mental retardation and heart defects in the United States. Affected individuals usually have a wide, flat skull, folds in the corner of the eyelids, and spots on the iris. They may have furrowed, large tongues that cause the mouth to remain partially open. Physical growth, behavior, and mental development are retarded, and approximately 40% of all Down syndrome children have congenital heart defects. In addition, these children are very susceptible to respiratory infections and contract leukemia at a rate far above the normal population. In the past two decades, improvements in medical care have increased survival rates dramatically, so that many people with Down syndrome survive into adulthood, although few reach the age of 50 years. In spite of these handicaps, many individuals who have Down syndrome lead rich, productive lives and can serve as an inspiration to us all.

■ **Trisomy 21** Aneuploidy involving the presence of an extra copy of chromosome 21, resulting in Down syndrome.

6.5 What Are the Risks for Autosomal Trisomy?

The causes of autosomal trisomy are unknown, but a variety of genetic and environmental factors have been proposed, including genetic predisposition, exposure to radiation, viral infection, and abnormal hormone levels. However, the only known risk factor for autosomal aneuploidy is advanced maternal age. In fact, a relationship between maternal age and Down syndrome was well established 25 years before the chromosomal basis for the condition was discovered.

Maternal age is the leading risk factor for trisomy.

Young mothers have a low probability of having trisomy 21 children, but the risk increases rapidly after the age of 35 years. At age 20, the risk of Down syndrome

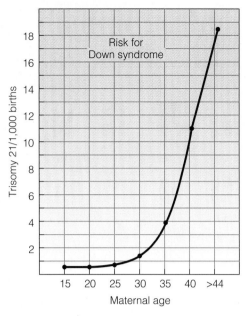

FIGURE 6.19 The relationship between maternal age and the frequency of trisomy 21 (Down syndrome). The risk increases rapidly after 35 years of age.

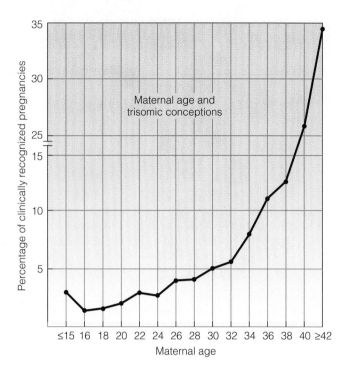

FIGURE 6.20 Maternal age is the major risk factor for autosomal trisomies of all types. By age 42, about one in three identified pregnancies is trisomic.

is 0.05%; by age 35 the risk has climbed to 0.9%; and at age 45, 3% of all newborns have trisomy 21 (▷ Figure 6.19). Maternal age is also a risk factor for other autosomal aneuploidies (▷ Figure 6.20). Paternal age has also been proposed as a factor in autosomal trisomy, but the evidence is weak, and no clear-cut link has been demonstrated.

Keep in mind

■ Age of the mother is the best known risk factor for trisomy.

Maternal age as a risk factor is supported by studies on the parental origin of nondisjunction. Occasionally, members of a chromosome pair have some minor differences in their banding patterns. By examining banded chromosomes from the trisomic child and the parents, the nondisjunction event can be traced to one or the other parent. For trisomy 21, nondisjunction occurs in the mother about 94% of the time and about 6% of the time in the father, and the great majority of these nondisjunction events take place at meiosis I in oocytes.

Why is maternal age a risk factor?

One idea about the relationship between maternal age and nondisjunction focuses on the duration of meiosis in females. Recall from Chapter 2 that primary oocytes are formed early in embryonic development and enter the first meiotic prophase well before birth. Meiosis I is not completed until ovulation, so that eggs produced at age 40 have been in meiosis I for more than 40 years. During this time, intracellular events or environmental agents may damage the cell so that aneuploidy results when meiosis resumes at ovulation. However, it is not yet known whether the age of the egg is directly related to the increased frequency of nondisjunction.

A second idea focuses on the interaction between the implanting embryo and the uterine environment. According to this idea, the embryo–uterine interaction normally results in the spontaneous abortion of chromosomally abnormal embryos, a process called maternal selection. As women age, maternal selection may become less effective, allowing more chromosomally abnormal embryos to implant and develop. There may well be other factors in addition to age of the egg and maternal selection that play a role in the relationship between maternal age and autosomal aneuploidy, and more research is needed to clarify the underlying mechanisms.

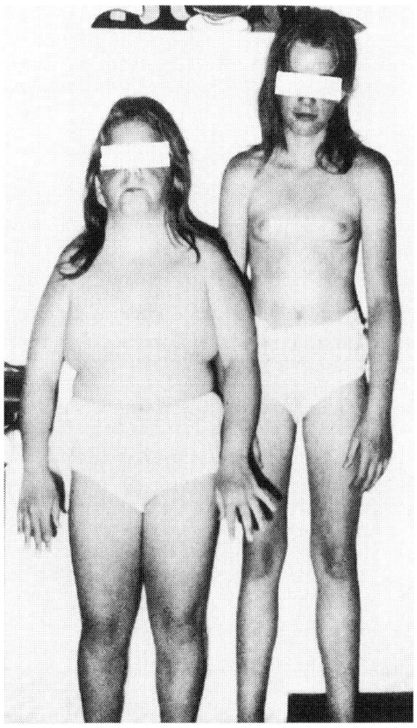

6.6 Aneuploidy of the Sex Chromosomes

Aneuploidy of the X and Y chromosomes is more common than autosomal aneuploidy. The overall incidence of sex chromosome anomalies in live births is 1 in 400 for males and 1 in 650 for females. Unlike autosomes, in which monosomy is always fatal, monosomy for the X chromosome is a viable condition. Monosomy for the Y chromosome (45,Y), however, is always lethal.

Keep in mind

■ Changes in the number of sex chromosomes have less impact than changes in autosomes.

Turner syndrome (45,X)

Females with **Turner syndrome** are short and wide-chested with rudimentary ovaries (▶ Figure 6.21). At birth, puffiness of the hands and feet is prominent, but this disappears in infancy. Many also have narrowing, or coarctation, of the aorta. There is no mental retardation associated with this syndrome. It is estimated that 1% of all conceptions are 45,X and that 95% to 99% of all 45,X embryos die before birth. Turner syndrome occurs with a frequency of 1 in 10,000 female births.

The phenotypic impact of the single X chromosome in Turner syndrome is strikingly illustrated in a case of identical twins, one of them 46,XX and the other 45,X (▶ Figure 6.22). In spite of being identical twins, they have significant differences in height, sexual development, hearing, and dental maturity. Although environmental factors may contribute to these differences, the major role of the second X chromosome in normal female development is apparent. Two X chromosomes are needed

From E. Weiss et al., American Journal of Medical Genetics 13:389–399

▲ **FIGURE 6.22** Monozygotic twins, one of which has Turner syndrome. The twin who has Turner syndrome (*left*) is 45,X; the healthy twin (*right*) is 46,XX.

■ **Turner syndrome** A monosomy of the X chromosome (45,X) that results in female sterility.

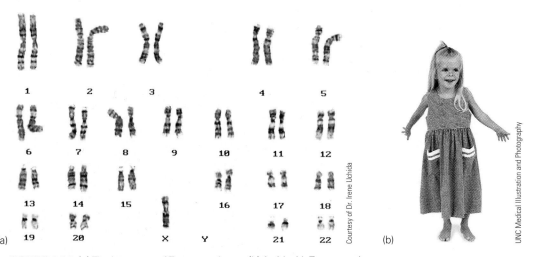

Courtesy of Dr. Irene Uchida

UNC Medical Illustration and Photography

▲ **FIGURE 6.21** (a) The karyotype of Turner syndrome. (b) A girl with Turner syndrome.

for normal development of the ovary, normal growth patterns, and development of the nervous system in females. Complete absence of an X chromosome in the absence or presence of a Y chromosome is always lethal, emphasizing that the X chromosome is an essential component of the karyotype.

Klinefelter syndrome (47,XXY)

Klinefelter syndrome Aneuploidy of the sex chromosomes involving an XXY chromosomal constitution.

The phenotype of **Klinefelter syndrome** was first described in 1942, and Patricia Jacobs and John Strong reported the XXY chromosomal condition in 1959. The frequency of Klinefelter syndrome is approximately 1 in 1,000 male births. The features of this syndrome do not develop until puberty (▶ Figure 6.23). Affected individuals are male but have very low fertility. Some men with Klinefelter syndrome have learning disabilities or subnormal intelligence. A significant number of Klinefelter males are mosaics with some cells having an XY chromosome combination and others having an XXY set of sex chromosomes. In these cases, nondisjunction occurred during mitosis of embryonic cells.

Overall, about 60% of the cases are the result of maternal nondisjunction, and advanced maternal age is known to increase the risk of affected offspring. Other forms of Klinefelter have XXYY, XXXY, and XXXXY sex chromosome sets. Additional

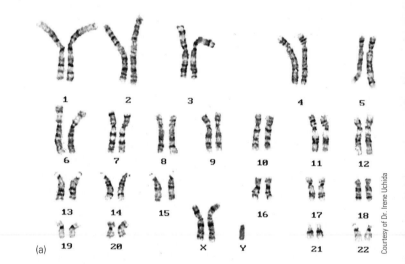

(a)

Courtesy of Dr. Irene Uchida

▶ FIGURE 6.23 (a) The characteristic karyotype of Klinefelter syndrome. (b) The young man in these photos has Klinefelter syndrome.

Stefan Schwarz

(b)

Stefan Schwarz

X chromosomes in these karyotypes increases the severity of the phenotypic symptoms and brings on clear-cut mental retardation.

XYY syndrome (47,XYY)

In 1965, a cytogenetic survey of 197 males institutionalized for violent and dangerous antisocial behavior aroused a great deal of interest in the scientific community and the popular press. The findings indicated that nine of these males (about 4.5% of the males in the survey) were **XYY** (▶ Figure 6.24). These individuals were above average in height, suffered personality disorders, and seven of the nine were of subnormal intelligence. Subsequent studies indicated that the frequency of XYY males in the general population is 1 in 1,000 male births (about 0.1% of the males in the general population) and that the frequency of XYY individuals in penal and mental institutions is significantly higher than in the population at large.

Early investigators associated the tendency to violent criminal behavior with the presence of an extra Y chromosome. In effect, this would mean that some forms of violent behavior were genetically determined. In fact, the XYY karyotype has been used on several occasions as a legal defense (unsuccessfully, so far) in criminal trials. The question is this: Is there really a direct link between the XYY condition and criminal behavior? There is no strong evidence to support such a link. In fact, the vast majority of XYY males lead socially normal lives. In the United States, long-term studies of the relationship between antisocial behavior and the 47,XYY karyotype were discontinued. Researchers feared that identifying children with potential behavioral problems might lead parents to treat them differently and result in behavioral problems as a self-fulfilling prophecy.

What are some conclusions about aneuploidy of the sex chromosomes?

Several conclusions can be drawn from the study of sex chromosome disorders. First, at least one copy of an X chromosome is essential for survival. Embryos without any X chromosomes (44,−XX and 45,Y) are not observed in studies of spontaneous abortions. They must be eliminated even before pregnancy is recognized, emphasizing the role of the X chromosome in normal development. The second general conclusion is that the addition of extra copies of either sex chromosome interferes with normal development and causes both physical and mental problems. As the number of sex chromosomes in the karyotype increases, the phenotype becomes more severe, indicating that a balance of sex chromosomes is essential to normal development in both males and females.

6.7 Structural Alterations within Chromosomes

Now that we have discussed changes in chromosome number, we will focus on structural changes within and between chromosomes that result in an abnormal phenotype. These changes can involve one, two, or more chromosomes and result from the breakage and reunion of chromosomal parts.

■ XYY karyotype Aneuploidy of the sex chromosomes involving XYY chromosomal constitution.

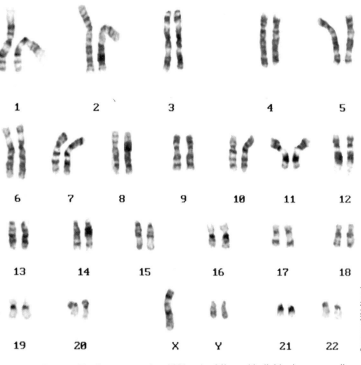

Courtesy of Ifti Ahmed

▲ FIGURE 6.24 The karyotype of an XYY male. Affected individuals are usually taller than normal, and some, but not all, suffer from personality disorders.

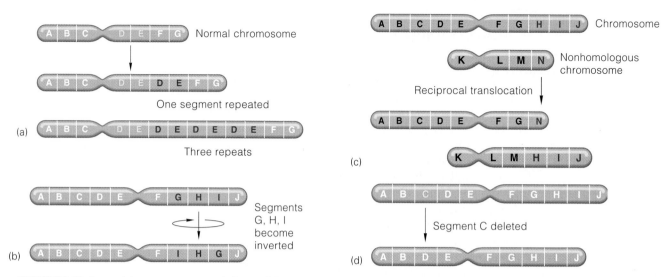

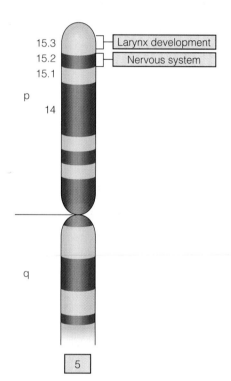

FIGURE 6.25 Some of the common structural abnormalities seen in chromosomes. (a) A duplication has a chromosomal segment repeated (in this example, genes *D* and *E* are duplicated). (b) In an inversion, the order of part of the chromosome is reversed. This does not change the amount of genetic information carried by the chromosome, only its arrangement. (c) In a translocation, chromosomal parts are exchanged. (d) In a deletion, part of the chromosome is lost. This can occur at the tip of the chromosome, or an internal segment can be lost, as shown here.

FIGURE 6.26 A deletion of part of chromosome 5 is associated with cri du chat syndrome. By comparing the region deleted with its associated phenotype, investigators have identified regions of the chromosome that carry genes involved in developing the larynx.

■ **Cri du chat syndrome** A deletion of the short arm of chromosome 5 associated with an array of congenital malformations, the most characteristic of which is an infant cry that resembles a meowing cat.

Breaks can occur spontaneously through errors in replication or recombination. Environmental agents, such as ultraviolet light, radiation, viruses, and chemicals can also produce them. Structural alterations that result from breaks include duplications (extra copies of a chromosome segment), translocations (moving a segment from one chromosome to another, nonhomologous chromosome), and deletions (loss of chromosome segments). These changes are summarized in ▶ Figure 6.25. Rather than considering how such aberrations are produced, we'll look at the phenotypic effects of these alterations and what they can tell us about the location and action of genes.

Keep in mind

■ Chromosomes can lose, gain, or rearrange genes.

Deletions involve loss of chromosomal material.

Deletion of more than a small piece of a chromosome is detrimental to a developing embryo, and deletion of an entire autosome is lethal. Consequently, only a few viable conditions are associated with large-scale deletions. Some of these are listed in ▶ Table 6.3.

Cri du chat syndrome is caused by a deletion in the short arm of chromosome 5 and occurs in 1 of 100,000 births. The loss of genes in the deleted chromosome produces the abnormal phenotype, not the presence of any mutant genes. Affected infants are mentally retarded, with defects in facial development, gastrointestinal malformations, and abnormal throat structures. Affected infants have a cry that sounds like a cat meowing, hence the name **cri du chat syndrome** (OMIM 123450). This deletion affects the motor and mental development of affected individuals but does not seem to be life-threatening.

By correlating phenotypes with chromosomal breakpoints, two regions associated with this syndrome have been identified on the short arm of chromosome 5 (▶ Figure 6.26). Loss of chromosome segments in 5p15.3 results in abnormal larynx development; deletions in 5p15.2 are associated with mental retardation and other

Table 6.3 Chromosomal Deletions

Deletion	Syndrome	Phenotype
5p-	Cri du chat syndrome	Infants have catlike cry, some facial anomalies, severe mental retardation
11q-	Wilms tumor	Kidney tumors, genital and urinary tract abnormalities
13q-	Retinoblastoma	Cancer of eye, increased risk of other cancers
15q-	Prader-Willi syndrome	Infants: weak, slow growth; children and adults: obesity, compulsive eating

phenotypic features of this syndrome. This indicates that genes controlling larynx development may be located in 5p15.3, and genes important in the development or function of the nervous system are located in 5p15.2.

Translocations involve exchange of chromosomal parts.

A translocation results when a chromosome segment is moved to a nonhomologous chromosome. There are two major types of translocations: reciprocal translocations and Robertsonian translocations. In a reciprocal translocation, two nonhomologous chromosomes exchange parts. No genetic information is gained or lost in the exchange, but genes are moved to new chromosomal locations. In some cases, there are no phenotypic effects, and the translocation is passed through a family for generations. In other cases, these translocations can produce genetically unbalanced gametes with duplicated or deleted chromosomal segments that can result in embryonic death or abnormal offspring.

About 5% of all cases of Down syndrome involve a Robertsonian translocation, most often between chromosomes 21 and 14. In this translocation, centromeres of two chromosomes fuse, and chromosomal material is lost from the short arms (▷ Figure 6.27). Someone who carries this translocation is phenotypically normal, even though they are actually aneuploid (they have only 45 chromosomes) and the short arms of both chromosomes are

▼ FIGURE 6.27 Segregation of chromosomes at meiosis in a 14/21 translocational carrier. Six types of gametes are produced. When these gametes fuse with those of a normal individual, six types of zygotes are produced. Of these, two (translocational carrier and normal) have a normal phenotype, one is Down syndrome, and three are lethal combinations.

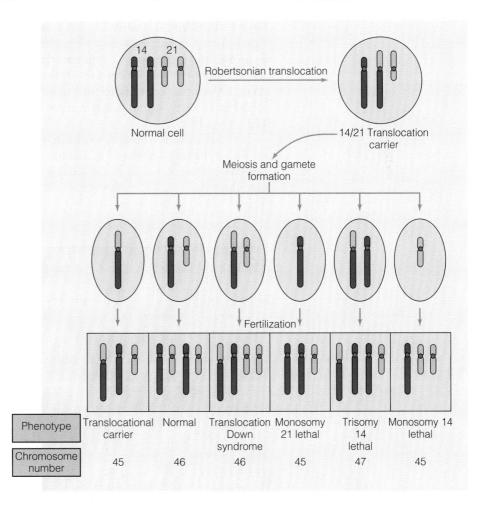

Phenotype	Translocational carrier	Normal	Translocation Down syndrome	Monosomy 21 lethal	Trisomy 14 lethal	Monosomy 14 lethal
Chromosome number	45	46	46	45	47	45

missing. These carriers have two copies of the long arm of chromosome 14 and two copies of the long arm of chromosome 21 (a normal 14, a normal 21, and a translocated 14/21), so there is no phenotypic effect. At meiosis the carrier produces six types of gametes in equal proportions (▶ Figure 6.27). Three of these produce zygotes that do not develop. Of the remaining three, one will produce a Down syndrome child, one is a translocation heterozygote, and one is chromosomally normal.

Although it might seem that translocation heterozygotes have a 33% risk of having a Down syndrome child, the observed frequency is somewhat lower. It is important to remember that this risk does not increase with maternal age. In addition, there is also a one in three chance of producing a phenotypically normal translocation carrier who is at risk of producing children with Down syndrome. For this reason it is important to cytogenetically analyze a Down syndrome child and the parents to determine whether a translocation is involved. This information is essential in counseling parents about future reproductive risks.

6.8 What Are Some Consequences of Aneuploidy?

Aneuploidy is the most common chromosomal abnormality in humans and has several important consequences. We will discuss three of these consequences: pregnancy loss, birth defects, and cancer.

Aneuploidy is a major cause of spontaneous abortions (see ▶ Figure 6.15). ▶ Table 6.4 summarizes some of the major chromosomal abnormalities found in miscarriages. These include triploidy, monosomy for the X chromosome (45,X), and trisomy 16. It is interesting to compare the frequency of chromosomal abnormalities found in spontaneous abortions with those in live births. Triploidy is found in 17 of every 100 spontaneous abortions but in only about 1 in 10,000 live births; 45,X is found in 18% of chromosomally abnormal miscarriages but in only 1 in 7,000 to 10,000 live births.

Comparison of the number of chromosomal abnormalities detected by CVS (performed at 10 to 12 weeks of gestation) versus amniocentesis (at 16 weeks of gestation) show that the abnormalities detected by CVS are two to five times more frequent than those detected by amniocentesis, which in turn are about two times higher than

Table 6.4 Chromosomal Abnormalities in Spontaneous Abortions	
Abnormality	**Frequency (%)**
Trisomy 16	15
Trisomies, 13, 81, 21	9
XXX, XXY, XYY	1
Other	27
45,X	18
Triploidy	17
Tetraploidy	6
Other	7

those found in newborns. This decrease in the frequency of chromosomal abnormalities during pregnancy is evidence that chromosomally abnormal embryos and fetuses are eliminated by spontaneous abortion throughout pregnancy (▶ Figure 6.28).

Birth defects are another consequence of chromosomal abnormalities. The frequencies of chromosomal aberrations in newborns are shown in ▶ Table 6.5. Trisomy 16, which is common in spontaneous abortions, is not found among infants, indicating that fetuses with this condition are not viable. Only trisomy 13, 18, and 21 occur with any frequency in live births. Trisomy 21 occurs with a frequency of about 1 in 900 births, but cytogenetic surveys of spontaneous abortions indicate that about two-thirds of such conceptions are lost by miscarriage. Similarly, over 99% of all 45,X conceptions are lost before birth. Overall, although selection against chromosomally abnormal embryos and fetuses is efficient, the high rate of nondisjunction in humans means there is a significant reproductive risk for chromosomal abnormalities. Over 0.5% of all newborns are affected with an abnormal karyotype.

The relationship between cancer and its accompanying chromosomal changes is a third consequence of chromosomal abnormalities. A significant number of cancers, especially leukemia, are associated with specific chromosomal translocations. Solid tumors have a wide range of chromosomal abnormalities, including aneuploidy, translocations, and duplications. Evidence suggests that these abnormalities may

Table 6.5 Chromosomal Abnormalities in Newborns	
Abnormality	**Approximate Frequency**
45,X	1/7,500
XXX	1/1,200
XXY	1/1,000
XYY	1/1,100
Trisomy 13	1/15,000
Trisomy 18	1/11,000
Trisomy 21	1/900
Structural abnormalities	1/400

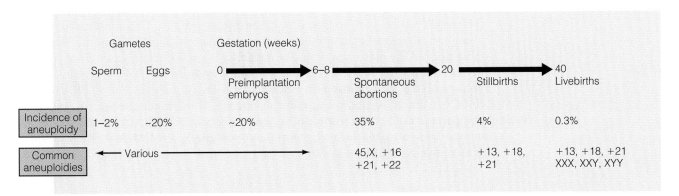

▲ FIGURE 6.28 The frequency of aneuploidy changes dramatically over developmental time. Between 6 to 8 weeks and 20 weeks, about 35% of spontaneous abortions are aneuploid. Around 20 weeks, the frequency falls by an order of magnitude to about 4% in stillbirths. The frequency decreases again by an order of magnitude, with about 0.3% of newborns being aneuploid.

arise during a period of genomic instability that precedes or accompanies the transition of a normal cell into a malignant cell. The chromosomal changes that accompany the development of cancer are discussed in Chapter 12.

6.9 Other Forms of Chromosomal Abnormalities

In some cases, the karyotype and individual chromosomes appear to be normal, but the phenotype is abnormal, and careful analysis reveals a subtle chromosome change. One of these situations is **uniparental disomy.** In this situation, both members of a chromosome pair are inherited from one parent. Fragile sites are another rare form of chromosome abnormality. These appear only when cells are grown in the laboratory and certain chemicals are added to the growth medium.

Uniparental disomy

Normally, meiosis ensures that one member of each chromosomal pair is derived from the mother, and the other member is from the father. On rare occasions, however, a child gets both copies of a chromosome from one parent, a condition known as uniparental disomy (UPD). UPD can arise in several ways, all of which involve two chromosomal errors in cell division. These errors can occur in meiosis (▶ Figure 6.29) or in mitotic divisions following fertilization.

UPD has been identified in some unusual situations. These include females affected with rare X-linked disorders such as hemophilia; father-to-son transmission of rare, X-linked disorders in which the mother is homozygous normal; and children who are affected with rare autosomal recessive disorders, but in which only one parent is heterozygous. Prader-Willi syndrome and Angelman syndrome (OMIM 105830) can be caused by deletions in region 15q11.12 or by UPD. If both copies of chromosome 15 are inherited from the mother, the child will have Prader-Willi syndrome. If both copies of chromosome 15 are inherited from the father, the child will have Angelman syndrome. The origin of these disorders by UPD is discussed in detail in Chapter 11.

Fragile sites appear as gaps or breaks in chromosomes.

Fragile sites appear as gaps or breaks at specific sites on a chromosome when cells are grown in the laboratory. Fragile sites are inherited as codominant traits. Over 100 fragile sites have been identified in the human genome. Chromosome breaks often occur at fragile sites, producing chromosome fragments, deletions, and other aberrations. The molecular nature of most fragile sites is unknown but is of great interest because they represent regions susceptible to breakage. Several fragile sites are located on the X chromosome (▶ Figure 6.30). Two of these rare sites, *FRAX E* and *FRAX A,* are associated with genetic disorders. A rare fragile site near the tip of the long arm of the X chromosome is associated with an X-linked form of mental retardation known as Martin-Bell syndrome, or **fragile-X** (OMIM 309500) syndrome. The fragile-X syndrome is caused by an alteration in the *FMR-1* gene and is discussed in Chapter 11.

▦ **Uniparental disomy** A condition in which both copies of a chromosome are inherited from a single parent.

▦ **Fragile X** An X chromosome that carries a nonstaining gap, or break, at band q27; associated with mental retardation in males.

Keep in mind

▦ Some fragile sites are associated with mental retardation.

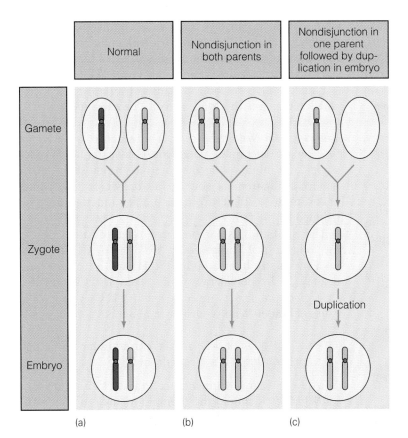

▲ FIGURE 6.29 Uniparental disomy can be produced by several mechanisms, involving nondisjunction in meiosis or nondisjunction in the zygote or early embryo. (a) Normally, gametes contain one copy of each chromosome, and fertilization produces a zygote carrying two copies of a chromosome, one derived from each parent. (b) Nondisjunction in both parents, in which one gamete carries both copies of a chromosome and the other gamete is missing a copy of that chromosome. Fertilization produces a diploid zygote, but both copies of one chromosome are inherited from a single parent. (c) Nondisjunction in one parent, resulting in the loss of a chromosome. This gamete fuses with a normal gamete to produce a zygote monosomic for a chromosome. An error in the first mitotic division results in duplication of the monosomic chromosome, producing uniparental disomy.

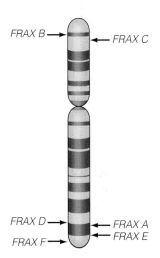

▲ FIGURE 6.30 The fragile sites on the human X chromosome. Sites B, C, and D are common sites and found on almost all copies of the X chromosome. A, E, and F are rare sites; expression of A is associated with fragile-X syndrome.

Genetics in Practice

Genetics in Practice case studies are critical thinking exercises that allow you to apply your new knowledge of human genetics to real-life problems. You can find these case studies and links to relevant websites at **http://biology.brookscole.com /cummings7**

CASE 1

Michelle was a 42-year-old Caucasian woman who had declined counseling and amniocentesis at 16 weeks of pregnancy but was referred for genetic counseling following an abnormal ultrasound at 20 weeks gestation. Following the ultrasound, a number of findings suggested a possible chromosome abnormality in the fetus. The ultrasound showed swelling under the skin at the back of the fetus's neck; shortness of the femur, humerus, and ear length; and underdevelopment of the middle section of the fifth finger. Michelle's physician performed an amniocentesis and referred her to the genetics program. Michelle and her husband did not want genetic counseling before receiving results of the cytogenetic analysis.

This was Michelle's third pregnancy; she and her husband, Mike, had a 6-year-old daughter and a 3-year-old son. At their next session, the counselor informed the couple that the results revealed trisomy 21, explored their understanding of Down syndrome, and elicited their expe-

(continued)

riences with people with disabilities. She also reviewed the clinical concerns revealed by the ultrasound and associated anomalies (mild to severe mental retardation, cardiac defects, and kidney problems). The options available to the couple were outlined. They were provided with a booklet written for parents making choices following the prenatal diagnosis of Down syndrome. After a week of careful deliberation with their family, friends, and clergy, they elected to terminate the pregnancy.

1. Do you think that this couple had the right to terminate the pregnancy given the prenatal diagnosis? If not, under what circumstance would a couple have this right? What other options were available to the couple?

2. Should physicians discourage a 42-year-old woman from having children because of an increased chance of a chromosomal abnormality?

CASE 2

A genetic counselor was called to the nursery for a consultation on a newborn that was described as "floppy with a weak cry." The counselor noted that the newborn's chart indicated that he was having feeding problems and had not gained weight since his delivery 15 days earlier. The counselor noted several other findings during his evaluation. The infant had almond-shaped eyes, a small mouth with a thin upper lip, downturned corners of the mouth, and a narrow face. The infant was born with undescended testes and a small penis. The counselor suspected that this child had the genetic disorder known as Prader-Willi syndrome.

Prader-Willi syndrome is caused by the absence of a small region on the long arm of chromosome 15. It is always the lack of the paternal copy of this region that causes Prader-Willi syndrome. This absence can occur in three ways: deletion of a segment of the paternal chromosome 15, a mutation on the paternal chromosome 15, or maternal uniparental disomy—in other words, both copies of chromosome 15 are from the mother and none are contributed by the father.

The child and his parents were tested for a deletion in the long arm of chromosome 15 (15q11–q13) by fluorescence in situ hybridization (FISH) and for uniparental disomy 15 by polymerase chain reaction (PCR). In this case, maternal disomy was detected by PCR—which is the cause of Prader-Willi syndrome in about 30% of the cases.

1. Why is a copy of the paternal chromosome 15 needed to prevent Prader-Willi syndrome?

2. Are there any treatments for Prader-Willi syndrome? What steps should the family now take to cope with the diagnosis?

3. Explain to the parents how maternal disomy happens during gamete formation and/or in mitosis following fertilization.

Summary

6.1 The Human Chromosome Set

- Human chromosomes are analyzed by the construction of karyotypes. A system of identifying chromosome regions allows any region to be identified by a descriptive address. Chromosome analysis is a powerful and useful technique in human genetics.

6.2, 6.3 Constructing and Analyzing Karyotypes

- The study of variations in chromosomal structure and number began in 1959 with the discovery that Down syndrome is caused by the presence of an extra copy of chromosome 21. Since then the number of genetic diseases related to chromosomal aberrations has steadily increased. The development of chromosome banding and techniques for identifying small changes in chromosomal structure have contributed greatly to the information that is now available.

6.4 Variations in Chromosome Number

- There are two major types of chromosomal changes: a change in chromosomal number and a change in chromosomal arrangement. Polyploidy and aneuploidy are major causes of reproductive failure in humans. Polyploidy is rarely seen in live births, but the rate of aneuploidy in humans is reported to be more than tenfold higher than in other primates and mammals. The reasons for this difference are unknown, but this represents an area of intense scientific interest.

6.5 What Are the Risks for Autosomal Trisomy?

- The loss of a single chromosome creates a monosomic condition, and the gain of a single chromosome is called a trisomic condition. Autosomal monosomy is eliminated early in development. Autosomal trisomy is selected against less stringently, and cases of partial development and live births of trisomic individuals are observed. Most cases of autosomal trisomy greatly shorten life expectancy, and only individuals who have trisomy 21 survive into adulthood.

6.6 Aneuploidy of the Sex Chromosomes

■ Aneuploidy of sex chromosomes involves both the X and Y chromosome. Studies of sex chromosome aneuploidies indicate that at least one copy of the X chromosome is required for development. Increasing the number of copies of the X or Y chromosome above the normal range causes progressively greater disturbances in phenotype and behavior, indicating the need for a balance in gene products for normal development.

6.7 Structural Alterations within Chromosomes

■ Changes in the arrangement of chromosomes include duplications, inversions, translocations, and deletions. Deletions of chromosomal segments are associated with several genetic disorders, including cri du chat and Prader-Willi syndromes. Translocations often produce

no overt phenotypic effects but can result in genetically imbalanced and aneuploid gametes. We discussed a translocation resulting in Down syndrome that in effect makes Down syndrome a heritable genetic disease, potentially present in one in three offspring.

6.8 What Are Some Consequences of Aneuploidy?

■ Aneuploidy is the leading cause of reproductive failure in humans, resulting in spontaneous abortions and birth defects. In addition, aneuploidy is associated with most cancers.

6.9 Other Forms of Chromosome Abnormalities

■ Uniparental disomy (UPD) is a condition in which both copies of a chromosome are inherited from a single parent. UPD is associated with several genetic diseases. Fragile sites appear as gaps, or breaks, in chromosome-specific locations. One of these fragile sites on the X chromosome is associated with a common form of mental retardation that affects a significant number of males.

Questions and Problems

Constructing and Analyzing Karyotypes

1. Originally, karyotypic analysis relied on size and centromere placement to identify chromosomes. Because many chromosomes are similar in size and centromere placement, the identification of individual chromosomes was difficult, and chromosomes were placed into eight groups, identified by letters A–G. Today, each human chromosome can be readily identified.
   a. What technical advances led to this improvement in chromosome identification?
   b. List two ways this improvement can be implemented.
2. What clinical information does a karyotype provide?
3. Given the adjacent karyotype, is this a male or female? Normal or abnormal? What would the phenotype of this individual be?

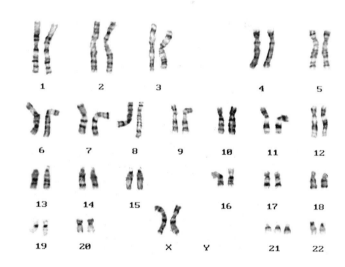

4. A colleague emails you a message that she has identified an interesting chromosome variation at 21q13. In discussing this discovery with a friend who is not a cytogeneticist, explain how you would describe the location, defining each term in the chromosome address 21q13.

5. What are the two prenatal diagnosis techniques used to detect genetic defects in an unborn baby? Which technique can be performed earlier, and why is this an advantage?

6. What are some conditions that warrant prenatal diagnosis?

Variations in Chromosome Number—Polyploidy

7. Discuss the following sets of terms:
   a. trisomy and triploidy
   b. aneuploidy and polyploidy

8. What chromosomal abnormality can result from dispermy?

9. Tetraploidy may result from:
   a. lack of cytokinesis in meiosis II
   b. nondisjunction in meiosis I
   c. lack of cytokinesis in mitosis
   d. nondisjunction in mitosis in the early embryo
   e. none of the above

10. A cytology student believes he has identified an individual suffering from monoploidy. The instructor views the cells under the microscope and correctly dismisses the claim. Why was the claim dismissed? What type of cells were being viewed?

11. An individual is found to have some tetraploid liver cells but diploid kidney cells. Be specific in explaining how this condition might arise.

12. A spermatogonial cell undergoes mitosis before entering the meiotic cell cycle en route to the production of sperm. However, during mitosis the cytoplasm fails to divide, and only one daughter cell is produced. A resultant sperm eventually fertilizes a normal ovum. What is the chromosomal complement of the embryo?

13. A teratogen is an agent that produces nongenetic abnormalities during embryonic or fetal development. Suppose a teratogen is present at conception. As a result, during the first mitotic division the centromeres fail to divide. The teratogen then loses its potency and has no further effect on the embryo. What is the chromosomal complement of this embryo?

14. As a physician, you deliver a baby with protruding heels and clenched fists with the second and fifth fingers overlapping the third and fourth fingers.
   a. What genetic disorder do you suspect the baby has?
   b. How do you confirm your suspicion?

Variations in Chromosome Number—Aneuploidy

15. Describe the process of nondisjunction and explain when it takes place during cell division.

16. A woman gives birth to monozygotic twins. One boy has a normal genotype (46,XY), but the other boy has trisomy 13 (47,13). What events—and in what sequence—led to this situation?

17. Assume that a meiotic nondisjunction event is responsible for an individual who is trisomic for chromosome 8. If two of the three copies of chromosome 8 are absolutely identical, at what point during meiosis did the nondisjunction event take place?

18. Two hypothetical human conditions have been found to have a genetic basis. Suppose a hypothetical genetic disorder responsible for condition 1 is similar to Marfan syndrome. The defect responsible for condition 2 resembles Edwards syndrome. One of the two conditions results in more severe defects, and death occurs in infancy. The other condition produces a mild phenotypic abnormality and is not lethal. Which condition is most likely lethal, and why?

19. What is the genetic basis and phenotype for each of the following disorders (use proper genetic notation)?
   a. Edwards syndrome   b. Patau syndrome
   c. Klinefelter syndrome   d. Down syndrome

20. The majority of nondisjunction events leading to Down syndrome are maternal in origin. Based on the duration of meiosis in females, speculate on the possible reasons for females contributing aneuploid gametes more frequently than males.

21. Name and describe the theory that deals with embryo-uterus interaction that explains the relationship between advanced maternal age and the increased frequency of aneuploid offspring.

22. If all the nondisjunction events leading to Turner syndrome were paternal in origin, what trisomic condition might be expected to occur at least as frequently?

Structural Alterations within Chromosomes

23. Identify the type of chromosomal aberration described in each of the following cases:
   a. Loss of a chromosome segment
   b. Extra copies of a chromosome segment
   c. Reversal in the order of a chromosome segment
   d. Movement of a chromosome segment to another, nonhomologous chromosome

24. Describe the chromosomal alterations and phenotype of cri du chat syndrome and Prader-Willi syndrome.

25. A geneticist discovers that a girl with Down syndrome has a Robertsonian translocation involving chromosomes 14 and 21. If she has an older brother who is phenotypically normal, what are the chances that he is a translocation carrier?

26. Albinism is caused by an autosomal recessive allele of a single gene. An albino child is born to phenotypically normal parents. However, the paternal grandfather is albino. Exhaustive analysis suggests that neither the mother nor her ancestors carry the allele for albinism. Suggest a mechanism to explain this situation.

Other Forms of Chromosomal Abnormalities

27. Fragile-X syndrome causes the most common form of inherited mental retardation. What is the chromosomal abnormality associated with this disorder? What is the phenotype of this disorder?

Internet Activities are critical thinking exercises using the resources of the World Wide Web to enhance the principles and issues covered in this chapter. For a full set of links and questions investigating the topics described below, log on to **http://biology.brookscole.com/cummings7**

1. *Identifying Chromosomes.* The University of Arizona's Biology Project provides a chromosome karyotyping activity. In this exercise, you have the opportunity to create part of a human karyotype. In the first part of the activity you will be arranging chromosomes onto a karyotyping sheet; once you have completed the karyotype, you will interpret the results of your efforts. Read the introductory material, and then proceed to "Patient Histories."
Further Exploration. To read more about the latest high-tech methods in karyotyping, go to The Biology Project's "New Methods for Karyotyping" web page.

2. *Exploring a Chromosomal Defect.* The chromosomal abnormality called fragile-X syndrome, discussed in this chapter, is a leading genetic cause of mental retardation.

Go to the *Your Genes, Your Health* website maintained by the Dolan DNA Learning Center at Cold Spring Harbor Laboratory, and click on the "Fragile X Syndrome" link. (If you want to find out about hemophilia or Marfan syndrome, there are links at this same site.) For this exercise, you should choose the "What causes it?" link. We'll continue to discuss various aspects of fragile-X syndrome in later chapters of this text. If you would like to investigate some of this information now, go to the fragile-X Internet Activities for Chapters 7 and 11.
Further Exploration. To find out more about general aspects of fragile-X syndrome, from current research to how to get involved with support groups, go to the *FRAXA* (Fragile X Research Foundation) website.

How would you vote now?

The most common chromosomal disorder in humans is Down syndrome, occurring in about 1 in every 800 births. The symptoms of Down syndrome are variable and cannot be accurately predicted before birth. Prenatal diagnostic testing can reveal whether a fetus has Down syndrome. More than 90% of couples learning such a diagnosis elect to terminate the pregnancy. The Fairchild family discussed in this chapter's opening story chose to continue the pregnancy of their Down syndrome child, Naia, who is now a loving child and an integral part of their family. Now that you know more about chromosomal abnormalities, risk factors, and outcomes, what do you think? Would you elect to terminate or continue a pregnancy after a diagnosis of Down syndrome? Would you consider adopting a Down syndrome child? Visit the Human Heredity Companion website to find out more on the issue, then cast your vote online at **http://biology.brookscole.com/cummings7**

For further reading and inquiry, log on to **InfoTrac College Edition,** your world-class online library, including articles from nearly 5,000 periodicals, at **http://infotrac.thomsonlearning.com**

7

Development and Sex Determination

IN the summer of 1965, Janet and Ron Reimer had twin sons, Bruce and Brian. A few months later, the boys were circumcised. For Bruce, the procedure went terribly wrong, and most of his penis was so badly burned that it could not be repaired. When he was 21 months of age, he was examined at a clinic in the United States, and his parents were advised to have reconstructive surgery and raise Bruce as a female. Bruce had the surgery, and went home with a new name, Brenda. His parents had instructions not to tell Brenda the truth, and to raise him as a girl.

Later, this case was hailed as proof that children are psychosexually neutral at birth, and that nurture has more to do with sexual roles than nature. While Bruce's case was the result of a surgical error, the apparent success of his transformation from male to female was used as a guideline in treating the 1 in 1,500 to 1 in 2,000 children born every year with genital structures that are not fully male or fully female, a condition known as ambiguous genitalia.

The treatment became focused on what was surgically possible and little attention was given to the psychological, social, or ethical consequences of these decisions. In general, males with a small or malformed penis were surgically altered into females because, in reconstructive genital surgery, it is easier to make a vagina than a penis.

In spite of the glowing reports about Bruce and his progress as Brenda, the reality was much different. As a child, Brenda refused to wear dresses, and preferred to play with boys. As a young teen, Brenda rebelled at having fur-

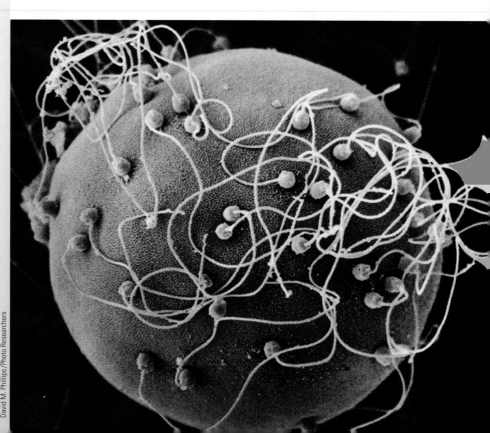

David M. Phillips/Photo Researchers

ther surgery to construct a vagina, and threatened suicide. One day, on the way home after a counseling session, Brenda's father told him the truth. Within weeks, Brenda demanded sex change surgery, and changed his name to David. After surgery to reconstruct a penis, he married and, through adoption, became the father of three children.

In an important follow-up of this case, two investigators concluded that it is wrong to assume that sexual identity is neutral at birth and that it can be shaped by the environment. This conclusion was confirmed by studies of children born as males with ambiguous genitals and surgically reassigned as females.

In this chapter, we will review the stages of human development and discuss the genetic and environmental factors that interact during prenatal sexual differentiation. We will also consider how differences in gene dosage between males and females are adjusted, and how the same gene can produce different phenotypes in males and females.

How would you vote?

The standard treatment for children born with genital abnormalities involves sex reassignment surgery, most often converting males into females. If you had a child with such a condition, would you consent to such surgery for your child, or would you allow the child to make that decision when he or she reaches puberty? Visit the Human Heredity Companion website to find out more on the issue, then cast your vote online at **http://biology.brookscole.com /cummings7**

7.1 The Human Reproductive System

We all begin as a single cell, the **zygote,** produced by the fusion of a **sperm** and an **oocyte.** The sperm (from a male) and the oocyte (produced by a female) are **gametes.** Males and females produce gametes in their **gonads,** paired organs that have associated ducts and accessory glands. The **testes** of males produce spermatozoa and sex hormones, and the **ovaries** of females produce oocytes and female sex hormones. Within the gonads, cells produced by meiosis mature into gametes, and by fertilization, gametes from two parents unite to form a zygote, from which a new individual develops.

The male reproductive system

Testes form in the abdominal cavity during male embryonic development and descend into the **scrotum,** a pouch of skin located outside the body cavity. In addition to the testes, the male reproductive system also includes (1) a duct system that trans-

■ Zygote   The fertilized egg that develops into a new individual.

■ Sperm   Male gamete.

■ Oocyte   Female gamete.

■ Gametes   Unfertilized germ cells.

■ Gonads   Organs where gametes are produced.

■ Testes   Male gonads that produce spermatozoa and sex hormones.

■ Ovaries   Female gonads that produce oocytes and female sex hormones.

■ Scrotum   A pouch of skin outside the male body that contains the testes.

■ **Seminiferous tubules**   Small, tightly coiled tubes inside the testes where sperm are produced.

■ **Spermatocytes**   Diploid cells that undergo meiosis to form haploid spermatids.

■ **Spermatogenesis**   The process of sperm production.

■ **Epididymis**   Where sperm are stored.

■ **Vas deferens**   A duct connected to the epididymis, which sperm travels through.

■ **Ejaculatory duct**   A short connector from the vas deferens to the urethra.

■ **Urethra**   A tube that passes from the bladder and opens to the outside. It functions in urine transport and, in males, also carries sperm.

■ **Seminal vesicles**   Glands that secrete fructose and prostaglandins into the sperm.

■ **Prostaglandins**   Locally acting chemical messengers that stimulate contraction of the female reproductive system to assist in sperm movement.

ports sperm out of the body, (2) three sets of glands that secrete fluids to maintain sperm viability and motility, and (3) the penis (▶ Active Figure 7.1).

The interior of the testis is divided into a series of lobes, each of which contains tightly coiled lengths of **seminiferous tubules,** where sperm are produced (▶ Active Figure 7.2). Altogether, about 250 m (850 ft.) of tubules are packed into the testes. In the tubules, cells called **spermatocytes** divide by meiosis to produce four haploid spermatids, which in turn differentiate to form mature sperm. This process of sperm production, also called **spermatogenesis,** begins at puberty and continues throughout life; each day, several hundred million sperm are in various stages of maturation. Once formed, sperm move from the seminiferous tubules to the **epididymis,** where they are stored.

Sperm move through the male reproductive system in stages. When a male is sexually aroused, sperm move from the epididymis into the **vas deferens,** a duct connected to the epididymis. The walls of the vas deferens are lined with muscles, which contract rhythmically to move sperm forward. The vas deferens from each testis joins to form a short **ejaculatory duct** that connects to the **urethra.** The urethra (which also functions in urine transport) passes through the penis and opens to the outside. In the second stage, sperm are propelled by muscular contractions accompanying orgasm from the vas deferens through the urethra and expelled from the body.

As sperm are transported through the duct system in the first stage, secretions are added from three sets of glands. The **seminal vesicles** contribute fructose, a sugar that serves as an energy source for the sperm, and **prostaglandins,** locally acting chemical messengers that stimulate contraction of the female reproductive system to

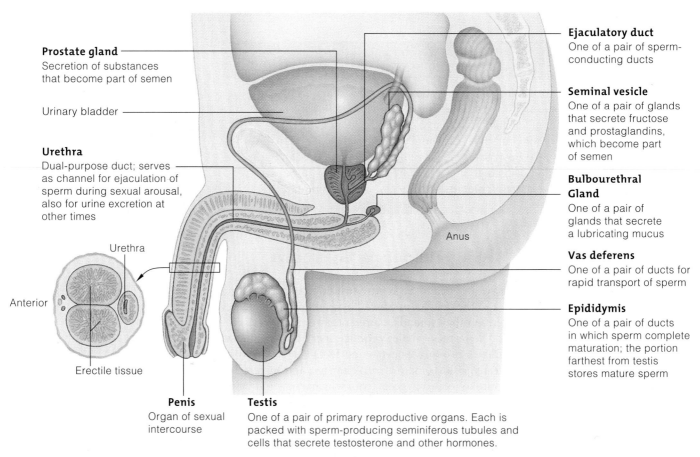

Prostate gland
Secretion of substances that become part of semen

Urinary bladder

Urethra
Dual-purpose duct; serves as channel for ejaculation of sperm during sexual arousal, also for urine excretion at other times

Urethra

Anterior

Erectile tissue

Penis
Organ of sexual intercourse

Testis
One of a pair of primary reproductive organs. Each is packed with sperm-producing seminiferous tubules and cells that secrete testosterone and other hormones.

Ejaculatory duct
One of a pair of sperm-conducting ducts

Seminal vesicle
One of a pair of glands that secrete fructose and prostaglandins, which become part of semen

Bulbourethral Gland
One of a pair of glands that secrete a lubricating mucus

Vas deferens
One of a pair of ducts for rapid transport of sperm

Epididymis
One of a pair of ducts in which sperm complete maturation; the portion farthest from testis stores mature sperm

Anus

▲ **ACTIVE FIGURE 7.1** The anatomy of the male reproductive system and the functions of its components.

HUMAN
Genetics ⑧ Now™   Learn more about the male reproductive system by viewing the animation on Human GeneticsNow at **http://now.ilrn.com/cummings7**

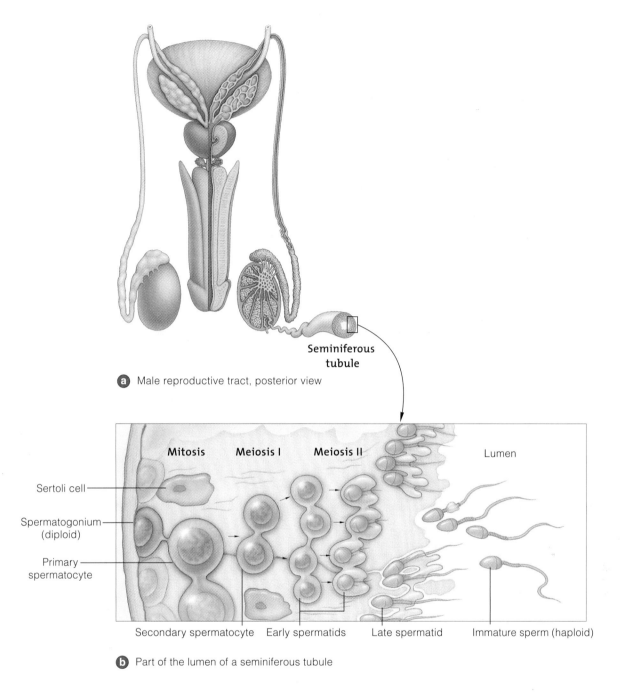

Seminiferous tubule

a Male reproductive tract, posterior view

Mitosis   Meiosis I   Meiosis II

Lumen

Sertoli cell

Spermatogonium (diploid)

Primary spermatocyte

Secondary spermatocyte   Early spermatids   Late spermatid   Immature sperm (haploid)

b Part of the lumen of a seminiferous tubule

Head (DNA in enzyme-rich cap)   Tail (with core of microtubules)

Midpiece with mitochondria

c Structure of a mature human sperm

▲ **ACTIVE FIGURE 7.2** (a) The male reproductive tract. (b) Cross section of the seminiferous tubule showing the process of sperm formation. Mitosis, meiosis, and incomplete cytokinesis produce haploid cells that differentiate into mature sperm. (c) A mature sperm and its components.

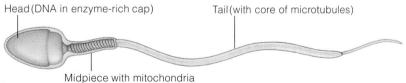 Learn more about sperm production by viewing the animation on Human GeneticsNow at
http://now.ilrn.com/cummings7

Table 7.1 The Male Reproductive System

Component	Function
Testes	Produce sperm and male sex steroids
Epididymis	Stores sperm
Vas deferens	Conducts sperm to urethra
Sex accessory glands	Produce seminal fluid that nourishes sperm
Urethra	Conducts sperm to outside
Penis	Organ of copulation
Scrotum	Provides proper temperature for testes

Prostate gland A gland that secretes a milky, alkaline fluid that neutralizes acidic vaginal secretions and enhances sperm viability.

Bulbourethral glands Glands that secrete a mucus-like substance that provides lubrication for intercourse.

Semen A mixture of sperm and various glandular secretions containing 5% spermatozoa.

Follicles A developing egg surrounded by an outer layer of follicle cells, contained in the ovary.

assist in sperm movement. The **prostate gland** secretes a milky, alkaline fluid that neutralizes acidic vaginal secretions and enhances sperm viability. The **bulbourethral glands** secrete a mucus-like substance that provides lubrication for intercourse. Together, the sperm and these various glandular secretions make up **semen,** a mixture that is about 95% secretions and about 5% spermatozoa. The components and functions of the male reproductive system are summarized in ▶ Table 7.1.

The female reproductive system

The female gonads are a pair of oval-shaped ovaries about 3 cm long, located in the abdominal cavity (▶ Active Figure 7.3). The ovary contains many **follicles,** consisting of a developing egg surrounded by an outer layer of follicle cells (▶ Active Fig-

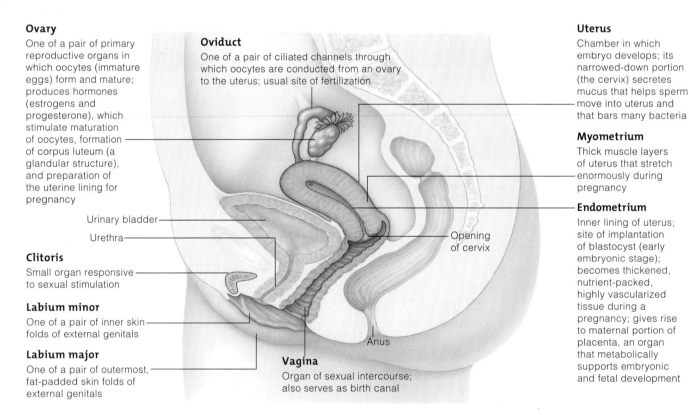

Ovary
One of a pair of primary reproductive organs in which oocytes (immature eggs) form and mature; produces hormones (estrogens and progesterone), which stimulate maturation of oocytes, formation of corpus luteum (a glandular structure), and preparation of the uterine lining for pregnancy

Oviduct
One of a pair of ciliated channels through which oocytes are conducted from an ovary to the uterus; usual site of fertilization

Urinary bladder

Urethra

Clitoris
Small organ responsive to sexual stimulation

Labium minor
One of a pair of inner skin folds of external genitals

Labium major
One of a pair of outermost, fat-padded skin folds of external genitals

Vagina
Organ of sexual intercourse; also serves as birth canal

Anus

Opening of cervix

Uterus
Chamber in which embryo develops; its narrowed-down portion (the cervix) secretes mucus that helps sperm move into uterus and that bars many bacteria

Myometrium
Thick muscle layers of uterus that stretch enormously during pregnancy

Endometrium
Inner lining of uterus; site of implantation of blastocyst (early embryonic stage); becomes thickened, nutrient-packed, highly vascularized tissue during a pregnancy; gives rise to maternal portion of placenta, an organ that metabolically supports embryonic and fetal development

▲ ACTIVE FIGURE 7.3 The anatomy of the female reproductive system and the functions of its components.

HUMAN Genetics Now™ Learn more about the female reproductive system by viewing the animation on Human GeneticsNow at **http://now.ilrn.com/cummings7**

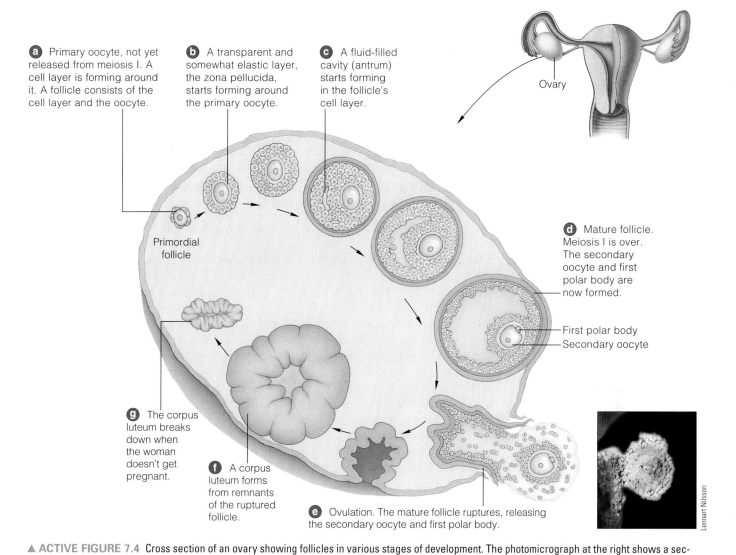

a Primary oocyte, not yet released from meiosis I. A cell layer is forming around it. A follicle consists of the cell layer and the oocyte.

b A transparent and somewhat elastic layer, the zona pellucida, starts forming around the primary oocyte.

c A fluid-filled cavity (antrum) starts forming in the follicle's cell layer.

Ovary

Primordial follicle

d Mature follicle. Meiosis I is over. The secondary oocyte and first polar body are now formed.

First polar body
Secondary oocyte

g The corpus luteum breaks down when the woman doesn't get pregnant.

f A corpus luteum forms from remnants of the ruptured follicle.

e Ovulation. The mature follicle ruptures, releasing the secondary oocyte and first polar body.

Lennart Nilsson

▲ **ACTIVE FIGURE 7.4** Cross section of an ovary showing follicles in various stages of development. The photomicrograph at the right shows a secondary oocyte being released from the surface of the ovary. This oocyte will enter the fallopian tube and move toward the uterus.

Genetics ⊛ **Now**™ Learn more about the development of oocytes by viewing the animation on Human GeneticsNow at **http://now.ilrn.com/cummings7**

ure 7.4). The developing egg is a primary oocyte and begins meiosis in the third month of female prenatal development. At birth, the female carries a lifetime supply of developing oocytes, each of which is in the prophase of the first meiotic division (▶ Active Figure 7.4). The first developing egg, called a secondary oocyte, is released from a follicle at puberty by **ovulation,** and over a female's reproductive lifetime, about 400 to 500 gametes will be produced.

The ovulated cell, called a secondary oocyte, is swept by ciliary action into the **oviduct** (also called the fallopian tube or uterine tube). At one end, the oviduct has fingerlike projections that partially surround the ovary.

The oviduct is connected to the **uterus,** a hollow, pear-shaped muscular organ about 7.5 cm (3 in.) long and 5 cm (2 in.) wide. The uterus consists of a thick, muscular outer layer called the myometrium and an inner membrane called the **endometrium.** This blood-rich inner lining is shed at menstruation if fertilization has not occurred. The lower neck of the uterus, the **cervix,** opens into the **vagina.** The vagina receives the penis during intercourse and also serves as the birth canal. The vagina opens to the outside of the body behind the urethra. The components and functions of the female reproductive system are summarized in ▶ Table 7.2.

▦ **Ovulation** The release of a secondary oocyte from the follicle; usually occurring monthly during a female's reproductive lifetime.

▦ **Oviduct** A duct with fingerlike projections partially surrounding the ovary and connecting to the uterus. Also called the fallopian or uterine tube.

▦ **Uterus** A hollow, pear-shaped muscular organ where a fertilized egg will develop.

▦ **Endometrium** The inner lining of the uterus that is shed at menstruation if fertilization has not occurred.

▦ **Cervix** The lower neck of the uterus opening into the vagina.

▦ **Vagina** The opening that receives the penis during intercourse and also serves as the birth canal.

Table 7.2	The Female Reproductive System
Component	**Function**
Ovaries	Produce ova and female sex steroids
Uterine tubes	Transport sperm to ova; transport fertilized ova to uterus
Uterus	Nourishes and protects embryo and fetus
Vagina	Site of sperm deposition, birth canal

What is the timing of meiosis and gamete formation in males and females?

The entire process of spermatogenesis takes about 48 days: 16 for meiosis I, 16 for meiosis II, and 16 to convert the spermatid into the mature sperm. Each of the four products of meiosis form sperm. The tubules within the testis contain many spermatocytes, and large numbers of sperm are always in production. A single ejaculate may contain 200 to 400 million sperm, and over a lifetime a male produces billions of sperm.

In females, cleavage of the cytoplasm in meiosis I does not produce cells of equal size. One cell, destined to become the oocyte, receives about 95% of the cytoplasm and is known as the secondary oocyte (see Spotlight on The Largest Cell). In the second meiotic division, the same disproportionate cleavage results in one cell retaining most of the cytoplasm. The large cell becomes the functional gamete, and the nonfunctional, smaller cells are known as polar bodies. Thus, in females, only one of the four cells produced by meiosis becomes a gamete. All oocytes contain 22 autosomes and an X chromosome.

The timing of meiosis and gamete formation in females is different than in males (▶ Table 7.3). In females this process is called **oogenesis**. In oogenesis, mitotically active cells in the ovaries, called **oogonia**, produce primary oocytes. These cells undergo meiosis I during embryonic development and then stop. They remain in meiosis I until the female undergoes puberty. At puberty, usually one oocyte per menstrual cycle completes the first meiotic division, is released from the ovary, and moves down the oviduct. If the egg is fertilized, it quickly completes meiosis II, producing a diploid zygote. Unfertilized eggs are sloughed off during menstruation, along with uterine

■ **Oogenesis**   The process of oocyte production.

■ **Oogonia**   Mitotically active cells that produce primary oocytes.

Table 7.3	A Comparison of the Duration of Meiosis in Males and Females			
Spermatogenesis		**Oogenesis**		
Begins at Puberty		**Begins During Embryogenesis**		
Spermatogonium ↓	}	Oogonium ↓	}	Forms at 2 to 3 months after conception
Primary spermatocyte ↓	} 16 days	Primary oocyte ↓	}	Forms at 2 to 3 months of gestation. Remains in meiosis I until ovulation, 12 to 50 years after formation.
Secondary spermatocyte ↓	} 16 days	Secondary oocyte ↓	}	
Spermatid ↓	} 16 days	Ootid	}	Less than 1 day, when fertilization occurs
Mature sperm Total time	48 days	Mature egg-zygote Total time		12 to 50 years

tissue. Each month until menopause, another oocyte completes meiosis I and is released from the ovary. Altogether, a female releases about 450 oocytes during the reproductive phase of her life.

In females, then, meiosis takes years to complete. It begins with prophase I, while she is still an embryo, and continues to the completion of meiosis II following fertilization. Depending on the time of ovulation, meiosis can take from 12 to 50 years in human females.

Keep in mind

■ There are important differences in the timing and duration of meiosis and gamete formation between males and females.

■ **Fertilization** The fusion of two gametes to produce a zygote.

7.2 A Survey of Human Development from Fertilization to Birth

Fertilization, the fusion of male and female gametes, usually occurs in the upper third of the oviduct (▷ Active Figure 7.5). Sperm deposited in the vagina swim through the cervix, up the uterus, and into the oviduct. About 30 minutes after ejaculation, sperm are present in the oviduct. Sperm travel this distance (about 7 inches) by swimming, using whip-like contractions of their tails, and are assisted by muscular contractions of the uterus.

Usually only one sperm fertilizes the egg, but many other sperm assist (▷ Active Figure 7.5) by helping to trigger chemical changes in the egg. During fertilization, a

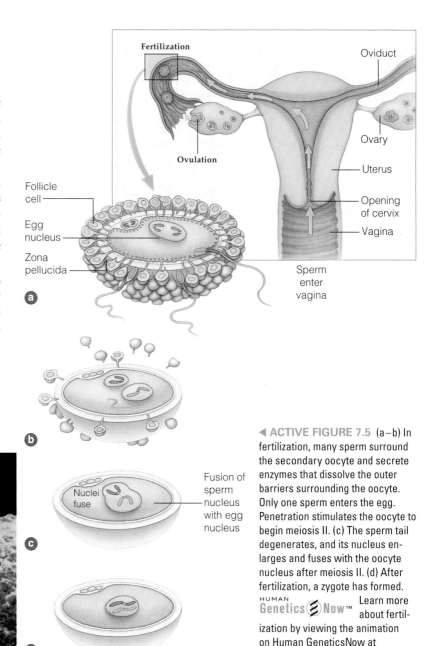

◀ **ACTIVE FIGURE 7.5** (a–b) In fertilization, many sperm surround the secondary oocyte and secrete enzymes that dissolve the outer barriers surrounding the oocyte. Only one sperm enters the egg. Penetration stimulates the oocyte to begin meiosis II. (c) The sperm tail degenerates, and its nucleus enlarges and fuses with the oocyte nucleus after meiosis II. (d) After fertilization, a zygote has formed.

HUMAN Genetics ℰ Now™ Learn more about fertilization by viewing the animation on Human GeneticsNow at **http://now.ilrn.com/cummings7**

sperm binds to receptors on the surface of the egg (technically, a secondary oocyte) and fuses with the cell's outer membrane. This fusion triggers a series of chemical changes in the membrane and prevents any other sperm from entering the oocyte. As a sperm enters the cytoplasm, its presence reinitiates meiosis in the egg, and the second meiotic division is completed. After meiosis, the haploid sperm nucleus fuses with the haploid oocyte nucleus, forming a diploid zygote.

The zygote is swept along by cilia lining the walls of the oviduct, and travels down the oviduct to the uterus over the next 3 to 4 days (▶ Active Figure 7.6). While in the oviduct, the zygote begins mitosis and becomes an embryo. The embryo, consisting of a small number of cells, descends into the uterus and floats unattached in the uterine interior for several days, drawing nutrients from the uterine fluids. Cell division continues during this time and the embryo enters a new stage of development; it is now called a **blastocyst** (▶ Active Figure 7.6).

A blastocyst, made up of about 100 cells, has several parts: the **inner cell mass,** a cyst-like internal cavity, and an outer layer of cells (the **trophoblast**). While the embryo is growing to form the blastocyst, the cells lining the uterus (called the endometrium) enlarge and differentiate, preparing for attachment of the embryo. During the weeklong process of implantation, the embryo's trophoblast sticks to the endometrium and releases enzymes that dissolve endometrial cells, allowing fingerlike growths from trophoblasts to lock the embryo into place (▶ Active Figure 7.6).

By about 12 days after fertilization, the embryo is firmly embedded and the trophoblast has formed a two-layered structure, called the **chorion.** Once formed, the chorion makes and releases a hormone called human chorionic gonadotropin (hCG). This hormone prevents breakdown of the uterine lining and stimulates endometrial cells to release hormones that help maintain the pregnancy. Excess hCG is eliminated in the urine. Home pregnancy tests work by detecting elevated hCG levels as early as the first day of a missed menstrual period.

As the chorion grows and expands, it forms a series of fingerlike projections called villi that extend into endometrial cavities filled with maternal blood. Capillaries from the embryo's developing circulatory system extend into the villi. The blood of the embryo and the maternal pools of blood are separated from each other only by a thin layer of cells. Food molecules and oxygen cross easily from the mother's blood into the embryo, and waste molecules and carbon dioxide move from the embryo into the mother's blood. The villi eventually form the placenta, a disc-shaped structure that will nourish the embryo throughout prenatal development. Membranes connecting the embryo to the placenta form the umbilical cord, which contains two umbilical arteries and a single umbilical vein as extensions of the embryo's circulatory system.

Development is divided into three trimesters.

Development in the period between fertilization and birth is divided into three trimesters, each of which lasts about 12 to 13 weeks. During the 36 to 39 weeks of development, the single-celled zygote undergoes 40 to 44 rounds of mitosis, producing trillions of cells that become organized into the tissues and organs of the fully developed fetus.

Organ Formation Occurs in the First Trimester. The first trimester is a period of radical change in the size, shape, and complexity of the embryo (▶ Figure 7.7). In the week after implantation, three basic tissue layers are formed, and by the end of the third week, organ systems are beginning to take shape. By 4 weeks, the embryo is about 5 mm long (about one-fifth of an inch), and much of the body is composed of paired segments.

■ **Blastocyst** The developmental stage at which the embryo implants into the uterine wall.

■ **Inner cell mass** A cluster of cells in the blastocyst that gives rise to the embryonic body.

■ **Trophoblast** The outer layer of cells in the blastocyst that gives rise to the membranes surrounding the embryo.

■ **Chorion** A two-layered structure formed from the trophoblast.

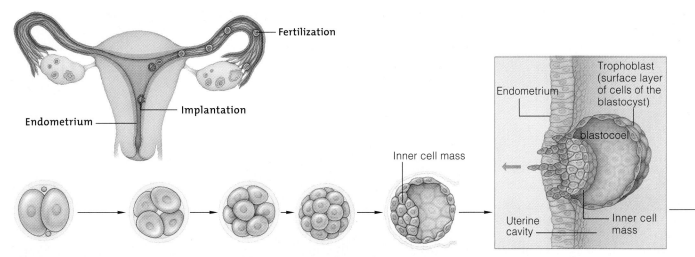

a **DAYS 1–2.** The first cleavage furrow extends between the two polar bodies. Later cuts are angled, so cells become asymmetrically arranged. Until the eight-cell stage forms, they are loosely organized, with space between them.

b **DAY 3.** After the third cleavage, cells abruptly huddle into a compacted ball, which tight junctions among the outer cells stabilize. Gap junctions formed along the interior cells enhance intercellular communication.

c **DAY 4.** By 96 hours there is a ball of sixteen to thirty-two cells shaped like a mulberry. It is a morula (after *morum*, Latin for mulberry). Cells of the surface layer will function in implantation and will function in implantation and will give rise to a membrane, the chorion.

d **DAY 5.** A blastocoel (fluid-filled cavity) forms in the morula as a result of surface cell secretions. By the thirty-two-cell stage, differentiation is occurring in an inner cell mass that will give rise to the embryo proper. This embryonic stage is the blastocyst.

e **DAYS 6–7.** Some of the blastocyst's surface cells attach themselves to the endometrium and start to burrow into it. Implantation has started.

Actual size

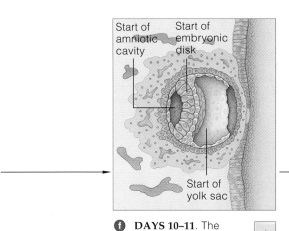

f **DAYS 10–11.** The yolk sac, embryonic disk, and amniotic cavity have started to form from parts of the blastocyst.

Actual size

g **DAY 12.** Blood-filled spaces form in maternal tissue. The chorionic cavity starts to form.

Actual size

h **DAY 14.** A connecting stalk has formed between the embryonic disk and chorion. Chorionic villi, which will be features of a placenta, start to form.

Actual size

▲ **ACTIVE FIGURE 7.6** From fertilization through implantation. A blastocyst forms and its inner cell mass gives rise to a disk-shaped early embryo. As the blastocyst implants into the uterus, cords of chorionic cells start to form. When implantation is complete, the blastocyst is buried in the endometrium.

HUMAN
Genetics Now™   Learn more about early development and implantation by viewing the animation on Human GeneticsNow at
http://now.ilrn.com/cummings7

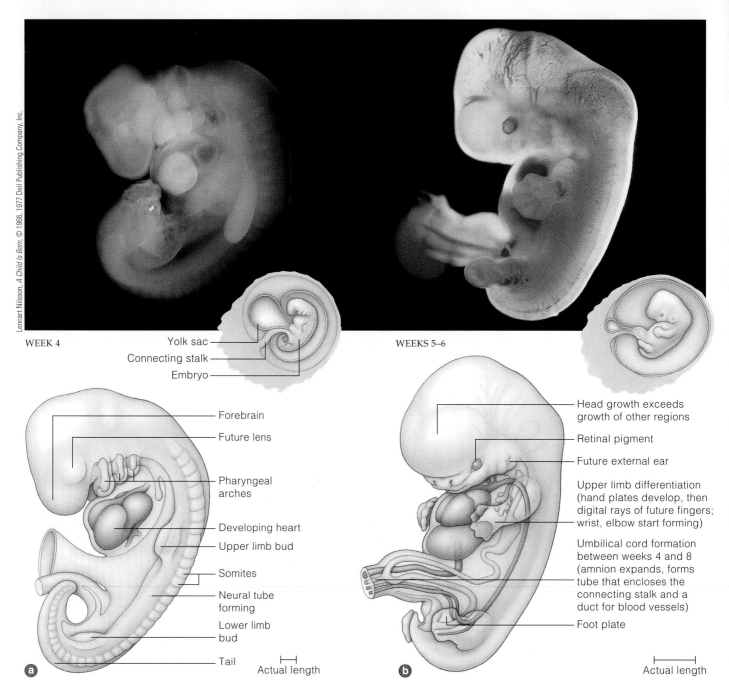

WEEK 4

Yolk sac
Connecting stalk
Embryo

WEEKS 5–6

Forebrain

Future lens

Pharyngeal arches

Developing heart

Upper limb bud

Somites

Neural tube forming

Lower limb bud

Tail

Actual length

Head growth exceeds growth of other regions

Retinal pigment

Future external ear

Upper limb differentiation (hand plates develop, then digital rays of future fingers; wrist, elbow start forming)

Umbilical cord formation between weeks 4 and 8 (amnion expands, forms tube that encloses the connecting stalk and a duct for blood vessels)

Foot plate

Actual length

(a)

(b)

▲ FIGURE 7.7 Stages of human development. (a) Human embryo 4 weeks after fertilization. (b) Embryo at 4 to 6 weeks of development. (c) Embryo at the transition to the fetal stage of development. (d) Fetus at 16 weeks of development.

During the second month, the embryo grows dramatically to a length of about 3 cm (about 1½ in.) and undergoes a 500-fold increase in size. Most of the major organ systems, including the heart, are formed. Limb buds develop into arms and legs, complete with fingers and toes. The head is very large in relation to the rest of the body because of the rapid development of the nervous system.

By about 7 weeks, the embryo is called a fetus. Although chromosomal sex (XX in females and XY in males) is determined at the time of fertilization, the fetus is neither male nor female at the beginning of the third month. During the third month, specific gene sets are activated, and sexual development is initiated. External sex organs can be seen in ultrasound scans between the twelfth and fifteenth week. Sex differentiation is discussed later in this chapter.

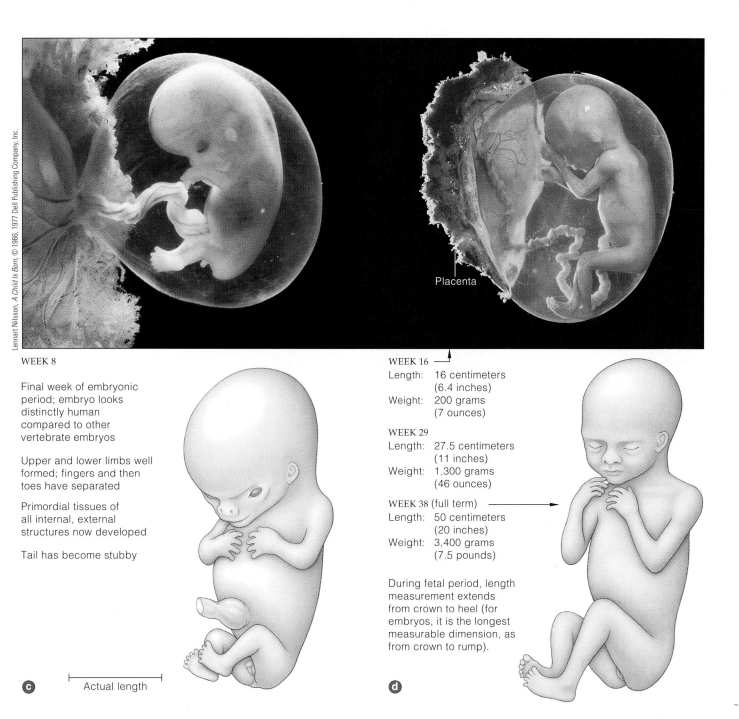

Lennart Nilsson, *A Child Is Born*, © 1966, 1977 Dell Publishing Company, Inc.

Placenta

WEEK 8

Final week of embryonic period; embryo looks distinctly human compared to other vertebrate embryos

Upper and lower limbs well formed; fingers and then toes have separated

Primordial tissues of all internal, external structures now developed

Tail has become stubby

c |— Actual length —|

WEEK 16
Length: 16 centimeters
        (6.4 inches)
Weight: 200 grams
        (7 ounces)

WEEK 29
Length: 27.5 centimeters
        (11 inches)
Weight: 1,300 grams
        (46 ounces)

WEEK 38 (full term)
Length: 50 centimeters
        (20 inches)
Weight: 3,400 grams
        (7.5 pounds)

During fetal period, length measurement extends from crown to heel (for embryos, it is the longest measurable dimension, as from crown to rump).

d

By the end of the first trimester, the fetus is about 9 cm (about 3½ in.) long and weighs about 15 g (about half an ounce). All the major organ systems have formed and are functional.

Keep in mind

■ Most of the important events in human development occur in the first trimester. The remaining months are mainly a period of growth.

The Second Trimester Is a Period of Organ Maturation.   In the second trimester, major changes include an increase in size and the further development of organ systems. Bony parts of the skeleton begin to form, and the heartbeat can be heard with

Table 7.4 Human Teratogens

Known

Radiation
  Fallout
  x-rays

Infectious Agents
  Cytomegalovirus
  Herpes virus II
  Rubella virus
  Toxoplasma gondii (spread in cat feces)

Maternal Metabolic Problems
  Phenylketonuria (PKU)
  Diabetes
  Virilizing tumors

Drugs and Chemicals
  Alcohol
  Aminopterins
  Chlorobiphenyls
  Coumarin anticoagulants
  Diethylstilbestrol
  Tetracyclines
  Thalidomide

Possible

Cigarette smoking
High levels of vitamin A
Lithium
Zinc deficiency

■ **Teratogen** Any physical or chemical agent that brings about an increase in congenital malformations.

a stethoscope. Fetal movements begin in the third month, and by the fourth month, the mother can feel movements of the fetus's arms and legs. At the end of the second trimester, the fetus weighs about 700 g (27 oz.) and is 30 to 40 cm (about 13 in.) long. It has a well-formed face, its eyes can open, and it has fingernails and toenails.

Rapid Growth Takes Place in the Third Trimester.   The fetus grows rapidly in the third trimester, and the circulatory system and the respiratory system mature to prepare for air breathing. During this period of rapid growth, maternal nutrition is important because most of the protein the mother eats will be used for growth and development of the fetal brain and nervous system. Similarly, much of the calcium in the mother's diet is used to develop the fetal skeletal system.

The fetus doubles in size during the last 2 months, and chances for survival outside the uterus increase rapidly during this time. In the last month, antibodies pass from the mother to the fetus, conferring temporary immunity on the fetus. In the first months after birth, the baby's immune system matures, and as it begins to make its own antibodies, the maternal antibodies disappear. At the end of the third trimester, the fetus is about 50 cm (19 in.) long and weighs from 2.5 to 4.8 kg (5.5 to 10.5 lb.).

Birth is hormonally induced.

Birth is a hormonally induced process. During the last trimester, the cervix softens and the fetus shifts downward, usually with its head pressed against the cervix. Mild uterine contractions start during the third trimester, but at the start of the birth process, they become more frequent and intense. Release of the hormone oxytocin from the pituitary gland helps stimulate uterine contractions. During labor, the cervical opening dilates in stages to allow passage of the fetus, and uterine contractions expel the fetus. The head usually emerges first. If any other body part enters the birth canal first, the result is a breech birth. A short time after delivery, a second round of uterine contractions begins delivery of the placenta. These contractions separate the placenta from the lining of the uterus, and the placenta is expelled through the vagina.

7.3 Teratogens Are a Risk to the Developing Fetus

Although about 97% of all babies are normal at birth, birth defects can be produced by genetic disorders or exposure to environmental agents (▷ Active Figure 7.8). Most birth defects are caused by disruptions of embryonic development, but the brain and nervous system can be damaged at any time during development, leading to mild to serious conditions such as learning disabilities or mental retardation.

Chemicals and other agents that produce embryonic and/or fetal abnormalities are called **teratogens.** Defects produced by teratogens are nongenetic and are not passed on to following generations. In 1960 only four or five agents were known to be teratogens. The discovery that thalidomide, a tranquilizer prescribed to help stop morning sickness, caused limb defects in unborn children helped focus attention on environmental factors that produce birth defects. Today, we know that 30 to 40 agents are teratogens, and another 10 to 12 chemicals are strongly suspected of causing birth defects. Some of these are listed in ▷ Table 7.4.

Radiation, viruses, and chemicals can be teratogens.

Radiation, especially medical x-rays, can be teratogens. Women of childbearing age should not have abdominal x-rays unless they know they are not pregnant. Pregnant

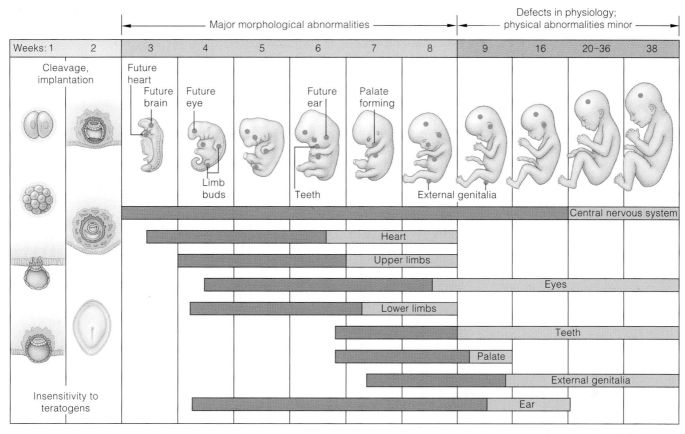

Major morphological abnormalities

Defects in physiology; physical abnormalities minor

Cleavage, implantation

Future heart

Future brain

Future eye

Future ear

Palate forming

Limb buds

Teeth

External genitalia

Central nervous system

Heart

Upper limbs

Eyes

Lower limbs

Teeth

Palate

External genitalia

Ear

Insensitivity to teratogens

▲ **ACTIVE FIGURE 7.8** Teratogens are chemical and physical agents that can produce deformities in the embryo and the fetus. The effect of most teratogens begins after 3 weeks of development. Dark blue represents periods of high sensitivity; light blue shows periods of development with less sensitivity to teratogens.

Genetics Now™ Learn more about the action of teratogens by viewing the animation on Human GeneticsNow at **http://now.ilrn.com/cummings7**

women should avoid all unnecessary x-rays, and all females should have abdominal shielding for most x-ray procedures.

Some viruses are teratogens. These include HIV, the measles virus, the German measles virus (rubella), and the virus that causes genital herpes. Fetuses infected with HIV are at risk for being stillborn or born prematurely and with low birth weight. The other viruses can cause severe brain damage and mental retardation in a developing fetus. Some infectious organisms, such as *Toxoplasma gondii*, which is transmitted to humans by cats, are teratogenic and can result in a stillborn child or a child with mental retardation or other disorders.

Many chemicals, including medications, such as the antibiotic tetracycline, are teratogens. Case 1 at the end of this chapter discusses drugs with teratogenic effects.

Fetal alcohol syndrome is a preventable tragedy.

A fetus's exposure to alcohol is one of the most serious and the most widespread teratogenic problems; it is also the leading preventable cause of birth defects. Alcohol consumption during pregnancy can result in spontaneous abortion, growth retardation, facial abnormalities (▶ Figure 7.9), mental retardation, and learning disabilities. This collection of defects is known as **fetal alcohol syndrome (FAS)**. In milder forms, the condition is known as fetal alcohol effects. The incidence of FAS is about 1.9 affected infants per 1,000 births, and the incidence for fetal alcohol effects is about 3.5 affected infants per 1,000 births.

The teratogenic effects of alcohol can occur at any time during pregnancy, but

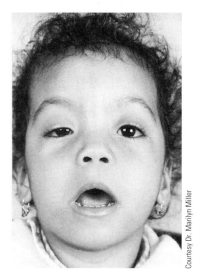

▲ **FIGURE 7.9** A child with fetal alcohol syndrome. The misshapen eyes, flat nose, and facial features are hallmarks of this condition.

■ **Fetal alcohol syndrome (FAS)** A constellation of birth defects caused by maternal alcohol consumption during pregnancy.

weeks 8 to 12 are particularly sensitive periods. Even in the third trimester, alcohol can seriously impair fetal growth. Most studies show that the consumption of one or more drinks per day is associated with an increased risk of having a child with growth retardation. However, because fetal damage is related to blood alcohol levels, thinking about averages can be misleading. Having six drinks in one day and no drinks the rest of the week may pose a greater risk to the fetus than having one drink each day of the week. To emphasize the risks, the U.S. Surgeon General requires that all alcohol containers carry this warning: "Drinking during pregnancy may cause mental retardation and other birth defects. Avoid alcohol during pregnancy." The American Academy of Pediatrics issues this policy statement: "Because there is no known safe amount of alcohol consumption during pregnancy, the Academy recommends abstinence from alcohol for women who are pregnant or who are planning a pregnancy."

The economic cost of FAS is enormous. The lifetime cost of caring for a child with FAS exceeds $1.4 million, and annual estimates for the overall costs to society range into billions of dollars. The mental retardation associated with FAS is estimated to account for 11% of the cost for treating all institutionalized, mentally retarded individuals. The emotional costs and social effects are difficult to estimate. Insight into the struggles of a family with an FAS child is recorded by Michael Dorris in his book, *The Broken Cord*.

Although the actions of alcohol as a teratogen are now well known, work is needed to resolve the degree of risk involved with other chemicals and substances that are suspected teratogens and to identify new teratogens among the thousands of chemicals currently used. More importantly, research is needed to investigate the genetic basis for susceptibility to teratogenic agents and to develop tests to identify those who are susceptible to teratogens.

7.4 How Is Sex Determined?

In humans, as in many other species, we can see obvious differences between the sexes, a condition known as sexual dimorphism. In humans, secondary sex characteristics such as body size, muscle mass, patterns of fat distribution, and amounts and distribution of body hair emphasize the differences between the sexes. These differences are the outcome of a long chain of events that begin early in embryonic development and involve a network of interactions between gene expression and the environment.

Chromosomes can help determine sex.

In humans whether someone is male or female is determined in stages beginning at fertilization, when the sex chromosomes carried by the gametes combine in the zygote. As discussed in Chapter 2, females have two X chromosomes (XX) and males have an X and a Y chromosome.

Although saying that females are XX and that males are XY seems straightforward, it does not provide all the answers to the question of what determines maleness and femaleness. Is a male a male because he has a Y chromosome or because he does *not* have two X chromosomes? Can someone be XY and develop as a female? Can someone be XX and develop as a male? These questions are still not completely resolved, but we began to discover answers about 40 years ago when individuals with only 45 chromosomes (45,X) were identified. Those with only one X chromosome are female. At about the same time, males who carry two X chromosomes along with a Y chromosome were discovered (47,XXY). From the study of people with abnormal numbers of sex chromosomes, it is clear that some females have only one X chromosome and that some males can have more than one X chromosome. Furthermore, anyone who has a Y chromosome is almost always male, no matter how many X chromosomes he may have. However, having an XX or XY chromosome set does not always mean someone is male or female. The outcome depends on interactions between genes on the X and Y chromosomes with many different factors.

Keep in mind

■ Chromosomal sex is determined at fertilization. Sexual differentiation begins in the seventh week and is influenced by a combination of genetic and environmental factors.

The sex ratio in humans changes with stages of life.

All gametes produced by human females carry an X chromosome, while males produce roughly equal numbers of gametes carrying an X chromosome and gametes carrying a Y chromosome. Because the male makes two different kinds of gametes, he is referred to as the heterogametic sex. The female is homogametic because she makes only one type of gamete. An egg fertilized by an X-bearing sperm results in an XX zygote that will develop as a female. Fertilization by a Y-bearing sperm will produce an XY, or male zygote (▶ Active Figure 7.10).

Because males produce approximately equal numbers of X- and Y-bearing sperm, males and females should be produced in equal proportions (▶ Active Figure 7.10). This proportion, known as the **sex ratio,** changes throughout the life cycle. At fertilization, the sex ratio (known as the primary sex ratio) should be 1:1. While direct determinations are impossible, estimates indicate that more males are conceived than females. The sex ratio at birth, known as the secondary sex ratio, is about 1.05 (105 males for every 100 females). The tertiary sex ratio is measured in adults. Between the ages of 20 and 25, the ratio is close to 1:1. After that, females outnumber males in ever-increasing proportions. Genetic and environmental factors are responsible for the higher death rate among males. The expression of deleterious X-linked recessive genes is one cause of male death in both prenatal and postnatal stages of life. Between the ages of 15 and 35, accidents are the leading cause of death in males.

■ **Sex ratio** The proportion of males to females, which changes throughout the life cycle. The ratio is close to 1:1 at fertilization but the female to male ratio increases as a population ages.

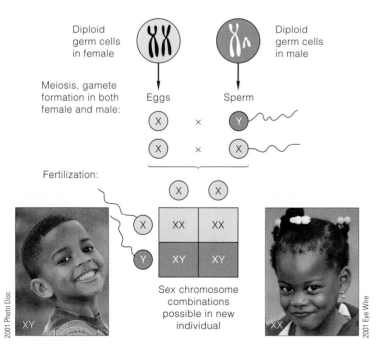

◀ ACTIVE FIGURE 7.10 The segregation of sex chromosomes and the random combination of X- or Y-bearing sperm with an X-bearing egg produces, on average, a 1:1 ratio of males to females.

HUMAN Genetics ⒮ Now™ Learn more about sex determination in humans by viewing the animation on Human GeneticsNow at **http://now.ilrn.com/cummings7**

▲ FIGURE 7.11 A cascade of gene action that begins in the seventh week of gestation results in the development of the male and female sexual phenotype.

7.5 Defining Sex in Stages: Chromosomes, Gonads, and Hormones

The XX-XY method of sex determination provides a genetic framework for developmental events that guide the embryo toward the male or female phenotypes (▶ Figure 7.11). The formation of male or female reproductive structures depends on several factors, including gene action, interactions within the embryo, interaction with other embryos that may be in the uterus, and interactions with the maternal environment. As a result of these interactions, the chromosomal sex (XX or XY) of an individual may differ from the phenotypic sex. These differences arise during embryonic and fetal development and can produce a phenotype opposite to the chromosomal sex, intermediate to the phenotypes of the two sexes, or a phenotype that has characteristics and genitalia of both sexes. The sex of an individual can be defined at several levels: chromosomal sex, gonadal sex, and phenotypic sex. In most cases, all of these definitions are consistent, but in others they are not (see Genetic Journeys: Sex Testing in the Olympics—Biology and a Bad Idea on page 170). To understand these variations and the interactions of genes with the environment, let's first consider what happens during normal sexual differentiation.

Sex differentiation begins in the embryo.

The first step in sex differentiation occurs at fertilization with the formation of a diploid zygote with an XX or XY chromosome pair. Although the chromosomal sex of the zygote is established at fertilization, the external genitalia of early embryos are neither male nor female for the first 7 or 8 weeks. During this time, two undifferentiated gonads are present, and both male and female reproductive duct systems develop. The two internal duct systems are the Wolffian (male) and Müllerian (female) ducts (▶ Figure 7.12a). At about 7 weeks, developmental pathways activate different sets of genes and cause the undifferentiated gonads to develop as testes or ovaries, establishing the gonadal sex of the embryo. This process takes place over the next 4 to 6 weeks. Although it is convenient to think of only two pathways, one leading to males and the other to females, there are many alternative pathways that produce in-

Umbilical cord (lifeline between the embryo and the mother's tissues)

Amnion (a protective, fluid-filled sac surrounding and cushioning the embryo)

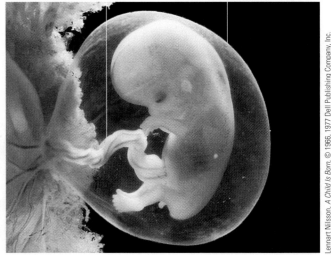

▶ FIGURE 7.12 (a) A human embryo at eight weeks, about the time sex differentiation begins. (b) Two duct systems (Wolffian and Müllerian) are present in the early embryo. They enter different developmental pathways in the presence and absence of a Y chromosome. (c) Steps in the development of phenotypic sex from an undifferentiated stage to the male or female phenotype. The male pathway of development takes place in response to the presence of a Y chromosome and production of the hormones testosterone and dihydrotestosterone (DHT). Female development takes place in the absence of a Y chromosome and without these hormones.

(a)

termediate outcomes in gonadal sex and in sexual phenotypes, some of which we consider in the following paragraphs.

If a Y chromosome is present, expression of genes on the Y chromosome causes the indifferent gonad to develop as a testis. A gene called the sex-determining region of the Y (*SRY*; OMIM 480000), located on the short arm of the Y chromosome, activates expression of other genes and plays a major role in testis development. Other genes on the Y chromosome and on autosomes also play important roles at this time.

Once testis development is initiated, cells in the testis secrete two hormones, **testosterone** and **Müllerian inhibiting hormone (MIH)**. Testosterone stimulates the Wolffian ducts to form the male internal duct system that will carry sperm. These ducts include the epididymis, seminal vesicles, and vas deferens. The MIH secreted by the developing testis stops further development of female duct structures and causes the Müllerian ducts to degenerate (▶ Figure 7.12b).

■ *SRY* A gene, called the sex-determining region of the Y, located near the end of the short arm of the Y chromosome, plays a major role in causing the undifferentiated gonad to develop into a testis.

■ **Testosterone** A steroid hormone produced by the testis; the male sex hormone.

■ **Müllerian inhibiting hormone (MIH)** A hormone produced by the developing testis that causes the breakdown of the Müllerian ducts in the embryo.

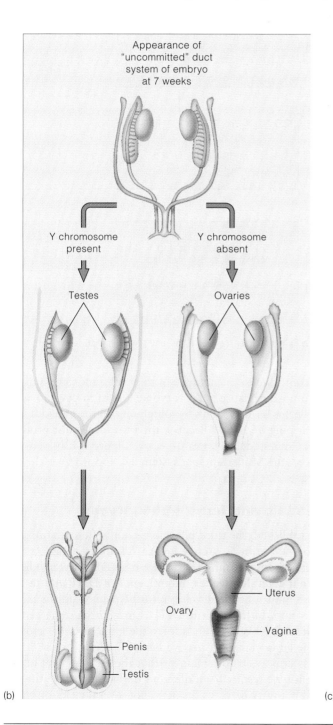

(b)

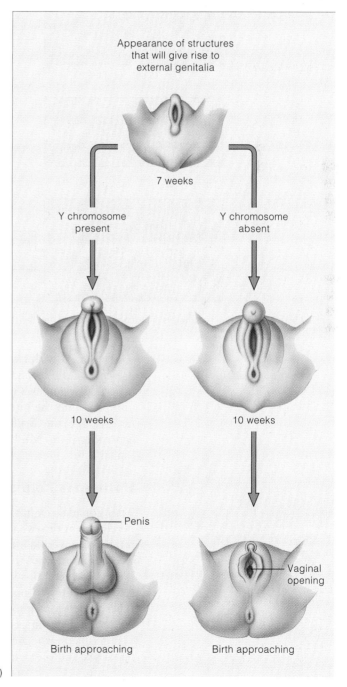

(c)

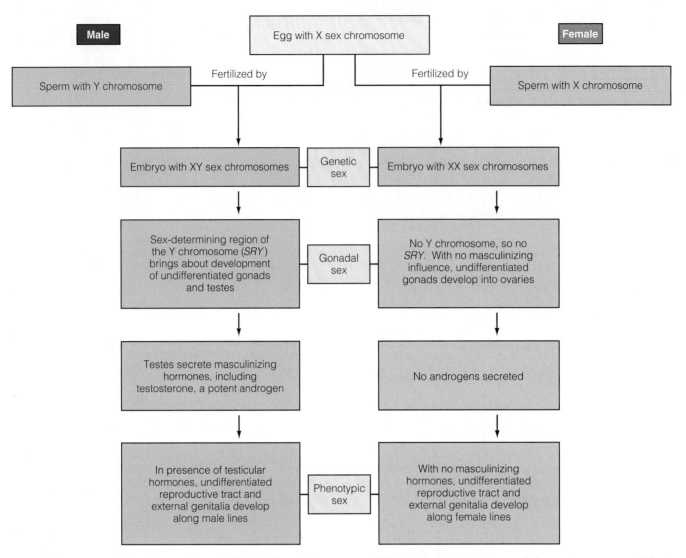

Male

Female

Egg with X sex chromosome

Sperm with Y chromosome

Fertilized by

Fertilized by

Sperm with X chromosome

Embryo with XY sex chromosomes

Genetic sex

Embryo with XX sex chromosomes

Sex-determining region of the Y chromosome (*SRY*) brings about development of undifferentiated gonads and testes

Gonadal sex

No Y chromosome, so no *SRY*. With no masculinizing influence, undifferentiated gonads develop into ovaries

Testes secrete masculinizing hormones, including testosterone, a potent androgen

No androgens secreted

In presence of testicular hormones, undifferentiated reproductive tract and external genitalia develop along male lines

Phenotypic sex

With no masculinizing hormones, undifferentiated reproductive tract and external genitalia develop along female lines

▲ **FIGURE 7.13** The major pathways of sexual differentiation and the stages at which genetic sex, gonadal sex, and phenotypic sex are established.

In embryos with two X chromosomes, the absence of the Y chromosome and the presence of the second X chromosome cause the embryonic gonad to develop as an ovary. Ovarian development begins as cells along the outer edge of the gonad divide and push into the interior, forming an ovary. Because the ovary does not produce testosterone, the Wolffian duct system degenerates (▶ Active Figure 7.12b), and because there is no MIH produced, the Müllerian duct system develops to form the fallopian tubes, uterus, and parts of the vagina.

Hormones help shape male and female phenotypes.

After gonadal sex has been established, the third phase of sexual differentiation, the development of sexual phenotype begins (▶ Active Figure 7.12c). In males, testosterone is converted into another hormone, dihydrotestosterone (DHT), which helps direct formation of the external genitalia. Under the influence of DHT and testosterone, the genital folds and genital tubercle develop into the penis, and the labio-scrotal swelling forms the scrotum. In females, no DHT is present, and the genital tubercle develops into the clitoris, the genital folds form the labia minora, and the labioscrotal swellings form the labia majora (▶ Active Figure 7.12c).

In terms of gene action, it is important to remember that the development of gonadal sex and the sexual phenotype results from different developmental pathways (▶ Figure 7.13). In males, this pathway involves the action of several genes on the Y chromosome, the presence of at least one X chromosome, and expression of several autosomal genes. In females, this pathway involves the presence of two X chromosomes, the absence of Y chromosome genes, and expression of a female-specific set of autosomal genes. These distinctions indicate that there may be important differences in the way genes in these pathways are activated, and they may provide clues in the search for genes that regulate these pathways.

7.6 Mutations Can Uncouple Chromosomal Sex from Phenotypic Sex

Developmental pathways that begin with the indifferent gonad often result in a gonadal and/or sexual phenotype that differs from the XX or XY chromosomal sex. These outcomes can result from several causes: chromosomal events that exchange segments of the X and Y chromosomes, mutations that affect the ability of cells to respond to the products of Y chromosome genes, or action of autosomal genes that control events on the X and/or Y chromosome.

Hermaphrodites have male and female gonads.

True hermaphrodites have both ovaries and testes and their associated duct systems. In some species, such as the earthworm, this condition is normal. In humans, a true hermaphrodite has ovarian and testicular tissue in separate gonads or in a single combined gonad. Most hermaphrodites are sex chromosome mosaics: some cells in the body are XX, and others are XY or XXY. In other cases, only XY cells are found. These individuals have mixed masculine and feminine genitals; the degree of maleness or femaleness depends on whether testicular or ovarian tissue predominates in gonads.

Androgen insensitivity can affect the sex phenotype.

The pattern of gene expression that leads from chromosomal sex to phenotypic sex can be disrupted at several stages. A mutation in an X-linked gene called the androgen receptor (*AR*) causes XY males to become phenotypic females (▶ Figure 7.14). This disorder is called **androgen insensitivity** (OMIM 313700).

In affected males, testis formation is normal and testosterone and MIH production begin as expected. MIH causes degeneration of the Müllerian duct system and no internal female reproductive tract is formed. However, because of the mutation, no testosterone receptors are produced, and cells cannot respond to testosterone or DHT. As a result, development proceeds as if there were no testosterone or DHT present. The Wolffian duct system degenerates and the genitalia develop as female structures. Individuals with this condition are chromosomal males but phenotypic females who do not menstruate and have well-developed breasts and very little pubic hair (see Genetics in Society: Joan of Arc—Was It Really John of Arc?).

Sex phenotypes can change at puberty.

Mutations in several different genes can produce a condition called pseudohermaphroditism. Affected individuals have both male and female structures, but at different times in their lives. At one stage, phenotypic sex does not match chromosomal sex, but later, the phenotypic sex changes. One autosomal form of **pseudohermaphroditism** (OMIM 264300) prevents conversion of testosterone to DHT. In this disorder, the Y chromosome initiates development of testes, and the Wolffian ducts form the male duct system. MIH secretion prevents the development of female ducts. However, the failure to produce DHT results in genitalia that are essentially female. The scrotum resembles the labia, a blind vaginal pouch is present, and the penis resembles a clitoris. Although chromosomally male, these individuals are identified and raised as females.

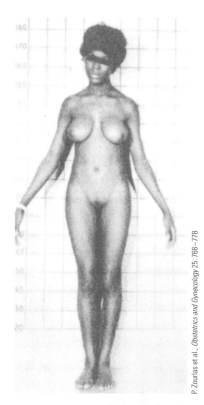

P. Zourlas et al., *Obstetrics and Gynecology* 25:768–778

▲ **FIGURE 7.14** A phenotypic female who has an XY chromosomal constitution and androgen insensitivity.

■ **Androgen insensitivity** An X-linked genetic trait that causes XY individuals to develop into phenotypic females.

■ **Pseudohermaphroditism** An autosomal genetic condition that causes XY individuals to develop the phenotypic sex of females.

At puberty, however, these females change into males. The testes move down into a developing scrotum, and what resembled a clitoris develops into a functional penis. The voice deepens, a beard grows, and muscle mass increases. In most cases, sperm production is normal. What causes these changes? This phenotype is altered by the increased levels of testosterone secretion that accompany puberty. This condition is rare, but in a group of small villages in the Dominican Republic, more than 30 such cases are known. The high incidence is the result of common ancestry through intermarriage. In 12 of the 13 families in these villages, a line of descent can be traced from a single individual.

Genetic Journeys

Sex Testing in the Olympics— Biology and a Bad Idea

Success in amateur athletics, including the Olympics, is often a prelude to financial rewards and acclaim as a professional athlete. Because the stakes are so high, several methods are used to guard against cheating in competition. Competitors in many international events are required to submit urine samples (collected while someone watches) for drug testing. In other cases, urine testing is done at random in an attempt to detect and thus eliminate the use of steroids or performance-enhancing drugs. In the 1960s, rumors about males attempting to compete as females led the International Olympic Committee (IOC) to require sex testing of all female athletes, beginning with the 1968 Olympic Games.

The IOC's test involved analysis of Barr bodies in cells collected by scraping the inside of the mouth. In genetic females (XX), the inactivated X chromosome forms a Barr body, which can be stained and seen under a microscope. Genetic males (XY) do not have a Barr body. The procedure is noninvasive, and females are not required to submit to a physical examination of their genitals. If sexual identity were called into question as a result of the test, a karyotype was required, and if necessary, a gynecological examination followed.

In both theory and practice, the IOC's test was a bad idea for several reasons. Barr body testing is unreliable and leads to both false positive and false negative results. It fails to take into account phenotypic females who are XY with androgen insensitivity and other conditions that result in a discrepancy between chromosomal and phenotypic sex. In addition, the test does not take into account the psychological, social, and cultural factors that enter into one's identity as a male or a female. Ironically, no men attempting to compete as women were identified, but the test unfairly prevented females from competition. Of the more than 6,000 women athletes tested, 1 in 500 had to withdraw from competition as a consequence of failing the sex test. The Spanish hurdler Maria Martinez Patino led a courageous fight against sex testing. She has complete androgen insensitivity, was raised as a female, and competed as a female.

In response to criticism, the IOC and the International Amateur Athletic Federation (IAAF) reconsidered the question of sex testing and instituted a new test, based on recombinant DNA technology, to detect the presence of the male-determining gene *SRY*, which is carried on the Y chromosome. This test was instituted at the 1992 Winter Olympics. A positive test makes the athlete ineligible to compete as a female. However, again the test was flawed because it fails to recognize several chromosomal combinations that result in a female phenotype, even though an *SRY* gene is present. At the 1996 Summer Olympic Games in Atlanta, 8 of 3,387 females were *SRY* positive. Of these, 7 of the 8 had partial or complete androgen insensitivity. Again, no males attempting to compete as females were identified.

Finally, in the face of criticism from medical professionals and athletes, in 1999, the IOC decided to abandon the use of genetic screening of female athletes at the 2000 Olympic Games in Australia. However, the IAAF still retains the option of testing a competitor should the question arise.

7.7 Equalizing the Expression of X Chromosomes in Males and Females

Because females have two copies of all genes on the X chromosome and males have only one copy of these genes, at first glance it would seem that females should have higher levels of all products encoded by these genes. Is this true, or is there a way to equalize the expression of genes on the X chromosome between males and females?

Dosage compensation equalizes the expression of genes on the X chromosome.

In Chapter 4, we discussed hemophilia A, an X-linked genetic disorder in which clotting factor VIII is missing. Because normal females have two copies of the clotting factor gene and normal males have only one, we can ask whether blood from females has twice as much of this clotting factor as blood from males. The answer is straightforward: careful measurements indicate that females have the same amount of this clotting factor as males. In fact, the same is true for all X chromosome genes tested: the amount of gene product is the same in males and females. A process called **dosage compensation** equalizes the amount of X chromosome gene products in both sexes. How that is accomplished in humans and how it came to be understood is an interesting story.

Mice, Barr bodies, and X inactivation can help explain dosage compensation.

The explanation of how dosage compensation works in female mammals leads from a physiologist working on cat nerves to a geneticist working on the inheritance of coat color in mice.

In the late 1940s, Murray Barr and his colleagues were studying nerve cells from cats. Under the microscope he saw a small, dense spot on the inside of the nuclear membrane in cells from female cats that did not appear in cells from male cats (▶ Figure 7.15). A geneticist, Susumo Ohno, suggested that this spot, now called the **Barr body,** is actually an inactivated X chromosome.

About a decade later, Mary Lyon was studying the inheritance of coat color in mice. In female mice heterozygous for X-linked coat-color genes, Lyon found that the coat color was unique. It was not the same as either homozygous parent, nor was it a blend of the parent's coat color. Instead, the fur had patches of the two parental colors in a random arrangement. Males, hemizygous for either gene, never showed such patches and had coats of uniform color. This genetic evidence suggested to Lyon that in heterozygous females, both alleles were active but not in the same cells.

Mary Lyon put her genetic results together with Ohno's suggestion about Barr

■ Dosage compensation A mechanism that regulates the expression of sex-linked gene products.

■ Barr body A densely staining mass in the somatic nuclei of mammalian females. An inactivated X chromosome.

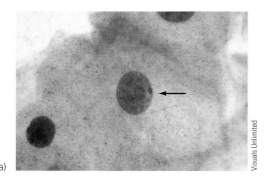

(a)

Visuals Unlimited

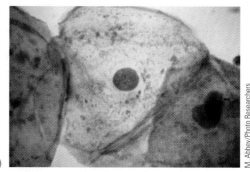

(b)

M. Abbey/Photo Researchers

▲ **FIGURE 7.15** (a) Nucleus from a female cell shows a Barr body (arrow). (b) Nucleus from a male cell shows no Barr body.

Genetics in Society

Joan of Arc—Was It Really John of Arc?

Joan of Arc, the national heroine of France, was born in a village in northeastern France in 1412, during the Hundred Years' War. At the age of 13 or 14, she began to have visions that directed her to help fight the English at Orleans. Following victory, she helped orchestrate the crowning of the new king, Charles VII. During a siege of Paris, the English captured Joan, and in 1431 she was tried for heresy. Although her trial was technically a religious one conducted by the English-controlled church, it was clearly a political trial. Shortly after being sentenced to life imprisonment, she was declared a relapsed heretic, and on May 30, 1431, she was burned at the stake in the marketplace at Rouen.

In 1455 Pope Callistus formed a commission to investigate the circumstances of her trial, and a Trial of Rehabilitation took place over a period of 7 months in 1456. The second trial took testimony from over 100 individuals who knew Joan personally. Extensive documentation from the original trial and the Trial of Re-

habilitation exists. This material has served as the source for the more than 100 plays and countless books written about her life. Although the story of her life is well known, perhaps more remains to be discovered. From an examination of the original evidence, R. B. Greenblatt proposed that Joan had phenotypic characteristics of androgen insensitivity. By all accounts, Joan was a healthy female who had well-developed breasts. Those living with her in close quarters testified that she never menstruated, and physical examinations conducted during her imprisonment revealed a lack of pubic hair. Although such circumstantial evidence is not enough for a diagnosis, it provides more than enough material for speculation. This speculation also provides a new impetus for those medicogenetic detectives who prowl through history, seeking information about the genetic makeup of the famous, infamous, the notorious, and the obscure.

■ **Lyon hypothesis** The proposal that dosage compensation in mammalian females is accomplished by partially and randomly inactivating one of the two X chromosomes.

bodies in cells of mammalian females and proposed her hypothesis (known as the **Lyon hypothesis**) about how dosage compensation works:

- One X chromosome is genetically active in the body cells (not the germ cells) of female mammals. The second X chromosome is inactivated and tightly coiled to form the Barr body.
- The inactivated chromosome can come from the mother or the father.
- Inactivation takes place early in development. After four to five rounds of cell division following fertilization, each cell of the embryo randomly inactivates one X chromosome.
- This inactivation is permanent (except in germ cells), and all descendants of a given cell will have the same X chromosome inactivated.
- The random inactivation of one X chromosome in females equalizes the activity of X-linked genes in males and females.

Females can be mosaics for X-linked genes.

The Lyon hypothesis means that female mammals are actually mosaics, constructed of two different cell types: some cells express genes from the mother's X chromosome, and some cells express genes from the father's X chromosome. The pattern of coat color that Lyon observed in the heterozygous mice is a result of this inactivation. In females heterozygous for X-linked coat-color genes, patches of one color are interspersed with patches of another color. According to the Lyon hypothesis, each patch represents a group of cells descended from a single cell in which the inactivation event occurred.

The tortoiseshell cat is an example of this mosaicism (▶ Active Figure 7.16). In cats, an X-linked gene for coat color has two alleles, a dominant allele (*O*) that produces an orange/yellow color and a recessive allele (*o*) that produces a black color. Heterozygous females (*O/o*) have patches of orange/yellow fur mixed with patches of black fur, called a tortoiseshell pattern. Cells expressing either the orange/yellow allele or the black allele cause this pattern. A cat with a tortoiseshell pattern on a white background is called a calico cat (white fur on the chest and abdomen in such cats is controlled by a different, autosomal gene). Therefore, tortoiseshell cats (and calico cats) are invariably female because males have only one X chromosome and would be either all orange/yellow or all black.

A mosaic pattern of gene expression can also be seen in human females. There is a gene on the X chromosome that controls the formation of sweat glands. A rare recessive mutant allele blocks the formation of sweat glands. This condition is called anhidrotic ectodermal dysplasia (OMIM 305100). Heterozygous women have patches of skin (▶ Figure 7.17) with sweat glands (cells in which the dominant allele is the active X chromosome) and patches of skin without sweat glands (cells in which the mutant recessive allele is on the active X chromosome).

▲ ACTIVE FIGURE 7.16 The differently colored patches of fur on this tortoiseshell cat result from X-chromosome inactivation.

HUMAN Genetics ⒮ Now™ Learn more about X-chromosome inactivation by viewing the animation on Human GeneticsNow at **http://now.ilrn.com/cummings7**

How and when are X chromosomes inactivated?

The process of X inactivation has presented researchers with several puzzling questions. How does the cell count the number of X chromosomes in the nucleus? If there are two X chromosomes in the nucleus, how is one X chromosome chosen to be turned off, but not the other? Finally, how is the chromosome inactivated? Detailed answers to these questions are not yet available, but we know that inactivation begins from a region on the X chromosome called the **X inactivation center (Xic)**. The Xic contains several genes, one of which is called *XIST*. If the *XIST* gene on an X chromosome is expressed, the chromosome becomes coated with *XIST* RNA (▶ Figure 7.18) and becomes coiled and inactivated. How the *XIST* gene on only one of the two X chromosomes in a female embryo is turned on is still a puzzle.

▨ **X inactivation center (Xic)** A region on the X chromosome where inactivation begins.

Keep in mind

▨ One X chromosome is randomly inactivated in all the somatic cells of human females. This event equalizes the expression of X-linked genes in males and females.

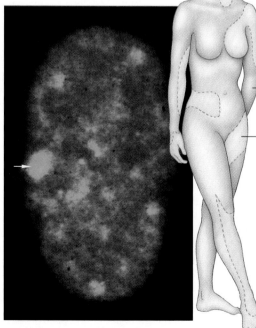

▶ FIGURE 7.17 (a) Photomicrograph of a Barr body (an inactive X chromosome) in a cell from a human female. (b) The mosaic pattern of sweat glands in a woman who is heterozygous for the X-linked recessive disorder anhidrotic ectodermal dysplasia.

Unaffected skin (X chromosome with recessive allele was condensed; its allele is inactivated. The dominant allele on other X chromosome is being expressed in this tissue.)

Affected skin with no normal sweat glands (In this tissue, the X chromosome with dominant allele has been condensed. The recessive allele on the other X chromosome is being transcribed.)

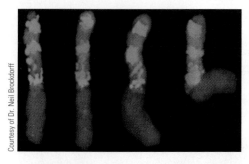

▲ FIGURE 7.18 In female mammals, expression of the *XIST* gene coats one X chromosome with *XIST* RNA (green) in the gene-rich regions between G bands as a first step in X inactivation. This photograph shows several X chromosomes coated with *XIST* RNA; other chromosome regions are stained red.

In humans, both X chromosomes are active in the zygote and all cells of the early embryo. Random inactivation of one X chromosome usually occurs when the embryo has about 32 cells. Because there are only a small number of cells in the embryo at the time of inactivation, is it possible that all or almost all the cells might randomly inactivate only the mother's or father's X chromosome? If this happens, heterozygous females would express recessive X-linked traits. In fact, this phenomenon has been seen in female monozygotic twins, one of whom expresses an X-linked recessive trait, whereas the other does not. In the pedigree shown in ▶ Figure 7.19, two female identical twins are heterozygotes for red-green color blindness through their color-blind father. One of the twins has normal color vision, and the other has red-green color blindness. The color-blind twin has three sons, two with normal vision and one who is color-blind (see pedigree).

Molecular testing of skin cells from the color-blind twin showed that almost all of the active X chromosomes were from her father and carry the allele for color blindness. In the twin with normal vision, the opposite situation is observed; almost all of the active X chromosomes are maternal in origin.

7.8 Sex-Influenced and Sex-Limited Traits

■ **Sex-influenced traits** Traits controlled by autosomal genes that are usually dominant in one sex but recessive in the other sex.

■ **Pattern baldness** A sex-influenced trait that acts like an autosomal dominant trait in males and an autosomal recessive trait in females.

■ **Sex-limited genes** Loci that produce a phenotype in only one sex.

■ **Precocious puberty** An autosomal dominant trait expressed in a sex-limited fashion. Heterozygous males are affected, but heterozygous females are not.

Sex-influenced traits are expressed in both males and females but are usually dominant traits in one sex, but recessive traits in the other sex. These traits are usually controlled by autosomal genes and illustrate the effect of hormonal differences on gene expression. **Pattern baldness** (OMIM 109200) is an example of sex-influenced inheritance (▶ Figure 7.20). This trait is expressed more often in males than in females. The allele for baldness behaves as an autosomal dominant trait in males and as an autosomal recessive trait in females. The pattern of expression is related to the different levels of testosterone present. In this case, the hormonal environment and the genotype interact in determining phenotypic expression of this gene.

Sex-limited genes are expressed only in one sex, whether they are inherited in an autosomal or sex-linked pattern. An autosomal dominant trait that controls **precocious puberty** (OMIM 176410) is expressed in heterozygous males but not in heterozygous females. Affected males undergo puberty at 4 years of age or earlier. Heterozygous females are unaffected but pass this trait on to half of their sons, making it hard to distinguish this trait from a sex-linked gene. Genes that deal with traits such as breast development in females and facial hair in males are other examples of sex-limited genes, as are virtually all other genes that deal with secondary sexual characteristics.

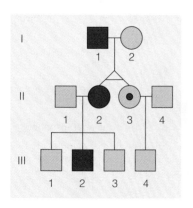

◀ FIGURE 7.19 Pedigree showing monozygotic female twins (II-2 and II-3) discordant for color blindness. The twins inherited the allele for color blindness from their father. Almost all of the active X chromosomes in the color-blind twin carry the mutant allele. Almost all of the active X chromosomes in the twin who has normal vision carry the allele for normal vision.

▲ FIGURE 7.20 Pattern baldness as an autosomal dominant trait in males. It is an autosomal recessive trait in females. The degree of baldness in both males and females is related to hormone levels and other environmental influences.

Genetics in Practice

 Genetics in Practice case studies are critical thinking exercises that allow you to apply your new knowledge of human genetics to real-life problems. You can find these case studies and links to relevant websites at **http://biology.brookscole.com/cummings7**

CASE 1

Melissa was referred to genetic counseling at 16 weeks in her pregnancy because of a history of epileptic seizures. She takes medication (valproic acid) for her seizures and has not had an attack for the last 3 years. Her physician became concerned when he learned that she was still taking this medication, against his advice, during her pregnancy. He wanted her to speak to a counselor about the possible effects of this medication on the developing fetus. The counselor took a detailed family history, which indicated that Melissa was the only family member with seizures and that no other genetic conditions were apparent in the family. The counselor asked Melissa why she continued to take valproic acid during her pregnancy. Melissa stated she was "afraid her child would be like her, if she didn't take her medicine." Melissa went on to say that she was teased as a child when she would have her "fits," and she wanted to prevent that from happening to her children.

With this in mind, the counselor reviewed the process of fetal development and why it is best that a physician carefully evaluate all medications that a woman takes while she is pregnant. Melissa's medication has been shown to cause spina bifida—affecting almost twice as many children who were exposed to it than children who are not exposed. Using illustrations, the counselor explained that spina bifida is a defect that occurs when the neural tube fails to completely close during embryonic development. The failure to fold exposes part of the spinal area when an infant is born. Valproic acid could also cause problems in the heart and in the genitals.

The counselor explained that prenatal diagnosis using ultrasound, and possibly amniocentesis, could help determine whether the baby's tube has closed properly.

Postscript: Melissa elected to have an ultrasound, which showed that the baby did not have a neural tube defect. However, she was offered an amniocentesis to rule out a possible false negative result of the ultrasound. She declined the amniocentesis, and Melissa delivered a healthy baby boy.

1. As a counselor, you have taken Melissa's family history. How can you address Melissa's fears that her child will develop epilepsy because she did?

2. From a genetics perspective, is Melissa at greater risk for having a child with epilepsy than someone without epilepsy?

3. Women taking valproic acid have a 1% to 2% risk of having a child with a neural tube defect. Does the fact that Melissa had a normal child increase the risk that her next child will be affected? Why or why not?

4. The neural tube forms and closes during the first trimester of pregnancy. What does this suggest about Melissa's medication program in future pregnancies?

Summary

7.1 The Human Reproductive System

- The human reproductive system consists of gonads (testes in males, ovaries in females), ducts to transport gametes, and genital structures for intercourse and fertilization.

7.2 A Survey of Human Development from Fertilization to Birth

- Human development begins with fertilization and the formation of a zygote. Cell divisions in the zygote form an early embryonic stage called the blastocyst. The embryo implants in the uterine wall, and a placenta develops to nourish the embryo.

7.3 Teratogens Are a Risk to the Developing Fetus

- The embryo and fetus are sensitive to chemical and physical agents that can produce birth defects. Fetal alcohol syndrome is a preventable form of birth defect.

7.4 How Is Sex Determined?

- Mechanisms of sex determination vary from species to species. In humans, the presence of a Y chromosome is associated with male sexual development, and the absence of a Y chromosome is associated with female development.

7.5 Defining Sex in Stages: Chromosomes, Gonads, and Hormones

- Chromosomal sex is established at fertilization, but other aspects of sex depend on the interaction of genes and environmental factors, especially hormones.

7.6 Mutations Can Uncouple Chromosomal Sex from Phenotypic Sex

- Early in development, the Y chromosome signals the indifferent gonad to begin development as a testis. Hormones secreted by the testis control later stages of male sexual differentiation, including the development of phenotypic sex.

7.7 Equalizing the Expression of X Chromosomes in Males and Females

- Human females have one X chromosome inactivated in all somatic cells to balance the expression of X-linked genes in males and females.

7.8 Sex-Influenced and Sex-Limited Traits

- In sex-influenced and sex-limited inheritance, the sex of the individual affects whether and the degree to which the trait is expressed. This is true for autosomal and sex-linked genes. Sex hormone levels modify expression of these genes, giving rise to altered phenotypic ratios.

Questions and Problems

HUMAN Genetics Now™ Preparing for an exam? Assess your understanding of this chapter's topics with a pre-test, a personalized learning plan, and a post-test at **http://now.ilrn.com/cummings7**

The Human Reproductive System

1. How many chromosomes are present in a secondary oocyte as it leaves the ovary during ovulation?
2. Discuss and compare the products of meiosis in human females and males. How many functional gametes are produced from the daughter cells in each sex?
3. A human female is conceived on April 1, 1979, and is born on January 1, 1980. Onset of puberty occurs on January 1, 1992. She conceives a child on July 1, 2004. How long did it take for the ovum that was fertilized on July 1, 2004, to complete meiosis?

A Survey of Human Development from Fertilization to Birth

4. The gestation of a fetus occurs over 9 months and is divided into three trimesters. Describe the major events that occur in each trimester. Is there a point at which the fetus becomes more "human"?
5. FAS is caused by alcohol consumption during pregnancy. It can result in spontaneous abortion, growth retardation, facial abnormalities, and mental retardation. How does FAS affect all of us, not just the unlucky children born with this syndrome? What steps need to be taken to prevent this syndrome?

How Is Sex Determined?

6. Describe, from fertilization, the major pathways of normal male sexual development; include the stages in which genetic sex, gonadal sex, and phenotypic sex are determined.
7. Which pathway of sexual differentiation (male or female) is regarded as the default pathway? Why?
8. The absence of a Y chromosome in an early embryo causes:
   a. the embryonic testis to become an ovary.
   b. the Wolffian duct system to develop.
   c. the Müllerian duct system to degenerate.
   d. the indifferent gonad to become an ovary.
   e. the indifferent gonad to become a testis.
9. Assume that human-like creatures exist on Mars. As in the human population on Earth, there are two sexes and even sex-linked genes. The gene for eye color is an example of one such gene. It has two alleles. The purple allele is dominant to the yellow allele. A purple-eyed female alien mates with a purple-eyed male. All of the male offspring are purple-eyed, whereas half of the female offspring are purple-eyed and half are yellow-eyed. Which is the heterogametic sex?

Mutations Can Uncouple Chromosomal Sex from Phenotypic Sex

10. Give an example of a situation in which genetic sex, gonadal sex, and phenotypic sex do not coincide. Explain why they do not coincide.
11. How can an individual who is XY be phenotypically female?
12. Discuss whether the following individuals are (1) gonadally male or female, (2) phenotypically male or female (discuss Wolffian/Müllerian ducts and external genitalia), and (3) sterile or fertile.
   a. XY, homozygous for a recessive mutation in the testosterone gene, which renders the gene nonfunctional
   b. XX, heterozygous for a dominant mutation in the testosterone gene, which causes continuous production of testosterone
   c. XY, heterozygous for a recessive mutation in the MIH gene
   d. XY, homozygous for a recessive mutation in the SRY gene

e. XY, homozygous for a recessive mutation in the *MIH* gene

Sex-Influenced and Sex-Limited Traits

13. It has been shown that hormones interact with DNA to turn certain genes on and off. Use this fact to explain sex-linked and sex-influenced traits.
14. What method of sex testing did the International Olympic Committee previously use? What method did they subsequently use? Do either of these methods conclusively test for "femaleness"? Explain.
15. Explain why pattern baldness is more common in males than in females yet the gene resides on an autosome.

Equalizing the Expression of X Chromosomes in Males and Females

16. Calico cats are almost invariably female. Why? (Explain genotype and phenotype of calico females and the theory of why calicos are female.)

17. How many Barr bodies would the following individuals have?
    a. normal male
    b. normal female
    c. Klinefelter male
    d. Turner female
18. Males have only one X chromosome and therefore only one copy of all genes on the X chromosome. Each gene is directly expressed, thus providing the basis of hemizygosity in males. Females have two X chromosomes, but one is always inactivated. Therefore females, like males, have only one functional copy of all the genes on the X chromosome. Again, each gene must be directly expressed. Why, then, are females not considered hemizygous, and why are they not afflicted with sex-linked recessive diseases as often as males?
19. Individuals with an XXY genotype are sterile males. If one X is inactivated early in embryogenesis, the genotype of the individual effectively becomes XY. Why will this individual not develop as a normal male?

Internet Activities

 Internet Activities are critical thinking exercises using the resources of the World Wide Web to enhance the principles and issues covered in this chapter. For a full set of links and questions investigating the topics described below, log on to **http://biology.brookscole.com /cummings7**

1. *Embryological Development. The Visible Embryo* website provides free images and descriptions of human developmental stages from conception to stage 23. (Descriptions only are available for stages beyond 10 weeks.) Follow the stages and read about the development of the embryo.
   Further Exploration. Check out the *Morphing Embryos* video at Nova Online's *Odyssey of Life* website.
2. *Further Exploration of a Chromosomal Defect.* Fragile-X syndrome, which you may have researched as part of the Chapter 6 Internet Activities, affects

males and females differently. Go to the *Your Genes, Your Health* website maintained by the Dolan DNA Learning Center at Cold Spring Harbor Laboratory, and click on the "Fragile X Syndrome" link. At this page, choose the "How is it inherited?" link and explore how males and females inherit and display fragile-X syndrome.
   Further Exploration. To explore the complexities of the genetics of coat color in cats, including the genetics of X-linked characteristics such as tortoiseshell coat patterns, you can try "The Cat Color FAQ."

How would you vote now?

The standard treatment for children born with genital abnormalities involves sex reassignment surgery, most often converting males into females. Now that you know more about how sex is determined and how sexual characteristics develop during pregnancy, what do you think? If you had a child with such a condition, would you consent to such surgery for your child, or would you allow the child to make that decision upon reaching puberty? Visit the Human Heredity Companion website to find out more on the issue, then cast your vote online at **http://biology.brookscole.com /cummings7**

 For further reading and inquiry, log on to **InfoTrac College Edition**, your world-class online library, including articles from nearly 5,000 periodicals, at **http://infotrac.thomsonlearning.com**

8

DNA Structure and Chromosomal Organization

I N February 2003, thousands of people in Asia became sick from a flu-like disease that began with a high fever, headaches, and respiratory problems. In the next few months, the disease spread to more than two dozen countries across Asia, Europe, North America, and South America. Scientists around the globe mobilized to identify the cause of this illness and quickly isolated a virus, called a cornavirus, from infected individuals. The disease, named severe acute respiratory syndrome (SARS), is spread by person-to-person contact, in the form of droplets produced by an infected person who sneezes or coughs. The SARS outbreak was contained by public health procedures such as quarantining patients and by screening travelers who might be infected. Despite these efforts, the World Health Organization (WHO) reported that just over 8,000 people became infected with SARS, and about 10% of those infected died from the virus. In the United States, eight people developed SARS, but all had previously traveled to parts of the world where SARS infections had been reported, and the disease did not spread.

By May of 2003, scientists using recombinant DNA technology and the techniques of genomics were able to derive the DNA sequence of the viral chromosome. The virus carries 11 genes and is a previously unknown strain of cornavirus. To help fight future outbreaks, development of a vaccine was a high priority. No vaccines against human cornaviruses had been successfully developed, so researchers turned to a new type of vaccine, a DNA vaccine.

In making a DNA vaccine against the SARS virus, one of the genes from the virus was isolated. Using a carrier molecule, the DNA carrying the gene was injected into mice. In a mouse muscle cell, the viral gene was switched on and directed the synthesis of a viral protein. The viral protein was recognized by the mouse's immune system and made antibodies against the protein, protecting the mouse against future infections with the SARS virus.

Courtesy of Dmitry Pruss, National Institutes of Health

No DNA vaccines have been approved for use in humans, and the DNA vaccine against SARS is now being tested in clinical studies using volunteers. Several concerns about DNA viruses are not yet resolved. Will the virus DNA insert itself into a human chromosome and disrupt a gene, perhaps causing cancer? Will the immune system make antibodies against the body, instead of the viral protein? Could injecting SARS DNA somehow increase susceptibility to SARS instead of preventing infection?

In this chapter we describe the structure of DNA and the events that led to the confirmation of DNA as the cellular molecule that carries genetic information. We also explore what is known about the way DNA is incorporated into chromosomes. In Chapter 17, we will discuss the immune system and its genetic components.

How would you vote?

DNA vaccines were developed quickly after the discovery of the virus that causes SARS. Although no DNA vaccines are currently approved for use in humans, clinical trials are underway to assess their safety and effectiveness. These trials will take several years to complete. Before the results of the clinical studies are in, if another outbreak of the deadly SARS virus occurs, or a bioterrorist attack releases anthrax or another potentially fatal disease-causing organism, would you agree to be treated with a DNA vaccine? Would you have members of your family injected with a DNA vaccine? Visit the Human Heredity Companion website to find out more on the issue, then cast your vote online at **http://biology.brookscole.com/cummings7**

8.1 DNA Carries Genetic Information

Early in the 1860s, a chemist named Frederick Miescher recovered a chemical substance from the nuclei of pus cells. To analyze this substance further, he broke open the cells by treating them with a protein-digesting substance called pepsin that he obtained from extracts of pig stomachs (a good source of pepsin, which functions in digestion). He treated the pus cells for several hours with the pepsin and found gray sediment collected at the bottom of the flask. Under the microscope this sediment turned out to be pure nuclei. Miescher was therefore the first person to isolate and purify a cellular organelle.

By chemically extracting the purified nuclei, Miescher obtained a substance he called nuclein. Chemical analysis revealed that it contained hydrogen, carbon, oxygen, and two uncommon substances, nitrogen and phosphorus. Nuclein was found in the nuclei of other cell types, including kidney, liver, sperm, and yeast. He regarded it as an important component of most cells. Many years later it was shown that his nuclein contained deoxyribonucleic acid (DNA).

Keep in mind as you read

■DNA is the macromolecular component of cells that encodes genetic information.

■Watson and Crick built models of DNA structure using information from x-ray diffraction studies and chemical analyses of DNA from various organisms.

■DNA is packaged into chromosomes by several levels of coiling and compaction.

■A newly replicated DNA molecule contains one old strand and one new strand.

Research in the first few decades of the twentieth century established that genes exist and are carried on chromosomes. But what chemical component of a cell is a gene? As is often the case in science, the answer to this question came from an unexpected direction, the study of an infectious disease.

At the beginning of the twentieth century, pneumonia was a serious public health problem and was the leading cause of death in the United States. Medical researchers of that era studied this infectious disease to develop an effective treatment, perhaps in the form of a vaccine. The unexpected outgrowth of this research was the discovery of the chemical nature of the gene.

DNA transfers genetic traits between bacterial strains.

By the 1920s, it was known that a bacterial infection could cause pneumonia. One form of pneumonia is caused by the bacterium *Streptococcus pneumoniae*. Two strains of this species were known: strain S formed a capsule, it was infective, and it caused pneumonia (that is, was a virulent strain). Strain R, on the other hand, did not form a capsule and was not infective. Fredrick Griffith, an English microbiologist, studied these strains, and the results of his experiment are straightforward and easily interpreted. He showed that mice injected with living cells of strain S developed pneumonia and soon died, but mice injected with live cells from strain R did not develop pneumonia and lived (▶ Active Figure 8.1).

In another experiment, Griffith found that if the strain S cells were killed with heat before injection, the mice survived and did not develop pneumonia. For Griffith, the most intriguing result was obtained when mice were injected with a mixture of heat-killed strain S cells and live cells from strain R. These mice developed pneumonia and died. Griffith recovered live strain S bacteria (with a capsule) from the bodies of the dead mice. Griffith concluded that in the bodies of the mice, living strain R cells were transformed into strain S cells. He proposed that hereditary information had somehow passed from the dead strain S cells into the living strain R cells, allowing them to make a capsule and become virulent. He called this process **transformation** and the unknown material the **transforming factor.**

In 1944, a team at the Rockefeller Institute in New York that included Oswald Avery, Colin MacLeod, and Maclyn McCarty discovered that the transforming factor is DNA. McCarty recounts the story of this discovery in a readable memoir: *The Transforming Principle: Discovering That Genes Are Made of DNA.*

In a series of experiments that stretched over 10 years, Avery and his colleagues extended Griffith's work on transformation. In their experiments, Avery and his col-

■ **Transformation** The process of transferring genetic information between cells by DNA molecules.

■ **Transforming factor** The molecular agent of transformation; DNA.

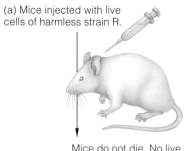 (a) Mice injected with live cells of harmless strain R.

Mice do not die. No live R cells in their blood.

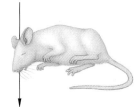 (b) Mice injected with live cells of killer strain S.

Mice die. Live S cells in their blood.

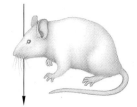 (c) Mice injected with heat-killed S cells.

Mice do not die. No live S cells in their blood.

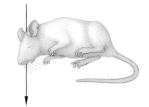 (d) Mice injected with live R cells *plus* heat-killed S cells.

Mice die. Live S cells in their blood.

▲ ACTIVE FIGURE 8.1 Griffith discovered that the ability to cause pneumonia is a genetic trait that can be passed from one strain of bacteria to another. (a) Mice injected with strain R do not develop pneumonia. (b) Mice injected with strain S develop pneumonia and die. (c) When the S strain cells are killed by heat treatment before injection, mice do not develop pneumonia. (d) When mice are injected with a mixture of heat-killed S cells and live R cells, they develop pneumonia and die. Griffith concluded that the live R cells acquired the ability to cause pneumonia from the dead S cells.

HUMAN Genetics ℰ Now™ Learn more about the process of transformation by viewing the animation on Human GeneticsNow at **http://now.ilrn.com/cummings7**

leagues separated the chemical components of S cells into chemical classes including carbohydrates, fats, proteins, and nucleic acids. Each component was mixed with live R cells and injected into mice. Mice only got pneumonia and died when injected with a mixture of S cell DNA and live R cells. Avery concluded that DNA from S cells was responsible for transforming the R cells into S cells. To confirm that DNA was responsible for transformation, they treated the DNA with enzymes that destroy protein and ribonucleic acid (RNA) before injection. This treatment removed any residual protein or RNA from the preparation but did not affect transformation. As a final test, the DNA preparation was treated with deoxyribonuclease, an enzyme that digests DNA, and the transforming activity was abolished.

The work of Avery and his colleagues produced two important conclusions:

■ DNA carries genetic information. Only DNA transfers heritable information from one bacterial strain to another strain.
■ DNA controls the synthesis of specific products. Transfer of DNA also results in the transferring of the ability to synthesize a specific gene product (in the form of a capsule).

Although the evidence was strong, many in the scientific community were not persuaded that DNA was the carrier of genetic information. They remained convinced that proteins were the only molecule complex enough to perform this task. A few years later, conclusive evidence for the idea that DNA encodes genetic information came from the study of viruses.

Replication of bacterial viruses involves DNA.

In the late 1940s and early 1950s scientists began working on a group of viruses that attack and kill bacterial cells (▶ Figure 8.2). These viruses, known as bacteriophages

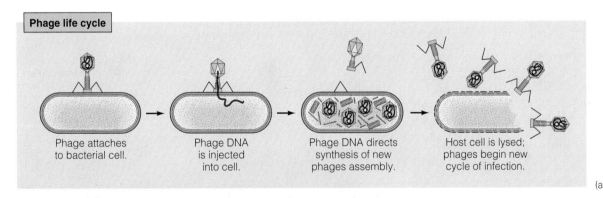

Phage life cycle

Phage attaches to bacterial cell. → Phage DNA is injected into cell. → Phage DNA directs synthesis of new phages assembly. → Host cell is lysed; phages begin new cycle of infection.

(a)

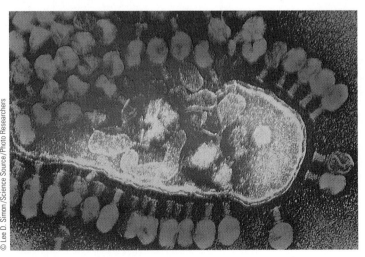

▲ FIGURE 8.2 Bacteriophages are viruses that attack and kill bacteria. (a) The virus attaches to the outside of the cell, and the viral DNA is injected. The phage DNA directs the synthesis and assembly of new phage particles that break open and destroy the bacterial cell, releasing new virus particles. (b) An electron micrograph of bacteriophages attacking a bacterial cell.

(b)

© Lee D. Simon/Science Source/Photo Researchers

(phages, for short), infect and replicate within *Escherichia coli,* the bacterium that inhabits the human intestinal tract. After injecting its DNA, a single virus reproduces rapidly, and in 20 to 25 minutes about a hundred new phages burst from the ruptured bacterium, ready to invade other cells.

Phages consist only of DNA and proteins, making them ideal candidates to help identify which of these molecules carry genetic information (▷ Active Figure 8.3). Alfred Hershey and Martha Chase grew phages with radioactive phosphorus (there is no phosphorus in proteins), making their DNA radioactive. They grew other phages with radioactive sulfur (there is no sulfur in DNA), making their protein coat radioactive. Then they did two experiments. In the first experiment, they added phages with radioactive DNA to a tube of bacteria. After waiting a few minutes for the viruses to attach to the bacterial cell, they put the mixture into a blender to separate the phages from the bacteria. They collected the bacteria and found that they were radioactive, but the protein coats of the phages were not. From this experiment, they concluded that after attaching to the bacteria, the phage DNA enters the cell, but the protein coat does not.

In a second experiment, they added phages with radioactive protein coats to a tube of bacteria, and after a few minutes, put the mixture in a blender. They discovered that the bacteria were not radioactive, but the phage protein coats pulled off the surface of the bacteria were radioactive. This confirmed that the phage protein coat remains on the outside of the bacterial cell during infection, and could not direct the synthesis of new phages.

From these simple experiments, Hershey and Chase concluded that only the phage DNA enters the bacterial cell and directs the production of new viruses, and that the phage DNA carries genetic information for this task, and not the protein coat.

DNA research has come a long way since the early experiments of Miescher,

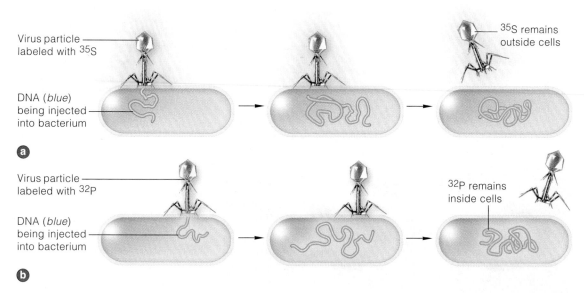

▲ **ACTIVE FIGURE 8.3** Phages contain only DNA and protein. Phage proteins contain sulfur, but not phosphorus. Phage DNA contains phosphorus, but not sulfur. Hershey and Chase designed two experiments to test whether DNA protein contained the genetic information needed to direct the replication of new phage particles. In one experiment, they used virus particles whose protein coat was labeled with radioactive sulfur to infect bacterial cells. In a second experiment, they infected bacteria with viruses whose DNA was labeled with radioactive phosphorus. They found that only the radioactively labeled DNA entered the bacterial cell and directed the synthesis of new virus particles. This was more evidence that DNA and not protein is the genetic material.

HUMAN Genetics *Now*™ Learn more about the Hershey-Chase experiment by viewing the animation on Human GeneticsNow at **http://now.ilrn.com/cummings7**

Griffith, Avery, and others. Today, DNA has entered the public purview and is so commonplace that it is even being used to sell products (see Genetics in Society: DNA for Sale).

Keep in mind

■ DNA is the macromolecular component of cells that encodes genetic information.

8.2 Watson, Crick, and the Structure of DNA

Recognition that DNA carries genetic information helped fuel efforts to understand the structure of DNA. From the mid-1940s through 1953, several laboratories made significant strides in unraveling the structure of DNA, culminating in a model for DNA structure proposed in 1953 by James Watson and Francis Crick. Watson doc-

Genetics in Society

DNA for Sale

The magazine ad for the perfume reads: "Where does love originate? Is it in the mind? Is it in the heart? Or in our genes?" A perfume named DNA is marketed in a helix-shaped bottle. There is no actual DNA in the fragrance, but the molecule is invoked to sell the idea that love emanates from the genes. Seem strange? Well, how about jewelry that actually contains DNA from your favorite celebrities? In this line of products, a process called the polymerase chain reaction (PCR) amplifies DNA in a single hair or cheek cell. The resulting solution contains millions of copied DNA molecules and is added to small channels drilled into acrylic earrings, pendants, or bracelets. The liquid can be colored to contrast with the acrylic and be more visible. Just as people wear T-shirts with pictures of Elvis or Einstein, they can now wear jewelry containing DNA from their favorite entertainer, poet, composer, scientist, or athlete. For dead heroes, the DNA can come from a lock of hair; in fact, a single hair will do.

Or, how about using DNA as a protection against counterfeit clothes? In the 2000 Summer Olympic Games in Sydney, DNA extracted from cheek cells swabbed from Australian athletes was amplified by PCR and mixed with the ink used to print souvenir shirts. More than 2,000 different types of items were created, and DNA testing of the labels was used to ensure that everything sold at the Games was genuine.

Want music composed from the base sequence of DNA? Composers have translated the four bases of DNA (adenine, guanine, cytosine, and thymine) into musical notes. Long sequences of bases, retrieved from computer databases, are converted into notes, transferred to sheet music, and played by instruments or synthesizers as the music of the genes. Those who have listened to this music say that DNA near chromosomal centromeres sounds much like the music of Bach or other Baroque composers, but that music from other parts of the genome has a contemporary sound.

From a scientific standpoint, this fascination with DNA may be a little difficult to understand, but DNA has clearly captured popular fancy and is being used to sell an ever-increasing array of products. DNA has name recognition. Over the past 40 years, DNA has moved from scientific journals and textbooks to the popular press and even to comic strips. The relationship between genes and DNA is now well known enough to be used in commercials and advertisements. In a few years, this fascination will probably fade and be replaced with another fad, but for now, if you want it to sell, relate it to DNA.

umented the scientific, intellectual, and personal intrigue that characterized the race to discover the structure of DNA in his book, *The Double Helix*. This book and others on the same topic provide a rare glimpse into the ambitions, jealousies, and rivalries that entangled scientists who were involved in the dash to a Nobel Prize.

Understanding the structure of DNA requires a review of some basic chemistry.

The structure of DNA and the structure of molecules, shown in later chapters, are drawn using chemical terms and symbols. For this reason, we will pause for a brief review of the terms and definition used in organic chemistry.

All matter is composed of atoms; the different types of atoms are known as elements (of which there are 114). In nature, atoms are combined into molecules, which are units of two or more atoms chemically bonded together. Molecules are represented by formulas that indicate how many atoms of each type are present. Each atom has a symbol for the element it represents: H for hydrogen, N for nitrogen, C for carbon, O for oxygen, and so forth. For example, a water molecule, composed of two hydrogen atoms and one oxygen atom, has its chemical formula represented as H_2O:

$$\text{two hydrogen atoms} \qquad H_2O$$

one oxygen atom

Many molecules in cells are large and have more complex formulas. A molecule of glucose contains 24 atoms and is written as

$$C_6H_{12}O_6$$

Atoms in molecules are held together by links called **covalent bonds.** In its simplest form, a covalent bond is a pair of electrons shared between two atoms. Sharing two or more electrons can form more complex covalent bonds. ▶ Figure 8.4a shows how such bonds are written in chemical structures. A second type of atomic interaction involves a weak attraction known as a **hydrogen bond.** In living systems, hydrogen bonds make an important contribution to the three-dimensional shape and functional capacity of biological molecules. Hydrogen bonds are weak interactions between two atoms (one of which is always hydrogen), that carry partial but opposite electrical charges. Hydrogen bonds are usually represented in structural formulas as dotted or dashed lines that connect two atoms (▶ Figure 8.4b).

Although individual hydrogen bonds are weak and easily broken, they hold molecules together by sheer force of numbers. As we see in a following section, hydrogen bonds hold together the two strands in a DNA molecule, and they are also responsible for the three-dimensional structure of proteins (Chapter 9).

■ **Covalent bonds** Chemical bonds that result from electron sharing between atoms. Covalent bonds are formed and broken during chemical reactions.

■ **Hydrogen bond** A weak chemical bonding force between hydrogen and another atom.

▶ FIGURE 8.4 Representations of chemical bonds. (a) Covalent bonds are represented as solid lines that connect atoms. Depending on the degree of electron sharing, there can be one (*left*) or more (*right*) covalent bonds between atoms. Once formed, covalent bonds are stable and are broken only in chemical reactions. (b) Hydrogen bonds are usually represented as dotted lines that connect two or more atoms. As shown, water molecules form hydrogen bonds with adjacent water molecules. These are weak interactions that are easily broken by heat and molecular tumbling and can be re-formed with other water molecules.

(a)

(b)

Nucleotides are the building blocks of nucleic acids.

Biological organisms contain two types of nucleic acids: **deoxyribonucleic acid (DNA)** and **ribonucleic acid (RNA).** Both are made up of subunits known as nucleotides. A **nucleotide** has three components: a **nitrogen-containing base** (either a **purine** or a **pyrimidine**), a **pentose sugar** (either ribose or deoxyribose), and a **phosphate group** (▷ Figure 8.5). The phosphate groups are strongly acidic and the reason DNA and RNA are called acids. Purines and pyrimidines have the same six-atom ring, but purines have an additional three-atom ring. The purine bases **adenine** (A) and **guanine** (G) are found in both RNA and DNA. The pyrimidine bases are **thymine** (T), found only in DNA; **uracil** (U), found only in RNA; and **cytosine** (C), found in both RNA and DNA. RNA has four bases (A, G, U, C), and DNA has four bases (A, G, T, C).

The sugars in nucleic acids contain five carbon atoms (which is why they are called pentoses). The sugar in RNA is known as **ribose,** and the sugar in DNA is **deoxyribose.** The difference between the two is a single oxygen atom; one is present in ribose and absent in deoxyribose.

The components of a nucleotide are assembled by covalently linking a base to a sugar, which in turn is covalently linked to a phosphate group (▷ Figure 8.6). Two or more nucleotides can be linked to each other by a covalent bond between the phosphate group of one nucleotide and the sugar of another nucleotide. Chains of nucleotides called polynucleotides can be formed in this way (▷ Figure 8.7).

Polynucleotides are directional molecules.

Polynucleotide chains have slightly different structures at either end. At one end is a phosphate group; this is the 5′ (pronounced "five prime") end. At the opposite end is an OH group attached to the sugar molecule; this is known as the 3′ ("three prime") end of the chain. By convention, nucleotide chains are written beginning with the 5′ end, such as 5′-CGATATGCGAT-3′. As we will see next, DNA is made up of two polynucleotide chains.

■ **Deoxyribonucleic acid (DNA)** A molecule consisting of antiparallel strands of polynucleotides that is the primary carrier of genetic information.

■ **Ribonucleic acid (RNA)** A nucleic acid molecule that contains the pyrimidine uracil and the sugar ribose. The several forms of RNA function in gene expression.

■ **Nucleotide** The basic building block of DNA and RNA. Each nucleotide consists of a base, a phosphate, and a sugar.

■ **Nitrogen-containing base** A purine or pyrimidine that is a component of nucleotides.

■ **Purine** A class of double-ringed organic bases found in nucleic acids.

■ **Pyrimidine** A class of single-ringed organic bases found in nucleic acids.

■ **Pentose sugar** A five-carbon sugar molecule found in nucleic acids.

■ **Phosphate group** A compound containing phosphorus chemically bonded to four oxygen molecules.

■ **Adenine and guanine** Nitrogen-containing purine bases found in nucleic acids.

■ **Thymine, uracil, and cytosine** Nitrogen-containing pyrimidine bases found in nucleic acids.

■ **Ribose and deoxyribose** Pentose sugars found in nucleic acids. Deoxyribose is found in DNA, ribose in RNA.

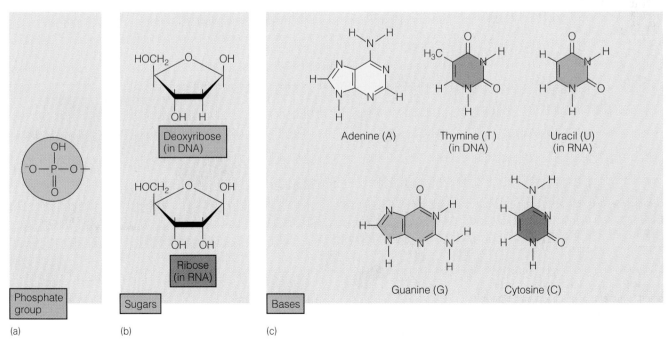

(a) (b) (c)

▲ FIGURE 8.5 DNA is made up of subunits called nucleotides. Each nucleotide is composed of (a) a phosphate group, (b) a sugar, and (c) a base.

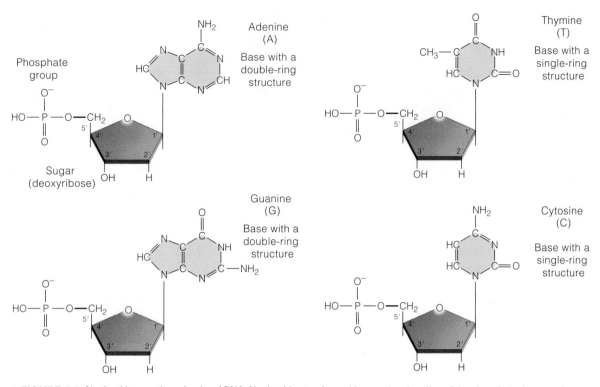

▲ FIGURE 8.6 Nucleotides are the subunits of DNA. Nucleotides are formed by covalent bonding of the phosphate, base, and sugar.

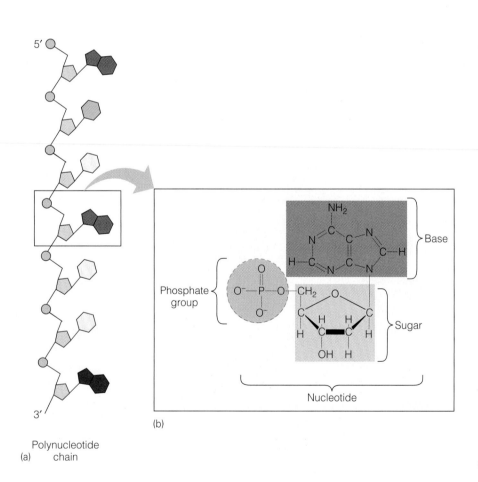

▶ FIGURE 8.7 (a) Nucleotides can be joined together to form chains called polynucleotides. Polynucleotides are polar molecules that have a 5′ end (at the phosphate group) and a 3′ end (at the sugar group). (b) The components of a nucleotide.

8.3 DNA Contains Two Polynucleotide Chains

In the early 1950s, James Watson and Francis Crick began to work out the structure of DNA. To build their model, they sifted through and organized the information that was already available about DNA. Their model is based on two types of information about DNA: x-ray crystallography, which provides information about the physical structure of the molecule, and chemical information about the nucleotide composition of DNA.

In x-ray crystallography, molecules are crystallized and placed in an x-ray beam. As the x-rays pass through the crystal, some hit the atoms in the crystal and are deflected at an angle. The pattern of x-rays emerging from the crystal can be recorded on photographic film and analyzed to produce information about the organization and shape of the crystallized molecule.

Working with Maurice Wilkins, Rosalind Franklin obtained x-ray crystallographic pictures from highly purified DNA samples. These pictures indicated that the DNA molecule has a helical shape with a constant diameter (▶ Figure 8.8). The x-ray films also suggested that the phosphates were on the outside of the helix, and also provided information about the distances between the stacked bases within the molecule.

Erwin Chargaff and his colleagues analyzed the base composition of DNA from a variety of organisms. Their results indicated that in DNA the amount of adenine equaled the amount of thymine, and the amount of guanine equaled the amount of cytosine. This relationship became part of what was known as Chargaff's rule.

Using the information from x-ray and chemical studies, Watson and Crick built a series of models of DNA using wire and cardboard. Eventually, they succeeded in producing a model that incorporated all of the information from the x-ray and chemical studies (▶ Active Figure 8.9). This model has the following features:

- DNA is composed of two polynucleotide chains running in opposite directions.
- The two polynucleotide chains are coiled to form a double helix.

These two features fit the x-ray results of Rosalind Franklin and Maurice Wilkins.

- In each chain, sugar and phosphate groups are linked together to form the backbone of the chain and are on the outside of the helix. The bases face inward, where

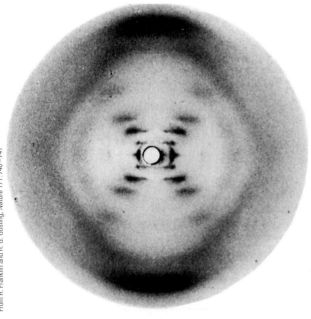

From R. Franklin and R. G. Gosling, *Nature* 171: 740–741

◀ FIGURE 8.8 An x-ray diffraction photograph of a DNA crystal. The central x-shaped pattern is typical of helical structures, and the darker areas at the top and bottom indicate a regular arrangement of subunits in the molecule. Watson and Crick used this and other photographs to construct their model of DNA.

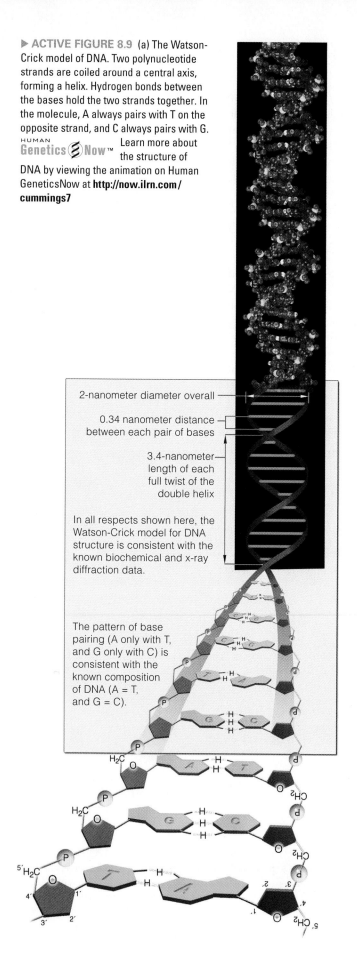

▶ **ACTIVE FIGURE 8.9** (a) The Watson-Crick model of DNA. Two polynucleotide strands are coiled around a central axis, forming a helix. Hydrogen bonds between the bases hold the two strands together. In the molecule, A always pairs with T on the opposite strand, and C always pairs with G.

Genetics ⦵ **Now**™ Learn more about the structure of DNA by viewing the animation on Human GeneticsNow at **http://now.ilrn.com/cummings7**

2-nanometer diameter overall

0.34 nanometer distance between each pair of bases

3.4-nanometer length of each full twist of the double helix

In all respects shown here, the Watson-Crick model for DNA structure is consistent with the known biochemical and x-ray diffraction data.

The pattern of base pairing (A only with T, and G only with C) is consistent with the known composition of DNA (A = T, and G = C).

they are paired by hydrogen bonds to bases in the opposite chain (▶ Active Figure 8.9).

- Base pairing is highly specific: A in one chain pairs only with T in the opposite chain, and C pairs only with G. Each set of hydrogen-bonded bases is called a base pair. The pairing of A with T and C with G fits the results obtained by Chargaff. Two hydrogen bonds link the A and T in opposite strands, and G and C are linked by three bonds.

- The base pairing of the model makes the two polynucleotide chains of DNA complementary in base composition (▶ Figure 8.10). If one strand has the sequence 5′-ACGTC-3′, the opposite strand must be 3′-TGCAG-5′, and the double-stranded structure would be written as

$$5'\text{-ACGTC-}3'$$
$$3'\text{-TGCAG-}5'$$

There are three important properties of this model:

- Genetic information is stored in the sequence of bases in the DNA. The linear sequence of bases has a high coding capacity. A DNA molecule n base pairs long has 4^n combinations. That means that a sequence of 10 nucleotides has 4^{10}, or 1,048,576, possible combinations of nucleotides. The complete set of genetic information carried by an organism (its genome) can be expressed as base pairs of DNA (▶ Table 8.1). Genomic sizes vary from a few thousand nucleotides (in viruses) that encode only a few genes to billions of nucleotides that encode 25,000 to 30,000 (as in humans). The human genome consists of about 3.2×10^9, or 3 billion, base pairs of DNA, distributed over 24 chromosomes (22 autosomes and two sex chromosomes).

- The model offers a molecular explanation for mutation. Because genetic information can be stored as a linear sequence of bases in DNA, any change in the order or number of bases in a gene can result in a mutation that produces an altered phenotype. This topic is explored in more detail in Chapter 11.

- As Watson and Crick noted, the complementary strands in DNA can be used to explain how DNA copies itself in S phase before each cell division. Each strand can be used as a **template** to reconstruct the base sequence in the opposite strand. This topic is discussed later in this chapter.

- Watson and Crick described their model in a brief paper in *Nature* in 1953. Although their model was based on the results of other workers, Watson and Crick correctly incorporated the physical and chemical data into a model that also could be used to explain the properties expected of the genetic material. Present-day applications, including genetic engineering, gene mapping, and gene therapy can be traced directly to this paper (see Spotlight on DNA Organization and Disease).

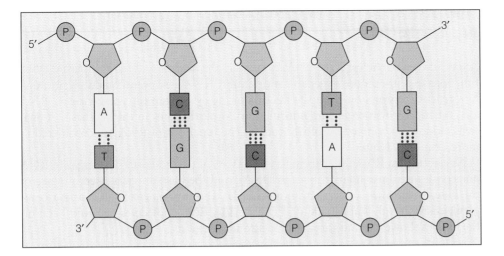

◀ FIGURE 8.10 The two polynucleotide chains in DNA run in opposite
directions. The top strand runs 5′ to 3′
and the bottom strand runs 3′ to 5′.
The base sequences in each strand
are complementary. An A in one strand
pairs with a T in the other strand, and a
C in one strand is paired with a G in the
opposite strand.

Table 8.1   Genome Size in Various Organisms

Organism	Species	Genome Size in Nucleotides
Bacterium	*E. coli*	4.6×10^6
Yeast	*S. cerevisiae*	1.2×10^7
Fruit fly	*D. melanogaster*	1.7×10^8
Tobacco plant	*N. tabacum*	4.8×10^9
Mouse	*M. musculus*	2.7×10^9
Human	*H. sapiens*	3.2×10^9

At the time there was no direct evidence to support the Watson-Crick model, but in subsequent years it was confirmed by experimental work in laboratories worldwide. The 1962 Nobel Prize for Medicine or Physiology was awarded to Watson, Crick, and Wilkins for their work on the structure of DNA. Although Rosalind Franklin provided much of the x-ray data for the Watson-Crick model, she did not receive a share of the prize. There has been some controversy over this, but because only living individuals are eligible, she could not have shared in the prize. Franklin died of cancer in 1958 at the age of thirty-seven, several years before the Nobel Prize was awarded to Watson, Crick, and Wilkins. You can read about her life in science and her role in the discovery of the structure of DNA in a recent biography, *Rosalind Franklin: The Dark Lady of DNA*. Often overlooked is the fact that although Erwin Chargaff made vital contributions to the Watson-Crick model, he did not receive a share of the Nobel Prize.

■ **Template**   The single-stranded DNA that serves to specify the nucleotide sequence of a newly synthesized polynucleotide strand.

Keep in mind

■ Watson and Crick built models of DNA structure using information from x-ray diffraction studies and chemical analyses of DNA from various organisms.

DNA Organization and Disease

Huntington disease (HD) is an autosomal dominant disease of the nervous system, characterized by involuntary movements, psychiatric and mood disorders, dementia, and death. HD is caused by an increase in the size of a cluster of nucleotide (CAG) repeats in the *HD* gene. The cluster tends to expand further when passed on by the father.

Measuring the number of CAG repeats can identify those at risk for HD. Normal individuals have 10 to 29 CAG repeats, and those with HD have 40 or more repeats. Those with 36 to 39 repeats are at risk for HD, and many of these individuals develop the disease. People with 30 to 35 repeats do not get HD, but males with 30 to 35 repeats may pass on an expanded HD gene to their offspring, who may become affected.

Individuals from families that have HD can have presymptomatic testing that follows recommendations established by the Huntington Disease Society and involves pretest and posttest visits with a neurologist, geneticist, and psychiatrist or psychologist.

8.4 RNA Is a Single-Stranded Nucleic Acid

A second type of nucleic acid, RNA, is found in the nucleus and the cytoplasm. DNA functions as a storehouse of genetic information. RNA has several functions: it transfers genetic information from the nucleus to the cytoplasm (in a few viruses, RNA also functions to store genetic information), it participates in the synthesis of proteins, and it is a component of ribosomes. The functions of RNA are considered in more detail in Chapter 9.

Nucleotides in RNA differ from those in DNA in two respects: The sugar in RNA nucleotides is ribose (deoxyribose in DNA), and the base uracil is used in place of the base thymine (▶ Table 8.2). In most cells, RNA is single-stranded, and a complementary strand is not made (▶ Figure 8.11). RNA molecules can fold back on themselves, however, and form double-stranded regions.

8.5 From DNA Molecules to Chromosomes

Although an understanding of DNA structure is an important development in genetics, it doesn't tell us how a chromosome is organized or what regulates the cycle of chromosome condensation as the cell moves from interphase to mitosis and back. This problem is significant because the spatial arrangement of DNA in the nucleus plays an important role in regulating the expression of genetic information. In addition, putting billions of nucleotides of DNA into the 46 human chromosomes requires packing a little more than 2 m (about 6.5 ft.) of DNA into a nucleus that measures about 5 μm in diameter. The length of DNA has to be compacted by a factor of almost 10,000 times to fit in the nucleus.

Within this cramped environment, the chromosomes unwind and become dispersed during interphase. In the nucleus, they undergo replication, gene expression, homologous pairing during meiosis, and contraction and coiling to become visible again during prophase. An understanding of chromosomal organization is necessary to understand these processes. In contrast to what is known about chromosomes in the nucleus, details of the structure and even the nucleotide sequence of the mitochondrial chromosome are well understood, and we will first consider this simpler system.

The mitochondrial chromosome is a circular DNA molecule.

As discussed in Chapter 2, mitochondria are cytoplasmic organelles involved in energy conversion. They contain DNA that encodes genetic information that is used within the mi-

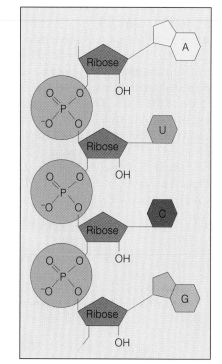

▶ FIGURE 8.11 RNA is a single-stranded polynucleotide chain. RNA molecules contain a ribose sugar instead of a deoxyribose and have uracil (U) in place of thymine.

Table 8.2	Differences between DNA and RNA	
	DNA	**RNA**
Sugar	Deoxyribose	Ribose
Bases	Adenine	Adenine
	Cytosine	Cytosine
	Guanine	Guanine
	Thymine	Uracil

tochondria. Because the sperm does not contribute any cytoplasm in fertilization, we inherit our mitochondria from our mothers. A number of genetic disorders resulting from mutations in mitochondrial DNA are known (see Chapter 4), and all offspring of affected mothers are also affected. In Case 2 at the end of this chapter, a family deals with a mitochondrial mutation in their child.

The DNA of the mitochondrial chromosome is not compacted and physically resembles the chromosomes of bacteria and other prokaryotes. This similarity reflects the evolutionary history of mitochondria and their transition from free-living prokaryotes to intracellular organelles.

Nuclear chromosomes have a complex structure.

A combination of biochemical, molecular, and microscopic techniques has provided a great deal of information about the organization and structure of human chromosomes, although we still do not know all the details. In humans and other eukaryotes, each chromosome contains a single DNA molecule. This DNA is compacted by binding with proteins to form **chromatin. Histones** are proteins that play a major role in chromosomal structure and gene regulation. Five types of histones bind to DNA and form small spherical bodies known as **nucleosomes,** connected to each other by thin threads of DNA (▶ Figure 8.12). Nucleosomes consist of DNA wound around a core of eight histone molecules.

Winding DNA around the histones shortens the length of the DNA molecule by a factor of 6 or 7. But because mitotic chromosomes are compacted by a factor of 5,000 to 10,000, there are more levels of organization between the nucleosome and the chromosome, each of which involves additional folding and/or compaction of DNA. Several models have been proposed to explain how nucleosomes are organized into more complex structures (▶ Active Figure 8.13). Most of these models are based on the idea that DNA/protein complexes (chromatin) fold into loops and fibers extending from a central protein scaffold or matrix.

The nucleus has a highly organized architecture.

The interphase nucleus is not a disorganized bag containing a diploid set of chromosomes and several nucleoli. Instead, the nucleus has an organized internal structure in which each chromosome occupies a distinct region, called a chromosome territory (▶ Figure 8.14). Chromosome territories do not overlap with one another; instead they are separated by spaces called interchromosomal domains. Nuclear organization is closely linked with function. As a result, these territories are not fixed; chromosomes move around in the nucleus at different times of the cell cycle. Some of these movements may be associated with DNA replication and chromosome duplication that takes place during S phase (review the cell cycle in Chapter 2). It has been proposed that DNA replication takes place at certain sites within the nucleus called "replication factories," and chromosomes move to these sites for replication. At other times, the chromosomes are located in territories where gene expression takes place. Much of what remains to be learned about how genes are turned on and off involves understanding the dynamics of chromosome organization in the nucleus.

Keep in mind

▪ DNA is packaged into chromosomes by several levels of coiling and compaction.

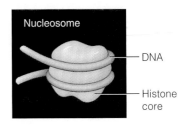

▲ FIGURE 8.12 A diagram showing how DNA coils around the outside of a histone cluster to form a nucleosome. This is the first level of DNA compaction.

▪ **Chromatin** The complex of DNA and proteins that makes up a chromosome.

▪ **Histones** DNA-binding proteins that help compact and fold DNA into chromosomes.

▪ **Nucleosomes** A bead-like structure composed of histones wrapped with DNA.

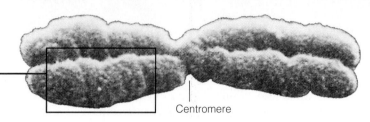

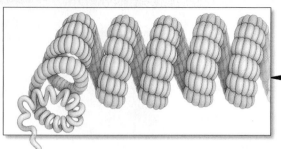

(b) At times when a chromosome is most condensed, the chromosomal proteins interact, which packages loops of already coiled DNA into a "supercoiled" array.

Centromere

(a) A duplicated human chromosome at metaphase, when it is most condensed.

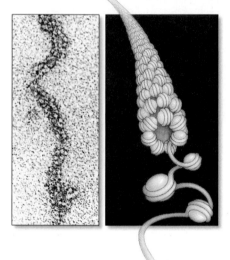

(c) At a deeper level of structural organization, the chromosomal proteins and DNA are organized as a cylindrical fiber.

8.6 DNA Replication Depends on Complementary Base Pairing

Between mitotic divisions, all cells replicate their DNA during the S phase of the cell cycle, so that each daughter cell will receive a complete set of genetic information. In their paper on the structure of DNA, Watson and Crick note that "It has not escaped our notice that the specific pairing we have postulated immediately suggests a possible copying mechanism for the genetic material." In a subsequent paper, they proposed a mechanism for DNA replication that depends on the complementary base pairing in the polynucleotide chains of DNA. If a DNA helix is unwound, each strand can serve as a template or pattern for synthesizing a new, complementary strand (▷ Active Figure 8.15). This process is known as **semiconservative replication** because one old strand is conserved in each new molecule, and one new strand is synthesized.

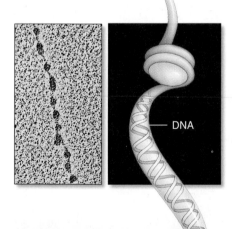

(d) Immerse a chromosome in saltwater and it loosens up to a beads-on-a-string organization. The "string" is one DNA molecule. Each "bead" is a nucleosome.

— DNA

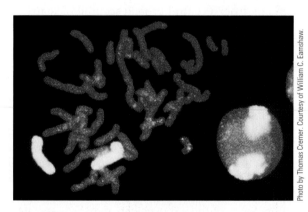

▲ FIGURE 8.14 Chromosome painting highlights both copies of human chromosome 4 in a metaphase spread (*left*), and shows the chromosome territories occupied by chromosome 4 in the interphase nucleus (*right*). In the nucleus, each chromosome occupies a distinct territory, separated from other chromosomes by a region called the interchromosome domain, a region that is free of chromosomes.

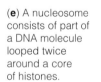

(e) A nucleosome consists of part of a DNA molecule looped twice around a core of histones.

Core of histone molecules —

◀ ACTIVE FIGURE 8.13 A model of chromosomal structure beginning with a double-stranded DNA molecule. The DNA is first coiled into nucleosomes. Then, the nucleosomes are coiled again and again into fibers that form the body of the chromosome. Chromosomes undergo cycles of coiling and uncoiling in mitosis and interphase, so their structure is dynamic.

HUMAN Genetics ⓢ Now™ Learn more about chromosome structure by viewing the animation on Human GeneticsNow at **http://now.ilrn.com/cummings7**

DNA replication in all cells, from bacteria to humans, is a complex, multistep process requiring the action of more than a dozen different enzymes. In humans, replication begins at sites called origins of replication that are present at intervals along the length of the chromosome. At these sites, proteins unwind the double helix by breaking the hydrogen bonds between bases in adjacent strands. This opens the molecule to the action of the enzyme **DNA polymerase.** DNA polymerase reads the sequence in the strand being copied and links complementary nucleotides together to form a newly synthesized strand. The completed DNA molecule contains one old strand (the strand that was copied) and one new strand (complementary to the old strand).

From the chromosomal perspective, recall that each chromosome contains one double-stranded DNA helix running from end to end. When replication is finished, the chromosome consists of two sister chromatids, joined at a common centromere. Each chromatid contains a DNA molecule that consists of one old strand and one new strand. When the centromeres divide at the beginning of anaphase, each chromatid becomes a separate chromosome that contains an accurate copy of the genetic information in the parental chromosome. In the next chapter, we will explore how the information encoded in the base sequence of DNA is converted into the amino acid sequence of proteins whose action produces phenotypes.

Semiconservative replication A model of DNA replication that provides each daughter molecule with one old strand and one newly synthesized strand. DNA replicates in this fashion.

DNA polymerase An enzyme that catalyzes the synthesis of DNA using a template DNA strand and nucleotides.

Keep in mind

■ A newly replicated DNA molecule contains one old strand and one new strand.

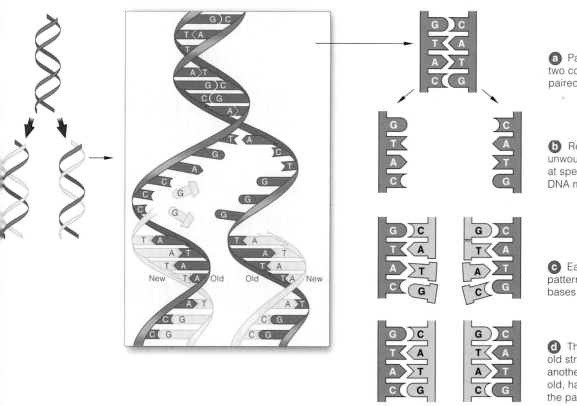

ⓐ Part of a parent DNA molecule: two complementary strands of base-paired nucleotides.

ⓑ Replication starts. The strands are unwound and separate from each other at specific sites along the length of the DNA molecule.

ⓒ Each "old" strand is the structural pattern, a template, for a sequence of bases that follows the base-pairing rule.

ⓓ The bases positioned on each old strand are hooked up to one another as a "new" strand. Each half-old, half-new DNA molecule is just like the parent molecule.

▲ **ACTIVE FIGURE 8.15** In DNA replication, the two polynucleotide strands uncoil, and each is a template for synthesizing a new strand. As the strands uncoil, complementary base pairing occurs with the template strand. These are linked together by DNA polymerase, assisted by other enzymes that help uncoil the DNA and seal up gaps in the new strands. A completed DNA molecule contains one new strand and one old strand.

Genetics ⓔ Now™ Learn more about DNA replication by viewing the animation on Human GeneticsNow at **http://now.ilrn.com/cummings7**

Genetics in Practice

CASE 1

Tune in to news programs regularly and you will probably become aware of the considerable debate over the patenting of genes. This controversy is largely fueled by the work of the Human Genome Project and the biotechnology industry. Despite numerous meetings and publications on the subject, Congress has not used U.S. patent laws to shape a policy that allows maximum innovation from biotech inventions. The first gene patents, issued in the 1970s, were granted for genes whose full nucleotide sequence was known, the protein product was known, and the protein's function was well understood.

Since then, genome projects have produced new ways of finding genes. Short sequences, only 25 to 30 nucleotides in length, called expressed sequence tags (ESTs), can be used to identify genes but provide no information about the entire gene, the product, the function of the product, or its association with any genetic disorder. Using gene-hunting software, researchers can take a short sequence of DNA and use it to search gene databases, turning up theoretical information about the sequence. For example, the sequence may belong to a gene encoding a plasma membrane protein or is similar to one in yeast that is involved in cell-cell signaling. At the present time, there are tens of thousands of ESTs and gene-hunting patent applications filed at the U.S. Patent Office.

The unresolved question, at the moment, is how much do you need to know about a gene and its usefulness to file a patent application. How should utility be defined? The diagnosis of disease certainly meets the definition of utility. Many discoveries have identified disease-causing genes such as for cystic fibrosis, fragile-X syndrome, breast cancer, colon cancer, and obesity. Many of these discoveries have patents based on diagnostic utility. An increasing number of patent applications are being filed for discoveries of hereditary disease-causing genes. These discoveries frequently lack immediate use for practical therapy, however, because gene discovery does not always include knowledge of gene function or a plan for developing a disease therapy.

The impact of a decision about gene patents is enormous. Pharmaceutical and biotech companies have invested hundreds of millions of dollars in identifying genes to be used in developing diagnostic tests and drugs. Without patents, it is unlikely that companies will invest in developing these drugs. On the other hand, patenting genes can lead to roy-alty-based gene testing with exorbitant fees, and licensing arrangements requiring payment to companies who own the patent on a particular gene. As the results of the Human Genome Project redefine health care, these issues are important to everyone.

1. What is a patent?

2. Is patenting a gene different from patenting another product or invention? Should patents be awarded for genes under any circumstances? Explain.

3. If patenting genes were not allowed, do you think it would slow gene research in a significant way?

CASE 2

A 34-year-old woman and her 1-month-old newborn were seen by a genetic counselor in the neonatal intensive care unit in a major medical center. The neonatologist was suspicious that the newborn boy had a genetic condition and requested a genetic evaluation. The newborn was very pale, was failing to thrive, had diarrhea, and had markedly increased serum cerebrospinal fluid lactate levels. In addition, he had severe muscle weakness with chart notes describing him as "floppy," and he had already had two seizures since birth. The neonatologist reported that the infant was currently suffering from liver failure, which would probably result in his death in the next few days. The panel of tests performed on the infant led the neonatologist and genetic counselor to the diagnosis of Pearson syndrome. The combination of marked metabolic acidosis and abnormalities in bone marrow cells is highly suggestive of Pearson syndrome.

Pearson syndrome is associated with a large deletion of the mitochondrial (mt) genome. The way the deletion-containing mtDNA molecules are distributed during mitosis is not known. However, it is assumed that during cell division daughter cells randomly receive mitochondria carrying wild type (WT) or mutant mtDNA. Mitochondrial DNA is, theoretically, transmitted only to offspring through the mother via the large cytoplasmic component of the oocyte. Nearly all cases of Pearson syndrome arise from new mutational events. Mitochondria have extremely poor DNA repair mechanisms, and mutations accumulate very rapidly. Most infants with Pearson syndrome die before age 3, often due to infection or liver failure.

(continued)

A diagnosis of Pearson syndrome results in an extremely grave prognosis for the patient. Unfortunately, at this point, treatment can be directed only toward symptomatic relief.

1. How would a large deletion in the mitochondrial genome cause a disease?

2. Why doesn't the mother have the disease if she has mutant mitochondrial DNA?

3. How would you react to hearing this diagnosis? How would you counsel a couple through this kind of situation?

Summary

8.1 DNA Carries Genetic Information

■ At the turn of the twentieth century, scientists identified chromosomes as the cellular components that carry genes. This discovery focused efforts to identify the molecular nature of the gene on the chromosomes and the nucleus. Biochemical analysis of the nucleus began around 1870 when Frederick Miescher first separated nuclei from cytoplasm and described nuclein, a protein/nucleic acid complex now known as chromatin.

■ Originally, proteins were regarded as the only molecular component of the cell with the complexity to encode genetic information. This changed in 1944 when Avery and his colleagues demonstrated that DNA is the genetic material in bacteria.

8.2 Watson, Crick, and the Structure of DNA

■ In 1953 Watson and Crick constructed a model of DNA structure that incorporated information from the chemical studies of Chargaff and the x-ray crystallographic work of Wilkins and Franklin. They proposed that DNA is composed of two polynucleotide chains oriented in opposite directions and held together by hydrogen bonding to complementary bases in the opposite strand. The two strands are wound around a central axis in a right-handed helix.

■ The mitochondrial chromosome, carrying genes that can cause maternally transmitted disorders, is a circular DNA molecule.

8.3 DNA Contains Two Polynucleotide Chains

■ Within chromosomes, DNA is coiled around clusters of histones to form structures known as nucleosomes. Supercoiling of nucleosomes may form fibers that extend at right angles to the axis of the chromosome. The structure of chromosomes must be dynamic to allow the uncoiling and recoiling seen in successive phases of the cell cycle, but the details of this transition are still unknown.

8.4 RNA Is a Single-Stranded Nucleic Acid

■ RNA is another type of nucleic acid. It contains a different sugar than DNA and uses the base uracil in place of thymine. RNA molecules are single-stranded but can fold back on themselves to produce double-stranded regions. RNA has a variety of functions in the cell.

8.5 From DNA Molecules to Chromosomes

■ Each human chromosome contains a single DNA molecule. Each DNA molecule is extensively coiled to allow it to fit into the nucleus.

8.6 DNA Replication Depends on Complementary Base Pairing

■ In DNA replication, strands are copied to produce semi-conservatively replicated daughter strands.

Questions and Problems

HUMAN Genetics Now™ Preparing for an exam? Assess your understanding of this chapter's topics with a pre-test, a personalized learning plan, and a post-test at **http://now.ilrn.com/cummings7**

DNA Carries Genetic Information
1. Until 1944, which cellular component was thought to carry genetic information?
   a. carbohydrate
   b. nucleic acid
   c. protein
   d. chromatin
   e. lipid

2. Why do you think nucleic acids were not originally considered to be carriers of genetic information?
3. The experiments of Avery and his coworkers led to the conclusion that:
   a. bacterial transformation occurs only in the laboratory.

b. capsule proteins can attach to uncoated cells.

c. DNA is the transforming agent and is the genetic material.

d. transformation is an isolated phenomenon in *E. coli*.

e. DNA must be complexed with protein in bacterial chromosomes.

4. In the experiments of Avery, what was the purpose of treating the transforming extract with enzymes?

5. Read the following experiment and interpret the results to form your conclusion. Experimental data: S bacteria were heat-killed and cell extracts were isolated. The extracts contained cellular components including lipids, proteins, DNA, and RNA. These extracts were mixed with live R bacteria and then injected together into mice along with various enzymes (proteases, RNAases, and DNAases). Proteases degrade proteins, RNAases degrade RNA, and DNAases degrade DNA.

S extract + live R cells	mouse dies
S extract + live R cells + 1 protease	mouse dies
S extract + 1 live R cell + 1 RNAase	mouse dies
S extract + live R cells + DNAase	mouse lives

Based on these results, what is the transforming principle?

6. Recently, scientists discovered that a rare disorder called polkadotism is caused by a bacterial strain, polkadotiae. Mice injected with this strain (P) develop polka dots on their skin. Heat-killed P bacteria and live D bacteria, a nonvirulent strain, do not produce polka dots when separately injected into mice. However, when a mixture of heat-killed P cells and live D cells were injected together, the mice developed polka dots. What process explains this result? Describe what is happening in the mouse to cause this outcome.

DNA Contains Two Polynucleotide Chains

7. Nucleosomes are complexes of:
   a. nonhistone protein and DNA.
   b. RNA and histone.
   c. histones, nonhistone proteins, and DNA.
   d. DNA, RNA, and protein.
   e. amino acids and DNA.

8. Discuss the levels of chromosomal organization with reference to the following terms:
   a. nucleotide   **b.** DNA double helix
   c. histones   **d.** nucleosomes
   e. chromatin

9. List the pyrimidine bases, the purine bases, and the base pairing rules for DNA.

10. In analyzing the base composition of a DNA sample, a student loses the information on pyrimidine content. The purine content is A = 27% and G = 23%. Using Chargaff's rule, reconstruct the missing data, and list the base composition of the DNA sample.

11. The basic building blocks of nucleic acids are:
   a. nucleosides.   **b.** nucleotides.
   c. ribose sugars.   **d.** amino acids.
   e. purine bases.

12. Adenine is a:
   a. nucleoside.   **b.** purine.
   c. pyrimidine.   **d.** nucleotide.
   e. base.

13. Polynucleotide chains have a 5′ and a 3′ end. Which groups are found at each of these ends?
   a. 5′ sugars, 3′ phosphates
   b. 3′ OH, 5′ phosphates
   c. 3′ base, 5′ phosphate
   d. 5′ base, 3′ OH
   e. 5′ phosphates, 3′ bases

14. DNA contains many hydrogen bonds. Describe a hydrogen bond and how this type of chemical bond holds DNA together.

15. Watson and Crick received the Nobel Prize for:
   a. generating x-ray crystallographic data of DNA structure.
   b. establishing that DNA replication is semiconservative.
   c. solving the structure of DNA.
   d. proving that DNA is the genetic material.
   e. showing that the amount of A equals the amount of T.

16. State the properties of the Watson-Crick model of DNA in the following categories:
   a. number of polynucleotide chains
   b. polarity (running in same or opposite directions)
   c. bases on interior or exterior of molecule
   d. sugar/phosphate on interior or exterior of molecule
   e. which bases pair with which
   f. right- or left-handed helix

17. Using ▶ Figure 8.7 as a guide, draw a dinucleotide composed of C and A. Next to this, draw the complementary dinucleotide in an antiparallel fashion. Connect the dinucleotides with the appropriate hydrogen bonds.

18. A beginning genetics student is attempting to complete an assignment to draw a base pair from a DNA molecule. The drawing is incomplete, and the student does not know how to finish. He asks for your advice. The assignment sheet shows that the drawing is to contain three hydrogen bonds, a purine, and a pyrimidine. From your knowledge of the pairing rules and the number of hydrogen bonds in A/T and G/C base pairs, what base pair do you help the student draw?

RNA Is a Single-Stranded Nucleic Acid

19. What is the purpose of making an RNA copy of the DNA in gene expression?

20. How does DNA differ from RNA with respect to the following characteristics?
   a. number of chains   **b.** bases used
   c. sugar used   **d.** function

21. RNA is ribonucleic acid, and DNA is *deoxy*ribonucleic acid. What exactly is deoxygenated about DNA?

DNA Replication Depends on Complementary Base Pairing

22. What is the function of DNA polymerase?
   a. It degrades DNA in cells.

b. It adds RNA nucleotides to a new strand.

c. It coils DNA around histones to form chromosomes.

d. It adds DNA nucleotides to a replicating strand.

e. none of the above

23. Which of the following statements is *not* true about DNA replication?

a. It occurs during the M phase of the cell cycle.

b. It makes a sister chromatid.

c. It denatures DNA strands.

d. It occurs semiconservatively.

e. It follows base pairing rules.

24. Make the complementary strand to the following DNA template, and label both strands as 5′ to 3′, or 3′ to 5′ (P = phosphate in the diagram). Draw an arrow showing the direction of synthesis of the new strand. How many hydrogen bonds are in this double strand of DNA?

template P—AGGCTCG—OH

new strand:

25. How does DNA replication occur in a precise manner to ensure that identical genetic information is put into the new chromatid? See ▶ Figure 8.15.

Internet Activities

Internet Activities are critical thinking exercises using the resources of the World Wide Web to enhance the principles and issues covered in this chapter. For a full set of links and questions investigating the topics described below, log on to **http://biology.brookscole.com/cummings7**

1. *Experimenting with the Structure of DNA.* The Genetic Science Learning Center (a joint project of the University of Utah and the Utah Museum of Natural History) provides general genetics information to students and the community. Go to the link on "How to Extract DNA from Anything Living." This is an activity that you *can* do at home, but even if you choose not to try the experiment, you can still learn from the experimental design and discussion.

Further Exploration. The Genetic Science Learning Center home page has a variety of review materials, interesting visuals, and fun activities.

2. *How Do Scientific Advances Occur? Access Excellence* is a website for "health and bioscience teachers and learners" run by the National Health Museum. Follow

the "About Biotech" link to *Biotech Chronicles* and then click on "Pioneer Profiles." Read the profiles of Rosalind Franklin and James Watson. Then, from the home page, follow the "Activities Exchange" link to *Classic Collection* and read "A Visit with Dr. Francis Crick."

Further Exploration. The "On-line Biology Book" has a good overview of DNA and molecular genetics that goes through the process of the discovery of DNA.

3. *Is The Pursuit of Science Always Objective and Unbiased? Access Excellence* was first developed in 1993 by the pioneering biotechnology company Genentech. In 1999, the website was donated to the nonprofit National Health Museum but is still partially funded and intellectually supported by Genentech.

How would you vote now?

No DNA vaccines are currently approved for use in humans; however, clinical trials of such vaccines are underway to assess their safety and effectiveness. These trials are of DNA vaccines developed quickly after the discovery of the SARS virus. Because the trials will last several years, another outbreak of deadly SARS virus could possibly occur before the results of the vaccine studies are in. There is also the threat of a bioterrorist attack releasing a potentially fatal disease-causing organism before the studies are complete. Now that you know more about the structure and organization of DNA, what do you think? If another SARS outbreak or a bioterrorist attack occurred, would you agree to be treated with a DNA vaccine? Would you allow members of your family to be injected with a DNA vaccine? Visit the Human Heredity Companion website to find out more on the issue, then cast your vote online at **http://biology.brookscole.com/cummings7**

For further reading and inquiry, log on to **InfoTrac College Edition,** your world-class online library, including articles from nearly 5,000 periodicals, at **http://infotrac.thomsonlearning.com**

Gene Expression: From Genes to Proteins

S IX months after earning a business degree from the University of Miami in 2001, Charlene Singh began to experience memory loss and behavior changes. A short time later, she had difficulty walking, and in 2002 she was diagnosed with variant Cruetzfeldt-Jakob disease (vCJD), the human form of mad-cow disease. The disease takes 10 to 15 years to develop, and younger people are more likely to develop vCJD than the elderly.

People get vCJD by eating meat from cows infected with a brain-wasting disease called bovine spongiform encephalopathy (BSE). Charlene was born in England and moved with her family to the United States in 1992. BSE first appeared in cows in England in the 1980s, and the first cases of vCJD developed in the early 1990s. Because of the circumstances, it is believed that Charlene became infected with vCJD while she lived in England. She died in June 2004, at age 25, in her home in Florida. Charlene was the first U.S. resident to die from vCJD, but within Great Britain, just over 140 people have already died from this disease, and during the next decade, several thousand additional cases may develop.

BSE, vCJD, and several other diseases are called prion diseases. They develop when normal proteins in the body refold into a new, infectious three-dimensional shape that kills cells of the brain and nervous system. Proteins perform tasks essential for life; they form most of the structures in a cell, transfer energy to drive all processes in a living cell, help copy chromosomes for cell division, and control which genes are switched on and off, relay signals, fight infection, and repair damage caused by environmental agents.

The information to make proteins is encoded in genes. In this chapter, we will examine the relationship between genes and proteins and the role of

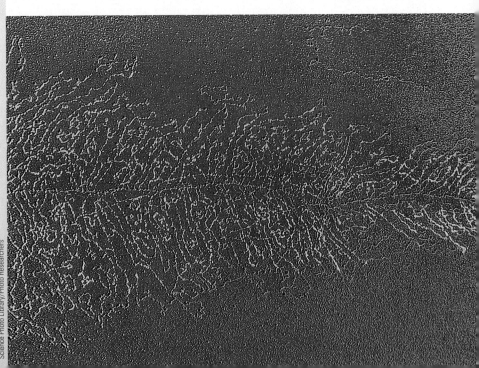

Science Photo Library/Photo Researchers

DNA as a carrier of genetic information. We will discuss the transfer of genetic information from the sequence of nucleotides in DNA to the sequence of amino acids in protein, and the relationship between protein structure and function.

How would you vote?

Most cases of prion diseases caused by eating infected beef have been reported in the United Kingdom, not in the United States. Prions have also been transmitted by contaminated surgical and dental instruments, and in other cultures, by cannibalism. There is no cure for a prion infection, and prions cannot be destroyed by sterilization. Some countries are now testing all beef used in human consumption, while others, such as the United States, are randomly testing only a small sample of cows. If you were traveling or living in a country with a history of infected cows, would you eat beef or allow your children to eat it? If not, what if infected beef was linked to human deaths in that country? Visit the Human Heredity Companion website to find out more on the issue, then cast your vote online at **http://biology.brookscole.com /cummings7**

9.1 The Link between Genes and Proteins

At the turn of the twentieth century, soon after the rediscovery of Mendel's work, Archibald Garrod recognized the relationship between genes, proteins, and phenotype in his studies of infants with a condition called **alkaptonuria** (OMIM 203500). Newborns with alkaptonuria can be identified because their urine turns black. Garrod chemically analyzed the urine of these children and found it contained large quantities of a compound he called alkapton (now called homogentisic acid). He reasoned that normally, homogentisic acid must be converted into other products because it does not build up in the urine of unaffected people. In infants with alkaptonuria, however, this conversion must be blocked, causing the buildup of homogentisic acid, which is excreted in the urine. He called this condition an "inborn error of metabolism." By doing a pedigree analysis, Garrod also discovered that alkaptonuria is a genetic disorder inherited as an autosomal recessive trait.

Genes are linked to metabolism.

Biochemical reactions, like the conversion of homogentisic acid, are organized into chains of chemical reactions known as metabolic pathways. In a pathway, an enzyme converts one compound into another. This second compound is converted into a third, and so forth (▶ Figure 9.1). Each reaction is carried out by a different enzyme. Enzymes are protein molecules that serve as biological catalysts to carry out specific

■ **Alkaptonuria** An autosomal recessive trait with altered metabolism of homogentisic acid. Affected individuals do not produce the enzyme needed to metabolize this acid, and their urine turns black.

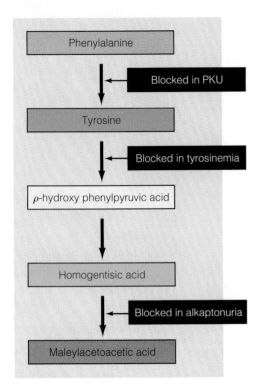

▲ FIGURE 9.1 A metabolic pathway beginning with the amino acid phenylalanine. The enzyme-catalyzed steps are shown as arrows. Mutations cause blocks in this pathway, leading to genetic disorders, including phenylketonuria (PKU), tyrosinemia, and alkaptonuria.

biochemical reactions. Loss of activity in a single enzyme can disrupt an entire pathway, causing an alteration in the cell's biochemistry and producing an altered phenotype. Garrod correctly predicted that genetic disorders could be caused by loss of enzyme activity, but the pathway from gene to protein to phenotype was not worked out for several decades. Case 1 at the end of the chapter deals with a disorder caused by the loss of a single enzyme by mutation.

What is the relationship between genes and enzymes?

George Beadle and Edward Tatum established the connection between genes, enzymes, and the phenotype in the late 1930s and early 1940s through their work on *Neurospora*, a common bread mold (▶ Figure 9.2). Using *Neurospora*, they demonstrated that mutation of a single gene caused loss of activity in a single enzyme, resulting in a mutant phenotype. This connection between a mutant gene, a mutant enzyme, and a mutant phenotype was a key step in understanding that genes produce phenotypes through the action of proteins. Beadle and Tatum received the Nobel Prize in 1958 for their work on the pathway from genes to proteins to phenotype.

9.2 Genetic Information Is Stored in DNA

Proteins are the intermediate between genes and phenotype. The phenotypes of a cell, tissue, and organism are all the result of protein function. When these functions are absent or changed, the result is a mutant phenotype, which we describe as a genetic disorder. Because proteins are the products of genes, and genes are made up of DNA, information encoded in DNA must control the kinds and amounts of proteins present in the cell. But just how is genetic information carried in DNA? Watson and Crick proposed that genetic information is encoded in the sequence of nucleotides in DNA. The amount of information stored in any cell is related to the number of nucleotides of DNA carried within the cell. This

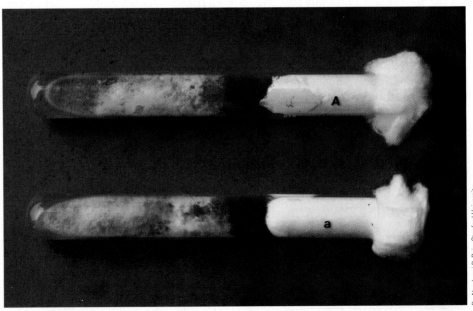

▶ FIGURE 9.2 A culture of the mold *Neurospora crassa*, an organism used by George Beadle and Edward Tatum in their work on biochemical genetics.

Dr. Namboori B. Raju, Stanford University

number ranges from a few thousand base pairs in some viruses to more than 3 billion base pairs in humans and more than twice that amount in some amphibians and plants. A gene typically consists of hundreds or thousands of nucleotides. Each gene has a beginning and end, marked by specific nucleotide sequences, and a molecule of DNA can contain thousands of genes.

Keep in mind

■ The information necessary to make proteins is encoded in the nucleotide sequence of DNA.

How do genes (in the form of DNA) control the production of proteins? Proteins are linear molecules assembled from subunits called amino acids. Twenty different types of amino acids are used to assemble proteins. The diversity of proteins found in nature results from the number of possible combinations of these 20 different amino acids. Because each amino acid in a protein can be any one of 20 different amino acids, the number of different combinations is 20^n, where n is the number of amino acids in the protein. In a protein composed of only 5 amino acids, 20^5, or 3,200,000, combinations are possible, each of which has a different amino acid sequence and a potentially different function. Most proteins are actually composed of several hundred amino acids, so literally billions and billions of combinations are possible.

If DNA has only four different nucleotides (A, T, C, and G), and there are 20 different amino acids in proteins, how can only four nucleotides encode the information for all 20 amino acids?

9.3 The Genetic Code: The Key to Life

Information that spells out the number and order of the amino acids in a protein is encoded in the nucleotide sequence of a gene. Because DNA is composed of only four different nucleotides, at first glance it seems difficult to envision how the information for literally billions of different combinations of 20 different amino acids can be carried in DNA. How exactly does DNA encode genetic information?

To answer this question, let's start with some hypothetical cases. If each nucleotide encoded the information for one amino acid, only four different amino acids could be inserted into proteins (four nucleotides, taken one at a time, or 4^1). If two nucleotides encoded the information for one amino acid, only 16 combinations would be possible (four nucleotides, taken two at a time, or 4^2). On the other hand, a sequence of three nucleotides allows 64 combinations (four nucleotides, taken three at a time, or 4^3), which doesn't seem right, because there are 44 more combinations than the 20 needed to encode amino acids.

The question of how many nucleotides are needed to encode one amino acid was answered in a series of experiments using mutants of a virus called T4. Francis Crick, Sidney Brenner, and their colleagues analyzed mutations in T4 genes and discovered that a sequence of three nucleotides encodes the information for one amino acid. They also found that some amino acids could be specified by more than one combination of three nucleotides. This built-in redundancy uses up most of the other 44 combinations in the code. This work established that in DNA a genetic code consists of a linear series of nucleotides, read three at a time, and that each triplet specifies an amino acid.

After Crick and Brenner's work, the question of which three nucleotides encode which amino acids was quickly worked out, and the coding information contained in all 64 triplet combinations was established (▶ Figure 9.3). By convention, the genetic code is written as it appears in an RNA copy of the information in DNA, and each group of three nucleotides is called a **codon.** ▶ Figure 9.3 shows that 61 of these com-

■ **Codon** Triplets of nucleotides in mRNA that encode the information for a specific amino acid in a protein.

First base	U	C	A	G	Third base
U	phenylalanine	serine	tyrosine	cysteine	U
	phenylalanine	serine	tyrosine	cysteine	C
	leucine	serine	STOP	STOP	A
	leucine	serine	STOP	tryptophan	G
C	leucine	proline	histidine	arginine	U
	leucine	proline	histidine	arginine	C
	leucine	proline	glutamine	arginine	A
	leucine	proline	glutamine	arginine	G
A	isoleucine	threonine	asparagine	serine	U
	isoleucine	threonine	asparagine	serine	C
	isoleucine	threonine	lysine	arginine	A
	methionine (or START)	threonine	lysine	arginine	G
G	valine	alanine	aspartate	glycine	U
	valine	alanine	aspartate	glycine	C
	valine	alanine	glutamate	glycine	A
	valine	alanine	glutamate	glycine	G

▲ FIGURE 9.3 The genetic code is read in blocks of three bases in mRNA called codons. The bases in the left column are the choices for the first base in a codon. The top column lists the second base, and the right-hand column lists the third base. Sixty-one codons code for amino acids. Three codons (UAA, UAG, UGA) are signals to stop translation. One codon (UAA) has two functions. It is the codon marking the beginning of an mRNA, and it codes for the amino acid methionine.

Messenger RNA (mRNA) A single-stranded complementary copy of the nucleotide sequence in a gene.

Transcription Transfer of genetic information from the base sequence of DNA to the base sequence of RNA, mediated by RNA synthesis.

Translation Conversion of information encoded in the nucleotide sequence of an mRNA molecule into the linear sequence of amino acids in a protein.

binations actually code for amino acids, but 3 (UAA, UAG, and UGA) do not encode amino acids, and they are called stop codons. The AUG codon has two functions. It encodes the information for the amino acid methionine and serves as the start codon, the first codon in a gene, marking the beginning of a coding sequence for a specific protein.

Aside from some exceptions, the same codons are used for the same amino acids in viruses and organisms, including bacteria, algae, fungi, and multicellular plants and animals. The universal nature of the genetic code means that the genetic code was established very early during the evolution of life on this planet. The existence of a universal code is strong evidence that all living things are closely related and may have evolved from a common ancestor.

Keep in mind

■ The three nucleotides in a codon are a universal language specifying the same amino acid in almost all organisms.

Now that we know how the information for proteins is encoded in DNA, let's turn our attention to another question: how is the linear sequence of nucleotides in a gene converted into the linear sequence of amino acids in a protein? In humans and other eukaryotes, almost all the cell's DNA is found in the nucleus (some is in the mitochondria), and almost all proteins are found in the cytoplasm. This means that the process of information transfer from gene to gene product must be indirect.

9.4 Tracing the Flow of Genetic Information from Nucleus to Cytoplasm

The transfer of genetic information from the linear sequence of nucleotides in a DNA molecule into the linear sequence of amino acids of a protein occurs in two steps. First, the information encoded in a gene is copied into an RNA molecule known as **messenger RNA (mRNA)**. This step is called **transcription** (▶ Figure 9.4). In humans and other eukaryotes, the mRNA moves to the cytoplasm, where the information encoded in the nucleotide sequence of the mRNA is converted into the amino acid sequence of a protein. This step is called **translation** (▶ Figure 9.4).

The amino acid sequence, in turn, determines the structural and functional characteristics of the protein and its role in phenotypic expression. In the next sections, we will examine the steps of protein synthesis in more detail.

Keep in mind

■ Genetic information for proteins, in the form of mRNA, moves from the nucleus to the cytoplasm where it is translated into the amino acid sequence of a polypeptide.

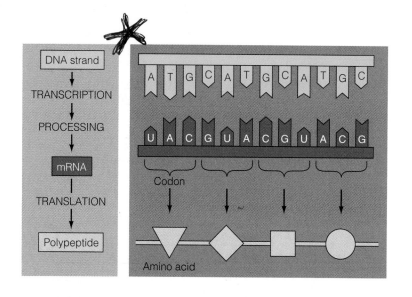

◀ **FIGURE 9.4** The flow of genetic information. One strand of DNA is transcribed into a strand of mRNA. The mRNA is processed and moves to the cytoplasm, where it is converted into the amino acid sequence of a protein.

9.5 Transcription Produces Genetic Messages

Transcription begins when the DNA in a chromosome unwinds, and one strand is used as a template to make an mRNA molecule (▶ Active Figure 9.5). Transcription has three stages: initiation, elongation, and termination. In initiation, an enzyme called RNA polymerase binds to a specific nucleotide sequence in the DNA adjacent to a gene. This sequence is called a **promoter region**. After the polymerase is bound,

■ **Promoter region**   A region of a DNA molecule to which RNA polymerase binds and initiates transcription.

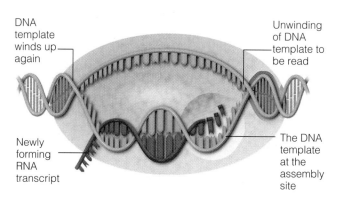

◀ **ACTIVE FIGURE 9.5** Transcription of a gene. An enzyme, RNA polymerase, uses one strand of DNA as a template to synthesize an mRNA molecule.

HUMAN Genetics ⊘ Now™   Learn more about transcription by viewing the animation on Human GeneticsNow at **http://now.ilrn.com/cummings7**

(a) RNA polymerase initiates transcription at a promoter region in the DNA. It will recognize the base sequence located downstream from that site as a template for linking together the nucleotides adenine, cytosine, guanine, and uracil into a strand of RNA.

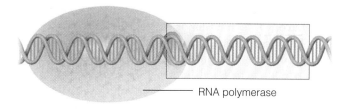

(b) All through transcription, the DNA double helix becomes unwound in front of the RNA polymerase. Short lengths of the newly forming RNA strand briefly wind up with its DNA template strand. New stretches of RNA unwind from the template (and the two DNA strands wind up again).

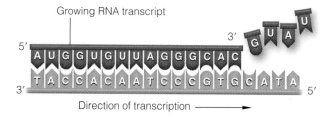

Direction of transcription ⟶

(c) What happened at the assembly site? The RNA polymerase catalyzed the base-pairing of RNA nucleotides, one after another, with exposed bases on the DNA template strand.

(d) At the end of the gene region, the last stretch of the new mRNA transcript is unwound and released from the DNA.

the two strands of DNA in the gene unwind, exposing the DNA strand that will be a template for RNA synthesis.

In the elongation stage of transcription, RNA polymerase links RNA nucleotides together, forming an RNA molecule called an mRNA transcript (▶ Active Figure 9.5). The rules of base pairing in transcription are the same as in DNA replication, with one exception: an A on the DNA template ends up as a U in the RNA transcript (recall from Chapter 8 that there is no T in RNA, so there is no A:T pairing in RNA). For example, if the nucleotide sequence in the DNA template strand is

CGGATCAT

the mRNA will have the sequence

GCCUAGUA

In humans, elongation proceeds at about 30 to 50 nucleotides per second. As the RNA polymerase moves along the DNA template, it eventually reaches the end of the gene. This region is marked by a nucleotide sequence called a **terminator region.** When the RNA polymerase reaches the terminator sequence, it falls off the DNA strand, the mRNA molecule is released, the DNA strands re-form a double helix, and transcription is terminated (▶ Active Figure 9.5). The length of the mRNA transcript depends on the size of the gene. Most transcripts in humans are about 5,000 nucleotides long, although lengths up to 200,000 nucleotides have been reported.

Most human genes have a complex internal organization.

Many, if not most, human genes contain nucleotide sequences that are transcribed but not translated into the amino acid sequence of a protein. These sequences, called **introns,** can vary in number from 0 to 75 or more. Introns also vary in size, ranging from about 100 nucleotides to more than 100,000.

The nucleotides in a gene that are transcribed and translated into the amino acid sequence of a protein are called **exons.** The internal organization of a typical human gene is shown in ▶ Figure 9.6. The combination of exons and introns determines the length of a gene, and often the exons are only a small fraction of the total nucleotides in a gene. For example, the dystrophin gene (see Chapter 4 for a discussion of muscular dystrophy and dystrophin) is more than 2 million nucleotides in length and contains 79 introns, which make up more than 99% of the gene. Most genes are not as long as the dystrophin and do not have as many introns. The alpha-globin gene (discussed in Chapter 10) is 800 nucleotides in length and has two introns that make up 30% of the gene.

Messenger RNA is processed and spliced.

In humans and other eukaryotes, transcription produces large mRNA precursor molecules called pre-mRNAs. These precursors are processed in the nucleus to remove introns and the exons are spliced together to form mRNA molecules (▶ Ac-

■ Terminator region The nucleotide sequence at the end of a gene that signals the end of transcription.

■ Introns DNA sequences present in some genes that are transcribed but are removed during processing and therefore are not present in mature mRNA.

■ Exons DNA sequences that are transcribed, joined to other exons during mRNA processing, and translated into the amino acid sequence of a protein.

■ Cap A modified base (guanine nucleotide) attached to the 5′ end of eukaryotic mRNA molecules.

■ Poly-A tail A series of A nucleotides added to the 3′ end of mRNA molecules.

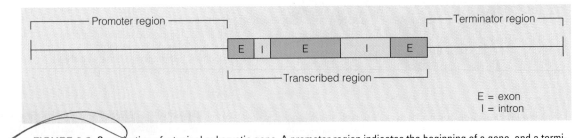

▲ FIGURE 9.6 Organization of a typical eukaryotic gene. A promoter region indicates the beginning of a gene, and a terminator region marks the end of a gene. The transcribed region contains introns and exons. Only the sequences in the exons appear in the mature mRNA and are translated into the amino acid sequence of a protein.

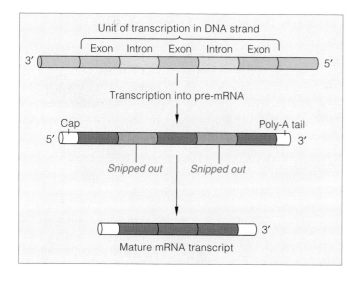

Unit of transcription in DNA strand

Exon | Intron | Exon | Intron | Exon

3' ⟶ 5'

Transcription into pre-mRNA

Cap   Poly-A tail

5'   3'

Snipped out   *Snipped out*

3'

Mature mRNA transcript

◄ **ACTIVE FIGURE 9.7** Steps in the processing and splicing of mRNA. The template strand of DNA is transcribed into a pre-mRNA molecule. The ends of this molecule are modified, and the introns are spliced out to produce a mature mRNA molecule. The mRNA is then moved to the cytoplasm for translation.

HUMAN
Genetics ⓔ Now™   Learn more about messenger RNA processing by viewing the animation on Human GeneticsNow at **http://now.ilrn.com/cummings7**

■ **Amino group**   A chemical group (NH_2) found in amino acids and at one end of a polypeptide chain.

■ **Carboxyl group**   A chemical group (COOH) found in amino acids and at one end of a polypeptide chain.

■ **R group**   A term used to indicate the position of an unspecified group in a chemical structure. They can be positively or negatively charged or neutral.

■ **Peptide bond**   A covalent chemical link between the carboxyl group of one amino acid and the amino group of another amino acid.

■ **Polypeptide**   A molecule made of amino acids joined together by peptide bonds.

tive Figure 9.7). The mature mRNAs are transported to the cytoplasm, where translation takes place.

Pre-mRNA molecules are processed by the addition of nucleotides to the 5' and the 3' ends. The sequence at the 5' end, known as a **cap**, helps attach the mRNA to ribosomes during translation. At the 3' end a string of 30 to 100 A nucleotides, called the **poly-A tail,** is added. The role of the tail is unclear, because some mRNAs lack this modification.

In addition to processing, the pre-mRNA molecules are cut and spliced to remove introns (see Spotlight on Mutations in Splicing Sites and Genetic Disorders). Enzymes cut the pre-mRNA at the junction between introns and exons. The exons are spliced together to form the mature mRNA, and the introns are discarded.

After processing and splicing, the mRNA is transported from the nucleus to the cytoplasm via the nuclear pores, where the encoded information is translated into the amino acid sequence of a protein.

9.6 Translation Requires the Interaction of Several Components

Translation converts the nucleotide sequence in mRNA into the amino acid sequence of a protein, a job that requires several different cytoplasmic components, each of which has a separate, specialized job. Before we examine the details of translation, let's look at the components.

First, we will examine the components of proteins. We have already explained that proteins are assembled from amino acids, and that twenty different amino acids can be used to make proteins. Each amino acid has three characteristic chemical groups: an **amino group** (NH_2), a **carboxyl group (COOH),** and an **R group** (▶ Figure 9.8a). R groups are side chains that are different for each amino acid. Some R groups are positively charged, some carry a negative charge, and others are electrically neutral. The 20 amino acids found in proteins and their abbreviations are listed in ▶ Table 9.1.

During protein synthesis, amino acids are linked together by the formation of covalent **peptide bonds** formed between the amino group of one amino acid and the carboxyl group of another amino acid (▶ Figure 9.8b). Two linked amino acids form a dipeptide, three form a tripeptide, and ten or more make a **polypeptide.** Each polypeptide (and protein) has a free

Table 9.1	Amino Acids Commonly Found in Proteins
Amino Acid	**Abbreviation**
Alanine	ala
Arginine	arg
Asparagine	asn
Aspartic acid	asp
Cysteine	cys
Glutamic acid	glu
Glutamine	gln
Glycine	gly
Histidine	his
Isoleucine	ile
Leucine	leu
Lysine	lys
Methionine	met
Phenylalanine	phe
Proline	pro
Serine	ser
Threonine	thr
Tryptophan	trp
Tyrosine	tyr
Valine	val

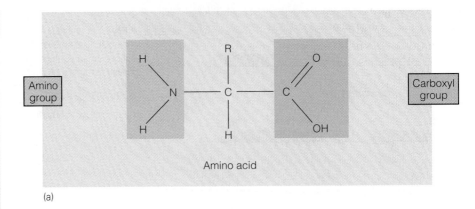

Amino group | Carboxyl group

Amino acid

(a)

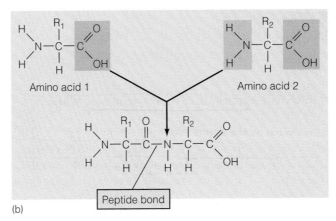

Amino acid 1 | Amino acid 2

Peptide bond

(b)

▲ **FIGURE 9.8** (a) An amino acid, showing the amino group, the carboxyl group, and the chemical side chain known as an R group. The R groups differ in each of the 20 amino acids used in protein synthesis. (b) Formation of a peptide bond between two amino acids.

■ **N-terminus** The end of a polypeptide or protein that has a free amino group.

■ **C-terminus** The end of a polypeptide or protein that has a free carboxyl group.

■ **Ribosomes** Cytoplasmic particles composed of two subunits that are the site of protein synthesis.

■ **Ribosomal RNA (rRNA)** RNA molecules that form part of the ribosome.

■ **Transfer RNA (tRNA)** A small RNA molecule that contains a binding site for a specific type of amino acid and a three-base segment known as an anticodon that recognizes a specific base sequence in messenger RNA.

■ **Anticodon** A group of three nucleotides in a tRNA molecule that pairs with a complementary sequence (known as a codon) in an mRNA molecule.

amino group at one end, known as the **N-terminus,** and a free carboxyl group, called the **C-terminus** at the other.

The nucleotide sequence of the mRNA is converted into the amino acid sequence of a protein with the help of two other components that involve RNA: ribosomes (▶ Figure 9.9) and transfer RNAs (tRNAs) (▶ Figure 9.10).

Ribosomes are cellular organelles with two subunits. Each subunit contains a type of RNA called **ribosomal RNA (rRNA) combined with proteins.** Ribosomes can float in the cytoplasm or attach to the outer membrane of the endoplasmic reticulum (ER) (review organelles in Chapter 2). At either location, ribosomes are the site of protein synthesis.

Transfer RNA (tRNA) molecules are adapters that recognize specific mRNA codons and their encoded amino acid. A tRNA molecule is a small (about 80 nucleotides) single-stranded molecule folded back on itself, forming a cloverleaf with several looped regions (▶ Figure 9.10). As adapters, tRNA molecules have two tasks: (1) bind to the appropriate amino acid, and (2) recognize the proper codon in mRNA. The folded structure of tRNA molecules allows them to perform both tasks. A loop

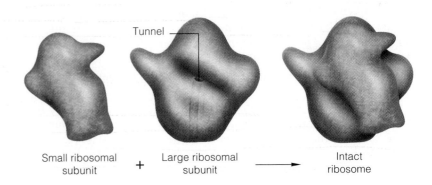

Tunnel

Small ribosomal subunit + Large ribosomal subunit ⟶ Intact ribosome

▶ **FIGURE 9.9** Three-dimensional models of the small and large subunits of ribosomes.

at one end of the molecule contains three nucleotides called an **anticodon.** The anticodon recognizes and pairs with a specific codon in an mRNA molecule. The stem region at the other end of the tRNA binds the amino acid specified by the codon (▶ Figure 9.10b).

There are 20 different amino acids, each matched by a different tRNA with its anticodon and amino acid binding site. The tRNA that matches glycine has CCC as its anticodon and binds glycine at its other end. However, tRNA molecules don't recognize and bind amino acids just by bumping into them. This task is carried out by an enzyme that binds a specific tRNA with its proper amino acid and links them together.

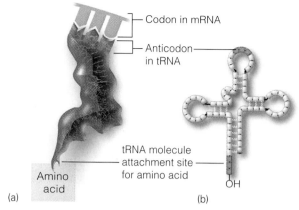

▲ FIGURE 9.10 Two models for transfer RNA (tRNA) molecules. (a) Three-dimensional model of tRNA showing the anticodon and the amino acid binding site. (b) A cloverleaf model for tRNA.

Genetic Journeys

Antibiotics and Protein Synthesis

Antibiotics are chemicals produced by microorganisms as defense mechanisms. The most effective antibiotics work by interfering with essential biochemical or reproductive processes. Many antibiotics block or disrupt one or more stages in protein synthesis. Some of these are listed here.

Tetracyclines are a family of chemically related compounds used to treat several types of bacterial infections. Tetracyclines interfere with the initiation of translation. The tetracycline molecule binds to the small ribosomal subunit and prevents binding of the tRNA anticodon in the first step in initiation. Both eukaryotic and prokaryotic ribosomes are sensitive to the action of tetracycline, but this antibiotic cannot pass through the plasma membrane of eukaryotic cells. Because it can enter bacterial cells to inhibit protein synthesis, it will stop bacterial growth, helping the immune system fight the infection.

Streptomycin is used in hospitals to treat serious bacterial infections. It binds to the small ribosomal subunit but does not prevent initiation or elongation; however, it does affect the efficiency of protein synthesis. When streptomycin binds to a ribosome, it changes the way codons in the mRNA interact with the tRNA anticodons. As a result, incorrect amino acids are incorporated into the growing polypeptide chain. In addition, streptomycin causes the ribosome to fall off the mRNA at random, preventing the synthesis of complete proteins.

Puromycin is not used clinically but has played an important role in studying the mechanism of protein synthesis in the research laboratory. The puromycin molecule has the same size and shape as a tRNA–amino acid complex. As a result, it enters the ribosome and is incorporated into a growing polypeptide chain. Once puromycin is added to the polypeptide, further synthesis is terminated because no peptide bond can be formed with an amino acid, and the shortened polypeptide falls off the ribosome.

Chloramphenicol was one of the first broad-spectrum antibiotics introduced. Eukaryotic cells are resistant to its actions, and it was widely used to treat bacterial infections. However, its use is now limited to external applications and serious infections. Chloramphenicol destroys cells in the bone marrow, the source of all blood cells. This antibiotic binds to the large ribosomal subunit in bacteria and inhibits the formation of peptide bonds. Another antibiotic, erythromycin, also binds to the large ribosomal subunit and inhibits the movement of ribosomes along the mRNA.

Almost every step of protein synthesis can be inhibited by one antibiotic or another. Work on designing new, synthetic antibiotics to fight infections is based on our knowledge of how the nucleotide sequence of mRNA is converted into the amino acid sequence of a protein.

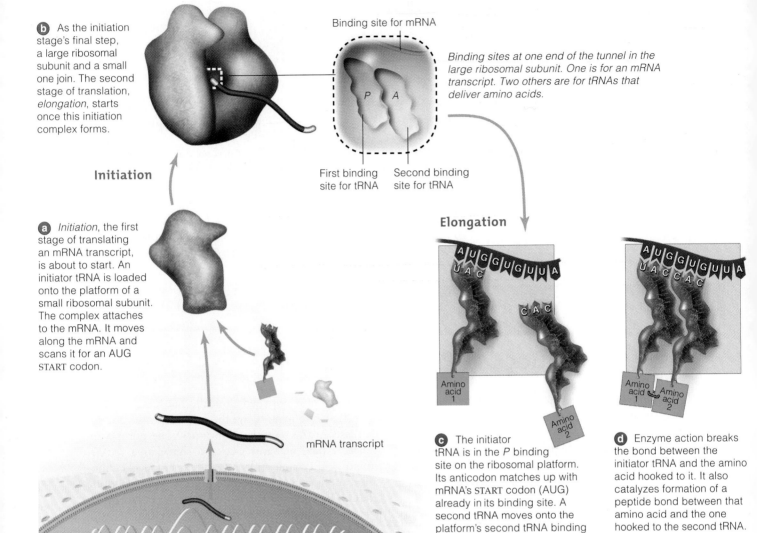

b As the initiation stage's final step, a large ribosomal subunit and a small one join. The second stage of translation, *elongation*, starts once this initiation complex forms.

Binding site for mRNA

Binding sites at one end of the tunnel in the large ribosomal subunit. One is for an mRNA transcript. Two others are for tRNAs that deliver amino acids.

Initiation

First binding site for tRNA

Second binding site for tRNA

a *Initiation*, the first stage of translating an mRNA transcript, is about to start. An initiator tRNA is loaded onto the platform of a small ribosomal subunit. The complex attaches to the mRNA. It moves along the mRNA and scans it for an AUG START codon.

mRNA transcript

Elongation

c The initiator tRNA is in the *P* binding site on the ribosomal platform. Its anticodon matches up with mRNA's START codon (AUG) already in its binding site. A second tRNA moves onto the platform's second tRNA binding site (*A*). It binds with the codon that follows the START codon.

d Enzyme action breaks the bond between the initiator tRNA and the amino acid hooked to it. It also catalyzes formation of a peptide bond between that amino acid and the one hooked to the second tRNA. Then the initiator tRNA is released from the ribosome.

▲ **ACTIVE FIGURE 9.11** Steps in the process of translation.

Human Genetics Now™ Learn more about translation by viewing the animation on Human GeneticsNow at **http://now.ilrn.com/cummings7**

Translation produces proteins from information in mRNA.

Translation requires mRNA, ribosomes, tRNA molecules linked to amino acids, and a variety of other proteins, some of which provide energy, some that help assemble the components, and others that disassemble components at the end of translation. Translation, just like transcription, has three steps: initiation, elongation, and termination. In the first step, mRNA, the small ribosomal subunit, and a tRNA that carries the first amino acid combine to form an **initiation complex** (▶ Active Figure 9.11). Because AUG is the **start codon** and also encodes methionine, this amino acid is inserted first in all human proteins. The initiation complex starts forming when the small ribosomal subunit binds to the start codon (AUG), and the anticodon (UAC) of a tRNA that carries methionine binds to the mRNA. Initiation is completed when a large ribosomal subunit binds to the small subunit. This complex is now ready to begin protein synthesis.

Elongation begins when amino acids are added to the growing protein. Ribosomes have two tRNA binding sites, the *P* site and the *A* site. During initiation, a tRNA car-

■ **Initiation complex** Formed by the combination of mRNA, tRNA, and the small ribosome subunit. The first step in translation.

■ **Start codon** A codon present in mRNA that signals the location for translation to begin. The codon AUG functions as a start codon.

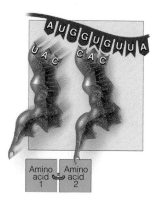

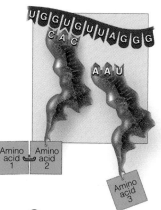

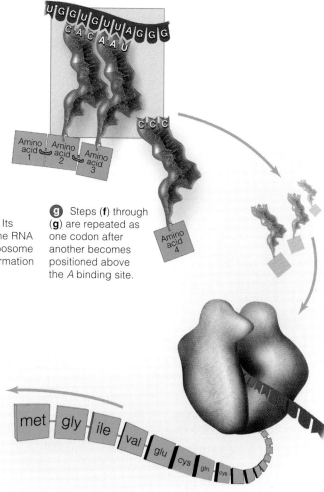

e The first amino acid is attached only to the second one, which is hooked to the second tRNA. The ribosome will move this tRNA into the *P* site and slide the mRNA along with it by one codon. This aligns the third codon in the *A* site.

f A third tRNA is moving into the vacated *A* site. Its anticodon can base-pair with the RNA transcript's third codon. The ribosome now catalyzes peptide bond formation between amino acids 2 and 3.

g Steps (**f**) through (**g**) are repeated as one codon after another becomes positioned above the *A* binding site.

Termination

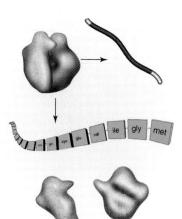

h A STOP codon moves into the area where the chain is being built. It is the signal to release the mRNA transcript from the ribosome.

i The new polypeptide chain is released from the ribosome. It is free to join the pool of proteins in the cytoplasm or to enter rough ER of the endomembrane system.

j The two ribosomal subunits now separate, also.

met — gly — ile — val — glu — cys — gln — cys

---

rying methionine binds to the *P* site. Elongation begins when a tRNA molecule that carries the second amino acid pairs with the mRNA codon next to the initiation codon in the *A* site (▷ Active Figure 9.11). When the second amino acid is in position, an enzyme forms a peptide bond between the two amino acids.

After this bond is formed, the first tRNA (the one in the *P* site) is released and moves out of the ribosome. Next, the ribosome moves down the mRNA to the next codon, and the tRNA with its two attached amino acids moves into the *P* site (▷ Active Figure 9.11). This places the third mRNA codon into the *A* site, where it is recognized by the anticodon of a tRNA carrying the third amino acid. A peptide bond is formed between the second and third amino acid, and the process repeats itself, adding amino acids to the growing polypeptide chain (▷ Active Figure 9.11).

Elongation continues until the ribosome reaches a **stop codon.** Stop codons (UAA, UAG, and UGA) do not code for amino acids, and there are no tRNA molecules with anticodons for stop codons. This is the termination point. Polypeptide synthesis is ended, and the polypeptide, mRNA, and tRNA are released from the ribosome (▷ Active Figure 9.11). Many antibiotics work by interfering with steps in protein synthesis, as described in Genetic Journeys: Antibiotics and Protein Synthesis.

■ **Stop codon** A codon present in mRNA that signals the end of a growing polypeptide chain. The codons UAG, UGA, and UAA function as stop codons.

9.7 Polypeptides Fold into Three-Dimensional Shapes to Form Proteins

After a polypeptide is synthesized, it folds into a three-dimensional shape that is determined by its amino acid sequence. Polypeptide folding is guided by proteins called molecular chaperones. Mutations in genes can alter folding and lead to genetic disorders, as discussed later. Polypeptides can be chemically modified after they are synthesized; this process is called post-translational modification. Over 200 different ways of modification have been identified. Some of these include attaching lipids or sugars to the polypeptide, chemically changing some of the amino acids in the polypeptide, or even removing some amino acids. Once a polypeptide is folded, is modified, and becomes functional, it is called a protein.

Polypeptides can have several different fates. Those made on the outer surface of the ER move inside the ER, where they are folded, chemically modified, and transported to the Golgi complex for packaging and secretion from the cell. Other polypeptides, made on cytoplasmic ribosomes, are folded, remain in the cell, and function in the cytoplasm or the nucleus.

How many different proteins can human cells make? The answer appears deceptively simple. Results of the Human Genome Project (discussed in Chapter 15) indicate that we carry between 25,000 and 30,000 protein coding genes. However, the set of proteins in a particular cell type, called its **proteome**, can be far greater than the number of genes in the genome. It is estimated that humans can make over 100,000 different proteins. Some of this diversity is produced by starting transcription at alternative sites, by processing out exons during mRNA maturation, and by other mechanisms we are only beginning to understand. These discoveries are one of the surprises of the Human Genome Project and are at the forefront of current research in human genetics.

Proteome The set of proteins present in a particular cell at a given time under a given set of environmental conditions.

Primary structure The amino acid sequence in a polypeptide chain.

Secondary structure The pleated or helical structure in a protein molecule that is brought about by the formation of bonds between amino acids.

Tertiary structure The three-dimensional structure of a protein molecule brought about by folding on itself.

Quaternary structure The structure formed by the interaction of two or more polypeptide chains in a protein.

Keep in mind

- Polypeptides fold into a three-dimensional shape that makes them functional. Mutations that prevent proper folding or that cause misfolding can cause disease.

9.8 Protein Structure and Function Are Related

The amino acid sequence of a protein determines its three-dimensional shape and function. There are four levels of protein structure. The sequence of amino acids in a polypeptide is the **primary structure** (▶ Active Figure 9.12). The next two levels are mostly determined by interactions among amino acids. The NH and CO groups of amino acids in different parts of the protein interact with each other via hydrogen bonds to form pleats or coils, called **secondary structure**. Most proteins have both pleats and coils. Folding helical or pleated regions back on themselves creates **tertiary structure**. Some functional proteins are composed of more than one polypeptide chain, and this level of interaction is known as the **quaternary structure**. When a polypeptide forms a functional, three-dimensional structure, it is called a protein.

It is this three-dimensional conformation, ultimately determined by its DNA-controlled primary structure, that determines a protein's function.

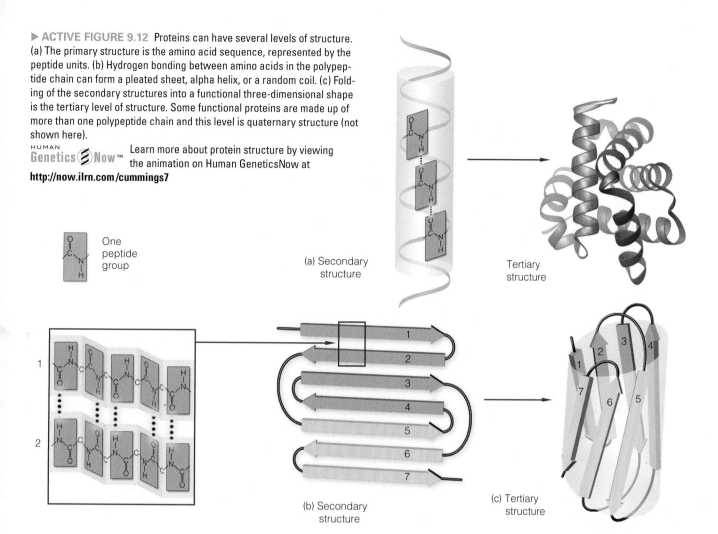

▶ ACTIVE FIGURE 9.12 Proteins can have several levels of structure. (a) The primary structure is the amino acid sequence, represented by the peptide units. (b) Hydrogen bonding between amino acids in the polypeptide chain can form a pleated sheet, alpha helix, or a random coil. (c) Folding of the secondary structures into a functional three-dimensional shape is the tertiary level of structure. Some functional proteins are made up of more than one polypeptide chain and this level is quaternary structure (not shown here).

HUMAN Genetics ⊜ Now™ Learn more about protein structure by viewing the animation on Human GeneticsNow at http://now.ilrn.com/cummings7

One peptide group

(a) Secondary structure

Tertiary structure

(b) Secondary structure

(c) Tertiary structure

Protein folding can be a factor in diseases.

Some mutations alter polypeptide folding and cause a genetic disorder. Several disorders, including Alzheimer disease (OMIM 104300 and other numbers), a metabolic disorder called MPS VI (OMIM 253200), and cystic fibrosis (OMIM 219700), are associated with defects in folding.

Defective folding prevents the formation of a functional protein, producing a mutant phenotype. The cystic fibrosis gene encodes a protein (called CFTR) of 1,440 amino acids (review cystic fibrosis in Chapter 4) that is embedded in the cell's plasma membrane where it controls the flow of chloride ions. The most common mutation in CF is the deletion of phenylalanine at position 508. This single amino acid change causes the polypeptide to fold improperly. As a result, the misfolded CFTR protein is identified as defective and is destroyed in the ER; it does not reach the plasma membrane.

Under certain conditions, some proteins can refold and change their three-dimensional shape, causing disease. Protein refolding diseases are called prion diseases (▶ Figure 9.13). In humans, Creutzfeldt-Jakob disease (CJD; OMIM 123400), Gerstmann-Straussler disease (OMIM 137440), and fatal familial insomnia (OMIM 600072) are prion diseases. In these disorders, a mutation changes one amino acid in the protein, predisposing it to refolding into a disease-causing shape.

In cattle, bovine spongiform encephalopathy (BSE), also known as mad-cow disease, is a prion disease. Prions cause degenerative changes in the nervous system, leading to early death. The disease begins when one or a small number of proteins

▨ Prion   A protein folded into an infectious conformation that is the cause of several disorders, including Creutzfeldt-Jakob disease and mad-cow disease.

▨ Mad-cow disease   A prion disease of cattle, also known as bovine spongiform encephalopathy, or BSE.

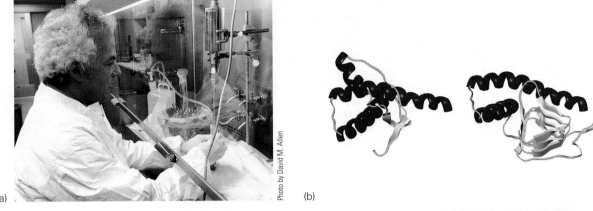

Photo by David M. Allen

(a)   (b)

▲ FIGURE 9.13 Misfolding or refolding of proteins can result in disease. (a) Stanley Prusiner won a Nobel Prize in 1997 for his discovery of prions and their role in disease. (b) At left, the normal folding pattern for a protein. Most of the protein is in a helical configuration. At right, the protein has refolded to form a disease-causing prion. This refolding has altered the secondary and tertiary levels of protein structure. In this form, most of the protein is in pleated sheets (the ribbon-like regions).

refold into a disease-causing shape (or when refolded proteins enter the body). These prions cause other proteins of the same type to refold into the disease-causing conformation. The process is slow, and the disease makes its appearance in about 5 to 15 years. Prion diseases like mad-cow disease are infectious, and the disease is transmitted when refolded proteins are transferred from one individual to another. Case 2 at the end of the chapter deals with CJD.

Proteins have many functions.

Proteins are the most abundant type of molecules found in any cell. They participate in a wide range of functions (summarized in ▶ Table 9.2), including muscle contraction (motion), transport, the immune response (protection), and receptors (nerve impulse transmission). Enzymes are one of the most important groups of proteins in the cell. They act as catalysts in biochemical reactions (▶ Figure 9.14a). Enzymes accelerate the rate of a chemical reaction by reducing the energy needed to carry out the reaction. The three-dimensional shape of the enzyme creates a region called the active site. Molecules that can fit into the active site are known as substrates. When they bind to the active site, they undergo a chemical change. Enzymes are usually named for their substrate, with the suffix _ase_ added. For example, the enzyme that catalyzes the breakdown of the sugar lactose is called lactase, and the enzyme that catalyzes the conversion of the amino acid phenylalanine to tyrosine is called phenylalanine hydroxylase.

The relationship between enzymes and genetic disorders is explored in Chapter 10. The function of all proteins depends ultimately on the amino acid sequence of the polypeptide chain. The nucleotide sequence of DNA determines the amino acid sequence of proteins. If protein function is to be maintained from cell to cell and from generation to generation, the nucleotide sequence of a gene must be maintained. Changes in the nucleotide sequence of DNA (a mutation) produce mutant genes that in turn produce mutant proteins with altered or impaired functions. These alterations result in an altered phenotype. Alkaptonuria, the condition described at the beginning of this chapter, is caused by a mutation that alters the function of an enzyme. The phenotypic consequences of mutational changes in DNA are discussed in Chapter 11.

Table 9.2 Some Biological Functions of Proteins

Protein Function	Examples	Occurrence or Role
Catalysis	Lactate dehydrogenase	Oxidizes lactic acid
	Cytochrome C	Transfers electrons
	DNA polymerase	Replicates and repairs DNA
Structural	Viral-coat proteins	Sheath around nucleic acid of viruses
	Glycoproteins	Cell coats and walls
	α-Keratin	Skin, hair, feathers, nails, and hoofs
	β-Keratin	Silk of cocoons and spider webs
	Collagen	Fibrous connective tissue
	Elastin	Elastic connective tissue
Storage	Ovalbumin	Egg-white protein
	Casein	A milk protein
	Ferritin	Stores iron in the spleen
	Gliadin	Stores amino acids in wheat
	Zein	Stores amino acids in corn
Protection	Antibodies	Form complexes with foreign proteins
	Complement	Complexes with some antigen-antibody systems
	Fibrinogen	Involved in blood clotting
	Thrombin	Involved in blood clotting
Regulatory	Insulin	Regulates glucose metabolism
	Growth hormone	Stimulates growth of bone
Nerve impulse transmission	Rhodopsin	Involved in vision
	Acetylcholine receptor protein	Impulse transmission in nerve cells
Motion	Myosin	Thick filaments in muscle fiber
	Actin	Thin filaments in muscle fiber
	Dynein	Movement of cilia and flagella
Transport	Hemoglobin	Transports O_2 in blood
	Myoglobin	Transports O_2 in muscle cells
	Serum albumin	Transports fatty acids in blood
	Transferrin	Transports iron in blood
	Ceruloplasmin	Transports copper in blood

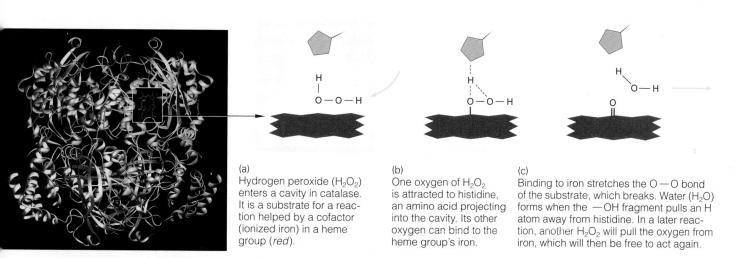

(a)
Hydrogen peroxide (H_2O_2) enters a cavity in catalase. It is a substrate for a reaction helped by a cofactor (ionized iron) in a heme group (*red*).

(b)
One oxygen of H_2O_2 is attracted to histidine, an amino acid projecting into the cavity. Its other oxygen can bind to the heme group's iron.

(c)
Binding to iron stretches the O—O bond of the substrate, which breaks. Water (H_2O) forms when the —OH fragment pulls an H atom away from histidine. In a later reaction, another H_2O_2 will pull the oxygen from iron, which will then be free to act again.

▲ ACTIVE FIGURE 9.14 The enzyme catalase has a quaternary structure and is composed of subunits. (a) The enzyme has an active site that binds hydrogen peroxide (the substrate). (b and c) The enzyme acts as a catalyst to carry out a chemical reaction, converting the substrate into a product (water). Mutation can change the folding pattern of an enzyme, making it nonfunctional.

HUMAN
Genetics 🄴 Now™   Learn more about enzyme action by viewing the animation on Human GeneticsNow at **http://now.ilrn.com/cummings7**

Genetics in Practice

Genetics in Practice case studies are critical thinking exercises that allow you to apply your new knowledge of human genetics to real-life problems. You can find these case studies and links to relevant websites at **http://biology.brookscole.com /cummings7**

CASE 1

A genetic counselor was called to the pediatric ward to examine a 3-week-old infant who was diagnosed with a genetic disorder of sugar metabolism, galactosemia. The infant was admitted to the hospital because of failure to thrive and severe jaundice (yellowing of the skin resulting from liver problems). Upon examination, the physician determined that the infant had an enlarged liver (hepatomegaly), cataracts, and constant diarrhea and vomiting when fed milk. *E. coli* infection is a common cause of death in infants who have galactosemia, and cultures were drawn on this infant. Laboratory results confirmed that the infant had a deficiency of the enzyme galactose-1-phosphate uridyltransferase and was infected with *E. coli*.

The counselor took a detailed family history and explained this condition to the parents. She indicated that this condition is due to the inheritance of a mutant gene from each parent (the trait is autosomal recessive) and that there is a 25%, or one in four, chance that with each pregnancy they have together that they will bear a child with this condition. The counselor explained that there is wide variability in phenotype, ranging from very mild to severe. A blood test could determine which variant of this disease they carry.

1. What exists in blood that can be tested for a variant of a disease-causing gene?

2. What are possible treatments for this disease?

CASE 2

Recurring cases of mad-cow disease have plagued the United Kingdom since the late 1990s. What started out as a topic of interest to a few cell biologists has blossomed into a huge public interest story. Mad-cow disease is caused by a prion, an infectious particle that consists only of protein. In 1986, the media began reporting that cows were dying all over England from a mysterious disease. Initially, however, there was little interest in determining whether humans could be affected. For 10 years, the British government maintained that this unusual disease could not be transmitted to humans. However, in March 1996, the government did an about-face and announced that bovine spongiform encephalopathy (BSE), commonly known as mad-cow disease, can be transmitted to humans. BSE and a similar condition known as Creutzfeldt-Jakob disease (CJD) eat away at the nervous system, destroying the brain and essentially turning it into a sponge. Victims suffer dementia; confusion; loss of speech, sight, and hearing; convulsions; coma; and finally succumb to death. Prion diseases are always fatal, and there is no treatment. Precautionary measures taken in Britain to prevent this disease in humans may have begun too late; many of the victims today may have contracted it over a decade ago when the BSE epidemic began, and the incubation period is long (CJD has an incubation period of 10 to 40 years).

1. How can a prion replicate itself without genetic material?

2. What measures have been taken to stop BSE?

3. If you were traveling in Europe, would you eat beef? Give sound reasons why or why not.

Summary

9.1 The Link between Genes and Proteins

- At the beginning of the last century, Garrod proposed that genetic disorders result from biochemical alterations.

- Using *Neurospora*, Beadle and Tatum showed that mutations can produce a loss of enzyme activity and a mutant phenotype. Beadle proposed that genes control the synthesis of proteins and that protein function is responsible for producing the phenotype.

9.2 Genetic Information Is Stored in DNA

■ In proposing their model, Watson and Crick maintained that DNA stores genetic information in its nucleotide sequence.

9.3 The Genetic Code: The Key to Life

■ The information transferred from DNA to mRNA is encoded in sets of three nucleotides, called codons. Of the 64 possible codons, 61 code for amino acids, and 3 are stop codons.

9.4 Tracing the Flow of Genetic Information from Nucleus to Cytoplasm

■ The processes of transcription and translation require the interaction of many components, including ribosomes, mRNA, tRNA, amino acids, enzymes, and energy sources. Ribosomes are the workbenches on which protein synthesis occurs. tRNA molecules are adapters that recognize amino acids and the nucleotide sequence in mRNA, the gene transcript.

9.5 Transcription Produces Genetic Messages

■ In transcription, one of the DNA strands is used as a template for making a complementary strand of RNA, called mRNA.

9.6 Translation Requires the Interaction of Several Components

■ Translation requires the interaction of tRNA molecules, amino acids, ribosomes, mRNA, and energy sources.

Within the ribosome, tRNA anticodons bind to complementary codons in the mRNA. The ribosome moves along the mRNA, linking amino acids and producing a growing polypeptide chain. At termination, this polypeptide is released from the ribosome and undergoes a conformational change to produce a functional protein.

9.7 Polypeptides Fold into Three-Dimensional Shapes to Form Proteins

■ After synthesis, polypeptides fold into a three-dimensional shape, often assisted by other proteins, called chaperones. Mutations in chaperones can cause genetic disorders. Polypeptides can be chemically modified in many different ways, producing functionally different proteins from one polypeptide.

9.8 Protein Structure and Function Are Related

■ Four levels of protein structure are recognized, three of which are results of the primary sequence of amino acids in the backbone of the protein chain.

■ Although proteins perform a wide range of tasks, enzyme activity is one of the primary tasks. Enzymes function by lowering the energy of activation required in biochemical reactions. The products of these biochemical reactions are inevitably involved in producing phenotypes.

Questions and Problems

HUMAN
Genetics ⓖ Now™ Preparing for an exam? Assess your understanding of this chapter's topics with a pre-test, a personalized learning plan, and a post-test at http://now.ilrn.com /cummings7

The Link between Genes and Proteins

1. The genetic material has to store information and be able to express it. What is the relationship between DNA, RNA, proteins, and phenotype?
2. Define replication, transcription, and translation. In what part of the cell does each process occur?

The Genetic Code: The Key to Life

3. If the genetic code used four bases at a time, how many amino acids could be encoded?
4. If the genetic code uses triplets, how many different amino acids can be coded by a repeating RNA polymer composed of UA and UC (UAUCUAUCUAUC . . .)

   a. one   b. two   c. three
   d. four   e. five

5. What is the start codon? What are the stop codons? Do any of these code for amino acids?

Transcription Produces Genetic Messages

6. The following segment of DNA codes for a protein. The uppercase letters represent exons. The lowercase letters represent introns. The lower strand is the template strand. Draw the primary transcript and the mRNA resulting from this DNA.

```
G C T A A A T G G C A a a a t t g c c g g a t g a c G C A C A T T G A C T C G G a a t c g a G G T C A G A T G C
C G A T T T A C C G T t t t a a c g g c c t a c t g C G T G T A A C T G A G C C t t a g c t C C A G T C T A C G
```

7. Is an entire chromosome made into an mRNA during transcription?

8. The 5′ promoter and the 3′ terminator regions of genes are important in:
   a. coding for amino acids.
   b. gene regulation.
   c. structural support for the gene.
   d. intron removal.
   e. anticodon recognition.

9. What are the three modifications made to pre-mRNA molecules before they become mature mRNAs ready to be used in protein synthesis? What is the function of each modification?

10. The pre-mRNA transcript and protein made by several mutant genes were examined. The results are given below. Determine where in the gene a likely mutation lies: the 5′ flanking region, exon, intron, cap on mRNA, or ribosome binding site.
    a. normal length transcript, normal length nonfunctional protein
    b. normal length transcript, no protein made
    c. normal length transcript, normal length mRNA, short nonfunctional protein
    d. normal length transcript, longer mRNA, longer nonfunctional protein
    e. transcript never made

Translation Requires the Interaction of Several Components

11. Briefly describe the function of the following in protein synthesis.
    a. rRNA   b. tRNA   c. mRNA

12. What is the difference between codons and anticodons?

13. Determine the percent of the following gene that will code for the protein product. Gene length is measured in kilobases (kb) of DNA. Each kilobase is 1,000 bases long.

14. How many kilobases of the DNA strand below will code for the protein product?

15. Write the anticodon(s) for the following amino acids:
    a. met   b. trp
    c. ser   d. leu

16. Given the following tRNA anticodon sequence, derive the mRNA and the DNA template strand. Also, make the protein that is encoded by this message.
    tRNA: UAC UCU CGA GGC
    mRNA:
    DNA:
    protein:

How many hydrogen bonds would be present in the DNA segment?

17. Given the following mRNA, write the double-stranded DNA segment that served as the template. Indicate both the 5′ and the 3′ ends of both DNA strands. Also make the tRNA anticodons and the protein that is encoded by the mRNA message.
    DNA:
    mRNA: 5′-CCGCAUGUUCAGUGGGCGUAAACACUGA-3′
    protein:
    tRNA:

18. The following is a portion of a protein:
    met-trp-tyr-arg-gly-pro-thr-
    Various mutant forms of this protein have been recovered. Using the normal and mutant sequences, determine the DNA and mRNA sequences that code for this portion of the protein, and explain each of the mutations.
    a. met-trp-
    b. met-cys-ile-val-val-leu-gln-
    c. met-trp-tyr-arg-ser-pro-thr-
    d. met-trp-tyr-arg-gly-ala-val-ile-ser-pro-thr-

19. Below is the structure of glycine. Draw a tripeptide composed exclusively of glycine. Label the N-terminus and C-terminus. Draw a box around the peptide bonds.

20. Indicate in which category, transcription or translation, each of the following functions: RNA polymerase, ribosomes, nucleotides, tRNA, pre-mRNA, DNA, A site, anticodon, amino acids.

Protein Structure and Function Are Related

21. Proteins have many critical functions in the human body. Some of these functions include:
    a. transporting oxygen.
    b. hormonal signaling.
    c. carrying out enzymatic reactions.
    d. destroying invading bacteria.
    e. all of the above.

22. Enzyme X normally interacts with substrate A and water to produce compound B.
    a. What would happen to this reaction in the presence of another substance that resembles substrate A and was able to interact with enzyme X?
    b. What if a mutation in enzyme X changed the shape of the active site?

23. Do mutations in DNA alter proteins all the time?

24. Can a mutation change a protein's tertiary structure without changing its primary structure? Can a mutation change a protein's primary structure without affecting its secondary structure?

Internet Activities are critical thinking exercises using the resources of the World Wide Web to enhance the principles and issues covered in this chapter. For a full set of links and questions investigating the topics described below, log on to **http://biology.brookscole.com/cummings7**

1. *Review of Gene Expression.* At the *Cell Biology Topics 1: Ribosome* website, review the basics of translation after the mRNA leaves the nucleus.

2. *Quiz Yourself.* At University of Arizona's *The Biology Project: Molecular Biology* website, click on the "Nucleic Acids" link to access quizzes on DNA replication, transcription, and translation. Correct answers are rewarded with brief overviews; if you answer incorrectly, you will be linked to a short tutorial that will help you solve the problem.

3. *Control of Gene Expression.* At the *On-line Biology: Control of Gene Expression* website, read about the control of gene expression in bacteria, viruses, and eukaryotes.

   a. How many different proteins and protein factors are involved in the various steps of gene expression? What would be the possible effects of a mutation that changed one of these proteins? Consequently, would you expect to see greater similarity or less similarity in the DNA sequences that code for these proteins in different organisms?

   b. In some cases the expression of multiple genes is controlled by a single protein factor, as in the operon model of transcriptional regulation proposed by Jacob and Monod. What might be the benefits of such a comparatively streamlined mechanism for the control of gene expression?

   c. Compare the genome sizes for various eukaryotes. What percentage of the average eukaryotic genome actually codes for protein? What percentage of the human genome codes for protein? What function, if any, does the noncoding portion of the genome serve?

How would you vote now?

Most cases of prion diseases caused by eating infected beef have been reported in the United Kingdom, not in the United States. Prions have also been transmitted by contaminated surgical and dental instruments, and in other cultures, by cannibalism. There is no cure for a prion infection, and prions cannot be destroyed by sterilization. Some countries are now testing all beef used in human consumption, while others, such as the United States, are randomly testing only a small sample of cows. Now that you know more about proteins and the relationship between protein structure and function, what do you think? Would you eat beef or allow your children to eat it if you were traveling or living in a country with a history of infected cows? What if infected beef was linked to human deaths in that country? Visit the Human Heredity Companion website to find out more on the issue, then cast your vote online at **http://biology.brookscole.com/cummings7**

For further reading and inquiry, log on to **InfoTrac College Edition,** your world-class online library, including articles from nearly 5,000 periodicals, at **http://infotrac.thomsonlearning.com**

10

From Proteins to Phenotypes

THE field of human biochemical genetics was founded in part by a young Norwegian mother, Borgny Egeland, who had two mentally retarded children. Her daughter, Liv, did not walk until nearly 2 years of age and spoke only a few words. She also had a musty odor that could not be washed away. Liv's younger brother, Dag, was also slow to develop and never learned to walk or talk. He had the same musty odor as his sister. Borgny was convinced that whatever was causing the odor might also be causing her children's mental retardation. To learn why both her children were retarded and had a musty odor, the mother went from doctor to doctor but to no avail. Finally, in the spring of 1934, the persistent woman took the two children, then aged 4 and 7 years, to Dr. Asbjorn Fölling, a biochemist and physician.

Because the children's urine had a musty odor, Fölling first tested the urine for signs of infection, but there was none. He discovered that the urine reacted with ferric chloride to produce a green color, indicting the presence of an unknown chemical. Beginning with 20 L of urine collected from the children, he worked to isolate and identify the unknown substance. Over the next 3 months, he managed to purify the compound and worked out its chemical structure. The chemical in the children's urine was a compound called phenylpyruvic acid. To confirm his finding, Fölling synthesized and purified phenylpyruvic acid from laboratory chemicals and showed that the compound from the urine and his synthetic phenylpyruvic acid had the same physical and chemical properties.

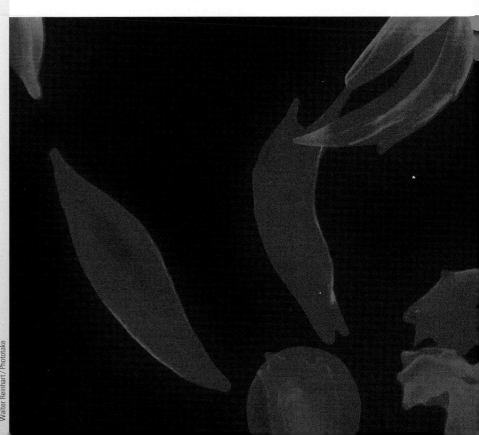

Walter Reinhart / Phototake

Fölling proposed that the phenylpyruvic acid in the urine was produced by a metabolic disorder that affected the breakdown of the amino acid phenylalanine. He further proposed that the accumulation of phenylpyruvic acid (the cause of the musty odor) in the bodies of the children caused their mental retardation. To confirm this, he examined the urine of several hundred retarded patients and normal individuals. He found phenylpyruvic acid in the urine of eight retarded individuals, but never in the urine of normal individuals. Less than 6 months after he began working on the problem, Fölling submitted a manuscript for publication that described this metabolic disorder called phenylketonuria (PKU). His work helped establish the relationship between a gene product and a phenotype, and PKU is now regarded as a prototype for metabolic genetic disorders.

As we discussed in the last chapter, DNA encodes information for the chemical structure of proteins. In this chapter we will show how protein function is related to the phenotype and how mutations that change or eliminate protein function produce an abnormal phenotype.

How would you vote?

All 50 states and the District of Columbia require testing of newborns for PKU. Genetic testing of newborns is becoming increasingly mandatory; however, as mentioned in Chapter 3, the number of genetic diseases newborns are tested for varies from state to state. Some states test for only 6 to 8 genetic diseases, while others test for more than 40. One of the rationales given for testing for a small number of disorders is that cost benefit analysis shows that it is not cost-efficient to do so. Some diseases are so rare that the costs of testing all newborns outweigh the health care costs for affected children. However, forgoing testing for those rare disorders means that some children may go undiagnosed or fail to receive proper treatment. Do you think that cost benefit analysis should be used as a determining factor in setting up and running newborn testing programs? Visit the Human Heredity Companion website to find out more on the issue, then cast your vote online at **http://biology.brookscole.com/cummings7**

10.1 Proteins Are the Link between Genes and the Phenotype

As outlined in Chapter 9, proteins are among the most important molecules in a cell. They are essential parts of all structures and biological processes carried out in cells. Proteins are part of membrane systems and the internal skeleton of cells. They are the glue that holds cells and tissues together. Proteins carry out biochemical reac-

Keep in mind as you read

- Phenotypes are the visible end product of a chain of events that starts with the gene, the mRNA, and the protein product.

- Phenylketonuria and several other metabolic disorders can be treated by dietary restrictions.

- Sickle cell anemia is caused by substitution of a single amino acid in beta globin.

- Small differences in proteins can have a large effect on our ability to taste, smell, and metabolize medicines.

▲ FIGURE 10.1 Portrait of a dwarf by Goya. Some genetic forms of dwarfism are caused by mutations in genes that encode proteins that act as growth hormones, receptors, and growth factors.

■ **Substrate** The specific chemical compound that is acted upon by an enzyme.

■ **Product** The specific chemical compound that is the result of enzymatic action. In biochemical pathways, a compound can serve as the product of one reaction and the substrate for the next reaction.

■ **Metabolism** The sum of all biochemical reactions by which cells convert and utilize energy.

tions; destroy invading microorganisms; and act as hormones (▶ Figure 10.1), receptors, and transport molecules. Even the replication of DNA and the expression of genes depend on the action of proteins.

The many different functions of proteins are matched by their enormous diversity. Because any one of 20 amino acids can occupy each position in a protein (amino acid 1, amino acid 2, and so forth), an enormous number of different proteins can be made. For example, in a protein containing only five amino acids, 20^5, or 3,200,000, different molecules, each with a unique amino acid sequence, can be made. When you consider that most proteins contain around 100 amino acids, the possible numbers of different proteins reaches astronomical proportions.

As we will see in this chapter, there is a direct link between a person's genotype, the proteins the person makes, and his or her phenotype. Mutations that alter the amino acid sequence of a protein can produce phenotypes that range from insignificant to lethal. We will examine this link using examples of proteins as enzymes and as transport molecules. In addition, we will explore how variations in the proteins we make affect our reactions to drugs and to environmental chemicals.

10.2 Enzymes and Metabolic Pathways

Enzymes are proteins that facilitate biochemical reactions. They convert molecules known as **substrates** into **products** by catalyzing chemical reactions (▶ Figure 10.2). In the cell, enzymatic reactions do not occur randomly; they are interconnected to form chains of reactions called *biochemical pathways* (▶ Figure 10.3a). The sum of all the biochemical reactions going on in a cell is called **metabolism,** and the biochemical reactions are called metabolic pathways.

Keep in mind

■ Phenotypes are the visible end product of a chain of events that starts with the gene, the mRNA, and the protein product.

In a metabolic pathway, the product of one reaction serves as the starting point (substrate) for the next reaction (see Spotlight on Why Wrinkled Peas Are Wrinkled). If a mutation shuts down an enzyme that performs one step in a pathway, all the reactions beyond that point are shut down, because there is no substrate for reactions beyond the one that is blocked (▶ Figure 10.3b). If one reaction is shut down, it also results in the accumulation of products in the pathway leading up to the block.

▶ FIGURE 10.2 Each step in a metabolic pathway is a separate chemical reaction catalyzed by an enzyme, in which a substrate is converted to a product.

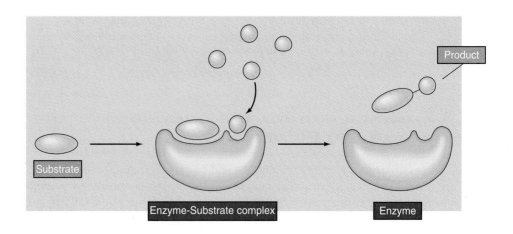

Substrate · Enzyme-Substrate complex · Product · Enzyme

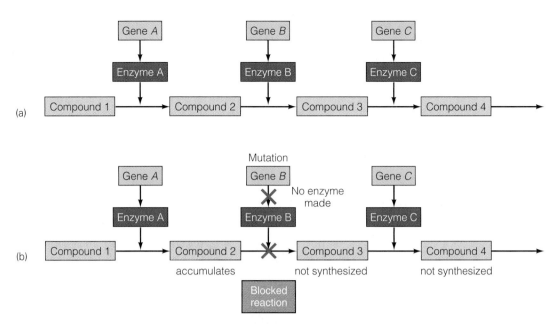

▲ FIGURE 10.3 (a) The sequence of reactions in a metabolic pathway. In this pathway, compound 1 is present in the diet and is converted in the body into compound 2, which is then converted into compound 3. Finally, compound 3 is converted into compound 4. A specific enzyme catalyzes each of these reactions. Each enzyme is the product of a gene. (b) In this pathway, a mutation in gene *B* leads to the production of a defective protein that cannot function as an enzyme. As a result, compound 2 cannot be converted into compound 3. Because no compound 3 is made, compound 4 will not be produced, even though enzyme C is present. Compound 1 is supplied by the diet and is converted into compound 2, which accumulates because it cannot be metabolized.

In the early years of the twentieth century, Sir Archibald Garrod was the first to propose that human genetic disorders and metabolism are related (see Genetic Journeys: Garrod and Metabolic Disease). Garrod studied a disorder called **alkaptonuria** (OMIM 203500), in which large amounts of a compound called homogentisic acid are excreted in urine. Garrod called this heritable condition an **inborn error of metabolism.** He proposed that people with alkaptonuria lacked activity of an enzyme needed to carry out a biochemical reaction, and that this trait was inherited. His work represented a pioneering study in applying Mendelian genetics to humans and in understanding the relationship between genes and biochemical reactions.

Mutations that destroy or alter the activity of an enzyme can cause phenotypic effects in several ways. First, the substrate for the blocked reaction may build up and reach toxic levels, causing an abnormal phenotype. Second, the enzyme may control a reaction that produces a molecule needed for some cellular function. If this product is not made, a mutant phenotype may result. Mutations that affect the action of enzymes can produce a wide range of phenotypes, ranging from inconsequential effects to those that are lethal prenatally or early in infancy.

■ **Alkaptonuria** An autosomal recessive trait with altered metabolism of homogentisic acid. Affected individuals do not produce the enzyme needed to metabolize this acid, and their urine turns black.

■ **Inborn error of metabolism** The concept advanced by Archibald Garrod that many genetic traits result from alterations in biochemical pathways.

■ **Essential amino acids** Amino acids that cannot be synthesized in the body and must be supplied in the diet.

10.3 Phenylketonuria: A Mutation That Affects an Enzyme

To make the proteins required to maintain life, we need all 20 amino acids. Humans can synthesize most of these amino acids from the food we eat. However, we cannot make all 20 amino acids; some must be included in our diet. The amino acids we cannot synthesize are called **essential amino acids.** Humans require nine essential amino acids: histidine, isoleucine, leucine, lysine, methionine, phenylalanine, threonine, tryptophan, and valine. In other words, our diet has to be varied enough to provide 9 of the 20 amino acids.

Why Wrinkled Peas Are Wrinkled

Wrinkled peas were one variety used by Mendel in his experiments. At the time, nothing was known about how peas became wrinkled or smooth. All Mendel needed to know was that a factor controls seed shape and that it had two forms—a dominant one for smooth shape and a recessive one for wrinkled shape.

Recently, scientists have discovered how peas become wrinkled, providing a connection between a gene and its phenotype. While the pea is developing, starch is synthesized and stored as a food source. Starch is a large, branched molecule made up of sugar molecules, and the ability to form branches in starch molecules is controlled by an enzyme. Normally, starch molecules are highly branched structures. This allows more sugar to be stored in each molecule. In peas that have the wrinkled genotype, the branching gene is inactive. Thus, the developing pea converts sugar into starch very slowly using other enzymes, and excess sugar accumulates. The excess sugar causes the pea to take up large amounts of water, and the seed swells. In a final stage of development, water is lost from the seed. In homozygous wrinkled peas, more water is lost than in the smooth seeds, causing the outer shell of the pea to become wrinkled.

Mendel's contribution was to show that a specific factor controlled a trait and that a given gene could have different forms. Now we know that genes exert their effect on phenotype through the production of a gene product.

■ **Phenylketonuria (PKU)** An autosomal recessive disorder of amino acid metabolism that results in mental retardation if untreated.

How is the metabolism of phenylalanine related to PKU?

Phenylalanine is one of the essential amino acids and is the starting point for a network of metabolic pathways. Here we will focus on what happens when the first step in the phenylalanine metabolic pathway is blocked by a mutation that prevents the conversion of phenylalanine to another amino acid, tyrosine. About two-thirds of the phenylalanine we eat gets converted to tyrosine; the rest is incorporated into proteins. A mutation that prevents the conversion of phenylalanine to tyrosine results in a genetic disorder called **phenylketonuria** (OMIM 261600), or PKU. This is the disorder described at the beginning of the chapter. About 1 in every 12,000 newborns has PKU. In almost all cases, this is caused by a mutation in a gene for the enzyme phenylalanine hydroxylase (PAH), which converts phenylalanine to tyrosine.

In people with PKU, the phenylalanine from protein-containing foods cannot be converted to tyrosine and builds up to high levels (▶ Figure 10.4). If untreated, newborns with high levels of phenylalanine become severely mentally retarded, have enhanced reflexes that cause their arms and legs to move in a jerky fashion, develop epileptic seizures, and never learn to talk. Because the skin pigment melanin is also a product of the blocked metabolic pathway (▶ Figure 10.4), most people with PKU usually have lighter hair and skin color than their siblings or other family members. Even though there are two pathways leading from phenylalanine, blockage of the pathway to tyrosine overloads the alternate pathway, also producing high levels of phenylacetic acid (responsible for the musty odor of affected individuals) and other compounds that contribute to the clinical phenotype.

How does the failure to convert phenylalanine to tyrosine produce mental retardation and the other aspects of the phenotype? These effects are not caused by a lack of tyrosine, because tyrosine is available from food. The problem is caused by high levels of phenylalanine and its metabolic by-products (▶ Figure 10.4) in infants during the time when the nervous system is maturing.

The human brain and nervous system continue to grow and develop after birth. New cells are produced, and nerve cells connect to each other during this period. This growth requires a constant supply of amino acids for protein synthesis. Transport proteins embedded in the plasma membrane of cells help move amino acids into nerve cells. Phenylalanine and seven other amino acids (called the neutral amino acids) are transported by one of these systems. As phenylalanine accumulates in the fluid outside cells of the maturing nervous system, phenylalanine molecules greatly outnumber those of the seven other amino acids, and the transport system takes in too much phenylalanine. It is not clear whether the damage to the nervous system is the result of transporting too much phenylalanine, not enough of other amino acids, or whether the breakdown products of phenylalanine accumulate in the nerve cells and cause the damage. The result, however, is brain damage, mental retardation, and the other neurological symptoms that result in the phenotype of PKU.

PKU can be treated with a diet low in phenylalanine.

Most people with PKU have heterozygous mothers and develop normally before birth because the mother has enough PAH enzyme in her body to break down any excess phenylalanine that accumulates during fetal development. After they are born, this safeguard is no longer present, and PKU homozygotes suffer neurological damage and become retarded when fed a normal diet containing protein.

PKU is a genetic disorder, but it is also an environmental disease. If phenylalanine is not present in the environment (diet), there is no abnormal phenotype. In the early 1950s, PKU was treated using a diet with very low levels of phenylalanine. This treatment is widely used today and has been successful in reducing the effects of this

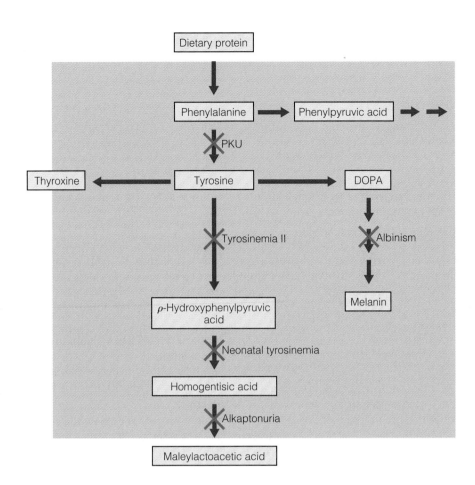

◀ FIGURE 10.4 The metabolic pathway that leads from the essential amino acid phenylalanine. Normally, phenylalanine is converted to tyrosine and from there to many other compounds. A metabolic block, caused by a mutation in the gene encoding the enzyme phenylalanine hydroxylase, prevents the conversion of phenylalanine to tyrosine and, in homozygotes, produces the phenotype of phenylketonuria (PKU). The diagram also shows other metabolic diseases produced by mutations in genes encoding enzymes in this pathway.

Genetic Journeys

Garrod and Metabolic Disease

Sir Archibald Garrod was a distinguished physician, Oxford professor, and physician to the royal family. Garrod was widely trained; he was skilled in biology, biochemistry, and clinical medicine. In studying alkaptonuria, he showed that the condition causes the excretion of abnormal amounts of a compound called homogentisic acid, which causes the urine to turn black upon exposure to air.

Garrod proposed that the presence of homogentisic acid in the urine of affected individuals was caused by a metabolic block, which prevented metabolism of the compound. He also speculated that the metabolic block was caused by an enzyme deficiency. Garrod observed that 60% of the affected individuals were the result of first-cousin marriages and that the parents were unaffected. This led

him to propose that alkaptonuria is inherited as a recessive Mendelian trait. Remember, Garrod did his work just a few years after Mendel's work was rediscovered and widely publicized. Garrod studied other metabolic disorders and published his findings in a book, *Inborn Errors of Metabolism*, which is widely regarded as a milestone in human genetics.

Why was Garrod's work, like that of Mendel, not immediately appreciated? From our perspective, it may seem strange, but Garrod worked in three areas that at the time were almost totally isolated from one another: genetics, medicine, and chemistry. Chemists had no interest in studying heredity, physicians regarded the conditions he studied as too rare to be important, and geneticists cared little about either metabolism or medicine.

disease. However, managing PKU by controlling dietary intake is both difficult and expensive.

■ Phenylketonuria and several other metabolic disorders can be treated by dietary restrictions.

One major problem is that phenylalanine is present in many protein sources, and it is impossible to eliminate all protein from the diet. The protein restriction means that meat, fish, milk, cheese, bread, cake, and nuts cannot be eaten. The diet is hard to follow; PKU children cannot eat hamburgers, chicken nuggets, pizza, or ice cream or many other favorite childhood foods. Instead, to get the amino acids needed to make proteins, they must drink a dietary supplement containing a synthetic mixture of amino acids (with very low levels of phenylalanine) along with vitamins and minerals. The supplement is foul-smelling and bad-tasting; in most cases, it must be continued for life.

The challenge of a PKU diet is maintaining blood levels of phenylalanine high enough to make proteins that allow normal development of the nervous system but low enough to prevent mental retardation. To avoid the consequences of PKU, dietary treatment must be started in the first month after birth. After 30 days the brain is damaged beyond repair and treatment is less effective.

In newborns, the first sign of PKU is abnormally high levels of phenylalanine in the blood and urine. Since the 1960s, newborns in the United States have been routinely tested for PKU by analyzing blood or urine for phenylalanine levels (▶ Figure 10.5). By the mid-1970s, many countries were testing newborns for PKU (see Chapter 14 for a discussion of genetic testing).

To date, more than 100 million infants have been screened in the United States, and over 10,000 cases of PKU have been detected and treated with a low phenylalanine diet. All states require screening of newborns for PKU, so the number of untreated cases is very low. Screening and treatment with a low phenylalanine diet allows PKU homozygotes to lead essentially normal lives.

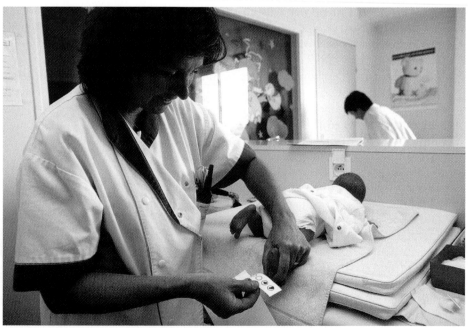

▶ FIGURE 10.5 A drop of blood from a newborn's heel will be used to test for phenylketonuria (PKU).

Photo Researchers

There is some controversy about how long a low-phenylalanine diet must be continued. Some studies suggest that PKU homozygotes can begin to eat a normal diet at about 10 to 14 years of age without any effects on intellect or behavior. Other treatment centers recommend that the treatment be continued for life. Recent findings indicate that parts of the brain continue to develop into adulthood. If confirmed, these results will probably require that the PKU diet be extended well into adulthood.

What happens when women with PKU have children of their own?

As PKU children treated with diet therapy have matured and reached reproductive age, the question arises: can a woman homozygous for PKU have an unaffected child? The answer to this seems straightforward. If she has a child with a homozygous normal man, the child would be heterozygous and unaffected. If she has a child with a heterozygote, the chances are 50% that the child will be unaffected. Only if she has a child with a homozygous PKU man will the child have a 100% chance of being affected.

The real answer is that all the children born to homozygous PKU women who eat a regular diet during pregnancy will be mentally retarded, no matter what their genotype. A pregnant PKU woman who eats a normal diet accumulates high levels of phenylalanine in her blood. This excess phenylalanine will cross the placenta and damage the nervous system of the developing fetus no matter what the genotype of the fetus.

To avoid this outcome, it is recommended that women with PKU stay on a low-phenylalanine diet throughout their reproductive years, or return to the diet for several months before becoming pregnant and stay on the diet throughout pregnancy. In addition, PKU females have other reproductive options, including *in vitro* fertilization and the use of surrogate mothers (see Chapter 16).

10.4 Other Metabolic Disorders in the Phenylalanine Pathway

The mutation that blocks a step in the conversion of phenylalanine to tyrosine is not the only mutation identified in this pathway. Several other genetic disorders are caused by mutations that block reactions leading from phenylalanine. For example, this pathway leads to the production of the thyroid hormones thyroxine and triiodothyronine. A mutation that blocks this part of the pathway leads to the autosomal recessively inherited disorder called genetic goitrous cretinism (▶ Figure 10.4). Newborn homozygotes are unaffected because during prenatal development, maternal thyroid hormones cross the placenta and promote normal growth. In the weeks following birth, physical development is slow, mental retardation begins, and the thyroid gland greatly enlarges. In this case, the phenotype is caused by the failure to synthesize an essential product, thyroid hormone, whereas in PKU, the problem is caused by toxic levels of a dietary amino acid and its breakdown products. If diagnosed early, infants with goitrous cretinism can be treated with thyroid hormone.

In this same network of metabolic pathways, a mutation in a gene encoding an enzyme leads to the buildup of homogentisic acid and causes alkaptonuria, an autosomal recessive condition. This is the disorder first investigated by Garrod at the beginning of the twentieth century. Mutations in other genes controlling enzymes in this network (▶ Figure 10.4) result in neonatal tyrosinemia (OMIM 276700), tyrosinemia II (OMIM 276600), albinism (OMIM 203100), and a number of other disorders.

10.5 Genes and Enzymes of Carbohydrate Metabolism

Mutations in genes encoding enzymes are not limited to amino acid metabolic pathways. Other pathways, including those of lipid metabolism, nucleic acid metabolism, and carbohydrate metabolism are also affected. We will briefly illustrate some mutations in carbohydrate metabolism.

Carbohydrates are organic molecules that include sugars, starches, glycogens, and celluloses. The simplest carbohydrates are sugars called monosaccharides (▶ Figure 10.6a). Fructose, glucose, and galactose are all monosaccharides and are important as energy sources for the cell. Chemically linking two monosaccharides produces a disaccharide (▶ Figure 10.6b). Some common disaccharides are sucrose (a glucose and fructose molecule; the sugar you buy at the store), maltose (two glucose molecules; a sugar used in brewing beer), and lactose (a glucose and a galactose molecule; the sugar found in milk). Long strings of many sugars linked together form polysaccharides; these include glycogen, starch, and cellulose.

Many different enzymes catalyze the chemical reactions that synthesize and break down sugars. Mutations that cause metabolic blocks in any of these reactions can have serious phenotypic consequences. Some genetic disorders associated with the metabolism of the polysaccharide glycogen are listed in ▶ Table 10.1. We will look at two examples of how mutations that affect enzymes of carbohydrate metabolism produce genetic disorders.

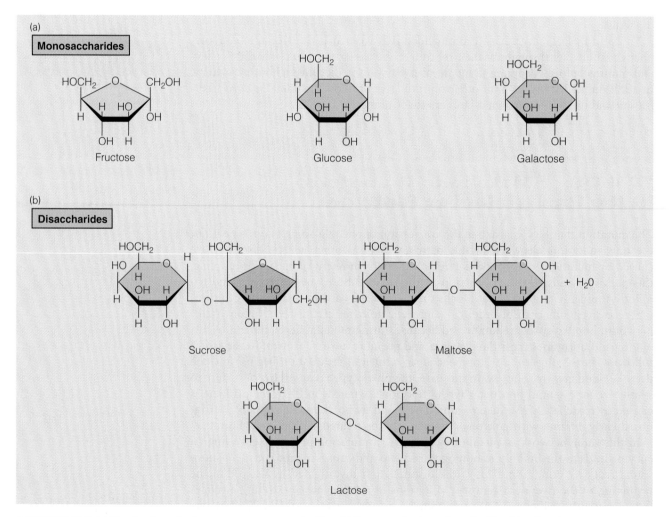

▲ FIGURE 10.6 (a) The structures for three common monosaccharides. (b) Structures for three disaccharides. Mutations in genes controlling metabolism of sugars are associated with genetic disorders.

Table 10.1   Some Inherited Diseases of Glycogen Metabolism

Type	Disease	Metabolic Defect	Inheritance	Phenotype	OMIM Number
I	Glycogen storage disease, Von Gierke disease	Glucose-6-phosphatase deficiency	Autosomal recessive	Severe enlargement of liver, often recognized in second or third decade of life; may cause death due to renal disease	232200
II	Pompe disease	Lysosomal glucosidase deficiency	Autosomal recessive	Accumulation of membrane-bound glycogen deposits. First lysosomal disease known. Childhood form leads to early death	232300
III	Forbes disease, Cori disease	Amylo-1,6-glucosidase deficiency	Autosomal recessive	Accumulation of glycogen in muscle, liver. Mild enlargement of liver, some kidney problems	232400
IV	Amylopectinosis, Andersen disease	Amylo-1,4-transglucosidase deficiency	Autosomal recessive	Cirrhosis of liver, eventual liver failure, death	232500

Galactosemia is caused by an enzyme deficiency.

Galactosemia (OMIM 230400) is an autosomal recessive disorder caused by the inability to break down galactose, one of the molecular components of lactose, the sugar found in milk (▶ Figure 10.7). Galactosemia occurs with a frequency of 1 in 57,000 births and is caused by lack of the enzyme galactose-1-phosphate uridyl transferase. When this enzyme is missing, a compound called galactose-1-phosphate accumulates and reaches toxic levels in the body. Like PKU, homozygous recessive individuals usually have a heterozygous mother and are unaffected before birth but develop symptoms a few days later. These include dehydration and loss of appetite; later the infants develop jaundice, cataracts, and mental retardation. In severe cases the condition is progressive and fatal. Seriously affected infants die within a few months, but mild cases may remain undiagnosed for many years. A galactose-free diet and the use of galactose- and lactose-free milk substitutes and foods lead to a reversal of symptoms. However, unless treatment is started within a few days of birth, mental retardation cannot be prevented.

Unlike PKU, dietary treatment in galactosemia does not prevent long-term complications. Many of those on a galactose-restricted diet develop difficulties with balance and have impaired motor skills, including problems with handwriting. It is not clear whether this is caused by low levels of damage to the nervous system that began during fetal development, or whether dietary treat-

■ **Galactosemia**   A heritable trait associated with the inability to metabolize the sugar galactose. If it is left untreated, high levels of galactose-1-phosphate accumulate, causing cataracts and mental retardation.

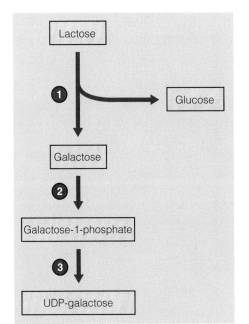

◀ **FIGURE 10.7** Metabolic pathway involving lactose and galactose. Lactose, the main sugar in milk, is enzymatically digested to glucose and galactose in step 1. In many adults, the enzyme for digesting lactose is not produced. Step 2 is the conversion of galactose into galactose-1-phosphate. In galactosemia, a mutation in the gene that controls step 3 prevents the conversion of galactose-1-phosphate into UDP-galactose. As a result, the concentration of galactose-1-phosphate rises in the blood, causing mental retardation and blindness.

ment is only partly effective. It is also not clear why or how the accumulation of galactose-1-phosphate is toxic.

Galactosemia is an example of a multiple-allele gene system. In addition to the normal allele, G, and the recessive mutant allele, g, a third allele, known as G^D (the Duarte allele, named after Duarte, California, the city in which it was discovered), has been found. Homozygous G^D/G^D individuals have only half of the normal enzymatic activity but show no symptoms of the disease. The existence of three alleles produces six possible genotypic combinations, and enzymatic activities ranging from 100% to 0% (▶ Table 10.2). This disease can be detected in newborns, and mandatory screening programs in many states test all newborns for galactosemia.

Lactose intolerance is a genetic variation.

Human milk is about 7% lactose, which is a major energy source for the nursing infant. The first step in breaking down lactose splits the molecule into two sugars, glucose and galactose (▶ Figure 10.7). This step is controlled by the enzyme lactase. In many parts of the world, lactase levels drop off during mid- to late childhood so that adults have less than 10% of the lactase activity found in infants. The decline in adult lactase levels is inherited as an autosomal recessive trait.

Adults with low lactase levels are unable to digest the lactose found in milk and other dairy products. If these adults eat lactose-containing foods, the result is a series of intestinal symptoms that include bloating, cramps, gas, and diarrhea. This condition is called lactose intolerance, and most lactase-deficient adults learn to avoid dairy products. Lactose intolerance is not considered a genetic disorder but only a variation in gene expression. Most human populations have low adult lactase levels, but the frequency of lactose intolerance varies from 0% to 100%. In Chapter 19, we will explore the role of natural selection in controlling the frequency of lactose intolerance in human populations.

10.6 Mutations in Receptor Proteins

Although many proteins function as enzymes, proteins have many other roles, including signal receptors and transducers. These proteins are usually embedded in the plasma membrane of the cell, and mutations in receptor function can have drastic consequences. For example, in androgen insensitivity (discussed in Chapter 7), a mutation in a gene encoding a receptor makes cells unable to respond to the presence of the hormone testosterone, causing a genotypic male to develop into a phenotypic female. Other genetic disorders associated with receptors, including familial hypercholesterolemia, are listed in ▶ Table 10.3.

Table 10.2   Multiple Alleles of Galactosemia

Genotype	Enzyme Activity (%)	Phenotype
G^+/G^+	100	Normal
G^+/G^D	75	Normal
G^D/G^D	50	Normal
G^+/g	50	Normal
G^D/g	25	Borderline
g/g	0	Galactosemia

Table 10.3   Some Heritable Traits Associated with Defective Receptors

Disease	Defective/Absent Receptor	Inheritance	Phenotype	OMIM Number
Familial hypercholesterolemia	Low-density lipoprotein (LDL)	Autosomal dominant	Elevated levels of cholesterol in blood, atherosclerosis, heart attacks; early death	144010
Pseudohypoparathyroidism	Parathormone (PTH)	X-linked dominant	Short stature, obesity, round face, mental retardation	300800
Diabetes insipidus	Vasopressin receptor defect	X-linked recessive	Failure to concentrate urine; high flow rate of dilute urine, severe thirst, dehydration; can produce mental retardation in infants unless diagnosed early	304800
Androgen insensitivity	Testosterone/ DHT receptor	X-linked recessive	Transformation of genotypic male into phenotypic female; malignancies often develop in intra-abdominal testes	313700

10.7 Defects in Transport Proteins: Hemoglobin

Hemoglobin, an iron-containing protein in red blood cells, transports oxygen from the lungs to the cells of the body. The hemoglobin molecule occupies a central position in human genetics. The study of hemoglobin variants led to an understanding of the molecular relationship between genes, proteins, and human disease in several ways.

- The discovery of variations in the amino acid composition of hemoglobin was the first example of inherited variations in protein structure.
- The altered hemoglobin in sickle cell anemia was the first direct proof that mutations result in a change in the amino acid sequence of proteins.
- The mutation in sickle cell anemia provided evidence that a change in a single nucleotide is enough to cause a genetic disorder.
- The molecular organization of the globin gene clusters has helped scientists understand how genes evolve and how gene expression is regulated.

Heritable defects in globin structure or synthesis are well understood at the molecular level and are truly "molecular diseases," as Linus Pauling called them. In this section we consider the structure of the hemoglobin molecule, the organization of the globin genes, and some genetic disorders related to globin structure and synthesis.

Hemoglobin is composed of four protein molecules called globins. Within each globin is a heme group. Heme is an organic molecule containing an iron atom (▶ Figure 10.8). In the lungs, oxygen enters the red blood cell and binds to the iron for transport to cells of the body. Although there are several different kinds of globin molecules (and hemoglobins), the heme group is the same in all cases.

Each adult hemoglobin molecule (called Hb A) is made up of two alpha globins and two beta globins (▶ Figure 10.9). Alpha globin is encoded in a gene cluster on chromosome 16 (▶ Figure 10.10); beta globin is encoded in a gene cluster on chromosome 11 (▶ Figure 10.11). Each red blood cell contains about 280 million molecules of hemoglobin, and there are between 4 and 6×10^{12} red blood cells in each liter of blood. Each red blood

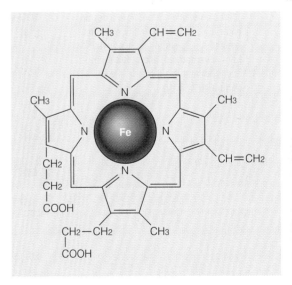

▲ FIGURE 10.8 A heme group is a flat molecule that inserts into the folds of a globin polypeptide. Each heme group carries an iron atom, which binds oxygen in the lungs for transport to the cells and tissues of the body.

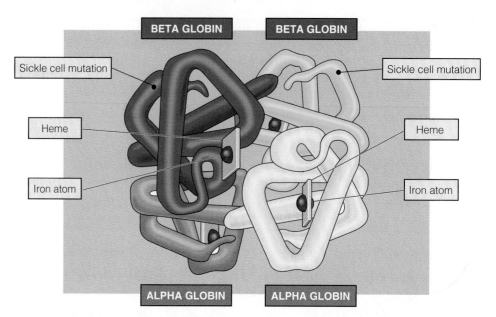

▶ FIGURE 10.9 A functional hemo-globin molecule is composed of two alpha-globin polypeptides and two beta-globin polypeptides. Each globin molecule carries a heme group within its folds. The location of the mutation in beta globin that is responsible for sickle cell anemia is shown near the start of each beta chain.

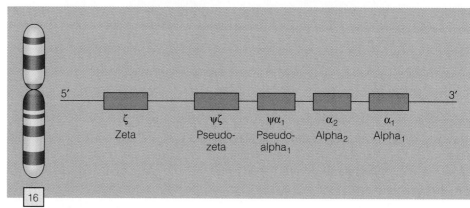

▶ FIGURE 10.10 The chromosomal location and organization of the alpha-globin cluster. Each copy of chromosome 16 contains two copies of the alpha-globin gene (alpha$_1$ and alpha$_2$), two nonfunctional versions (called pseudogenes), and a zeta gene, which is active only during early embryonic development.

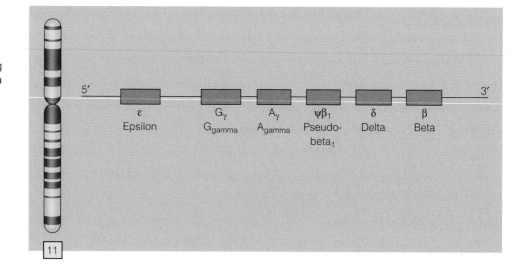

▶ FIGURE 10.11 The chromosomal location and organization of the beta-globin complex on chromosome 11. Each copy of chromosome 11 has an epsilon gene, active during embryonic development; two gamma genes (G$_{gamma}$ and A$_{gamma}$), active in fetal development; and a delta gene and a beta gene, which are transcribed after birth.

cell is replaced every 120 days, so hemoglobin production is a major metabolic process in the body, with millions of new hemoglobin molecules produced each second of each day.

Unlike most genes we carry, there are two copies of the alpha-globin gene (designated alpha$_1$ and alpha$_2$) in the alpha-gene cluster on each chromosome 16 (▶ Figure 10.10), along with three related genes: the zeta gene, pseudozeta, and pseudoal-

Table 10.4 Beta-Globin Chain Variants with Single Amino Acid Substitutions

Hemoglobin	Amino Acid Position	Amino Acid	Phenotype
A_1	6	glu	Normal
S	6	val	Sickle cell anemia
C	6	lys	Hemoglobin C disease
A_1	7	glu	Normal
Siriraj	7	lys	Normal
San Jose	7	gly	Normal
A_1	58	tyr	Normal
Hb M Boston	58	his	Reduced O_2 affinity
A_1	145	cys	Normal
Bethesda	145	his	Increased O_2 affinity
Fort Gordon	145	asp	Increased O_2 affinity

pha$_1$ genes. **Pseudogenes** are nonfunctional copies of genes whose nucleotide sequence is similar to a functional gene but with mutations that prevent their expression.

Genetic disorders of hemoglobin fall into two categories: the **hemoglobin variants,** which involve changes in the amino acid sequence of the globin polypeptides, and the **thalassemias,** characterized by imbalances in globin synthesis. More than 400 hemoglobin variants have been identified, each of which is caused by a different mutation. More than 90% of all variants are caused by the substitution of one amino acid for another in the globin chain, and more than 60% of these are found in beta globin (▶ Table 10.4). Some hemoglobin variants have no visible phenotype, whereas others produce mild symptoms, and still others result in lethal conditions.

Sickle cell anemia is an autosomal recessive disorder.

Sickle cell anemia (OMIM 141900) is inherited as an autosomal recessive trait. Affected individuals have a wide range of symptoms, including weakness, abdominal pain, kidney failure, and heart failure (▶ Active Figure 10.12), which lead to early death if left untreated.

This painful and disabling condition is caused by a mutation in the beta-globin gene. After oxygen is unloaded and transferred to cells in the body, hemoglobin molecules containing mutant beta-globin subunits come out of solution. The insoluble hemoglobin molecules stick together and form long tubular structures inside the cell (▶ Figure 10.13). These tubes distort and harden the membrane of the red blood cell, twisting the cell into a characteristic sickle shape. The deformed blood cells break easily. The lowered number of red blood cells reduces the oxygen-carrying capacity of the blood and results in anemia. The sickled cells also clog capillaries and small blood vessels, producing pain and tissue damage.

Linus Pauling and his colleagues showed that hemoglobin from sickle cell patients is abnormal. Later, Vernon Ingram showed that the only difference between normal hemoglobin and sickle cell hemoglobin is a change in the amino acid at position 6 in the beta chain. This change in a single amino acid is the molecular basis of sickle cell anemia (▶ Active Figure 10.14). All of the symptoms of the disease and its inevitably fatal outcome if left untreated derive from this alteration of one amino acid out of the 146 found in beta globin.

▦ **Pseudogenes** Nonfunctional genes that are closely related (by DNA sequence) to functional genes present elsewhere in the genome.

▦ **Hemoglobin variants** Alpha and beta globins with variant amino acid sequences.

▦ **Thalassemias** Disorders associated with an imbalance in the production of alpha or beta globin.

Keep in mind

▦ Sickle cell anemia is caused by substitution of a single amino acid in beta globin.

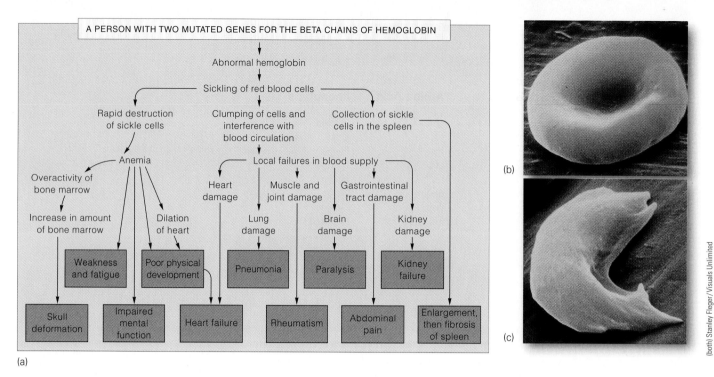

A PERSON WITH TWO MUTATED GENES FOR THE BETA CHAINS OF HEMOGLOBIN

(a)

(b)

(c)

▲ **ACTIVE FIGURE 10.12** (a) The cascade of phenotypic effects resulting from the mutation that causes sickle cell anemia. Affected homozygotes suffer effects at the molecular, cellular, and organ levels, all resulting from the substitution of a single amino acid in the beta-globin polypeptide chain. (b) Normally shaped red blood cell. (c) Sickled red blood cell.

HUMAN
Genetics ◯ Now™   Learn more about sickle cell anemia by viewing the animation on Human GeneticsNow at **http://now.ilrn.com/cummings7**

▶ **FIGURE 10.13** (a) A computer-generated image of the stages in the polymerization of sickle cell beta globin to form rods. *Upper:* A pair of intertwined fibers formed from stacked hemoglobin molecules. *Middle:* Seven pairs of fibers form the polymer responsible for distorting red blood cells. *Lower:* A large fiber composed of many smaller fibers. (b) An electron micrograph of a ruptured sickled red blood cell, showing the internal fibers of polymerized hemoglobin.

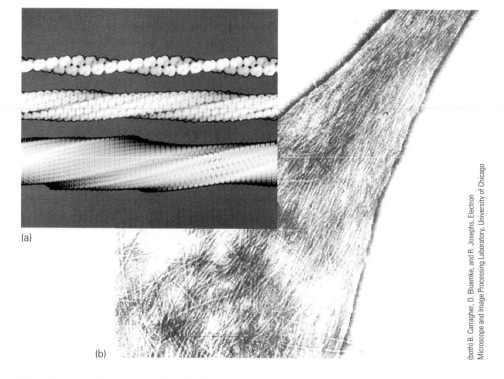

(a)

(b)

Thalassemias are also inherited hemoglobin disorders.

The thalassemias are a group of inherited hemoglobin disorders in which a mutant phenotype results from an imbalance in the relative amounts of alpha and beta globins produced. Normally, equal amounts of alpha and beta globin are produced, and

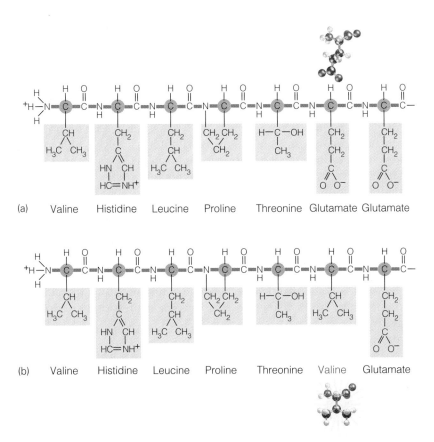

◀ ACTIVE FIGURE 10.14 (a) The normal sequence of amino acids at the start of a beta-globin chain. (b) A single amino acid substitution is present in the beta chains of Hb S molecules. Valine is present as the amino acid at position 6 instead of glutamate. This single amino acid difference is responsible for all the symptoms of sickle cell anemia.

HUMAN Genetics ⓔ Now™ Learn more about the molecular basis of sickle cell anemia by viewing the animation on Human GeneticsNow at **http://now.ilrn.com/cummings7**

(a) Valine Histidine Leucine Proline Threonine Glutamate Glutamate

(b) Valine Histidine Leucine Proline Threonine Valine Glutamate

normal hemoglobin molecules contain two molecules of each type of globin. In thalassemia, the synthesis of alpha or beta globin is reduced or absent, causing the formation of hemoglobin molecules with an abnormal number of alpha or beta globins. These hemoglobin molecules do not bind oxygen efficiently and can have serious and even fatal effects.

Thalassemias are common in several parts of the world, especially the Mediterranean region and Southeast Asia, where up to 20% or 30% of the population can be affected. The name "thalassemia" is derived from the Greek word *thalassa*, for "sea," emphasizing that this condition was first described in people living around the Mediterranean Sea.

There are two types of thalassemia: **alpha thalassemia** (OMIM 141800), in which the synthesis of alpha globin is reduced or absent, and **beta thalassemia** (OMIM 141900), which affects the synthesis of beta chains (▶ Table 10.5). Both conditions

◼ **Alpha thalassemia** Genetic disorder associated with an imbalance in the ratio of alpha and beta globin caused by reduced or absent synthesis of alpha globin.

◼ **Beta thalassemia** Genetic disorder associated with an imbalance in the ratio of alpha and beta globin caused by reduced or absent synthesis of beta globin.

Table 10.5 Summary of Thalassemias

Type of Thalassemia	Nature of Defect
α-Thalassemia-1	Deletion of two alpha-globin genes/haploid genome
α-Thalassemia-2	Deletion of one alpha-globin gene/haploid genome
β-Thalassemia	Deletion of beta and delta genes/haploid genome
Nondeletion α-thalassemia	Absent, reduced, or inactive alpha-globin mRNA
β^{0}-Thalassemia	Absent, reduced, or inactive beta-globin mRNA. No beta-globin produced
β^{+}-Thalassemia	Absent, reduced, or inactive beta-globin mRNA. Reduced beta-globin production

have more than one cause, and although inherited as autosomal recessive traits, both alpha and beta thalassemia have phenotypic effects in the heterozygous condition.

Alpha thalassemia is caused by the deletion of one or more alpha-globin genes. Six genotypes are possible, five of which have symptoms ranging from mild to lethal (▶ Figure 10.15). There are several forms of beta thalassemia, but most do not involve deletions of the gene. In some forms of beta thalassemia mutations lower the efficiency of beta-globin pre-mRNA processing. In β^0-thalassemia, a mutation at the junction between an intron and an exon interferes with normal mRNA splicing, resulting in very low levels of functional mRNA and, in turn, low levels of beta globin. Low levels of beta globin result in the formation of hemoglobin molecules with more alpha globins than beta globins.

Hemoglobin disorders can be treated through gene switching.

If untreated, sickle cell anemia is a fatal disease, and most affected individuals die by the age of 2 years. Even with an understanding of the molecular basis of the disease, treatments are only partially successful in relieving the symptoms. Recently, a discovery that certain anticancer drugs change patterns of gene expression has created a new and effective treatment for sickle cell anemia.

The drug hydroxyurea shuts off cell division and is used to treat cancer patients. As a side effect, patients have elevated levels of a hemoglobin type usually seen in developing fetuses. This fetal hemoglobin is a combination of two alpha globins and two gamma globins. Gamma-globin genes are part of the beta cluster and are switched off at birth, when the beta gene is activated (▶ Figure 10.16).

Treatment with hydroxyurea reactivates the gamma genes and makes fetal hemoglobin reappear in the red blood cells. Because sickle cell anemia is caused by a defect in beta globin, switching on a normal member of the beta cluster (gamma globin) produces fetal hemoglobin and reduces the amount of hemoglobin carrying mutant beta globins. This in turn reduces the number of sickled red blood cells, relieving many of the disorder's symptoms. Other drugs, including sodium butyrate, also switch on the synthesis of fetal hemoglobin. In some patients treated with sodium

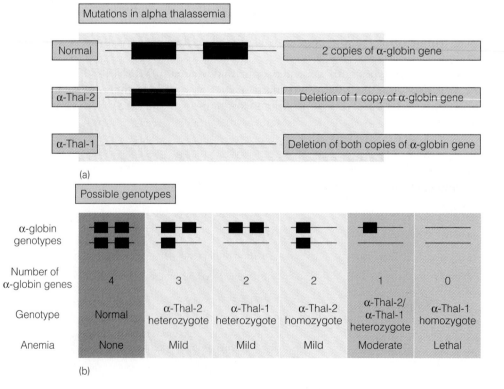

▶ FIGURE 10.15 Deletions of alpha-globin genes in alpha thalassemia. (a) Normally, each copy of chromosome 16 carries two copies of the alpha-globin gene (normal). One copy is deleted in the *alpha-thal-2* allele, and both copies are deleted in the *alpha-thal-1* allele. (b) These three alleles can be combined to form six genotypic combinations that have zero to four copies of the alpha-globin gene. Genotypes that have one copy deleted have moderate anemia and other symptoms, and genotypes that have no copies of the gene are lethal.

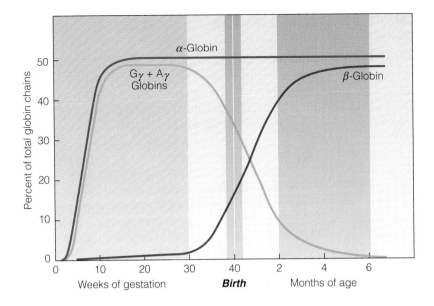

Percent of total globin chains

α-Globin

Gγ + Aγ
Globins

β-Globin

| | | | | | | | | |
0   10   20   30   40   2   4   6

Weeks of gestation   **Birth**   Months of age

◄ FIGURE 10.16 Patterns of globin gene expression during development. The alpha genes are switched on early in development and continue throughout life. The G$_{gamma}$ and A$_{gamma}$ genes, members of the beta family, are active during fetal development and switch off just before birth. The beta-globin gene is switched on at birth and is active throughout life. Sickle cell anemia and beta thalassemia are caused by mutations that affect beta globin. Research aimed at treating these conditions is directed at switching on the gamma genes, producing fetal hemoglobin to correct the conditions.

butyrate, up to 25% to 30% of the hemoglobin in the blood is fetal hemoglobin. Because sodium butyrate and related chemicals are less toxic than hydroxyurea, they are used to treat both sickle cell anemia and beta thalassemia by switching on genes that are normally turned off at birth.

10.8 Pharmacogenetics

As we have seen in previous sections of this chapter, variations in the type and amount of proteins produced by an individual can result in genetic disorders of metabolism. We are also discovering that variations in the amino acid sequence of proteins affect the way individuals react to prescription drugs and chemicals in the environment. For example, why is it that some people smoke cigarettes for years and never develop lung cancer? The answer may be in their genes. Alleles of genes for a family of enzymes called the P450 enzymes control the metabolism of carcinogens in cigarette smoke. Certain combinations of these alleles convert the carcinogens into less harmful compounds, offering protection against lung cancer.

Like some metabolic disorders, phenotypic differences in drug reactions or exposure to environmental chemicals appear only when an individual is exposed to the drug or chemical. These reactions are often the result of heritable variations in proteins and can be dominant or recessive traits. A branch of genetics known as **pharmacogenetics** studies the genetic variations that underlie drug responses. A branch of genetics called **ecogenetics** studies differences in the reactions to environmental agents. We will first describe some of the advances in pharmacogenomics and then discuss how ecogenetics is revealing how each of us is genetically unique.

Differences in drug responses can produce a range of phenotypic responses: drug resistance, toxic sensitivity to low doses, development of cancer after prolonged exposure, or an unexpected reaction to a combination of drugs. Some of these variations are harmless, whereas others can be life-threatening. In this section, we consider how exposure to drugs produces a wide range of phenotypes and describe the role of specific proteins in generating these phenotypes (if known).

■ **Pharmacogenetics** A branch of genetics concerned with the inheritance of differences in the response to drugs.

■ **Ecogenetics** A branch of genetics that studies genetic traits related to the response to environmental substances.

Keep in mind

■ Small differences in proteins can have a large effect on our ability to taste, smell, and metabolize medicines.

Taste and smell differences: we live in different sensory worlds.

Shortly after Garrod proposed that we are all biochemically unique individuals because of our genotypes, researchers began to demonstrate differences in the way people respond to chemicals. The discovery that we all have different abilities to taste and smell chemicals and that these differences are inherited was the first indication that there are important genetic differences in how people respond to drugs used to treat diseases.

The first pharmacogenetic trait was discovered in the 1930s as a by-product of work on artificial sweeteners. In searching for sugar substitutes, workers at DuPont discovered that some people cannot taste the chemical phenylthiocarbamide (PTC), whereas others find it to be very bitter. Shortly thereafter, it was found that the ability to taste PTC depends on a single pair of alleles and that genotypes *TT* and *Tt* represent tasters, whereas those who have genotype *tt* are nontasters. The ability to taste PTC varies from population to population. In the United States, about 30% of adult whites are nontasters, whereas only about 3% of U.S. blacks are nontasters. Later work showed that the ability to taste PTC is more complex than originally thought. When PTC solutions at various dilutions are used, a wide range of tasters can be detected. It appears that modifying genes affect the threshold of taste sensitivity.

How does such a discovery impact us? Some foods contain compounds similar to PTC and a related compound, PROP. These plants, including kale, cabbage, broccoli, and Brussels sprouts (▶ Figure 10.17) are bitter tasting to some people. So, if you don't like broccoli or Brussels sprouts, you may be able to blame it on your genotype.

Other evidence indicates that PTC/PROP tasters may live in a different taste world than nontasters. For example, capsaicin, the compound that makes hot peppers hot, has a more intense taste to PTC/PROP tasters; sucrose (table sugar) and artificial sweeteners are more intensely sweet to tasters. In addition, tasters have more food dislikes than nontasters and usually do not like foods such as black coffee, dark beer, anchovies, and strong cheeses.

Are there relationships between our genotypes, taste preferences, and our overall diets? For example, do tasters choose fruits and vegetables lower in cancer-*fighting* compounds, or do they choose foods that are lower in cancer-*causing* compounds? Is there a relationship between genotype, diet preference, and obesity? More research is needed to answer these and other questions related to taste preferences.

▶ FIGURE 10.17 Genetic differences can make some vegetables such as broccoli, kale, and Brussels sprouts taste bitter to some people.

Grant Heilman Photography

◀ FIGURE 10.18 Pink and red verbena flowers. Many people can smell the fragrance from the pink flowers but not the red ones. Others can smell the fragrance from the red flowers but not the pink ones.

The ability to smell is mediated by a family of 100 to 1,000 different membrane proteins. These proteins are present on the surface of cells in the nose and sinuses. There are many combinations of alleles for these proteins, so that each of us lives in a slightly different world of smell. In fact, our sensory worlds can be so different that some people cannot smell the odor released by skunks (OMIM 270350).

The garden flower *Verbena* comes in a variety of colors, including red and pink (▶ Figure 10.18). Blakeslee discovered that people differ in their ability to smell these flowers. About two-thirds of the people he tested could detect a fragrance in the pink flowers but not the red ones. The remaining one-third could detect a smell in the red flowers but not the pink ones.

Although the genetics of taste and smell demonstrate that different genotypes may be responsible for our food preferences and the ability to smell flowers, the importance of pharmacogenetics lies in determining the genetic foundations for the wide range of reactions to therapeutic drugs.

Drug sensitivities are genetic traits.

During the past 50 years, tens of thousands of new drugs have been developed. As these were tested on human volunteers and put into general use, distinctive patterns of response to these chemicals were identified. Subsequent work has shown that many of the differences people experience in response to drugs are genetically controlled. We will examine two examples of drug sensitivities.

Succinylcholine Sensitivity   Succinylcholine is used as a muscle relaxant and as a short-acting anesthetic (called suxamethonium). Soon after its introduction about 50 years ago, it became apparent that some people took hours rather than minutes to recover from a small dose of the drug. Normally, the drug is broken down to an inactive form by the enzyme serum cholinesterase. Those who take a long time to recover from the drug have a form of serum cholinesterase that breaks down the drug very slowly, prolonging the effect of the anesthetic (OMIM 177400). Pedigree analysis indicates that this trait is inherited in an autosomal recessive manner. In a study of Canadians, the frequency of heterozygotes was 3% to 4%, and about 1 in 2,000 people were sensitive recessive homozygotes. Use of succinylcholine on sensitive individuals can lead to paralysis of the respiratory muscles and death.

Primaquine Sensitivity and Favism   During World War II, several new drugs were developed to protect military personnel from malaria. Two of these drugs were primaquine and pamaquine. When these drugs were given to members of certain ethnic groups, including U.S. blacks, African blacks, and those from the European shores of the Mediterranean, such as Greeks, the result was a massive destruction of red blood cells (hemolytic anemia). A similar response to eating fava beans (*Vicia fava*) has been known since ancient times in Greece, Italy, and nearby countries. Both responses are caused by a deficiency in an X-linked enzyme, glucose-6-phosphate dehydrogenase (G6PD). The link between antimalarial drugs and fava beans is that both produce peroxides in the blood. The enzyme G6PD breaks down and inactivates the peroxides. If G6PD activity is reduced or absent, the high concentration of peroxides causes red blood cells to burst, producing a life-threatening anemia.

There are many other drugs and chemicals that cause lysis of red blood cells in G6PD-deficient individuals, including naphthalene, the active ingredient in mothballs. More than 400 million people worldwide are affected by G6PD deficiency.

10.9 Ecogenetics

The scope of pharmacogenetics has expanded to study genetic differences in reactions to chemicals in food, occupational exposure, and industrial pollution, leading to the development of ecogenetics. It is well known that the health risks from environmental chemicals involve the properties of the chemical itself, as well as the dose and the length of exposure. It is now clear that the overall risks to environmental chemicals also depends upon genetically determined variations in the proteins involved in transport, metabolism, and excretion of these chemicals.

What is ecogenetics?

Ecogenetics is the study of genetic variation that affects responses to environmental chemicals. Although there are more than 500,000 different chemicals used in manufacturing and agriculture, only a few have been tested for their toxicity or ability to cause cancer. The recognition that some members of a population may be sensitive or resistant to environmental chemicals has important consequences for research, medicine, and public policy. In this section, we will focus on the ecogenetics of pesticides.

Pesticide metabolism is linked to resistance to nerve gases.

Insects, weeds, fungi, and other pathogens destroy about 35% of the world's crops. After harvesting, another 10% to 20% is destroyed in storage. Chemical agents, including herbicides, insecticides, and fungicides, are used to control these pests. In the United States, around 65% of the insecticides used each year are applied to two crops, cotton and corn.

Agricultural insecticides include a group of chemicals called organophosphates, which includes parathion, an insecticide used for more than 50 years. Exposure to parathion and other organophosphates can occur on the job (agricultural workers and forestry workers) or from eating contaminated food. In the human body, parathion is chemically inert, but is enzymatically converted to a compound called paraoxon. Paraoxon is a toxic chemical that disrupts the transmission of signals in the nervous system. Paraoxon is broken down by paraoxonase, an enzyme found in blood serum.

The gene for paraoxonase (*PON1*) has two alleles (*A* and *B*). The *A* allele encodes a protein with high levels of enzymatic activity that detoxifies paraoxon and other organophosphate pesticides ten times faster than the enzyme encoded by the *B* allele. The two proteins differ in a single amino acid at position 192. People homozygous for the *A* allele (*A/A*) are resistant to the effects of pesticides like parathion. Conversely, those homozygous for low activity (*B/B*) are highly sensitive to parathion poisoning (OMIM 168820).

Paraoxonase is also involved in metabolizing toxic nerve gas agents, such as somain and sarin (sarin is the nerve gas released in the Tokyo subway in 1995 by the Aum Shinrikyo cult). However, in this case, the allele effects are reversed. The *B/B* genotype has high activity, and the *A/A* genotype has low activity. Asian populations have a high frequency of the *B* allele (about 25%), while Caucasians have a lower frequency (about 10%) for this allele. The wide range in the frequency of the *A* and *B* alleles in different populations suggests that ethnic groups may differ in their sensitivity to poisoning by insecticides and nerve gases, a rather grim reminder of the role of proteins in producing a phenotype.

The constellation of genes present within each person is the result of the random combination of parental genes and the sum of changes brought about by recombination and mutation. This genetic combination confers a distinctive phenotype upon each of us. Garrod referred to this metabolic uniqueness as chemical individuality. Understanding the molecular basis for this individuality remains one of the great challenges of human biochemical genetics.

Genetics in Practice

Genetics in Practice case studies are critical thinking exercises that allow you to apply your new knowledge of human genetics to real-life problems. You can find these case studies and links to relevant websites at **http://biology.brookscole.com/cummings7**

CASE 1

A couple was referred for genetic counseling because they wanted to know the chances of having a child with dwarfism. Both the man and woman had achondroplasia, the most common form of short-limbed dwarfism. The couple knew that this condition is inherited as an autosomal dominant trait, but they were unsure what kind of physical manifestations a child would have if it inherited both genes for the condition. They were each heterozygous for the *FGFR3* gene that causes achondroplasia, and they wanted information on the chances of having a child homozygous for the *FGFR3* gene. The counselor briefly reviewed the phenotypic features of individuals who have achondroplasia. These include the facial features (large head with prominent forehead; small, flat nasal bridge; and prominent jaw), very short stature, and shortening of the arms and legs. Physical examination and skeletal x-ray films are used to diagnosis this condition. Final adult height is approximately 4 feet.

Because achondroplasia is an autosomal dominant condition, a person with this condition has a 1 in 2, or 50%, chance of having children with this condition. However, approximately 75% of individuals with achondroplasia are born to parents of average size. In these cases, achondroplasia is due to a new mutation. This couple is at risk for having a child with two copies of the changed gene or double homozygosity. Infants with homozygous achondroplasia are either stillborn or die shortly after birth. The counselor recommended prenatal diagnosis via serial ultrasounds. In addition, a DNA test is available to detect the homozygous condition prenatally. Achondroplasia occurs in 1 in every 14,000 births.

1. What is the chance this couple will have a child with two copies of the dominant mutant gene? What is the chance that the child will have normal height?

2. Should the parents be concerned about the heterozygous condition as well as the homozygous mutant condition?

3. Why would the achrondroplasia gene be more susceptible to mutation than other genes?

CASE 2

Tina is 12 years old. Although symptomatic since infancy, she was not diagnosed with acid maltase deficiency (AMD) until she was 10 years old. The progression of her disease has been slow and insidious. She has great difficulty walking and breathing due to severe muscle weakness. She relies on a respirator to assist her breathing. Tina has severe scoliosis (curvature of the spine), which further restricts her breathing and causes even greater difficulty in walking. She is extremely tired and suffers from constant muscle pain. Although she is very bright and thinks like a normal teenager, her body won't let her function like one. She can no longer attend school. The future is bleak for Tina and other children like her. The cause of death in the childhood form of AMD is frequently due to complications from respiratory infections, which are a constant threat. Life expectancy in this form of AMD is only to the second or third decade of life.

AMD, also called glycogen storage disease type II (or Pompe disease), is an autosomal recessive condition that is genetically transmitted from carrier parents to their child. *(continued)*

When both parents are carriers (that is, they are heterozygous), there is a 25% chance during each pregnancy that the child will have two abnormal genes and be affected.

Normally, glycogen is synthesized from sugars and is stored in the muscle cells for future use. The acid maltase enzyme breaks down the glycogen in the muscle cells. Someone with AMD lacks this enzyme, and glycogen is not broken down but gradually builds up in the muscle tissues, leading to progressive muscle weakness and degeneration. There is no treatment or cure for AMD. Enzyme replacement and gene therapy are tools that may be useful in the future but have been unsuccessful in the past. However, a new treatment for enzyme replacement is being tested and offers new hope for children like Tina.

1. Should researchers continue with gene therapy even if it has not worked in the past? Who should fund this work?

2. Tina is 12 years old and her life expectancy is 20 to 30 years. What accommodations are needed to help her live as fulfilling and comfortable a life as possible?

Summary

10.1 Proteins Are the Link between Genes and the Phenotype

- In the early part of the twentieth century, Sir Archibald Garrod's studies on the metabolic diseases cystinuria, albinism, and alkaptonuria provided the first hints that gene products control the phenotype. These diseases and many others result from mutations that block biochemical reactions. In alkaptonuria, Garrod argued, the normal reaction is blocked by the lack of a needed enzyme. He speculated that the inability to carry out this reaction results from a gene inherited as a recessive trait.

10.2 Enzymes and Metabolic Pathways

- Biochemical reactions in the cell are linked together to form metabolic pathways. Mutations that block one reaction in a pathway can produce a mutant phenotype in several ways.

10.3 Phenylketonuria: A Mutation That Affects an Enzyme

- Phenylalanine is an essential amino acid and the starting point for a network of metabolic reactions. A mutation in a gene encoding the enzyme that controls the first step in this network causes phenylketonuria (PKU). The phenotype is caused by the buildup of phenylalanine and the products of secondary reactions.

10.4 Other Metabolic Diseases in the Phenylalanine Pathway

- The mutation causing PKU is only one of several genetic disorders caused by the mutation of genes in the phenylalanine pathway. Others include defects of thyroid hormone, albinism, and alkaptonuria, the disease investigated by Garrod.

10.5 Genes and Enzymes of Carbohydrate Metabolism

- Mutations in genes encoding enzymes can affect metabolic pathways of other biological molecules, including carbohydrates. Galactosemia is a genetic disorder caused by lack of an enzyme in sugar metabolism. Lactose intolerance is not a genetic disorder, but a genetic variation that affects millions of adults worldwide.

10.6 Mutations in Receptor Proteins

- Defects in receptor proteins, transport proteins, structural proteins, and other nonenzymatic proteins can cause phenotypic effects in the heterozygous state, and many show an incompletely dominant or dominant pattern of inheritance. Mutations in receptor proteins cause familial hypercholesterolemia.

10.7 Defects in Transport Proteins: Hemoglobin

- In 1949, James Neel identified sickle cell anemia as a recessively inherited disease. This disorder is caused by a mutation in a gene encoding beta globin, a protein that transports oxygen from the lungs to cells and tissues of the body. Other mutations cause thalassemia, an imbalance in the production of globins, affecting the transport of oxygen within the body.

10.8 Pharmacogenetics

- Differences in the reactions to therapeutic drugs represents a "hidden" set of phenotypes that are not revealed until exposure occurs. Understanding the genetic basis for these differences is the concern of pharmacogenetics and may lead to customized drug treatment for infections and other diseases.

10.9 Ecogenetics

■ Ecogenetics is the study of genetic variation that affects responses to environmental chemicals. The fact that some members of a population may be sensitive or resistant to environmental chemicals including pesticides has important consequences for research, medicine, and public policy.

Questions and Problems

Human Genetics Now™ Preparing for an exam? Assess your understanding of this chapter's topics with a pre-test, a personalized learning plan, and a post-test at **http://now.ilrn.com/cummings7**

Enzymes and Metabolic Pathways

1. Many individuals with metabolic diseases are normal at birth but show symptoms shortly thereafter. Why?
2. List the ways in which a metabolic block can have phenotypic effects.
3. Enzymes have all of the following characteristics, except:
   a. they act as biological catalysts
   b. they are proteins
   c. they carry out random chemical reactions
   d. they convert substrates into products
   e. they can cause genetic disease

Questions 4 through 6 refer to the following hypothetical pathway in which substance A is converted to substance C by enzymes 1 and 2. Substance B is the intermediate produced in this pathway:

$$\begin{array}{ccc} \text{enzyme} & \text{enzyme} \\ 1 & 2 \\ A \text{---------}> & B \text{---------}> & C \end{array}$$

4. a. If an individual is homozygous for a null mutation in the gene that codes for enzyme 1, what will be the result?
   b. If an individual is homozygous for a null mutation in enzyme 2, what will be the result?
   c. What if an individual is heterozygous for a dominant mutation where enzyme 1 is overactive?
   d. What if an individual is heterozygous for a mutation that abolishes the activity of enzyme 2 (a null mutation)?
5. a. If the first individual in Question 4 married the second individual, would their children be able to convert substance A into substance C?
   b. Suppose each of the aforementioned individuals were heterozygous for an autosomal dominant mutation. List the phenotypes of their children with respect to compounds A, B, and C. (Would the compound be in excess, not present, and so on?)
6. An individual is heterozygous for a recessive mutation in enzyme 1 and heterozygous for a recessive mutation in enzyme 2. This individual marries an individual of the same genotype. List the possible genotypes of their children. For every genotype, determine the activity of enzyme 1 and 2, assuming that the mutant alleles have 0% activity and the normal alleles have 50% activity. For every genotype, determine if compound C will be made. If compound C is not made, list the compound that will be in excess.

Questions 7 to 11 refer to a hypothetical metabolic disease in which protein E is not produced. Lack of protein E causes mental retardation in humans. Protein E's function is not known, but it is found in all cells of the body. Skin cells from eight individuals who cannot produce protein E were taken and were grown in culture. The defect in each of the individuals is the result of a single recessive mutation. Each individual is homozygous for her or his mutation. The cells from one individual were grown with the cells from another individual in all possible combinations of two. After a few weeks of growth, the mixed cultures were assayed for the presence of protein E. The results are given in the following table. A plus sign means that the two cell types produced protein E when grown together (but not separately); a minus sign means that the two cell types still could not produce protein E.

	1	2	3	4	5	6	7	8
1	−	+	+	+	+	−	+	+
2	−		+	+	+	+	−	+
3		−		+	+	+	+	−
4			−		+	+	+	+
5				−		+	+	+
6					−		+	+
7						−		+
8								−

7. a. Which individuals seem to have the same defect in protein E production?
   b. If individual 2 married individual 3, would their children be able to make protein E?
   c. If individual 1 married individual 6, would their children be able to make protein E?
8. a. Assuming that these individuals represent all possible mutants in the synthesis of protein E, how many steps are there in the pathway to protein E production?
   b. Compounds A, B, C, and D are known to be intermediates in the pathway for production of protein E. To determine where the block in protein E production occurred in each individual, the various intermediates were given to each individual's cells in culture. After a few weeks of growth with the intermediate,

the cells were assayed for the production of protein E. The results for each individual's cells are given in the following table. A plus sign means that protein E was produced after the cells were given the intermediate listed at the top of the column. A minus sign means that the cells still could not produce protein E even after being exposed to the intermediate at the top of the column.

			Compounds		
Cells	A	B	C	D	E
1	−	−	+	+	+
2	−	+	+	+	+
3	−	−	−	+	+
4	−	−	−	−	+
5	+	+	+	+	+
6	−	−	+	+	+
7	−	+	+	+	+
8	−	−	−	+	+

9. Draw the pathway leading to the production of protein E.
10. Denote the point in the pathway in which each individual is blocked.
11. a. If an individual who is homozygous for the mutation found in individual 2 and heterozygous for the mutation found in individual 4 marries an individual who is homozygous for the mutation found in individual 4 and heterozygous for the mutation found in individual 2, what will be the phenotype of their children?
    b. List the intermediate that would build up in each of the types of children who could not produce protein E.

Phenylketonuria: A Mutation That Affects an Enzyme

12. Essential amino acids are:
    a. amino acids the human body can synthesize
    b. amino acids humans need in their diet
    c. amino acids in a box of Frosted Flakes
    d. amino acids that include arginine and glutamic acid
    e. amino acids that cannot harm the body if not metabolized properly

13. Suppose that in the formation of phenylalanine hydroxylase mRNA, the exons of the pre-mRNA fail to splice together properly, and the resulting enzyme is nonfunctional. This produces an accumulation of high levels of phenylalanine and other compounds, which causes neurological damage. What phenotype and disease would be produced in the affected individual?

14. PKU is an autosomal recessive disorder that causes mental retardation. In individuals afflicted with PKU, high levels of the essential amino acid phenylalanine are present because of a deficiency in the enzyme phenylalanine hydroxylase. If phenylalanine was not an essential amino acid, would diet therapy (the elimination of phenylalanine from the diet) work?

15. Phenylketonuria and alkaptonuria are both autosomal recessive diseases. If a person with PKU marries a person with AKU, what will be the phenotype of their children?

Genes and Enzymes of Carbohydrate Metabolism

16. The normal enzyme required for converting sugars into glucose is present in cells, but the conversion never takes place, and no glucose is produced. What could have occurred to cause this defect in a metabolic pathway?

17. Knowing that individuals who are homozygous for the G^D allele show no symptoms of galactosemia, is it surprising that galactosemia is a recessive disease? Why?

Mutations in Receptor Proteins

18. Familial hypercholesterolemia is caused by an autosomal dominant mutation in the gene that produces the LDL receptor. The LDL receptor is present in the plasma membrane of cells and binds to cholesterol and helps remove it from the circulatory system for metabolism in the liver. What is the phenotype of the following individuals?

    HH
    Hh
    hh

19. Suppose the gene for the LDL receptor has been isolated by recombinant DNA techniques. Could you treat this disease by producing an LDL receptor and injecting it into the bloodstream of affected individuals? Why or why not?

20. If a chromosomal male has a defect in the cellular receptor that binds the hormone testosterone, what condition results? What is the genotype and phenotype of this individual?

Defects in Transport Proteins: Hemoglobin

21. Describe the quaternary structure of the blood protein hemoglobin.

22. A person was found to have very low levels of functional beta-globin mRNA and therefore very low levels of the beta-globin protein. Name this person's disease and explain what mutation may have occurred in the conversion of pre-mRNA into mRNA.

23. If an extra nucleotide is present in the first exon of the beta-globin gene, what effect would it have on the amino acid sequence of the globin polypeptides? Would the globin most likely be fully functional, partly functional, or nonfunctional? Why?

24. Transcriptional regulators are proteins that bind to promoters (the 5′-flanking regions of genes) to regulate their transcription. Assume a particular transcription regulator normally promotes transcription of gene X, a transport protein. If a mutation makes this regulator gene nonfunctional, would the resulting phenotype be similar to a mutation in gene X itself? Why?

25. Mutations in the alpha thalassemia genes can result in a variety of abnormal phenotypes. If a heterozygous alpha thalassemia-1 man marries a heterozygous alpha thalassemia-2 woman, what will be the phenotypes of their offspring? (Refer to ▶ Figure 10.15.)

Pharmacogenetics

26. Explain why there are variant responses to drugs and why these responses act as heritable traits.

Ecogenetics

27. Ecogenetics is a branch of genetics that deals with genetic variation that underlies reactive differences to drugs, chemicals in food, occupational exposure, industrial pollution, and other substances. Cases have arisen in which workers claim that exposure to a certain agent has made them feel ill (whereas other workers are unaffected). Although claims like these are not always justified, what are some concrete examples that prove that variation to certain substances exist in the human population?

Internet Activities

 Internet Activities are critical thinking exercises using the resources of the World Wide Web to enhance the principles and issues covered in this chapter. For a full set of links and questions investigating the topics described below, log on to **http://biology.brookscole.com/cummings7**

1. *Sickle Cell Anemia.* At the *Sickle Cell Case Study* site, read about the genetics of sickle cell disease and the relationship between sickle cell disease and malaria. Read, too, about current research in sickle cell anemia.

2. *Enzyme Replacement Therapy and Pompe Disease.* At Applied Biosystems's *Biobeat* site, access and read the article on enzyme replacement therapy in the treatment of Pompe disease.

How would you vote now?

PKU and other metabolic disorders can be tested for in newborns, thus allowing for early treatment. All 50 states and the District of Columbia require testing for PKU, and several other prevalent genetic disorders. However, the exact number of genetic diseases newborns are tested for varies from state to state, from as few as 6 genetic disorders to more than 40. One of the rationales given for testing for a small number of disorders is that cost benefit analysis shows that it is not cost-efficient to do so. That is, some diseases are so rare that the costs of testing all newborns outweigh the health care costs for affected children, regardless of the severity of problems caused by the disorder. Now that you know more about metabolic disorders, and the relationships among genes, proteins, and phenotypes, what do you think? Should cost benefit analysis be used as a determining factor in setting up and running newborn testing programs? Visit the Human Heredity Companion website to find out more on the issue, then cast your vote online at **http://biology.brookscole.com/cummings7**

 For further reading and inquiry, log on to **InfoTrac College Edition,** your world-class online library, including articles from nearly 5,000 periodicals, at **http://infotrac.thomsonlearning.com**

Mutation: The Source of Genetic Variation

11

DURING the past 40 years, research has demonstrated that radiation can help preserve food and kill contaminating microorganisms. Irradiation prevents sprouting of root crops, such as potatoes; extends the shelf life of many fruits and vegetables; destroys bacteria and fungi in meat, fish, and grain; and kills insects and other pests in spices.

For irradiation, food is placed on a conveyor and moved to a sealed, heavily shielded chamber where it is exposed to a radioactive source. An operator views the process on video camera and delivers the dose. The food itself does not come in contact with the radioactive source, and the food is not made radioactive. Relatively low doses are used to inhibit sprouting of potatoes and to kill parasites in pork. Intermediate doses are used to retard spoilage in meat, poultry, and fish, and high doses can be used to sterilize foods, including meats. The amount of food irradiated varies from country to country, ranging from a few tons of spices to hundreds of thousands of tons of grain.

NASA has routinely used irradiated food to feed astronauts in space since 1972, and irradiated foods are sold in more than 40 countries, including the United States. The Food and Drug Administration (FDA) approved the first application for food irradiation in 1964, and approval has been granted for the irradiation of spices, herbs, fruits and vegetables, pork, beef, lamb, chicken, and eggs. All irradiated food sold in the United States must be labeled with an identifying logo (see inset).

Public concern about radiation has prevented the widespread sale of irradiated food in this country. Advocates point out that irradiation can eliminate the use of many chemical preservatives, lower food costs by preventing spoilage, and reduce the inci-

Lior Rubin/Peter Arnold, Inc.

dence of the 76 million cases of food-borne illnesses and 5,000 deaths caused by contaminated food each year in the United States. Those opposed to food irradiation argue that irradiation produces mutation-causing and cancer-causing compounds in food, and that the testing of irradiated food to detect any cancer-causing effects is inadequate. Opponents also point out that treatment may select for radiation-resistant microorganisms.

In this chapter we will consider the nature of mutations, how mutations are detected, the rate of mutation, and the role of radiation and chemicals in causing mutations.

How would you vote?

E. coli contamination in meat causes 20,000 illnesses and 500 deaths per year in the United States. Irradiation to kill *E. coli* in beef and *Salmonella* in poultry has been approved by the U.S. Food and Drug Administration and the U.S. Department of Agriculture. The World Health Organization and the American Medical Association have endorsed irradiation as an effective means of preventing disease and deaths. In spite of this approval, irradiated meat is not widely available. If such products were available in the supermarket, would you buy them? Visit the Human Heredity Companion website to find out more on the issue, then cast your vote online at **http://biology.brookscole.com /cummings7**

11.1 Mutations Are Heritable Changes

Mutation is the source of all genetic variation in humans and other organisms. The results of mutations can be classified in a number of ways. From a genetic perspective, mutations that produce dominant alleles are expressed in the heterozygous condition; mutations to recessive alleles are expressed only when homozygous. Mutations can also be ranked in categories, such as the severity of the phenotype or the age of onset.

Keep in mind

■ Mutation can occur spontaneously as the result of errors in DNA replication or be induced by exposure to radiation or chemicals.

For our purposes, two general categories of mutations are the most useful: mutations that affect chromosomes and mutations that change the nucleotide sequence of a gene. Chromosomal aberrations were discussed in Chapter 6. In this chapter we

Keep in mind as you read

■ Mutation can occur spontaneously as the result of errors in DNA replication or be induced by exposure to radiation or chemicals.

■ Mutations in DNA can occur in several ways, including nucleotide substitution, deletion, or insertion.

■ Damage to DNA can be repaired during DNA synthesis, and by enzymes that repair damage to DNA caused by radiation or chemicals.

focus on mutational changes in single genes, that is, changes in the sequence or number of nucleotides in DNA. First, we consider how mutations are detected and then investigate at what rate these mutations take place. Finally, we examine how mutation works at the molecular level.

11.2 Mutations Can Be Detected in Several Ways

How do we know that a mutation has taken place? In humans, the sudden appearance of a dominant mutation in a family can be observed in a single generation. But mutation from a dominant allele to a recessive allele can be detected only in the homozygous condition, posing a challenge for human geneticists, because its phenotype may appear only after many generations.

If an affected individual appears in an otherwise unaffected family, the first question is whether the trait is caused by genetic or nongenetic factors. For example, if a mother is exposed to the rubella virus (which causes a form of measles called German measles) early in pregnancy, the fetus may have a phenotype similar to those produced in some genetic disorders. The phenotype caused by rubella infection isn't produced by mutation but by the effect of the virus on the developing fetus. To determine whether an abnormal phenotype is caused by a genetic disorder, geneticists depend on pedigree analysis and the study of births over several generations (a family history).

If a mutant allele is dominant, is fully penetrant (expressed in all who carry the mutant allele), and appears in a family with no history of this condition in previous generations, geneticists usually presume that a mutation has taken place. In the pedigree shown in ▶ Figure 11.1, severe blistering of the feet appeared in one out of six children, although the parents were unaffected. The trait was transmitted by the affected female to six of her eight children and was passed to the next generation as an autosomal dominant condition. A reasonable explanation for this pedigree is that a mutation to a dominant allele causing foot blisters appeared in individual II-5. On the other hand, a number of uncertainties can affect this conclusion. For example, if the child's father is not the husband in the pedigree but is an affected male, then it would only seem that a mutational event had taken place.

If mutation results in a recessive sex-linked allele, it often can be detected by examining males in the family line. But it can be difficult to determine whether a het-

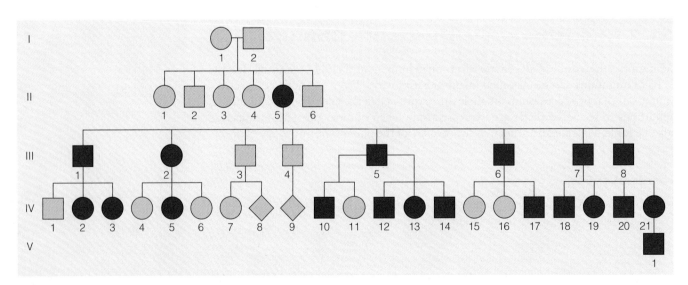

▲ FIGURE 11.1 A dominant trait, foot blistering, appeared (II-5) in a family that had no previous history of this condition. The trait is transmitted through subsequent generations in an autosomal dominant fashion.

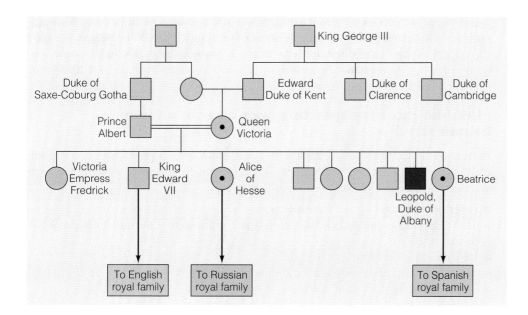

◀ FIGURE 11.2 Pedigree of Queen Victoria of Great Britain, showing her immediate ancestors and children. Because she passed the mutant allele for hemophilia on to three of her children, she was probably a heterozygote rather than the source of the mutation.

erozygous female who transmits a trait to her son is the source of the mutation or is only passing on a mutation that arose in an ancestor. The X-linked form of hemophilia that spread through the royal families of Western Europe and Russia in the nineteenth and twentieth centuries probably originated with Queen Victoria (▶ Figure 11.2; see Genetics in Society: Hemophilia and History in Chapter 4). None of the males in previous generations had hemophilia, but one of Victoria's sons was affected, and at least two of her daughters were heterozygous carriers.

Because Victoria transmitted the mutant allele to a number of her children, it is reasonable to assume that she was a heterozygous carrier. Her father was not affected, and there is nothing in her mother's pedigree to indicate that she was a carrier. It is therefore likely that Victoria received a newly mutated allele from one of her parents. We can only speculate as to which parent gave her the mutant gene.

If an autosomal recessive trait appears suddenly, it is usually difficult or impossible to trace the mutant allele through a family to identify the person or even the generation in which the mutation first occurred, because only homozygotes are affected. Such a new mutation can remain undetected for generations, passed from heterozygote to heterozygote.

11.3 Measuring Spontaneous Mutation Rates

Pedigree analysis reveals that mutation does take place in the human genome. The available evidence suggests that it is a rare event, but given the problems outlined previously, can we hope to accurately measure the rate of spontaneous mutations? If we knew the overall rate of mutation, we can monitor it over time to see if it is increasing, decreasing, or remaining the same.

Geneticists define **mutation rates** as the number of mutated alleles per gene per generation. Suppose that for a certain gene, 4 out of 100,000 births show a mutation from a recessive to a dominant allele. Because each of these 100,000 individuals carries two copies of the gene, we have sampled 200,000 copies of the gene. The four births represent four mutated genes (we are assuming that the newborns are heterozygotes for a dominant mutation and carry only one mutant allele). In this case, the mutation rate is 4/200,000, or 2/100,000. In scientific notation this would be written as 2×10^{-5} per allele per generation.

■ **Mutation rate** The number of events that produce mutated alleles per locus per generation.

If the gene was X-linked and if 100,000 male births were examined and four mutants were discovered, this would represent a sampling of 100,000 copies of the gene (because the males have only one copy of the X chromosome). Excluding contributions from female carriers, the mutation rate in this case would be 4/100,000, or 4×10^{-5} per allele per generation.

Mutation rates for specific genes can sometimes be measured.

Is there a way to measure directly the rate of mutation for a gene? For certain dominant alleles, the answer is yes, but only under certain conditions. To ensure the measurement is accurate, the mutant phenotype must

- never be produced by recessive alleles,
- always be fully expressed and completely penetrant so that mutant individuals can be identified,
- have clearly established paternity,
- never be produced by nongenetic agents such as drugs or infection, and
- be produced by dominant mutation of only one gene.

One dominant mutation, achondroplasia, fulfills these requirements. Achondroplasia (OMIM 100800) is a dominant form of dwarfism that produces short arms, short legs, and an enlarged skull (▶ Figure 11.3). A diagnosis by x-ray examination can be done shortly after birth. Several population surveys have used mutations in this gene to estimate the overall mutation rate in humans. One survey found 7 achondroplastic children of unaffected parents in a total of 242,257 births. From these data, the mutation rate for achondroplasia has been calculated at 1.4×10^{-5}, or one mutation in every 100,000 copies of the gene.

Although the mutation rate for achondroplasia can be measured directly, it is not clear whether this gene's mutation rate is typical for all human genes. Maybe this gene has an inherently high rate of mutation, or an unusually low rate of mutation. To get an accurate picture of the mutation rate in humans, it is important to measure the mutation rate in a number of different genes before making any general statements. As it turns out, two other dominant mutations have widely different rates of mutations.

Neurofibromatosis (OMIM 162200), an autosomal dominant condition, is characterized by pigmentation spots and tumors of the skin and nervous system (described in Chapter 4). About 1 in 3,000 births are affected. Many of these births (about 50%) occur in families that have no previous history of neurofibromatosis, indicating that this locus has a high mutation rate. In fact, the calculated mutation rate in this disease is 1 in 10,000 (1×10^{-4}), one of the highest rates so far discovered in humans. On the other end of the spectrum, the mutation rate for Huntington disease (OMIM 143100) has been calculated as 1×10^{-6}, a rate 100-fold lower than that of neurofibromatosis and 10-fold lower than that of achondroplasia.

Measurements of the mutation rate in several human genes are listed in ▶ Table 11.1. The average rate is approximately 1×10^{-5}. All the genes listed in the table are inherited as autosomal dominant or X-linked traits. It is almost impossible to measure directly the mutation rates in autosomal recessive alleles by pedigree analysis, but population surveys using recombinant DNA methods are now providing estimates of the rate and type of mutations found in many human genes, including those with autosomal recessive patterns of inheritance. Still, many geneticists feel that to reduce any potential bias, a more conservative estimate of the mutation rate in hu-

▲ FIGURE 11.3 The painting *Las Meninas* by Diego Velasquez shows Infanta Margarita of the seventeenth-century Spanish court accompanied by her maids and others, including an achondroplasic woman at right. Achondroplasia (OMIM 100800) is a form of dwarfism caused by a dominant mutation.

Table 11.1 Mutation Rates for Selected Genes

Trait	Mutants/Million Gametes	Mutation Rate	OMIM Number
Achondroplasia	10	1.4×10^{-5}	100800
Aniridia	2.6	2.6×10^{-6}	106200
Retinoblastoma	6	6×10^{-6}	180200
Osteogenesis imperfecta	10	1×10^{-5}	166200
Neurofibromatosis	50–100	$0.5–1 \times 10^{-4}$	162200
Polycystic kidney disease	60–120	$6–12 \times 10^{-4}$	173900
Marfan syndrome	4–6	$4–6 \times 10^{-6}$	154700
Von Hippel–Landau syndrome	<1	1.8×10^{-7}	193300
Duchenne muscular dystrophy	50–100	$0.5–1 \times 10^{-4}$	310200

mans should be used, and by convention 1×10^{-6} is used as the average mutation rate for human genes.

Why do genes have different mutation rates?

Several factors influence the mutation rate and contribute to the wide range we observe:

- *Size of the gene.* Larger genes are bigger targets for mutation. The gene for neurofibromatosis (*NF-1*) is an extremely large gene and has a high mutation rate. The *NF-1* gene extends over 300,000 base pairs of DNA. The gene for Duchenne and Becker muscular dystrophy, the largest gene identified to date in humans, contains more than 2 million base pairs. These large genes have high mutation rates.
- *Nucleotide sequence.* In some genes, short nucleotide repeats are present in the DNA. In the gene for fragile-X syndrome, the sequence CGG is repeated some 6 to 50 times in normal individuals. Those who have more than 230 copies experience symptoms of the disorder, and these become more severe as the number of CGG repeats increases. The presence of nucleotide repeats may predispose a gene to mutate at a higher rate.
- *Spontaneous chemical changes.* Among the bases in DNA, cytosine is especially susceptible to chemicals that can change the nucleotide sequence in DNA. These and other chemical changes are discussed in a later section of this chapter. Genes rich in G/C base pairs are more likely to undergo chemical changes than those rich in A/T pairs.

11.4 Environmental Factors Influence Mutation Rates

In general, mutations can occur as the result of mistakes that occur during normal cellular functions such as DNA replication or by the action of agents that attack DNA or cellular functions from outside the cell. These agents include chemicals and radiation.

Radiation is one source of mutations.

Radiation is a process by which energy travels through space. For example, the heat from a fire in a fireplace travels through space and warms a room. There are two

Radiation The process by which electromagnetic energy travels through space or a medium such as air.

main types of radiation: waves of energy (electromagnetic radiation) and particles of energy (corpuscular radiation). Waves are electrical or magnetic energy, whereas atomic and subatomic particles move through space at high speeds as a form of radiation. Both forms of radiation are known as **ionizing radiation** because they form chemically reactive ions when they collide with molecules in cells. However, some forms of radiation can cause mutations without producing ions. For example, ultraviolet (UV) light causes mutations in DNA without producing ions. The energy in UV light is absorbed directly by DNA and results in mutations.

Remember that exposure to radiation is unavoidable. Everything in the physical world contains sources of radiation. This includes our bodies, the air we breathe, the food we eat, and the bricks in our houses. Some of this radiation is left over from the birth of the universe, and some has been created by interaction of atoms on Earth with cosmic radiation. These natural sources of radiation are called **background radiation.** We are also exposed to radiation that results from human activity, including medical testing, nuclear testing, nuclear power, and consumer goods.

Radiation can cause biological damage at several levels. As radiation strikes the molecules in cells, it creates ions and charged atoms. Such ionized molecules are highly reactive and can cause mutations in DNA. Because cells are about 80% water, radiation often generates free radicals, by splitting water molecules into hydrogen ions (H^+) or hydroxyl radicals (OH^-). These can produce mutations in DNA. Often, the cell can repair these mutations. However, if too many mutations accumulate in a cell, the repair system can be overwhelmed, and cell death or cancer can result. If the mutations are not repaired, cancer is a possible outcome. In germ cells, mutations that are not repaired are transmitted from generation to generation as newly mutated alleles. As a result of the Chernobyl nuclear power plant accident (▶ Figure 11.4), the U.S. Nuclear Regulatory Commission estimates that 14,000 additional cancer deaths can be anticipated in Western Europe within 70 years following the accident. However, given the size of the population and the rate of cancer deaths, this number of additional deaths may be impossible to detect.

How much radiation are we exposed to?

A dose of radiation can be measured in several ways: the amount a person is exposed to, the amount absorbed by the body, or the amount of damage caused. Most often, the dose is expressed in terms of damage. A **rem** (radiation equivalent in man or Roentgen-equivalent man, for the scientist who discovered x-rays) is the amount of

▦ **Ionizing radiation** Radiation that produces ions during interaction with other matter, including molecules in cells.

▦ **Background radiation** Radiation in the environment that contributes to radiation exposure.

▦ **Rem** The unit of radiation exposure used to measure radiation damage in humans. It is the amount of ionizing radiation that has the same effect as a standard amount of x-rays.

▦ **Millirem** Each rem is equal to 1,000 millirems.

▶ FIGURE 11.4 In April 1986 the nuclear reactor at Chernobyl, Ukraine, exploded—spreading radioactive material across the Northern Hemisphere.

radiation that causes the same damage as a standard amount of x-rays. Because people are usually exposed to very small amounts of radiation, the dose is usually expressed in **millirems** (1,000 mrem equals 1 rem). At doses in the millirem range, cells can repair most, if not all, of the radiation damage. At doses of approximately 100 rem (100,000 mrem), cells begin to die, and radiation sickness results. At a dosage of 400 rem, about 50% of people will die within 60 days if they are not treated. Case 1 at the end of this chapter discusses the impact of the Chernobyl reactor explosion on the surrounding populations.

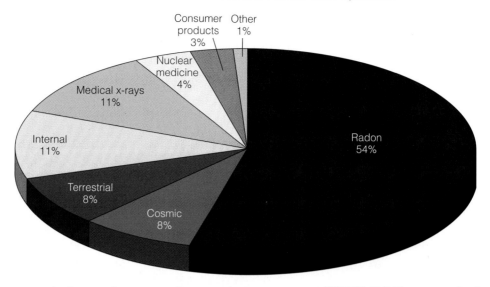

Sources of Radiation Exposure to the U.S. Population

▲ **FIGURE 11.5** The sources of radiation individuals are exposed to in the United States. The average dose is 360 mrem, 82% of which is from background radiation.

In the United States, the average person is exposed to about 360 mrem/year, 82% of which is from nature (▶ Figure 11.5). Because most people are exposed to much less than 100 rem/year and will not get radiation sickness, what are the major risks from radiation exposure? At levels below 5 rem (5,000 mrem), the major risk is mutations in somatic cells that increase susceptibility to cancer. Overall, this effect is very small, and risk analysis suggests that radiation is a small risk compared to others in our daily lives.

Genetics in Society

Rise of the Flame Retardants

Flame retardants are almost a universal feature of consumer goods in the United States. They are found in polyurethane foam, carpets, mattresses, upholstery, televisions, computer monitors, printers, cell phones, and a long list of other items. One class of flame retardants, PBDEs (short for polybrominated diphenyl ethers), is chemically related to well known and persistent environmental contaminants such as PCBs that cause cancer and other health problems in humans. PBDEs have not yet been directly linked to cancer but have been shown to cause brain damage in developing mice. The use of PBDEs is banned in some European countries but permitted in the United States. Recent measurements of PDBEs in human blood serum and milk in the United States show some disturbing trends. Analysis of serum samples collected between 1985 and 2002 showed a

sixfold increase in levels of PBDEs. Samples of human milk collected from a milk bank and analyzed for PBDEs shows levels 10 to 100 times greater than human serum and milk from European countries. PDBEs are also found in food, land sludge, house dust, freshwater fish, and even seals living above the Arctic Circle.

Several states have outlawed some forms of PDBEs, but these laws will not take effect for several years, and other states have taken no action. Other halogen-containing organic compounds such as PCBs persist in the environment for decades, and PDBEs will probably be around long after any global bans are put into effect. The link between PDBEs and cancer remains circumstantial, but as concentrations in human tissues continue to rise, the health consequences of exposure to these compounds needs to be explored.

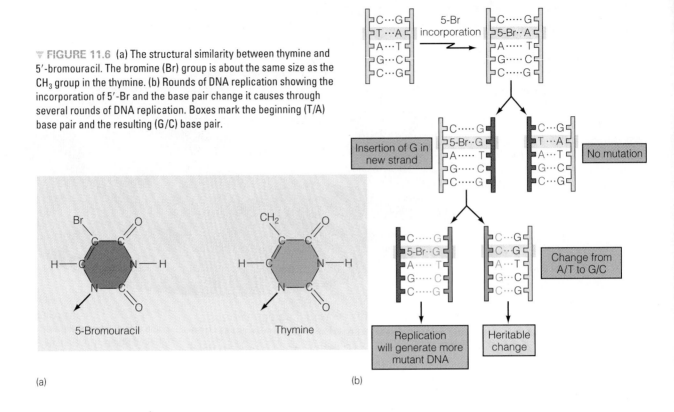

▼ FIGURE 11.6 (a) The structural similarity between thymine and 5′-bromouracil. The bromine (Br) group is about the same size as the CH₃ group in the thymine. (b) Rounds of DNA replication showing the incorporation of 5′-Br and the base pair change it causes through several rounds of DNA replication. Boxes mark the beginning (T/A) base pair and the resulting (G/C) base pair.

(a)

(b)

Chemicals can cause mutations.

We know of over 6 million chemical compounds, and almost 500,000 of these are used in manufacturing processes and are part of everyday life in many ways: in packaging, in food, in building materials, and so forth. Unfortunately, we know little or nothing about whether most of the chemicals cause mutations (see Genetics in Society: Rise of the Flame Retardants). Chemicals cause mutations in several ways, and they are often classified by the type of damage they cause to DNA. Some chemicals cause nucleotide substitutions or change the number of nucleotides in DNA, whereas others structurally change the bases in DNA, causing a base pair change after replication. Some of the ways chemicals act as mutagens are discussed here.

Base Analog   Mutagenic chemicals that structurally resemble nucleotides that are incorporated into DNA or RNA during synthesis are called **base analogs.** The structure of the base analog 5′-bromouracil is similar to that of thymine (▶ Figure 11.6a). If 5′-bromouracil is present in cells, and inserted into DNA in place of thymine (T), it generates a 5′-Br/A base pair (▶ Figure 11.6b). When the DNA replicates the next time, the 5′-bromouridine can serve as a template for the insertion of guanine in the new strand, creating a 5′-Br/G base pair. After another round of replication, the net result is that an A/T base pair is changed into a G/C base pair. Although most people will not come in contact with 5′-bromouridine, other common chemicals are base analogs. For example, the caffeine in your morning coffee is a base analog (intensive investigation has shown that caffeine is rapidly broken down in the body and does not pose much of a threat as a mutagen).

Chemical Modification of Bases   Some mutagens attack the bases in a DNA molecule, changing one base into another. Treatment of DNA with nitrous acid removes a chemical group from cytosine, converting it to uracil (▶ Figure 11.7). What was a G/C base pair is converted into a G/U pair. Uracil has the base pairing properties of thymine (T). In the next round of DNA replication, the U will act as a template for

■ **Base analogs**   A purine or pyrimidine that differs in chemical structure from those normally found in DNA or RNA.

the insertion of A in the newly synthesized strand, and the G/U base pair is converted to an A/U pair. After another round of replication, an A/T pair will be created, replacing the original G/C pair. Nitrates and nitrites used in the preservation of meats, fish, and cheese are converted into nitrous acid in the body. Although studied extensively, how much mutation is caused by these dietary chemicals has been difficult to assess.

Chemicals That Bind to DNA   Chemicals that bind directly to DNA generally produce frameshift mutations (described in a later section). These chemicals, called *intercalating agents*, generally insert themselves into the DNA, distorting the double helix. The distortion can cause a mistake during DNA replication, resulting in the addition or deletion of a base pair. The structure of one of these chemicals, acridine orange is shown in ▶ Figure 11.8. This molecule is about the same size as a purine/pyrimidine base pair, and wedges itself into DNA, distorting the shape of the double helix. When replication takes place in this distorted region, deletion or insertion of bases can take place, resulting in a frameshift mutation. Some components and breakdown products of commonly used pesticides are intercalating agents.

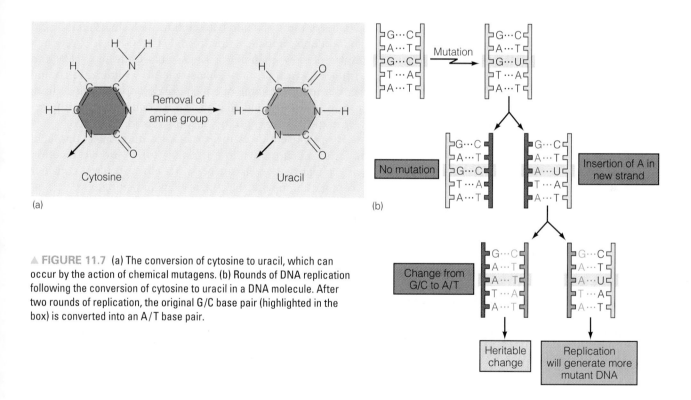

(a)

(b)

▲ FIGURE 11.7 (a) The conversion of cytosine to uracil, which can occur by the action of chemical mutagens. (b) Rounds of DNA replication following the conversion of cytosine to uracil in a DNA molecule. After two rounds of replication, the original G/C base pair (highlighted in the box) is converted into an A/T base pair.

◀ FIGURE 11.8 The molecular structure of acridine orange, an intercalating agent that inserts into the helical structure of DNA, distorting its shape. Replication in the distorted region can lead to the insertion or deletion of base pairs, producing a mutation.

■ **Nucleotide substitutions** Mutations that involve substitutions of one or more nucleotides in a DNA molecule.

■ **Frameshift mutations** Mutational events in which a number of bases (other than multiples of three) are added to or removed from DNA, causing a shift in the codon reading frame.

■ **Missense mutations** Mutations that cause the substitution of one amino acid for another in a protein.

■ **Sense mutations** Mutations that change a termination codon into one that codes for an amino acid. Such mutations produce elongated proteins.

11.5 Mutations at the Molecular Level: DNA as a Target

At the molecular level, mutations can involve substitutions, insertions, or deletions of one or more nucleotides in a DNA molecule. Mutations that alter the sequence but not the number of nucleotides in a gene are called **nucleotide substitutions.**

Generally, these substitutions involve only one or a small number of nucleotides. A second type of mutation, called **frameshift mutations,** causes the *insertion* or *deletion* of bases into a DNA molecule. Because codons are composed of three bases, changing the number of bases in a codon changes the sequence of all following codons and results in large-scale changes in the amino acid sequence of the protein product. We will begin by examining nucleotide substitutions and then consider frameshift mutations.

Keep in mind

■ Mutations in DNA can occur in several ways, including nucleotide substitution, deletion, or insertion.

Many hemoglobin mutations are caused by nucleotide substitutions.

Hemoglobin is one of the most studied molecules in the human body. Scientists have found several hundred variants of the alpha and beta globins with single amino acid substitutions (see Spotlight on Population Genetics of Sickle Cell Genes). These variants provide many examples of how changing a single nucleotide in a gene can change the structure and function of a protein. Nucleotide substitutions in coding regions can have a number of outcomes, some of which are described in the following section. In this discussion, keep in mind that the term *codon* refers to the sequence of three nucleotides in mRNA that specifies an amino acid.

Missense mutations are single nucleotide changes that substitute one amino acid for another in a protein. These substitutions do not always affect protein function and often do not have any phenotypic consequences. To illustrate, let's consider amino acid number 6 in beta globin, one of the components of hemoglobin (▶ Figure 11.9). Normal beta globin (Hb A) mRNA has the codon GAG at position 6, specifying the amino acid glutamine. A single nucleotide substitution from GAG (glu) to GUG (val) results in Hb S (sickle cell anemia), a condition with a potentially lethal phenotype. A condition known as Hb C (hemoglobin C disease) also has a nucleotide substitution at position 6. In this case, GAG is changed into AAG, and the beta globin has lysine as the sixth amino acid instead of glutamine. This change causes a mild set of clinical conditions that are less serious than sickle cell anemia. In yet another substitution, a change in codon 6 from GAG (glu) to GCG (ala) produces a hemoglobin variant known as Hb Makassar, a substitution that causes no clinical symptoms and is regarded as harmless.

In all three cases, the proteins differ only in amino acid number 6: Hb A has glutamic acid (glu), Hb S has valine (val), Hb C has lysine (lys), and Hb Makassar has alanine (ala). The other 145 amino acids in the protein are unchanged. In these examples, single nucleotide changes in the sixth codon of the beta-globin gene result in phenotypes that range from harmless (Hb Makassar), to the mild clinical symptoms of Hb C, to the serious and potentially life-threatening consequences of Hb S (sickle cell anemia) (see Genetic Journeys: The First Molecular Disease).

Other nucleotide substitutions produce proteins that are longer or shorter than normal. **Sense mutations** produce longer-than-normal proteins by changing a termination codon (UAA, UAG, or UGA), into one that codes for amino acids. Several hemoglobin variants with longer-than-normal globin molecules are shown in

▶ Table 11.2. In each case, the extended polypeptide chain can be explained by a single nucleotide substitution in the normal termination codon. In hemoglobin Constant Springs-1, the mRNA codon 142 is changed from UAA to CAA, and a glycine codon replaces a stop codon. Thirty more amino acids are inserted before another stop codon is reached.

Nonsense mutations change codons for amino acids into one of the three termination codons: UAA, UAG, or UGA (see ▶ Figure 9.3). This shortens the protein product. In the McKees Rock variant of beta globin, a change in codon 144 from UAU (tyr) to UAA (termination) results in a beta chain that is 143 amino acids instead of 145. This change has little or no effect on the function of the beta-globin molecule as a carrier of oxygen. However, other nucleotide substitutions can produce more drastic changes in polypeptide length and have more serious phenotypic effects.

Mutations can be caused by nucleotide deletions and insertions.

Nucleotide deletions and insertions within a gene can range from the deletion or duplication of one nucleotide to that of an entire gene. As more genes are sequenced in genome projects, deletions and insertions are emerging as a major cause of genetic disorders and in humans account for 5% to 10% of all known mutations. As defined earlier, the insertion or deletion of nucleotides within the coding sequence of a gene causes frameshift mutations. Because codons consist of groups of three bases, adding or subtracting a base from one codon changes the coding sense of all following codons in the gene. A frameshift mutation changes the amino acid sequence of the protein from the site of the mutation to the end of the protein. Suppose that a codon series reads as the following sentence:

THE FAT CAT ATE HIS HAT

A nucleotide (in this case, an A) inserted in the second codon destroys the sense of the remaining message:

insertion

THE FAA TCA TAT EHI SHA T

Similarly, a deletion in the second codon can also generate an altered message:

THE FTC ATA TEH ISH AT

deletion

In nucleotide substitutions, the number of nucleotides in the gene remains the same, and usually only one amino acid in the protein is altered (▶ Active Figure 11.10). Frameshift mutations change the number of nucleotides in the gene and usually cause large-scale changes in the amino acid sequence of the protein (▶ Active Figure 11.11). Usually, these changes result in a nonfunctional gene product. A he-

		1	2	3	4	5	6	7	8
Normal Hb A	DNA	CAC	GTG	GAC	TGA	GGA	CTC	CTC	TTC
	mRNA	GUG	CAC	CUG	ACU	CCU	GAG	GAG	AAG
	Amino acid	val	his	leu	thr	pro	glu	glu	lys
Sickle Hb S	DNA	CAC	GTG	GAC	TGA	GGA	CAC	CTC	TTC
	mRNA	GUG	CAC	CUG	ACU	CCU	GUG	GAG	AAG
	Amino acid	val	his	leu	thr	pro	val	glu	lys
Hb C	DNA	CAC	GTG	GAC	TGA	GGA	TTC	CTC	TTC
	mRNA	GUG	CAC	CUG	ACU	CCU	AAG	GAG	AAG
	Amino acid	val	his	leu	thr	pro	lys	glu	lys
Hb Makassar	DNA	CAC	GTG	GAC	TGA	GGA	CGC	CTC	TTC
	mRNA	GUG	CAC	CUG	ACU	CCU	GCG	GAG	AAG
	Amino acid	val	his	leu	thr	pro	ala	glu	lys

▲ FIGURE 11.9 The DNA code word, mRNA codon, and the first eight amino acids of normal adult hemoglobin (Hb A), sickle cell hemoglobin (Hb S), hemoglobin C (Hb C), and hemoglobin Makassar (Hb Mk). A single nucleotide substitution in codon 6 is responsible for the changes in the two variant forms of hemoglobin.

Table 11.2 Alpha Globins with Extended Chains Produced by Nucleotide Substitutions	
Alpha Globins	**Abnormal Chains**
Constant Springs-1	gln (142) + 30 amino acids
Icaria	lys (142) + 30 amino acids
Seal Rock	glu (142) + 30 amino acids
Koya Dora	ser (142) + 30 amino acids

■ **Nonsense mutations** Mutations that change an amino acid specifying a codon to one of the three termination codons.

Genetic Journeys

The First Molecular Disease

Linus Pauling, a two-time Nobel Prize winner, once recalled that when he first heard a description of how red blood cells change shape in sickle cell anemia, he had the idea that sickle cell anemia is really a molecular disease. He thought this disorder must involve an abnormality of the hemoglobin molecule caused by a mutated gene.

Early in 1949, Linus Pauling and his student Harvey Itano began a series of experiments to determine whether there is a difference between normal hemoglobin and sickle cell hemoglobin. They obtained blood samples from people who had sickle cell anemia and from unaffected individuals. They prepared hemoglobin from these blood samples, placed it in a tube with an electrode at each end, and passed an electrical current through the tube. Hemoglobin from individuals with sickle cell anemia migrated toward the cathode, indicating that it had a positive electrical charge. Samples of normal hemoglobin migrated in the opposite direction (toward the anode), indicating that it had a net negative electrical charge. In the same year, James Neel, working with sickle cell patients in the Detroit area, demonstrated that sickle cell anemia is an autosomal recessive trait.

Pauling and his colleagues published a paper on their results and incorporated Neel's findings into their discussion. They concluded that a mutant gene involved in the synthesis of hemoglobin caused sickle cell anemia (and the heterozygous condition known as sickle cell trait). The idea that a genetic disorder could be caused by a defect in a single molecule was revolutionary. Pauling's idea about a molecular disease helped start the field of human biochemical genetics and played a key role in understanding the molecular nature of mutations.

After Watson and Crick worked out the structure of DNA, Crick was anxious to prove that mutant genes produce mutant proteins whose amino acid sequences differ from those of the normal protein. He persuaded Vernon Ingram to look for such differences. Ingram settled on hemoglobin as the protein he would analyze because of Pauling's work. Ingram cut hemoglobin into pieces using the enzyme trypsin and separated the 30 resulting fragments. He noticed that normal hemoglobin and sickle cell hemoglobin differed in only one fragment, a peptide about 10 amino acids long. Ingram then worked out the amino acid sequence in this fragment. In 1956, he reported that there is a difference of only a single amino acid (glutamine in normal hemoglobin and valine in sickle cell hemoglobin) between the two proteins. This finding confirmed the relationship between a mutant gene and a mutant gene product and established a way of thinking about mutations and disease that changed human genetics.

moglobin variant with an extended chain will serve as an example of an altered protein produced by a frameshift mutation. In this case, the frameshift occurs near the end of the gene and has a minimal impact on the function of the gene product.

In normal alpha hemoglobin, the mRNA codons for the last few amino acids are as follows:

Position number	138	139	140	141	TER
mRNA codon	UCC	AAA	UAC	CGU	UAA
Amino acid	ser	lys	tyr	arg	

In a variant of alpha hemoglobin called Hb Wayne, the last nucleotide (A) in codon 139 is deleted, producing a frameshift:

Position number	138	139	140	141	142	143	144	145	146	TER
mRNA Codon	UCC	AAU	ACC	GUU	AAG	CUG	GAG	CCU	CGG	UAG
Amino acid	ser	asn	thr	val	lys	leu	gln	pro	arg	

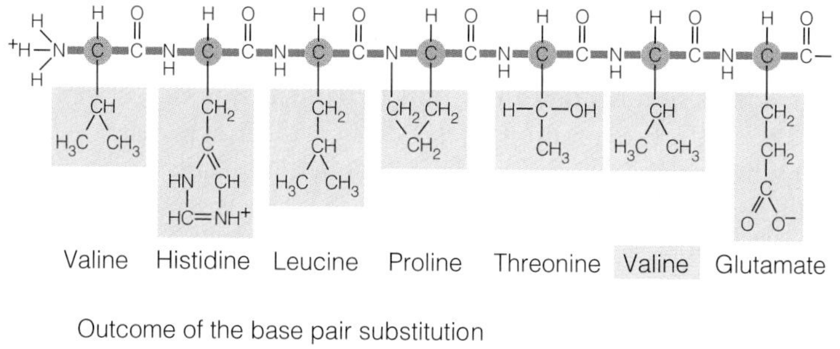

◄ ACTIVE FIGURE 11.10 A change in a single base pair is responsible for the replacement of glutamate with valine in the beta globin in sickle cell anemia.

HUMAN
Genetics ⊘ Now™ Learn more about the mutation in sickle cell anemia by viewing the animation on Human GeneticsNow at **http://now.ilrn.com/cummings7**

Valine Histidine Leucine Proline Threonine Valine Glutamate

Outcome of the base pair substitution

In this case, deletion of a single nucleotide causes a shift in the codon reading frame so that the termination sequence UAA that follows codon 141 is split into two codons, causing new amino acids to be added until another stop codon (generated by the deletion) is reached. The result is an alpha chain variant with 146 amino acid residues instead of 141.

Trinucleotide repeats and gene expansions are types of mutations.

Trinucleotide repeats are a recently discovered class of mutations associated with a number of genetic disorders. This type of mutation is a sequence of three nucleotides repeated several times in consecutive order within a gene (▶ Table 11.3). Mutations that increase the number of repeats are responsible for several genetic disorders. This expansion involves only one of the two alleles of a gene, and the phenomenon is called **allelic expansion.** The potential for expansion is a characteristic of a specific allele and occurs only within that allele. The study of allelic expansion in fragile-X syndrome has explained some aspects of the way this condition is inherited (see Chapter 6 for a review of fragile sites).

In fragile-X syndrome, the phenotype includes mental retardation. About 1% of all males institutionalized for mental retardation have this syndrome. Mothers of affected males are heterozygous carriers and pass the fragile-X chromosome to 50% of their offspring. These carrier mothers are phenotypically normal. In 20% to 50% of all cases, the mutant allele has a low degree of penetrance in males, and these males, with a normal phenotype, are called transmitter males. As discussed next, daughters of transmitter males have a high risk of bearing affected children, producing the type of pedigree shown in ▶ Figure 11.12.

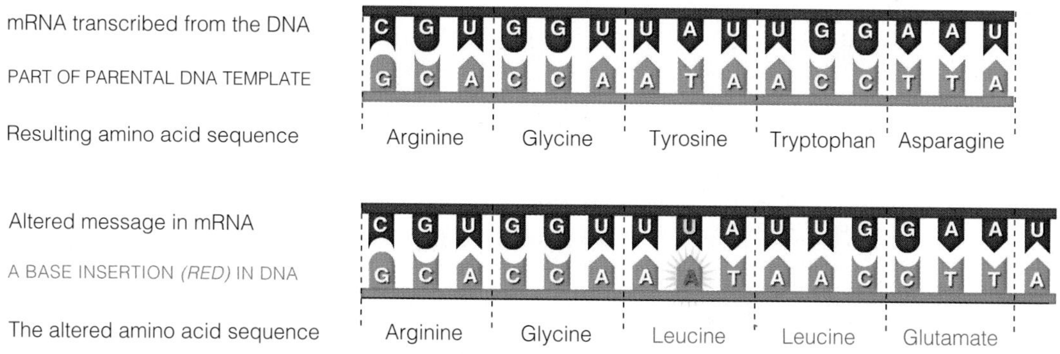

▦ **Trinucleotide repeats** A form of mutation associated with the expansion in copy number of a nucleotide triplet in or near a gene.

▦ **Allelic expansion** Increase in gene size caused by an increase in the number of trinucleotide sequences.

mRNA transcribed from the DNA

PART OF PARENTAL DNA TEMPLATE

| C G U | G G U | U A U | U G G | A A U |
| G C A | C C A | A T A | A C C | T T A |

Resulting amino acid sequence

Arginine | Glycine | Tyrosine | Tryptophan | Asparagine

Altered message in mRNA

A BASE INSERTION *(RED)* IN DNA

| C G U | G G U | U U A | U U G | G A A | U |
| G C A | C C A | A A T | A A C | C T T | A |

The altered amino acid sequence

Arginine | Glycine | Leucine | Leucine | Glutamate

▲ ACTIVE FIGURE 11.11 Insertions are mutations in which extra bases insert into genes. Insertions change the reading frame of the DNA code and mRNA codons, causing the wrong amino acids to be inserted into the protein product.

HUMAN
Genetics ⊘ Now™ Learn more about frameshift mutations by viewing the animation on Human GeneticsNow at **http://now.ilrn.com/cummings7**

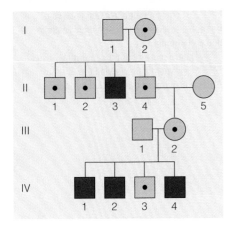

▲ FIGURE 11.12 A pedigree illustrating the inheritance of fragile-X syndrome. Mothers (I-2) of phenotypically normal but transmitting males (II-4) are phenotypically normal but have some offspring with fragile-X syndrome (II-3). Daughters (III-2) of transmitting males are at high risk of having affected children. Allelic expansion of premutation alleles is more likely when inherited from a female. III-2 has inherited such an allele, which is likely to undergo expansion and affect her children.

■ Anticipation   Onset of a genetic disorder at earlier ages and with increasing severity in successive generations.

Alleles of the *FMR-1* gene contain CGG repeats. Normal alleles contain 6 to 52 copies of the CGG repeat. These normal alleles are stable and do not undergo expansion. Premutation alleles have an expanded number of repeats (ranging from 60 to 200). Premutation alleles do not affect expression of the *FMR-1* gene, so males with premutation alleles are phenotypically normal, but are carriers. Children and grandchildren of transmitter males and carrier females are at high risk of being affected. Fully mutant *FMR-1* alleles have more than 230 CGG repeats that prevent expression of the *FMR-1* gene, producing mental retardation in all males and in 60% of females with this allele.

A surprising factor influencing the expansion of the fragile-X gene is the sex of the parent transmitting the mutant allele. When males transmit premutation alleles, the number of CGG repeats is likely to remain constant or even decrease. The change from a premutation allele to a fully mutant allele occurs only when females transmit the premutation alleles (▶ Figure 11.13), resulting in fragile-X syndrome. This transition occurs only after the premutation allele passes through oogenesis, but the mechanism has not been fully explained.

More than a dozen genetic disorders are associated with expansion of trinucleotide repeats. Some of these are listed in ▶ Table 11.3. In myotonic dystrophy (DM), the mutation is caused by expansion of a CTG repeat. As with fragile-X syndrome, there is a correlation between the size of the expanded repeat, the age of onset, and the severity of symptoms. In contrast to the fragile-X syndrome, expansion of trinucleotide repeats in DM and increased probability of clinical symptoms are more likely when transmitted by males than by females.

Gene expansion is related to anticipation.

Five disorders (the first five in ▶ Table 11.3) associated with expansion of trinucleotide repeats have some common characteristics. All are progressive, degenerative disorders of the nervous system that are inherited as autosomal dominant traits. All have expanded CAG repeats, and all show a correlation between the increasing size of the repeat and earlier age of onset. In these disorders, mildly affected parents have more seriously affected offspring, who develop symptoms at an earlier age than their parents.

The appearance of more severe symptoms at earlier ages in succeeding generations, called **anticipation,** was first noted for myotonic dystrophy. Although carefully documented by clinicians, geneticists discounted the phenomenon of anticipation because genes were regarded as highly stable entities that have only occasional mutations. The discovery of anticipation means that the concept of mutation must now

Table 11.3	Some Mutations with Expanded Trinucleotide Repeats			
Gene	Triplet Repeat	Normal Copy	Copy in Disease	OMIM Number
Spinal and bulbar muscular atrophy	CAG	12–34	40–62	313200
Spinocerebellar ataxia type 1	CAG	6–39	41–81	164400
Huntington disease	CAG	6–37	35–121	143100
Haw-River syndrome	CAG	7–34	54–70	140340
Machado-Joseph disease	CAG	13–36	68–79	109150
Fragile-X syndrome	CGG	5–52	230–72,000	309550
Myotonic dystrophy	CTG	5–37	50–72,000	160900
Friedreich ataxia	GAA	10–21	200–900	229300

recognize the existence of unstable genomic regions that undergo dynamic changes. In other words, when regions of the genome containing trinucleotide repeats undergo an expansion, this event enhances the chance of further expansions and the development of a disease phenotype. Forms of genome instability similar to trinucleotide expansion might explain genetic disorders that do not show simple Mendelian inheritance (for example, those showing incomplete penetrance or variable expression).

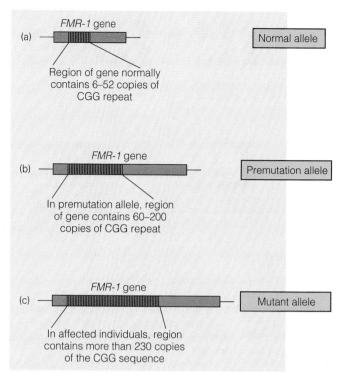

▲ FIGURE 11.13 Allelic expansion in the *FMR-1* gene at the fragile-X locus. (a) The gene normally contains 6 to 52 copies of a CGG trinucleotide repeat. (b) In premutation alleles, this region expands to include 60 to 200 copies of this repeat. (c) Affected individuals have more than 230 copies of the repeat.

11.6 Mutations and DNA Damage Can Be Repaired

Not every mutation is a permanent genomic change. All cells have enzyme systems that repair damage to DNA. ▶ Table 11.4 estimates the rate of spontaneous DNA damage in a typical mammalian cell at 37°C (body temperature). If not repaired, the accumulated damage would destroy much of the DNA in the cell. Fortunately, humans (and other organisms) have a number of highly efficient DNA repair systems (▶ Table 11.5). However, because the rate of background damage is so high, it is easy to overload the repair systems. One type of damage, the formation of **thymine dimers,** is a major cause of cell death, mutation, and cancer.

Keep in mind

- Damage to DNA can be repaired during DNA synthesis, and by enzymes that repair damage to DNA caused by radiation or chemicals.

■ **Thymine dimer** A molecular lesion in which chemical bonds form between a pair of adjacent thymine bases in a DNA molecule.

Cells have several DNA repair systems.

One repair system corrects errors made during DNA replication. In humans, replication proceeds rapidly, with 10 to 20 nucleotides added each second to a DNA strand at each replication site. About 3 billion nucleotides are copied in each round of cell division, and it is little wonder that mistakes occur. Sometimes, the wrong nucleotide is inserted into the newly synthesized strand, resulting in a potential spontaneous mutation. However, DNA polymerase, the enzyme involved in replication,

Table 11.4	Rates of DNA Damage in a Mammalian Cell
Damage	**Events/Hour**
Depurination	580
Depyrimidation	29
Deamination of cytosine	8
Single-stranded breaks	2,300
Single-stranded breaks after depurination	580
Methylation of guanine	130
Pyrimidine (thymine) dimers in skin (noon Texas sun)	5×10^4
Single-stranded breaks from background ionizing radiation	10^{-4}

Table 11.5	Maximum DNA Repair Rates in a Human Cell	
Damage		**Repairs/ Hour**
Single-stranded breaks		2×10^5
Pyrimidine dimers		5×10^4
Guanine methylation		10^4–10^5

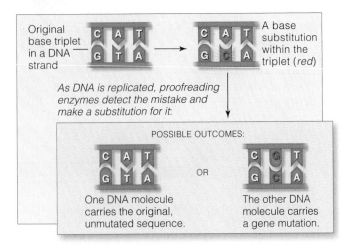

Original base triplet in a DNA strand → A base substitution within the triplet (*red*)

As DNA is replicated, proofreading enzymes detect the mistake and make a substitution for it:

POSSIBLE OUTCOMES:

One DNA molecule carries the original, unmutated sequence. OR The other DNA molecule carries a gene mutation.

Example of a base pair substitution

▲ ACTIVE FIGURE 11.14 Base pair substitutions have two possible outcomes. The mutation can be detected by enzyme proofreading and corrected (*bottom left*) or remain undetected and become a mutation.

HUMAN **Genetics⊜Now**™ Learn more about base pair substitutions by viewing the animation on Human GeneticsNow at **http://now.ilrn.com/cummings7**

corrects many of these mistakes. In addition to directing DNA replication, the enzyme has a proofreading function. If an incorrect nucleotide is inserted by mistake, the enzyme can detect the mistake and move backward, removing nucleotides until the incorrect nucleotide is eliminated (▶ Active Figure 11.14). Then the enzyme inserts the correct nucleotide and moves forward, resuming replication. Those few mistakes that elude the proofreading function of DNA polymerase remain as true spontaneous mutations.

Other repair systems recognize and repair damage to DNA in other phases of the cell cycle. These systems fall into several categories, each controlled by several genes. For example, exposure of DNA to UV light (from sunlight and tanning lamps) causes adjacent thymine molecules in the same DNA strand to pair with each other, forming thymine dimers (▶ Figure 11.15). Thymine dimers distort the DNA molecule and can interfere with normal replication. These dimers are corrected by several different DNA repair mechanisms.

Genetic disorders can affect DNA repair systems.

Because DNA repair is under genetic control, it, too, can undergo mutation. Several genetic disorders, including xeroderma pigmentosum (XP; OMIM 278700), are caused by mutations in genes that repair DNA. XP is an autosomal recessive disorder with a frequency of 1 in 250,000. Affected individuals are extremely sensitive to sunlight (which contains UV light). Even short exposure to the sun causes dry, flaking skin and pigmented spots that can develop into skin cancer (▶ Figure 11.16). Skin cancers are about 1,000 times more common in XP individuals. Early death from cancer is the usual fate of XP individuals who do not take extraordinary measures to protect themselves from UV light.

Mutations in at least eight different genes can cause XP, and all are defective in repairing DNA damaged by UV light.

▲ FIGURE 11.15 Thymine dimers are produced when ultraviolet light cross-links two adjacent thymine bases in the same strand of DNA. This structure causes a distortion in the DNA, and errors in replication are likely to occur unless corrected.

This disease illustrates a functional correlation between mutation and cancer, a topic we consider in detail in Chapter 12.

Several other genetic disorders are characterized by unusual sensitivity to sunlight and/or to other forms of radiation resulting from defects in DNA repair. These include Fanconi anemia (OMIM 227650), ataxia telangiectasia (OMIM 208900), and Bloom syndrome (OMIM 210900). The range of phenotypes seen in these disorders indicates that DNA repair is a complex process that involves many different genes.

11.7 Mutations, Genotypes, and Phenotypes

Sickle cell anemia was the first genetic disorder to be analyzed at the molecular level. In this disorder, a nucleotide substitution in codon 6 changes the amino acid in the beta-globin polypeptide at position 6 and produces a distinctive set of clinical symptoms. All affected individuals and all heterozygotes have the same nucleotide substitution.

As it turns out, sickle cell anemia is probably an exception rather than the rule. Analysis of mutations in other genes reveals that more often than not a number of different mutations in a single gene can produce the phenotype associated with a genetic disorder (▶ Figure 11.17).

11.8 The Position of a Mutation in a Gene Is Important

Cystic fibrosis provides a clear example of the types and numbers of mutations that can occur in a single gene, all of which result in a disease phenotype. More than 1,300 different mutations have been identified in the cystic fibrosis gene. These include single nucleotide substitutions and deletions, as well as larger deletions that involve one or more regions of the gene. In addition, there are frameshift mutations and splice-site mutations. Mutations are distributed in all regions of the *CFTR* gene (▶ Figure 11.17), strengthening the idea that any mutational event that interferes with expression of a gene produces an abnormal phenotype.

People with cystic fibrosis have a wide range of clinical symptoms. The relationship between the type and location of the *CFTR* mutation and the clinical phenotype has been investigated for a number of mutant alleles. In some mutations, such as the Δ508 deletion (present in 70% of all cases of CF), the CFTR protein is produced but is not inserted into the plasma membrane. As a result, there is no chloride ion transport and clinical symptoms are severe. In other mutations, such as R117, R334, and R347 (▶ Figure 11.18), the CFTR protein is produced and inserted into the membrane but is only partially functional. These mutations are associated with a milder form of cystic fibrosis. With over 1,300 different mutations known in the CF gene, it is possible for someone with CF to carry two different mutant alleles. This genetic variability further contributes to the phenotypic variability seen in this disease.

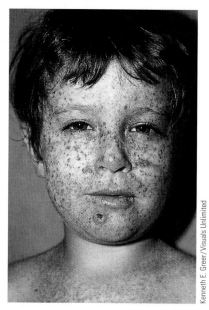

▲ FIGURE 11.16 Child affected with xeroderma pigmentosum. Affected individuals cannot repair damage to DNA caused by ultraviolet light from the sun and other sources.

Kenneth E. Greer/Visuals Unlimited

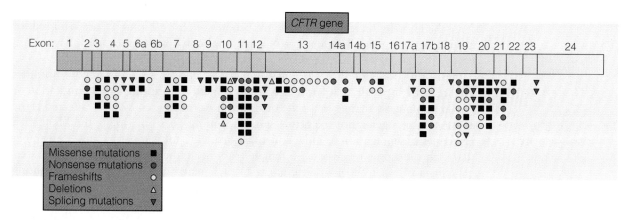

▲ FIGURE 11.17 Distribution of mutations in the cystic fibrosis gene, *CFTR*. More than 500 different mutations have now been discovered. The mutations shown here include nucleotide substitutions, deletions, and frameshift mutations. Any of these mutations in the homozygous condition or in combination with each other (that is, a compound heterozygote) results in the phenotype of cystic fibrosis.

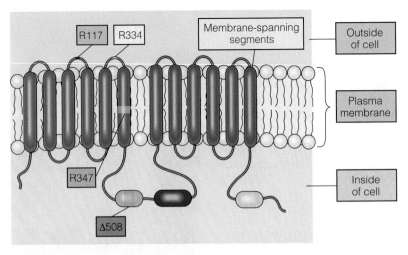

▲ FIGURE 11.18 Mutations in the *CFTR* gene differ in their phenotypic effects. When homozygous, mutations R117, R334, or R347 allow between 5% and 30% of normal activity for the gene product and produce only mild symptoms. These nucleotide substitution mutations are not common: Together they account for about 2% of all cases of cystic fibrosis. The most common mutation in European populations, Δ508, causes an amino acid deletion in a cytoplasmic region of the protein and accounts for 70% of all mutations in the *CF* gene. This mutation inactivates the CFTR protein and is associated with severe symptoms.

■ **Genomic imprinting** Phenomenon in which the expression of a gene depends on whether it is inherited from the mother or the father. Also known as genetic or parental imprinting.

■ **Uniparental disomy** A condition in which both copies of a chromosome are inherited from one parent.

11.9 Genomic Imprinting Is a Reversible Alteration of the Genome

We all carry two copies of each gene, one from mom and one from dad. Normally, there is no difference in the expression of the two copies. For certain genes, however, only the copy from one parent is expressed. This pattern of differential expression is called **genomic imprinting** and involves marking genes during gamete formation or early embryonic development. The mechanism of imprinting is not yet completely understood.

The first evidence for imprinting in mammals was discovered in mice when haploid nuclei were transplanted into eggs to produce zygotes carrying an all-female or all-male genome rather than the usual combination of one male and one female genome. Experimental embryos with only a male genome develop abnormal embryonic structures but have normal placentas. Embryos with only a female genome develop normal embryonic structures but have abnormal placentas. Both conditions are lethal, and we can conclude that both a maternal and a paternal genome are required for normal development.

Genomic imprinting plays a role in several genetic disorders, including Prader-Willi syndrome (PWS) and Angelman syndrome (AS). Most cases of PWS, an autosomal recessive disorder characterized by obesity, uncontrollable appetite, and mental retardation, are associated with a small deletion in the long arm of chromosome 15. In about 40% of all cases, however, no deletion can be detected. Using molecular markers, these nondeletion cases are found to have inherited both copies of chromosome 15 from the mother. This condition, in which both copies of a given chromosome are inherited from a single parent, is called **uniparental disomy**. In this situation, there are two copies of chromosome 15 present. In PWS, both copies come from the mother, and there is no copy of the chromosome from the father.

Severe mental retardation, uncontrollable puppet-like movements, and seizures of laughter characterize AS. About 50% of affected individuals have a small deletion in the long arm of chromosome 15 (the same region that is deleted in PWS). Molecular studies indicate that AS is a genetic mirror image of PWS. In nondeletion cases of AS, both copies of chromosome 15 come from the father. These two disorders reinforce the idea that paternal and maternal copies of chromosome 15 are required for normal development. The absence of a paternal copy of chromosome 15 results in PWS; the absence of a maternal copy results in AS.

Imprinting does not affect all genes. Only genes in certain regions of chromosomes 4p, 8q, 15q, 17p, 18p, 18q, and 22q are imprinted. Imprinting is not a mutation or even a permanent change in a gene or a chromosome region. What is affected is the expression of a gene, not the gene itself. Imprinting does not violate the Mendelian principles of segregation or independent assortment, nor is it permanent. Remember that a chromosome received by a female from her father is transmitted as a maternal chromosome in the next generation. In each generation, the previous imprinting is erased, and a new pattern of imprinting defines the chromosome as either paternal or maternal. Imprinting events are thought to take place during gamete formation or early embryonic development, and the effects are transmitted to all tissues of the offspring. Although imprinting is not strictly a mutational event, it does involve modification of DNA.

Genetics in Practice

 Genetics in Practice case studies are critical thinking exercises that allow you to apply your new knowledge of human genetics to real-life problems. You can find these case studies and links to relevant websites at **http://biology.brookscole.com/cummings7**

CASE 1

On April 26, 1986, one of the four reactors at the Chernobyl generating station in the Soviet Union melted down. It has been reported that the plant was running with disconnected safety measures. The result was fire, chaos, fear, a cloud of radioactive isotopes spreading across vast reaches of Eastern Europe, and the radioactive contamination of thousands of people.

Unfortunately, this human tragedy is not being investigated as it should be, according to scientists who are trying to learn from it. Cancer prevention specialists claim that there is a lack of resources available to look at cancer cases in the population exposed to radiation released from Chernobyl.

Scientists who want to study the aftermath of Chernobyl face many obstacles: (1) the dissolution of the Soviet Union split administrative, record-keeping, and medical responsibilities among Belarus, Ukraine, and Russia; (2) a general decline in living standards has reduced the level of medical care in the area; (3) individual doses cannot be reconstructed accurately; estimates on doses received by individuals have been complicated by the fact that some of the dose was external through exposure to radioactive dust and part of it was internal, through eating contaminated food; and (4) little money is available for the kind of large studies that are needed for extracting the best data.

Here are some of the facts—as best we know—about the Chernobyl meltdown. The accident released 1.85×10^{18} (1,850,000,000,000,000,000) international units of radioactive material. The releases contaminated an estimated 17 million people to some degree. The exact amount of exposure depended on location, wind direction, length of exposure, eating habits, and whether the person was a "liquidator."

These unfortunate heroes were pressed into service in a crude cleanup effort after the accident. One hundred and thirty-four people showed signs of acute radiation sickness immediately after the accident. Many of the 28 people who died from acute radiation sickness had skin lesions covering 50% or more of their bodies. After the fire, 135,000 people evacuated the area around the reactor,

and 800,000 liquidators moved in to try to decontaminate the area. Approximately 17% to 45% of the liquidators received doses between 10 and 25 rads. (For comparison, U.S. safety guidelines permit an annual dose to the general public of 0.1 rads; nuclear workers are permitted 5 rads.)

Despite the early confusion, some medical information is now available in Chernobyl's aftermath. The most compelling data involves radiation exposure and thyroid cancer, particularly in children. According to the International Chernobyl Conference (April 1996), radiation exposure caused "a substantial increase in reported cases of thyroid cancer in Belarus, Ukraine, and some parts of Russia, especially in young children." This is thought to be due to exposure to radioactive iodine during the early phases of the accident in 1986. By the end of 1995, approximately 800 cases of thyroid cancer had been reported in children who were under age 15 at the time of diagnosis. To date, three of these thyroid cancer victims have died, and several thousand more cases of thyroid cancer are expected. Ironically, most of the thyroid cancers could have been prevented if people in the contaminated areas had taken iodine tablets immediately after the accident. Most iodine in the body goes to the thyroid gland; if enough normal iodine is available, only a small amount of radioactive isotope irradiates the little gland, whose hormones help regulate growth. More information on the long-term effects of Chernobyl's explosion will be learned in the coming decades, as scientists and health care workers assess the full impact of this disaster.

1. How do you think radiation causes cancer?

2. The liquidators were unfortunately exposed to large amounts of radiation. Are their families at risk even though they were not part of the cleanup effort?

3. What kind of compensation, if any, should liquidators receive from the government for the exposure to radiation?

4. An increase in thyroid cancer was reported after Chernobyl. Are these people at risk of passing a mutant cancer gene to their offspring?

11.1 Mutations Are Heritable Changes

- The study of mutations is an essential part of genetics. Without mutations, it would be difficult to determine whether a trait is under genetic control and impossible to determine its mode of inheritance.

11.2 Mutations Can Be Detected in Several Ways

- Mutations can be classified in a variety of ways using criteria such as pattern of inheritance, phenotype, biochemistry, and degrees of lethality.

- Dominant mutations are the easiest to detect because they are expressed in the heterozygous condition. Accurate pedigree information can often identify the individual in whom the mutation arose. It is more difficult to determine the origin of sex-linked recessive mutations, but an examination of the male progeny is often informative. If the mutation in question is autosomal recessive, it is almost impossible to identify the original mutant individual.

11.3 Measuring Spontaneous Mutation Rates

- Studies of mutation rates in a variety of dominant and sex-linked recessive traits indicate that mutations in the human genome are rare events, occurring about once in every 1 million copies of a gene. The impact of mutation is diminished by several factors, including the redundant nature of the genetic code, the recessive nature of most mutations, and the lowered reproductive rate or early death associated with many genetic diseases.

11.4 Environmental Factors Influence Mutation Rates

- Environmental agents including radiation and chemicals can cause mutations.

11.5 Mutations at the Molecular Level: DNA as a Target

- Molecular analysis of mutations has shown a direct link between gene, protein, and phenotype. Mutations arise spontaneously as the result of errors in DNA replication or as the result of structural shifts in nucleotide bases. Environmental agents, including chemicals and radiation, also cause mutations. Frameshift mutations cause a change in the reading frame of codons, often producing dramatic alterations in the structure and function of proteins.

11.6 Mutations and DNA Damage Can Be Repaired

- Not all mutations cause genetic damage. Cells have a number of DNA repair systems that correct errors in replication and repair damage caused by environmental agents such as ultraviolet light, radiation, or chemicals.

11.7 Mutations, Genotypes, and Phenotypes

- In most genes associated with a genetic disorder, many different types of mutations can cause a mutant phenotype. In the cystic fibrosis gene, more than 500 different mutations have been identified, including deletions, nucleotide substitutions, and frameshift mutations.

11.8 The Position of a Mutation in a Gene Is Important

- The wide range of mutations found in genetic disorders leads to wide variation in clinical symptoms. Depending on the mutation, symptoms can range from very mild to very severe.

11.9 Genomic Imprinting Is a Reversible Alteration of the Genome

- Genomic imprinting alters expression of normal genes, depending on whether they are inherited maternally or paternally. Imprinting has been implicated in a number of disorders, including Prader-Willi and Angelman syndromes. Not all regions of the genome are affected, and only segments of chromosomes 4, 8, 17, 18, and 22 are imprinted. Genes are not permanently altered by imprinting, but are reimprinted in each generation.

Questions and Problems

HUMAN Genetics Now™ Preparing for an exam? Assess your understanding of this chapter's topics with a pre-test, a personalized learning plan, and a post-test at http://now.ilrn.com/cummings7

Measuring Spontaneous Mutation Rates

1. Define mutation rate.

2. Achondroplasia is an autosomal dominant form of dwarfism caused by a single gene mutation. Calculate

the mutation rate of this gene given the following data: 10 achondroplastic births to unaffected parents in 245,000 births.

3. Why is it almost impossible to measure directly the mutation rates in autosomal recessive alleles?

4. What are the factors that influence mutation rates of human genes?

5. Achondroplasia is a rare dominant autosomal defect resulting in dwarfism. The unaffected brother of an individual with achondroplasia is seeking counsel on the likelihood of his being a carrier of the mutant allele. What is the probability that the unaffected client is carrying the achondroplasia allele?

Environmental Factors Influence Mutation Rates

6. Although it is well known that x-rays cause mutations, they are routinely used to diagnose medical problems including potential tumors, broken bones, or dental cavities. Why is this done? What precautions need to be taken?

7. You are an expert witness called by the defense in a case in which a former employee is suing an industrial company because his son was born with muscular dystrophy, an X-linked recessive disorder. The employee claims that he was exposed to mutagenic chemicals in the workplace that caused his son's illness. His attorney argues that neither the employee, his wife, nor their parents have this genetic disorder, and, therefore, the disease in the employee's son represents a new mutation. How would you analyze this case? What would you say to the jury to refute this man's case?

8. Bruce Ames and his colleagues have pointed out that, although detailed toxicological analysis has been conducted on synthetic chemicals, almost no information is available about the mutagenic or carcinogenic effects of the toxins produced by plants as a natural defense against fungi, insects, and animal predators. Tens of thousands of such compounds have been discovered, and he estimates that in the United States adults eat about 1.5 g of these compounds each day, levels that are approximately 10,000 times higher than those levels of synthetic pesticides present in the diet. For example, cabbage contains 49 natural pesticides and metabolites, and only a few of these have been tested for their carcinogenic and mutagenic effects.

a. With the introduction of new foods into the U.S. diet over the last 200 years (mangoes, kiwi fruit, tomatoes, and so forth), has there been enough time

for humans to evolve resistance to the mutagenic effects of toxins present in these foods?

b. The natural pesticides present in plants constitute more than 99% of the toxins we eat. Should diet planning, especially for vegetarians, take into account the doses of toxins present in the diet?

Mutations at the Molecular Level: DNA as a Target

9. Define and compare the following types of nucleotide substitutions. Which is likely to cause the most dramatic mutant effect?
   a. missense mutation   b. nonsense mutation
   c. sense mutation

10. If the coding region of a gene (the exons) contains 2,100 base pairs of DNA, would a missense mutation cause a protein to be shorter, longer, or the same length as the normal 700 amino acid proteins? What would be the effect of a nonsense mutation? A sense mutation?

11. Two types of mutations discussed in this chapter are (1) nucleotide changes and (2) unstable genome regions that undergo dynamic changes. Describe each type of mutation.

12. What is a frameshift mutation?

13. A frameshift mutation is caused by:
   a. a nucleotide substitution
   b. a three-base insertion
   c. a premature stop codon
   d. a one-base insertion   e. a two-base deletion

14. In the gene coding sequence shown, which of the following events will produce a frameshift after the last mutational site?
   normal mRNA: UCC AAA UAC CGU CGU UAA
                ser lys tyr arg arg stop
   a. insertion of an A after the first codon
   b. deletion of the second codon (AAA)
   c. insertion of TA after the second codon and deletion of CG in the fourth codon
   d. deletion of AC in the third codon.

15. Trinucleotide repeats cause serious neurodegenerative disorders such as Huntington disease, fragile-X syndrome, and myotonic dystrophy (DM). The process of anticipation causes the appearance of symptoms at earlier ages in succeeding generations. Describe the current theory on how anticipation works.

16. Familial retinoblastoma, a rare autosomal dominant defect, arose in a large family that had no prior history of the disease. Consider the following pedigree (the darkly colored symbols represent affected individuals):

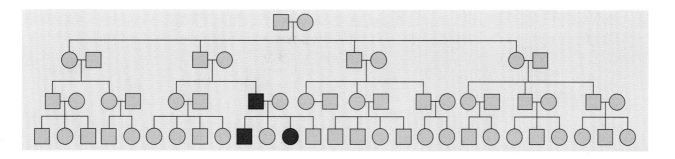

a. Circle the individual(s) in which the mutation most likely occurred.

b. Is this individual affected by the mutation? Justify your answer.

c. Assuming that the mutant allele is fully penetrant, what is the chance that an affected individual will have an affected child?

17. Tay-Sachs disease is an autosomal recessive disease. Affected individuals do not often survive to reproductive age. Why has Tay-Sachs persisted in humans?

Mutations and DNA Damage Can Be Repaired

18. Replication involves a period of time during which DNA is particularly susceptible to the introduction of mutations. If nucleotides can be incorporated into DNA at a rate of 20 nucleotides/second and the human genome contains 3 billion nucleotides, how long would replication take? How is this time reduced so that replication can take place in a few hours?

19. Our bodies are not defenseless against mutagens that alter our genomic DNA sequences. What mechanisms are used to repair DNA?

Mutations, Genotypes, and Phenotypes

20. The cystic fibrosis gene encodes a chloride channel protein necessary for normal cellular functions. Let us assume that if at least 10% normal channels are present, the affected individual has mild symptoms of cystic fibrosis. Less than 10% normal channels produces severe symptoms. At least 50% of the channels must be expressed for the individual to be phenotypically normal. This gene has various mutant alleles:

Allele	Molecular defect	% Functional Channels	Symptoms
CF100	deletion in exon	0%	severe
CF1	missense mutation in 5′ flanking region	25%	mild
CF2	nonsense mutation in exon	0%	severe
CF3	missense mutation in exon	5%	mild

Predict the percent of functional channels and severity of symptoms for the following genotypes:

a. heterozygous for CF100

b. homozygous for CF100

c. heterozygous, with one copy of CF100 and one of CF3

d. heterozygous, with one copy of CF1 and one copy of CF3

Internet Activities

Internet Activities are critical thinking exercises using the resources of the World Wide Web to enhance the principles and issues covered in this chapter. For a full set of links and questions investigating the topics described below, log on to **http://biology.brookscole.com/cummings7**

1. *Mutant Sequences*. Review the information on Pompe disease from the Internet Activities in Chapter 10. How many mutations have been found in humans for this one enzyme?

2. *Mutation Review*. Work through the Gene Action/Mutation worksheet at the *Access Excellence Activities Exchange* to reinforce the concepts related to the various types of mutations and their consequences.

3. *Using a Mutation Database*. The *Human Gene Mutation Database* at the Institute of Medical Genetics at Cardiff, Wales, contains information about mutations identified in human genes. This information includes nucleotide substitutions, missense and nonsense muta-

tions, splicing mutations, and small insertions and deletions. The data provided includes the name and symbol for the gene, its chromosomal location, the mutant sequence codon number, and a reference to the paper that first identified the mutation.

a. At the website, search on "breast cancer." From the list of breast cancer genes, select "*BRCA1*" and scroll through the list of mutations, noting both the type of mutation and its location.

b. Do a second search, on the gene for cystic fibrosis. How do the results for this entry compare to the information available for *BRCA1*?

How would you vote now?

E. coli contamination in meat causes 20,000 illnesses and 500 deaths per year in the United States. Irradiation to kill *E. coli* in beef and *Salmonella* in poultry has been approved by the U.S. Food and Drug Administration and the U.S. Department of Agriculture, and endorsed by the World Health Organization and the American Medical Association as an effective means of preventing disease and deaths. In spite of this approval, there is public concern about possible mutation-causing compounds in irradiated meat, and so it is not widely available. Now that you know more about mutations and the effects of radiation, what do you think? If such products were available in the supermarket, would you buy them? Visit the Human Heredity Companion website to find out more on the issue, then cast your vote online at
http://biology.brookscole.com /cummings7

 For further reading and inquiry, log on to **InfoTrac College Edition,** your world-class online library, including articles from nearly 5,000 periodicals, at **http://infotrac.thomsonlearning.com**

12

Genes and Cancer

JULIE, a 23-year-old with a family history of breast cancer, sought counseling at the family cancer risk assessment clinic. She had been going to her local breast center to be treated for fibrocystic breast disease. At her last mammogram, she requested an operation to remove her breasts. Julie came to the counseling session with a maternal aunt who had undergone this surgery several years earlier. The genetic counselor explored the reasons that had prompted Julie's request for surgery and reviewed her family history. Three of Julie's six maternal aunts had breast cancer in their late 30s or early 40s. One maternal first cousin had also been diagnosed with breast cancer in her 30s. Julie's mother was in her 50s and had no history of cancer.

Discussion revealed that the women in Julie's family believed that the only way to avoid this disease was to have their healthy breasts removed before cancer developed. Julie went on to say that her mother had her breasts removed at age 34, and the aunt attending the session had the surgery last year, when she turned 40. The aunt told stories of caring for her dying sisters. She also stated, "It would be the same as a death sentence not to do this." This led the remaining aunts to seek surgery, and now Julie was being encouraged to take this step.

The genetic counselor explained that one of the maternal aunts could take a genetic test to determine if a mutant gene is contributing to the development of breast cancer in the family. The counselor explained that if her aunt carries such a mutation, Julie could be tested for this same mutation. If Julie has the mutant gene, then surgery might be a reasonable option. However, if Julie did not inherit the mutation, her risk for breast cancer would be the same as for the general population (or 10%), and breast removal would be unnecessary. One of Julie's aunts with cancer was tested, and it was found that she carries a mutation for a predisposition to breast cancer. Julie and her mother

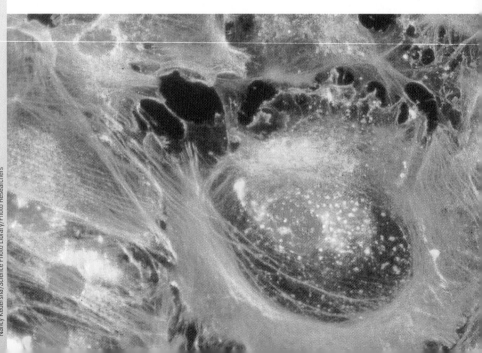

Nancy Kedersha/Science Photo Library/Photo Researchers

were tested for this same mutation and neither of them had the mutation. Julie continues to be followed by the local breast center for her fibrocystic breast disease; however, she has decided not to have surgery.

How would you vote?

Women who carry mutations in either of two cancer genes have a significantly higher risk of developing breast cancer. One of these mutant genes is present in members of Julie's family, but she and her mother do not carry this gene. On the basis of this information, Julie decided not to have surgery to remove her breasts. If a female member of your family carried one of these two cancer genes and asked your advice, would you recommend that she have her breasts removed, even though not everyone carrying this mutation will develop breast cancer? Visit the Human Heredity Companion website to find out more on the issue, then cast your vote online at **http://biology.brookscole.com/cummings7**

12.1 Cancer Is a Genetic Disorder

Cancer is a complex disease that affects many different cells and tissues in the body (▷ Figure 12.1). It is characterized by uncontrolled cell division and by the ability of these cells to spread, or metastasize, to other sites within the body. Unchecked, growth and metastasis result in death, making cancer a devastating and feared disease. Improvements in medical care have reduced deaths from infectious disease and have led to increases in life span, but these benefits have also helped make cancer a major cause of illness and death in our society. The risk of many cancers is age-related, and because more Americans are living longer, they are at greater risk of developing cancer.

The link between cancer and genetic mutations was forged early in the last century by Theodore Boveri, who proposed that normal cells mutate into malignant cells because of changes in chromosome constitution. Four lines of evidence support the idea that cancer has a genetic origin:

1. More than 50 forms of cancer have some degree of inherited predisposition.
2. It has been shown that most carcinogens are also mutagens.
3. Some viruses carry mutant genes, known as oncogenes that promote and maintain the growth of a tumor.
4. As Boveri proposed, specific chromosomal changes are found in some cancers, especially leukemia.

Mutation is a common feature of all cancers. In most cases, these mutations take place in somatic cells, and the mutant alleles are not passed on to offspring. In other

**Keep in mind
as you read**

■ Cancer can be caused by an inherited susceptibility or a sporadic event.

■ Cancer cells bypass cell cycle checkpoints and divide continuously.

■ Colon cancer is a multistep process that involves oncogenes and mutant tumor suppressor genes.

■ Several types of translocations associated with cancer create hybrid genes.

■ Diet and behavior are two major factors in cancer prevention.

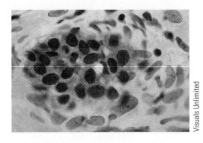

▲ FIGURE 12.1 The molecular differences between cancer cells and normal cells can be revealed by staining tissues.

■ **Metastasis** A process whereby cells detach from the primary tumor and move to other sites, forming new malignant tumors in the body.

cases, mutations occur in germ cells and are passed on to succeeding generations as a predisposition to cancer. Cancer, then, is a genetic disorder that acts at the cellular level.

Because mutation is the ultimate cause of cancer and because there is a constant background of spontaneous mutations, there will always be a baseline rate of cancer. The environment (ultraviolet light, chemicals, and viruses) and behavior (diet and smoking) can also play a significant role in cancer risks by increasing the rate of mutation.

12.2 Cancer Begins in a Single Cell

Tumors are abnormal growths but not all tumors are cancers. Benign tumors are self-contained growths that do not spread to other tissues and are not invasive. Benign tumors, including cysts, usually cause problems by increasing in size until they interfere with the function of neighboring organs.

Cancers are malignant tumors with several characteristics.

■ Cancer begins in a single cell. All the cells in a cancerous tumor are direct descendents of that original cell, and are clones.

■ Most cancers develop after a cell accumulates a number of specific mutations over a long time period.

■ Once formed, cancer cells divide continuously. Mutations continue to accumulate, and the cancer grows more aggressively with each mutation.

■ Cancer cells are invasive. Cells detach from the primary tumor and move to other sites in the body, forming new malignant tumors. This process is called **metastasis** (▶ Active Figure 12.2). The ability to invade new tissues results from new mutations in cancer cells.

In the following sections we consider the genetic changes that take place within cells that lead to cancer.

12.3 Inherited Susceptibility and Sporadic Cancers

Families with high rates of cancer have been known of for hundreds of years, but in most cases no clear-cut pattern of inheritance can be identified. How is it, then, that some families have a rate of cancer that is much higher than average? Many explanations have been offered, including inheritance, environmental agents, or chance alone. Studies of these families have helped identify specific genes involved in cancer.

In these families, some individuals inherit a mutant gene that causes a predisposition to cancer. As a result, these individuals carry this mutant gene in their germ cells and all the somatic cells of the body (▶ Table 12.1). If one or more additional mutations accumulate spontaneously or by exposure to mutagenic environmental agents in any somatic cell, the result is cancer. In some cases, people carrying a mutant allele causing a predisposition to cancer have a 100,000-fold increased risk of cancer.

Sporadic cases are the most common form of cancer. These cases arise when a single dominant mutation occurs in a somatic cell, or when the somatic cell acquires a certain

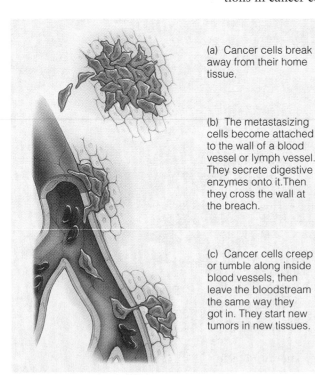

(a) Cancer cells break away from their home tissue.

(b) The metastasizing cells become attached to the wall of a blood vessel or lymph vessel. They secrete digestive enzymes onto it. Then they cross the wall at the breach.

(c) Cancer cells creep or tumble along inside blood vessels, then leave the bloodstream the same way they got in. They start new tumors in new tissues.

▲ ACTIVE FIGURE 12.2 Cancer cells can move to new locations, a process called metastasis.

HUMAN Genetics Now™ Learn more about metastasis by viewing the animation on Human GeneticsNow at

http://now.ilrn.com/cummings7

Table 12.1 Heritable Predispositions to Cancer

Disorder	Chromosome	OMIM Number
Early-onset familial breast cancer	17q	113705
Familial adenomatous polyposis	5q	175100
Hereditary nonpolyposis colorectal cancer	2p	120435
Li-Fraumeni syndrome	17p	151623
Multiple endocrine neoplasia type 1	11q	131100
Multiple endocrine neoplasia type 2	10	171400
Neurofibromatous type 1	17q	162200
Neurofibromatous type 2	22q	101000
Retinoblastoma	13	180200
Von Hippel-Lindau disease	3p	193300
Wilms tumor	11p	194070

number of recessive mutations. The exact number of mutations needed to cause cancer is specific for different forms of cancer, but two mutations may be the minimum number needed (▶ Figure 12.3). In most cases, cancer-causing mutations accumulate over a period of years, explaining why age is a leading risk factor for many cancers.

Keep in mind

▪ Cancer can be caused by an inherited susceptibility or a sporadic event.

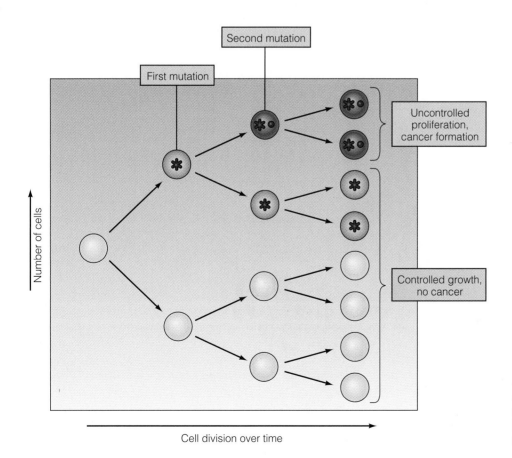

▸ FIGURE 12.3 Over time, cells acquire mutations. Two independently acquired mutations in the same cell may be enough to cause uncontrolled cell growth and cancer.

12.4 Cancer Is a Disease of the Cell Cycle

Cancer cells are characterized by uncontrolled cell division and have abnormal shapes (▶ Figure 12.4). These cells bypass checkpoints in the cell cycle that regulate cell division. As a result, studies of the cell cycle are now an important part of cancer research. The 2001 Nobel Prize for Physiology or Medicine was awarded to three scientists who helped establish the link between the cell cycle and cancer.

Recall from Chapter 2 that events in interphase and mitosis define the cell cycle (▶ Figure 12.5). The cycle is regulated at several points. For our discussion, two checkpoints are important: a point in G1 just before cells enter S, known as the G1/S transition and the transition between G2 and M (the G2/M transition). Mutations in the genes regulating these checkpoints are an important step in cancer development. Two classes of genes regulate these checkpoints: (1) genes that turn off or de-

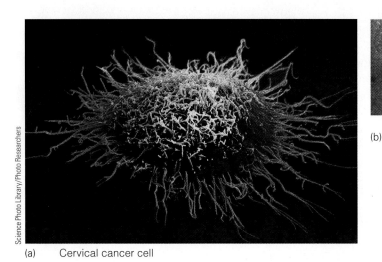

(a) Cervical cancer cell

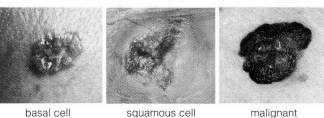

(b) basal cell squamous cell malignant
carcinoma carcinoma melanoma

◀ FIGURE 12.4 (a) Cancer cells have abnormal shapes. This cell can move to new locations using its cytoplasmic extensions. (b) Skin cancers. Basal cell carcinomas are slow growing and noninvasive. Squamous cell carcinomas are fast growing and can be invasive. Melanomas are dark, fast growing, invasive, and can be fatal if not caught early.

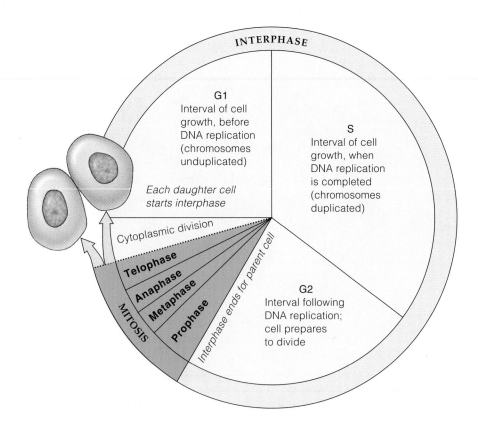

▶ FIGURE 12.5 The eukaryotic cell cycle consists of two parts: interphase and mitosis. Interphase is divided into three stages, G1, S, and G2, which make up the major part of the cycle. Mitosis involves partitioning cytoplasm and replicated chromosomes to daughter cells.

crease the rate of cell division and (2) genes that turn on or increase the rate of cell division. The first class is known as **tumor suppressor genes.** These genes act at either the G1/S or the G2/M control points to inhibit cell division. If these genes mutate, cell division cannot be switched off, and cells divide in an uncontrolled manner. The second class of checkpoint genes, called **proto-oncogenes,** turn on or maintain cell division. When these genes are active, cells undergo division. If these genes become permanently switched on or overproduce their products, uncontrolled cell division results. Mutant forms of proto-oncogenes are called **oncogenes.**

The following examples describe how mutations in tumor suppressor genes control the development of cancer. After we discuss oncogenes and their involvement in cancer, a genetic model of colon cancer will illustrate a third class of genes, the DNA repair genes that are involved in cancer.

Keep in mind

■ Cancer cells bypass cell cycle checkpoints and divide continuously.

Retinoblastoma is caused by mutation of a tumor suppressor gene.

Retinoblastoma (*RB1*; OMIM 180200) is a cancer that affects the retina, the light-sensitive layer of the eye. It occurs in 1 in 14,000 to 1 in 20,000 births, and is most often diagnosed between the ages of 1 and 3 years. There are two forms of retinoblastoma. Hereditary retinoblastoma (accounting for 40% of all cases) is an autosomal dominant trait in which a predisposition to retinoblastoma is inherited. In families with this trait, children have a 50% chance of inheriting the mutant *RB* allele, and 90% of these individuals will develop retinoblastoma, usually in both eyes. In addition, those carrying the mutant allele are at high risk for other cancers, especially osteosarcoma and fibrosarcoma. The second form (60% of all cases) is sporadic retinoblastoma. In these cases, no mutant retinoblastoma genes are inherited. Instead, sometime early in childhood, mutations in both copies of the *RB1* gene occur in a retinal cell and cause retinoblastoma. In sporadic cases, tumors usually develop in only one eye, and patients are not at high risk for other cancers because the cancer-causing mutations are only found in the tumor (▶ Figure 12.6).

Alfred Knudson and his colleagues proposed a two-step model that explains both forms of this disease. The model proposes that retinoblastoma develops when two mutant copies of *RB1* gene are present in a single retinal cell. According to the model, retinoblastoma can arise in two ways. In inherited retinoblastoma, a child inherits a mutant *RB1* allele, and all cells in the child's body, including all retinal cells, carry this mutation. If the normal copy of the *RB1* gene becomes mutated in any retinal cell, the child will develop retinoblastoma (▶ Figure 12.7a). Because all retinal cells already carry one mutant gene, only a small number of mutations of the normal allele are needed to cause cancer in both eyes (bilateral cases) at an early age.

In the sporadic form of retinoblastoma (▶ Figure 12.7b), a child inherits two normal copies of the *RB1* gene, and both copies of the *RB* gene must become mutated in a single retinal cell for a tumor to develop. Because the chance that two *RB1* mutations will occur in the same cell is low, sporadic cases are more likely to involve only one eye (unilateral cases). Because more time is required for two mutations to occur in a single cell, the sporadic form of retinoblastoma arises later in childhood.

Surveys of children with retinoblastoma have confirmed the model's predictions. Pedigree analysis shows that most bilateral cases involve an inherited mutant allele, but the vast majority of cases that involve only one eye are sporadic and result from acquired mutations. In addition, most cases of retinoblastoma involving one eye occur at a later age (although still in childhood) than bilateral cases.

■ **Tumor suppressor genes** Genes encoding proteins that suppress cell division.

■ **Proto-oncogenes** Genes that initiate or maintain cell division and that may become cancer genes (oncogenes) by mutation.

■ **Oncogenes** Genes that induce or continue uncontrolled cell proliferation.

■ **Retinoblastoma** A malignant tumor of the eye arising in retinoblasts (embryonic retinal cells that disappear at about 2 years of age). Because mature retinal cells do not transform into tumors, this is a tumor that usually only occurs in children. It is associated with a deletion on the long arm of chromosome 13.

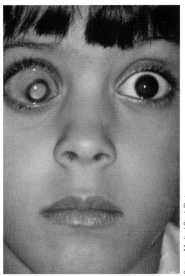

Custom Medical Stock Photo

▲ FIGURE 12.6 A child with retinoblastoma in one eye.

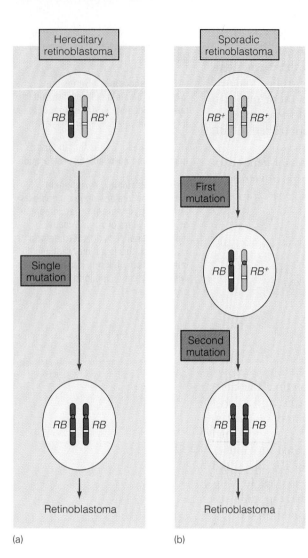

Hereditary retinoblastoma

RB   RB⁺

Single mutation

RB   RB

Retinoblastoma

(a)

Sporadic retinoblastoma

RB⁺   RB⁺

First mutation

RB   RB⁺

Second mutation

RB   RB

Retinoblastoma

(b)

◀ FIGURE 12.7 A model of retinoblastoma. In hereditary cases, one mutation is inherited. (a) Inheriting a single mutant allele of the *RB* gene causes a predisposition to retinoblastoma. If the normal *RB* allele mutates, retinoblastoma results. (b) Sporadic retinoblastoma requires two independent mutations in the *RB* gene in a single cell.

The *RB1* gene is located on chromosome 13 at 13q14 (▶ Figure 12.8) and encodes a protein (pRB) found only in the nucleus. pRB is present in retinal cells and in almost all other cell types in the body. pRB is found in the nucleus at all stages of the cell cycle, but its *activity* is regulated synchronously with the cell cycle. The protein is a molecular switch that controls progression through the cell cycle. If pRB is active during G1, the cell will not move from G1 into S and will not move through the cycle and divide. On the other hand, if pRB protein is inactive in G1, the cell moves into S phase, through G2, and on to mitosis. If both copies of the *RB1* gene are deleted or become mutated in a retinal cell, there is no pRB available to regulate cell division, and the cell begins to divide in an uncontrolled manner, forming a tumor.

Mutations in oncogenes cause cancer.

Oncogenes are mutant alleles of genes that normally turn on or maintain cell division and are turned off when division stops. These normal genes are called proto-oncogenes. When mutated, these genes are called oncogenes. In mutant form, the genes are permanently switched on, causing the uncontrolled cell division associated with cancer. The existence of oncogenes was first discovered by work on a virus that causes cancer in chickens.

How do proto-oncogenes become oncogenes?

Although oncogenes were discovered in viruses, only a few rare human forms of cancer are caused by virally transmitted oncogenes. In most cases, the conversion from proto-oncogene to oncogene takes place in the nucleus of a somatic cell without the action of a virus.

What is the difference between a proto-oncogene in a normal cell and its mutant allele (an oncogene) in a cancer cell? Many differences are possible. These can include a single base change that produces an altered gene product, mutations that cause underproduction or overproduction of the normal gene product, or mutations that increase the number of copies of the normal gene. In fact, all these types of mutations have been identified in human oncogenes or their adjacent regulatory regions.

Let's look at one oncogene and how mutation affects it. The *ras* gene encodes a protein of 189 amino acids that is embedded on the cytoplasmic side of the plasma membrane. The ras protein is involved in transmitting signals to switch on cell division and cycles between an active and an inactive state. In its active state, ras transfers growth-promoting signals from the plasma membrane to molecules in the cytoplasm, and then to the nucleus, where changes in gene expression begin cell division. *ras* genes isolated from many different human tumors show that in each case, a single base change is the only difference between the mutant oncogene in tumor cells and the proto-oncogene in normal cells. In all cases, the base change causes an amino acid change in the ras protein. The mutant gene and the mutant oncoprotein are found only in tumor cells and not in the normal tissue of the patient.

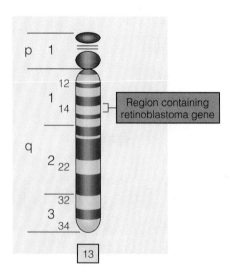

p   1

    12
1
    14   Region containing
                      retinoblastoma gene

q

2   22

    32
3
    34

13

▲ FIGURE 12.8 A diagram of chromosome 13 showing the retinoblastoma locus.

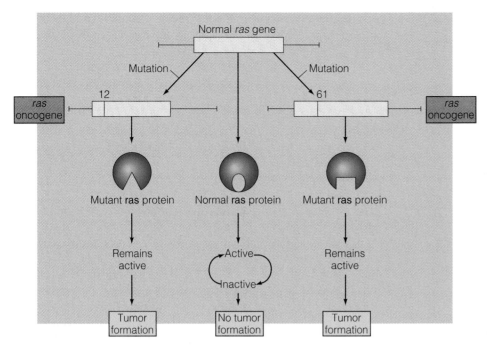

▲ FIGURE 12.9 The *ras* proto-oncogene encodes a protein that receives and transfers signals needed for cell division. Mutations at amino acid positions 12 or 61 cause the formation of an oncoprotein that permanently signals for cell division.

The amino acid change in all mutant *ras* genes occurs at one of two places: amino acid 12 or amino acid 61 (▶ Figure 12.9). Changing glycine at position 12 disrupts the structure of the protein and prevents it from folding to form an inactive signal molecule. As a result, the mutant protein is locked into the active state; it is constantly signaling for cell growth. Cells carrying this mutation escape from growth control and become cancerous.

Mutations in specific genes can predispose women to breast cancer.

Breast cancer is the most common form of cancer in women in the United States (see Spotlight on Male Breast Cancer). Each year, more than 40,000 women die from breast cancer, and more than 215,000 new cases are diagnosed. Although environmental factors may be involved in breast cancer, geneticists struggled for years with the question, Is there a genetic predisposition to breast cancer? After more than 20 years of work, the answer is clearly yes. Mutations in two different genes can predispose women to breast cancer and ovarian cancer.

One of these genes, *BRCA1* (OMIM 113705), is on the long arm of chromosome 17. Although mutant *BRCA1* alleles are present in only about 5% of all breast cancers, this gene is responsible for breast cancer susceptibility in women younger than 40 years of age. Approximately 1 in 200 females inherit the mutant allele, and of these approximately 90% will develop breast cancer.

The search for *BRCA1* began in the 1970s when Mary-Claire King and her colleagues analyzed familial patterns of breast cancer. They searched for families with a clear history of breast cancer and found that approximately 15% of the 1,500 families they studied had multiple cases of breast cancer. A genetic model predicted that about 5% (or 75/1500) of these cases were genetic, but it was impossible to know which families had a genetic predisposition and which had no predisposition, making further research on the gene difficult.

Instead of being discouraged, King decided on a brute force approach. She and her team began testing as many families as possible, knowing that finding a genetic marker for breast cancer was a long shot. They looked for linkage between certain forms of

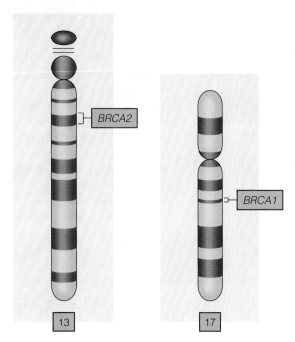

▲ FIGURE 12.10 The chromosome locations for *BRCA1* and *BRCA2*. Together these genes account for the majority of cases of breast cancer associated with a genetic predisposition.

proteins and breast cancer. In the 1980s, as recombinant DNA techniques became widely available, the team switched to DNA markers and the polymerase chain reaction (PCR) technique for screening. Finally, in 1990, after testing hundreds of families using hundreds of markers, they found linkage between breast cancer and a genetic marker. The 183rd marker they used, a variable-number tandem repeat (VNTR) locus on chromosome 17 called D17S74, was tested on family members from 23 pedigrees with a history of breast cancer. This marker was coinherited with breast cancer and was clearly linked to the disease. Other laboratories quickly confirmed their results and found that this marker was also linked to familial cases of ovarian cancer.

The investigators formed an international consortium and began studying members of 214 families with multiple cases of breast cancer or ovarian cancer. They used molecular markers to narrow the search for the gene to a small region on the long arm of chromosome 17 (▶ Figure 12.10). The *BRCA1* gene was finally identified and cloned in 1994, some 20 years after King began the search. The mutant allele of *BRCA1* is associated with an autosomal dominantly inherited predisposition to breast cancer. Not all women who inherit a mutation in the *BRCA1* gene will develop breast cancer. Approximately 82% of women inheriting one mutant *BRCA1* allele will develop a mutation in the other *BRCA1* allele and get breast cancer. Mutations in this gene are responsible for about half of all cases resulting from hereditary predisposition to breast cancer. Women with a *BRCA1* mutation are also at higher risk for ovarian cancer.

A second breast cancer predisposition gene, *BRCA2* (OMIM 600185), was discovered in 1995. Mutant alleles of this gene cause breast cancer susceptibility inherited as an autosomal dominant trait. *BRCA2* maps to q12–13 of chromosome 13 and may be responsible for the majority of inherited predispositions not caused by *BRCA1* (▶ Figure 12.10). Although mutations in *BRCA1* and *BRCA2* account for two-thirds of inherited breast cancers (totaling 10% to 15% of all cases of breast cancer), 85% to 90% of breast cancers are sporadic.

Overall, mutations in *BRCA1* and *BRCA2* are rare (0.12% of the general population carry these genes), but in some populations, the frequency of these mutant alleles is much higher than the general population. In women of eastern European Jewish ancestry (Ashkenazi Jews), the combined frequency of *BRCA1* and *BRCA2* is 2.5%. In Chapter 19, we will explore the factors that explain such population differences.

BRCA1 and *BRCA2* are DNA repair genes.

Many questions about the functions of both genes remain unanswered, as do questions about how mutations in these genes lead to breast cancer. Each gene encodes a large protein found only in the nucleus. In rapidly dividing cells, expression of *BRCA1* and *BRCA2* is highest at the G1/S boundary and into S phase. The BRCA1 protein is activated when DNA is damaged and stops DNA replication and participates in DNA repair. The BRCA2 protein has similar functions, and both proteins bind to a protein involved in the repair of double-stranded breaks in DNA molecules.

Because the predisposition to breast cancer is inherited as an autosomal dominant trait and both alleles must be mutant for cancer to develop, these genes are regarded as tumor suppressor genes.

12.5 Colon Cancer Is a Genetic Model for Cancer

As discussed earlier, cancer requires a number of mutations in specific genes. A mutation in only one of the two copies of a proto-oncogene can cause cancer, giving oncogenes a dominant phenotype. In retinoblastoma, a tumor suppressor gene, two mutational steps are required to convert a normal cell into a cancerous one, and the phenotype is recessive. In other cases, a half dozen or more mutations are required to initiate formation of a cancer cell.

Colon cancer is one of those latter types. The development of colon cancer is a useful model to study the number and order of mutations necessary to change a normal cell into a cancer cell. Colon/rectal cancer is one of the most common forms of cancer in the United States (▶ Table 12.2). This cancer starts as a benign tumor that later becomes malignant. In addition to spontaneous cases, two forms of genetic predisposition to colon cancer are known: an autosomal dominant trait, **familial adenomatous polyposis (FAP**; OMIM 175100), and **hereditary nonpolyposis colon cancer (HNPCC**; OMIM 120435 and 120436). FAP accounts for only about 1% of all cases of colon cancer but has been useful in deriving the main features of a genetic model for colon cancer described later. HNPCC accounts for approximately 15% of all cases and is associated with a form of genomic instability.

To clarify the role of inheritance in colorectal cancer, Randall Burt and his colleagues studied a large pedigree with more than 5,000 members covering six generations. This family has clusters of siblings and relatives with colon cancer. Like many other families, this one shows no definite pattern of inheritance for this cancer. However, as part of this study, about 200 family members were examined for intestinal growths (▶ Figure 12.11). These benign tumors, known as **polyps,** usually precede or accompany colon cancer and are regarded as the first step in colon cancer. When polyps and colon cancer are considered a single phenotype, an autosomal dominant pattern of inheritance is clear. This trait, called FAP, is caused by a mutation in the *APC* gene. The results also show that the dominant mutant allele for polyps and cancer (FAP) has a high frequency in the general population (3/1,000).

FAP can be related to colon cancer.

By studying mutations that cause benign tumors and the additional mutations that transform these growths into cancer, researchers have defined the number and order of steps that change a normal intestinal cell into a cancer cell. The genetic model for colon cancer, shown in ▶ Figure 12.12, has two important features: (1) Development of colon/rectal cancer requires five to seven mutations. If fewer mutations are present, benign growths or intermediate stages of malignant tumor formation result. (2) The order of mutations usually follows that shown in the figure, indicating that *both* the number and order of mutations are important in tumor formation.

In FAP-associated colon cancer, a mutation in the *APC* gene on chromosome 5 is the first step. In a normal homozygote, no growth of polyps occurs. In spontaneous cases, mutation of a single copy of *APC* takes place in a single intestinal cell. The cell carrying this mutation partially escapes cell cycle control and divides to form a polyp (a benign tumor). In familial cases, heterozygotes carry a mutant copy of *APC* and develop hundreds or thousands of polyps in the colon and rectum. In either case, the polyps are benign tumors made up of clones of cells, each of which carries a mutant *APC* gene. Because mutation of one copy of *APC* causes polyp formation, the *APC* gene is regarded as a tumor suppressor gene. A single mutation in *APC* is not enough to cause cancer, which develops only after mutation of other genes causes the transition from polyp to colon cancer.

Stages between polyp formation and colon cancer carry an intermediate number of mutations (▶ Figure 12.12). Mutation of one copy of the *k-ras* proto-oncogene in a polyp cell transforms the polyp into an adenoma, an intermediate stage

Table 12.2 Colon and Rectal Cancer in the United States

Estimated new cases, 2004	
Colon	106,370
Rectum	40,570
Total	146,940
Mortality (estimated deaths, 2004)	
Colon and rectum	56,730
	(16% of cancer deaths)
5-year survival rate (early detection)	
Colon	90%
Rectum	85%

■ **Familial adenomatous polyposis (FAP)** An autosomal dominant trait resulting in the development of polyps and benign growths in the colon. Polyps often develop into malignant growths and cause cancer of the colon and/or rectum.

■ **Hereditary nonpolyposis colon cancer (HNPCC)** A form of colon cancer associated with genomic instability of microsatellite DNA sequences.

■ **Polyps** Growths attached to the substrate by small stalks. Commonly found in the nose, rectum, and uterus.

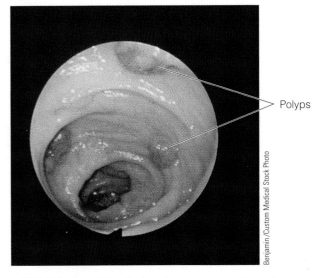

Benjamin/Custom Medical Stock Photo

Polyps

▲ **FIGURE 12.11** Polyps in the colon are a precursor to colon cancer. If a cell in one of these polyps acquires enough mutations, it will transform into a cancer cell, leading to colon cancer.

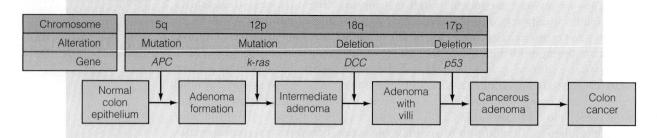

▲ FIGURE 12.12 A model for colon cancer. In this multiple-step model, the first mutation occurs in the *APC* gene, leading to the formation of polyps. Subsequent mutations in genes on chromosomes 12, 17, and 18 cause the transformation of the polyp into a tumor. In this model, the sum of the changes is more important than the order in which the changes take place.

tumor with fingerlike projections. To progress further, *both* alleles of the downstream genes shown in the figure must be mutated in a polyp cell. The 18q region contains a number of genes involved in colon cancer (all of them are tumor suppressor genes), including *DCC*, *DPC4*, and *JV18-1*. Mutation in both alleles of any of these genes forms late-stage adenomas. In the last stage, mutations in both alleles of the *p53* gene on chromosome 17 cause the late-stage adenoma to become cancerous. Mutations in the *p53* gene are pivotal in the formation of other cancers, including lung, breast, and brain cancer. The *p53* gene is a tumor suppressor gene that is active in regulating the passage of cells from the G1 to the S phase of the cell cycle.

In sum, the model for FAP-associated colon cancer requires a series of mutations that accumulate over time in a single cell. Each mutation confers a slight growth advantage on the cell, first allowing it to grow and divide, forming a benign tumor, then enlarging and transforming it in later stages as it gradually breaks away from cell cycle controls. Eventually, one cell accumulates enough mutations to escape completely from cell cycle controls to form a malignant tumor. Later, additional mutations allow tumor cells to become metastatic and break away to form tumors at remote sites.

Recombinant DNA techniques have been used to identify other cases in which cancer involves a number of mutations at specific chromosomal sites, often on different chromosomes (▶ Table 12.3).

HNPCC can be related to colon cancer.

Most cancers are caused by mutations in two or more genes that accumulate over time. If one of these mutations is inherited, fewer mutations are required to cause cancer, resulting in a genetic predisposition to cancer. But if the mutation rate is low (as we have seen in Chapter 11), how do the multiple mutations needed for cancer formation accumulate in a single cell? Work on a second form of colon cancer has provided a partial answer to this question.

Table 12.3	Number of Mutations Associated with Specific Forms of Cancer	
Cancer	**Chromosomal Sites of Mutations**	**Minimal Number of Mutations Required**
Retinoblastoma	13q14	2
Wilms tumor	11p13	2
Colon cancer	5q, 12p, 17p, 18q	5 to 7
Small-cell lung cancer	3p, 11p, 13q, 17p	10 to 15

There are two forms of HNPCC (*HPNCC1* and *HNPCC2*). The gene for *HNPCC1* is on chromosome 2 (2p16), and *HNPCC2* is on chromosome 3 (3p23–21.3). HNPCC-associated colon cancer may be one of the most common genetic disorders, affecting approximately 1 in 200 individuals. Mutations of either gene destabilizes the genome, generating a cascade of mutations in DNA sequences (microsatellites) that are repeated thousands of times and located on many chromosomes. It has been estimated that cells from *HNPCC* tumors can carry more than 100,000 mutations in microsatellites scattered throughout the genome.

Proteins encoded by *HNPCC1* and *HNPCC2* repair errors made during DNA replication. When these genes are inactivated by mutation, DNA repair is defective, and microsatellite mutation rates increase by at least 100-fold. This genomic instability promotes mutations in other genes, including the *APC* gene, eventually leading to colon cancer, as well as other forms of cancer.

Keep in mind

■ Colon cancer is a multistep process that involves oncogenes and mutant tumor suppressor genes.

Gatekeeper genes and caretaker genes have provided insights from colon cancer.

There are at least two pathways to colon cancer; one begins with a mutation in the *APC* gene, and the other (*HNPCC*) begins with a mutation in a DNA repair gene. Together, these two mechanisms provide insight into which genes can cause normal cells to become cancerous. Mutations in *APC* cause formation of hundreds or thousands of benign tumors. These benign growths progress slowly to cancer by accumulating mutations in other genes. Because there are thousands of polyps, there is a good chance that at least one of these will progress to colon cancer. In *HNPCC*, polyps accumulate slowly, forming only a small number of benign tumors. However, mutations in these polyps accumulate at a rate two to three times faster than in normal cells, making it almost certain that at least one benign growth will progress to colon cancer.

The different pathways to colon cancer show that two different types of genes are involved in cancer: **gatekeeper genes** and **caretaker genes**. Gatekeeper genes inhibit cell growth and control the cell cycle. Mutation in these genes opens the gate to uncontrolled cell division. Tumor suppressor genes are one type of gatekeeper gene. Many types of oncogenes are also gatekeeper genes.

Caretaker genes, on the other hand, encode proteins that maintain the integrity of the genome, for example, DNA repair genes. Some genetic disorders caused by mutations in DNA repair genes are listed in ▶ Table 12.4. Proteins encoded by these

■ **Gatekeeper genes** Genes that regulate cell growth and passage through the cell cycle; for example, tumor suppressor genes.

■ **Caretaker genes** Genes that help maintain the integrity of the genome; for example, DNA repair genes.

Table 12.4 Human Genetic Disorders Associated with Chromosome Instability and Cancer Susceptibility

Disorder	Inheritance	Chromosome Damage	Cancer Susceptibility	Hypersensitivity
Ataxia telangiectasia	Autosomal recessive	Translocations on 7, 14	Lymphoid, others	X-rays
Bloom syndrome	Autosomal recessive	Breaks, translocations	Lymphoid, others	Sunlight
Fanconi anemia	Autosomal recessive	Breaks, translocations	Leukemia	X-rays
Xeroderma pigmentosum	Autosomal recessive	Breaks	Skin	Sunlight

genes repair DNA mutations caused by mistakes in DNA replication or by environmental agents (ultraviolet light, for example). Mutation of a caretaker gene does not directly lead to tumor formation but does increase the mutation rate of all genes, including gatekeeper genes by a failure to repair DNA. This insight about gene types and genomic instability may also explain why many forms of cancer become associated with chromosomal instability and aneuploidy.

12.6 Chromosome Changes, Hybrid Genes, and Cancer

Changes in the number and structure of chromosomes are a common feature of cancer cells (▶ Figure 12.13). In some cases, the relationship between a single chromosome change and the development of cancer is not clear. For example, Down syndrome is caused by the presence of an extra copy of chromosome 21. This quantitative change in genetic information is not associated with any known gene mutation. In addition to defects in cardiac structure and in the immune system, children with Down syndrome are 18 to 20 times more likely to develop leukemia than those in the general population. How extra copies of genes on chromosome 21 predispose to cancer is not yet known, but it may be a by-product of increasing the dosage of some proto-oncogenes. In a limited number of cases, more direct information is available about the way chromosomal rearrangements are associated with developing and/or maintaining the cancerous condition.

Chromosome rearrangements can be related to leukemia.

The connection between chromosome rearrangements and cancer is evident in leukemias. In these cancers (involving uncontrolled division of white blood cells), specific chromosome changes are well defined and diagnostic (▶ Table 12.5).

One of the best-established links between cancer and a chromosomal aberration is the translocation between chromosome 9 and chromosome 22 in chronic myelogenous leukemia (CML) (▶ Figure 12.14). Originally called the **Philadelphia chromosome** (the city in which it was first found), this was the first example of a chromosome translocation accompanying a human disease.

Other cancers, including acute myeloblastic leukemia, Burkitt's lymphoma, and multiple myeloma, are associated with specific translocations (▶ Table 12.5). The finding that certain forms of cancer are consistently associated with specific chromosomal abnormalities suggests that these aberrations are related to the develop-

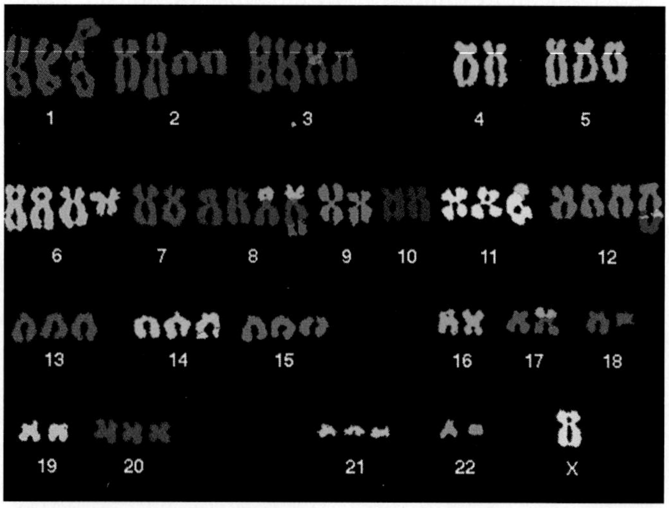

▦ Philadelphia chromosome   An abnormal chromosome produced by translocation of parts of the long arms of chromosomes 9 and 22.

▶ FIGURE 12.13 Chromosome aberrations including translocations, deletions, and aneuploidy result from the genomic instability of cancer cells.

Drs. Michael Speicher and David Ward, Yale University

Table 12.5 Chromosomal Translocation Associated with Human Cancers

Chromosomal Translocation	Cancer
t(9;22)	Chronic myelogenous leukemia (Philadelphia chromosome)
t(15;17)	Acute promyelocytic leukemia
t(11;19)	Acute monocytic leukemia, acute myelomonocytic leukemia
t(1;9)	Pre-B-cell leukemia
t(8;14),t(8;22),t(2;8)	Burkitt's lymphoma, acute lymphocytic leukemia of the B-cell type
t(8;21)	Acute myelogenous leukemia, acute myeloblastic leukemia
t(11;14)	Chronic lymphocytic leukemia, diffuse lymphoma, multiple myeloma
t(4;18)	Follicular lymphoma
t(4;11)	Acute lymphocytic leukemia
t(11;14)(p13;q13)	Acute lymphocytic leukemia

ment of the cancer. There is strong evidence that chromosome rearrangements are not by-products of malignancy but are important steps in the development of certain cancers. The genetic and molecular basis for this role is now becoming clear as the field of cancer cytogenetics merges with the molecular biology of oncogenes.

Translocations and hybrid genes can lead to leukemias.

As oncogenes were identified, cytogeneticists systematically mapped their chromosomal locations. It soon became clear that many of these genes are located at or very close to the breakpoints of chromosomal translocations involved with specific forms of leukemia. In fact, chromosome breaks can convert proto-oncogenes to oncogenes, initiating cancer formation.

In CML the translocation between chromosomes 9 and 22 creates a hybrid gene. The oncogene c-*abl* is on chromosome 9 at the breakpoint, and the *bcr* gene is on chromosome 22 at the breakpoint. The translocation produces a hybrid gene with

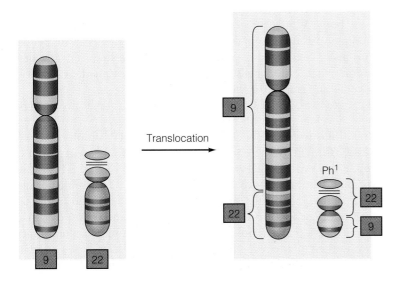

Translocation

◀ FIGURE 12.14 A reciprocal translocation between chromosomes 9 and 22 results in the formation of a chromosome involved in chronic myelogenous leukemia (CML).

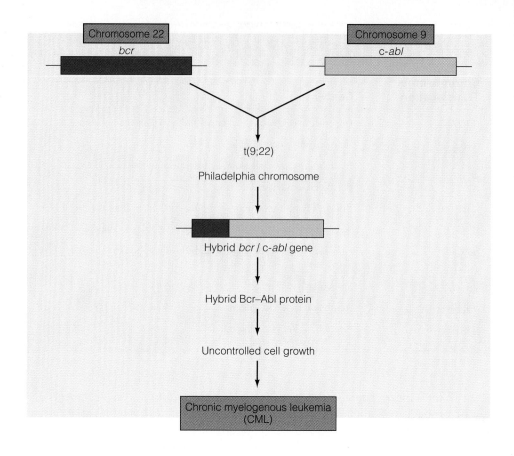

► FIGURE 12.15 Gene fusion associated with the 9;22 translocation in chronic myelogenous leukemia (CML). The *bcr* gene on chromosome 22 is fused to the c-*abl* proto-oncogene on chromosome 9. The hybrid gene is transcribed, and the resulting hybrid Bcr−Abl protein stimulates cell division in white blood cells. Overproduction of these cells results in CML.

bcr sequences at the beginning of the gene and c-*abl* sequences at the end of the gene (► Figure 12.15). The hybrid gene is transcribed and translated producing a hybrid protein that acts as a signal for cell division, causing CML.

Keep in mind

▪ Several types of translocations associated with cancer create hybrid genes.

New cancer drugs are being designed.

Cancer therapy has traditionally used radiation and chemicals to target and kill rapidly dividing cells in the body. While cancer cells are rapidly dividing, so are other cells in the body, including cells in bone marrow (making red blood cells), in the intestine (replacing worn away cells), and many other tissues. All these cells are destroyed or damaged along with cancer cells during radiation treatment or chemotherapy, often with serious side effects for the patient.

The fact that only cancerous CML cells contain the hybrid Bcr−Abl protein offered an opportunity to develop a chemotherapy drug that targets only the cancer cells. The hybrid protein folds to form a pocket for binding of ATP, a molecular energy source (► Figure 12.16). ATP binding is required for Bcr−Abl protein activity. Using this information, a drug called Gleevec was designed to fit into the Bcr−Abl pocket. When Gleevec binds to the Bcr−Abl protein, it prevents ATP from entering the pocket, making the protein inactive (► Figure 12.16). With the signal from the Bcr−Abl protein turned off, the cancer cell stops dividing. More than 90% of CML patients treated with this drug go into remission and show a dramatic reduction in white blood cells carrying the Philadelphia chromosome.

Knowledge of the molecular organization of oncogenes and the three-dimensional

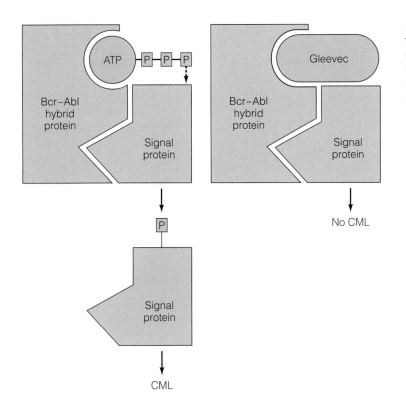

◀ FIGURE 12.16 (a) ATP binds to the Bcr–Abl protein and transfers a phosphate group (P) to a signal molecule. This signal moves to the nucleus and switches on cell division, leading to CML. (b) Binding of Gleevec in the ATP site blocks ATP binding and keeps the protein inactive, preventing the cancer cell from dividing.

structure of their products is being used to develop diagnostic tests for cancer. For example, if mutant proteins fold differently than the protein encoded by the normal proto-oncogene, antibodies that bind to the mutant proteins can be used to detect cancer at a very early stage. Tests are already available for a mutant protein released into the bloodstream by breast cancer cells, and others are being tested in clinical trials.

New strategies for treatment may also be derived from knowledge about oncogenes. In some cases, many copies of an oncogene are present in lung cancer, breast cancer, and cervical cancer. The number of copies of the mutant gene can be measured with recombinant DNA techniques, and these tumors receive more aggressive treatment in order to counteract the increased amount of the mutant protein in these tumors. Other strategies focus on how to turn off oncogene expression. In the laboratory, studies of cancer cells show switching off oncogenes converts cells into noncancer cells. Learning more about how to turn off genes may result in new anticancer drugs that act on the regulatory regions of specific genes.

Success in the development of cancer treatment drugs based on genetics has changed the way anticancer drugs are developed. In the past, such drugs were found by screening hundreds or thousands of chemicals for their ability to slow or stop the growth of cancer cells. With an understanding of the molecular events linked to cancer, it is now possible to design drugs for treatment of specific cancers without the side effects of other treatments.

12.7 Cancer and the Environment

The relationship between environmental factors and cancer has been studied for more than 50 years. During that time, sophisticated methods for gathering and analyzing data have provided solid evidence for the relationship between environmental factors and cancer.

Epidemiology is the study of factors that control the presence or absence of a disease. It is an indirect and inferential science that provides correlations between envi-

▪ **Epidemiology** The study of the factors that control the presence, absence, or frequency of a disease.

Table 12.6 Age-Adjusted Cancer Death Rates per 100,000 Population

Country	All Sites	
	Male	Female
United States	165.3 (27)*	111.1 (18)
Australia	158.5 (28)	100.2 (20)
Austria	171.6 (20)	105.6 (16)
Denmark	178.7 (17)	138.1 (1)
Germany	177.3 (18)	108.2 (11)
Hungary	258.7 (1)	135.2 (2)
Japan	149.8 (32)	75.2 (43)
Latvia	206.1 (6)	98.7 (23)
Mauritius	85.4 (47)	63.8 (46)
Mexico	81.6 (48)	77.6 (41)
Poland	204.2 (8)	107.6 (13)
Romania	140.2 (36)	84.5 (38)
Slovenia	203.9 (9)	108.0 (12)
Switzerland	167.2 (24)	96.5 (26)
Trinidad, Tobago	120.0 (42)	91.4 (31)
United Kingdom	179.1 (16)	124.6 (5)

* Number in parentheses refers to rank order.

ronmental agents and the existence of a disease, such as cancer. These correlations provide working hypotheses that must be confirmed in laboratory experiments on animal models and then in carefully controlled clinical trials with humans. Typically, an epidemiological cancer study measures the incidence of cancer in several different populations (▶ Table 12.6). If a statistically significant difference is found, further studies seek to identify factors correlated with this difference. The rates of many forms of cancer in the United States are related to our physical surroundings, personal behavior, or both. Estimates indicate that at least 50% of all cancer can be attributed to environmental factors.

What are some environmental factors for cancer?

The American Cancer Society estimates that 85% of the lung cancer cases in men and 75% of the cases in women are related to smoking. Smoking produces cancers of the oral cavity, larynx, esophagus, and lungs and accounts for 30% of all cancer deaths. Most of these cancers have very low survival rates. Lung cancer, for example, has a 5-year survival rate of 13%. Cancer risks associated with tobacco are not limited to smoking; the use of snuff or chewing tobacco carries a 50-fold increased risk of oral cancer.

Keep in mind

▪ Diet and behavior are two major factors in cancer prevention.

The Skin Cancer Epidemic

Skin cancer is the most common form of cancer in the United States. According to the American Academy of Dermatology, skin cancer is an undeclared epidemic. In 1967, about 100,000 cases of skin cancer were reported. This year, more than 1 million new cases will be diagnosed, accounting for half of all cancer cases, more than the combined number of breast, lung, colon, and prostate cancers. Americans now have a 1 in 5 lifetime risk of skin cancer. Ten thousand people will die from skin cancer this year, most from melanoma, the most deadly form of skin cancer. The incidence of melanoma among whites tripled between 1980 and 2003. There will be more than 100,000 new cases of melanoma this year, and at the current rate, 1 in 37 Americans have a lifetime risk of this cancer. While many young women worry about the future risk of breast cancer, few seem to worry about the present risk of skin cancer. Melanoma is more common than any non-skin cancer in women between the ages of 25 and 29.

Why has there been a more than tenfold increase in skin cancer since 1967, and why are skin cancer cases increasing by 3% to 5% per year? The answers are complex. Ozone depletion of the atmosphere contributes to increased levels of ultraviolet radiation, and there is more outdoor activity than in decades past, but attitudes and behavior are also important contributing factors. More than 80% of lifetime skin damage occurs by the age of 18. In spite of this, many Americans think suntans are healthy, and only 25% consistently use sunscreen lotions or oils.

About 1 million new cases of skin cancer are reported in the United States every year, almost all related to ultraviolet light from the sun or tanning lamps (see Genetics in Society: The Skin Cancer Epidemic). Skin cancer cases are increasing rapidly in the population, presumably as a result of an increase in outdoor recreation. Epidemiological surveys show that lightly pigmented people are at much higher risk for skin cancer than heavily pigmented individuals. This supports the idea that genetic characteristics can affect the susceptibility of individuals or subpopulations to environmental agents that cause a specific form of cancer.

Genetics in Practice

Genetics in Practice case studies are critical thinking exercises that allow you to apply your new knowledge of human genetics to real-life problems. You can find these case studies and links to relevant websites at **http://biology.brookscole.com/cummings7**

CASE 1

Mike was referred for genetic counseling because he was concerned about his extensive family history of colon cancer. His family history is highly suggestive of an inherited form of colon cancer, known as hereditary nonpolyposis colon cancer (HNPCC). This is an autosomal dominant predisposition to colon cancer, and those who carry the altered gene have a 75% chance of developing colon cancer by age 65. Mike was counseled about the inheritance of this condition, the associated cancers, and the possibility of genetic testing (on an affected family member). Mike's aunt elected to be tested for one of the genes that may be altered in this condition and discovered that she does have

an altered *MSH2* gene. Other family members are in the process of being tested for this mutation.

1. Seventy-five percent of people who carry the altered gene will get colon cancer by age 65. This is an example of incomplete penetrance. What could cause this?

2. Once a family member is tested for the gene, is it hard for other family members to remain unaware of their own fate, even if they did not want this information? How could family dynamics help (or hurt) this situation?

3. Is colon cancer treatable? What are the common treatments, and how effective are they?

Summary

12.1 Cancer Is a Genetic Disorder

- Cancers are malignant tumors. The primary risk factor for cancer is age. Heritable predispositions to cancer usually show a dominant pattern of inheritance.

12.2 Cancer Begins in a Single Cell

- Cancer begins when a single cell acquires mutations over time allowing it to escape control of the cell cycle and begin uncontrolled division. Cancer cells are clonal descendants from a single mutant cell. Because mutations accumulate slowly, age is the primary risk factor for cancer.

12.3 Inherited Susceptibility and Sporadic Cancers

- Cancer can be caused by inheriting genes that cause a predisposition to cancer (inherited cancer) or can be caused by the accumulation of mutations in somatic cells (sporadic cancer).

12.4 Cancer Is a Disease of the Cell Cycle

- The study of two classes of genes, tumor suppressor genes and oncogenes, has established the relationship

between cancer, the regulation of cell growth and division, and the cell cycle. The discovery of tumor suppressor genes that normally act to inhibit cell division has provided insight into the regulation of the cell cycle. These gene products act at control points in the cell cycle at G1/S or G2/M. Deletion or inactivation of these products causes cells to continuously divide.

12.5 Colon Cancer Is a Genetic Model for Cancer

- Cancer is a multistep process that requires a number of specific mutations. Colon cancer has been studied to provide insight into the number and order of steps involved in transforming normal cells into cancer cells. Two pathways to colon cancer illustrate that some genes are gatekeepers, controlling the cell cycle, whereas others are caretakers, repairing DNA damage to prevent genomic instability.

12.6 Chromosome Changes, Hybrid Genes, and Cancer

- Other human disorders, including Down syndrome, are associated with high rates of cancer. This predisposition may result from the presence of an initial mutation or genetic imbalance that moves cells closer to a cancerous state. Other cancers, including leukemia, are caused by translocation events, some of which create hybrid genes that activate cell division.

12.7 Cancer and the Environment

■ It is now apparent that many cancers are environmentally induced. Occupational exposure to minerals and chemicals poses a cancer risk to workers in a number of industries. The widespread dissemination of these materials poses an undefined but potentially large risk to the general population. Social behavior contributes to approximately 50% of all cancer cases in the United States; most, if not all, of which are preventable.

Questions and Problems

HUMAN Genetics ❷ Now™ Preparing for an exam? Assess your understanding of this chapter's topics with a pre-test, a personalized learning plan, and a post-test at http://now.ilrn.com/cummings7

Cancer Is a Genetic Disorder

1. Theodore Boveri predicted that malignancies would often be associated with chromosomal mutation. What lines of evidence substantiate this prediction?
2. Distinguish between a carcinogen and a mutagen.
3. Benign tumors:
   a. are noncancerous growths that do not spread to other tissues
   b. do not contain mutations
   c. are malignant and clonal in origin
   d. metastasize to other tissues
   e. none of the above
4. What does it mean to have a malignant tumor?
5. Metastasis refers to the process in which:
   a. tumor cells die
   b. tumor cells detach and move to secondary sites
   c. cancer does not spread to other tissues
   d. tumors become benign
   e. cancer can be cured

Inherited Susceptibility and Sporadic Cancers

6. Cancer is now viewed as a disease that develops in stages. What are the stages that this statement is referring to?
   a. Nonmalignant tumors become malignant.
   b. Proto-oncogenes become tumor suppressor genes.
   c. Younger people get cancer more than older people.
   d. Small numbers of individual mutational events exist that can be separated by long periods of time.
   e. None of the above.
7. It is often the case that a predisposition to certain forms of cancer is inherited. An example is familial retinoblastoma. What does it mean to have inherited an increased probability to acquire a certain form of cancer? What subsequent event(s) must occur?

Cancer Is a Disease of the Cell Cycle

8. A proto-oncogene is a gene that:
   a. normally causes cancer
   b. normally suppresses tumor formation
   c. normally functions to promote cell division
   d. is involved in forming only benign tumors
   e. is only expressed in blood cells
9. What is the difference between an oncogene and a tumor suppressor gene?
10. Distinguish between dominant inheritance and recessive inheritance in retinoblastoma.
11. Describe the likelihood of developing bilateral (both eyes affected) retinoblastoma in the inherited versus the sporadic form of the disease.
12. Parents of a 1-year-old boy are concerned that their son may be susceptible to retinoblastoma because the father's brother had bilateral retinoblastoma. Both parents are normal (no retinoblastoma). They have their son tested for the *RB* gene and find that he has inherited a mutant allele. The father is then tested and is also found to carry a mutant allele.
   a. Why didn't the father develop retinoblastoma?
   b. What is the chance that the couple will have another child carrying the mutant allele?
   c. Is there a benefit to knowing their son may develop retinoblastoma?
13. The search for the *BRCA1* breast cancer gene discussed in this chapter was widely publicized in the media (for example, *Newsweek*, Dec. 6, 1993). Describe the steps taken by Mary-Claire King and her colleagues to clone this gene. How long did this process take?
14. Viral oncogenes are often transduced copies of normal cellular genes—they become oncogenic when their gene product or its regulation is subtly altered. What are the roles of cellular proto-oncogenes, and how is this role consistent with their implication in oncogenesis?
15. Which of the following mutations will result in cancer?
   a. homozygous recessive mutation in a tumor suppressor gene coding for a nonfunctional protein
   b. dominant mutation in a tumor suppressor gene in which the normal protein product is overexpressed
   c. homozygous recessive mutation in which there is a deletion in the coding region of a proto-oncogene, leaving it nonfunctional

Questions and Problems ■ 287

d. dominant mutation in a proto-oncogene in which the normal protein product is overexpressed

16. In DNA repair, what are caretaker genes? How do they work?

17. The following family has a history of inherited breast cancer. Betty (grandmother) does not carry the gene. Don, her husband, does. Don's mother and sister had breast cancer. One of Betty and Don's daughters (Sarah) has breast cancer; the other (Karen) does not. Sarah's daughters are in their 30s. Dawn, 33, has breast cancer; Debbie, 31, does not. Debbie is wondering if she will get the disease because she looks like her mother. Dawn is wondering if her 2-year-old daughter (Nicole) will get the disease.

  a. Draw a pedigree indicating affected individuals, and identify all individuals.

  b. What is the most likely mode of inheritance of this trait?

  c. What are Don's genotype and phenotype?

  d. What is the genotype of the unaffected women (Betty and Karen)?

  e. An RFLP marker has been found that maps very close to the gene. Given the following RFLP data for chromosomes 4 and 17, which chromosome does this gene map to?

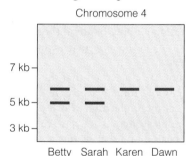 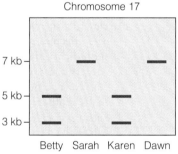

  f. Using the same RFLP marker, Debbie and Nicole were tested. The results are shown in the following figure. Based on their genotypes, is either of them at increased risk for breast cancer?

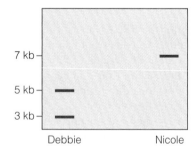

18. You are in charge of a new gene therapy clinic. Two cases have been referred to you for review and possible therapy.

Case 1. A mutation in the promoter of an oncogene causes the gene to make too much of its normal product, a receptor protein that promotes cell division. The uncontrolled cell division has caused cancer.

Case 2. A mutation in an exon of a tumor suppressor gene makes this gene nonfunctional. The product of this gene normally suppresses cell division. The mutant gene cannot suppress cell division and has led to cancer.

Given the current state of knowledge, in which case is gene therapy a viable option? Why?

19. Explain how the *APC* gene starts the progress toward colon cancer.

Chromosome Changes, Hybrid Genes, and Cancer

20. Can you postulate a reason or reasons why children with Down syndrome are 20 times more likely to develop leukemia than children in the general population?

21. What mutational event is typically associated with Burkitt's lymphoma? Which chromosomes are involved in this tumor?

Cancer and the Environment

22. What are some factors that epidemiologists have associated with a relatively high risk of developing cancer?

23. Smoking cigarettes has been shown to be associated with the development of lung cancer. However, a direct correlation between how many cigarettes one smokes and the onset of lung cancer does not exist. A heavy smoker may not develop lung cancer, although a light smoker may develop the disease. Explain why this may be.

24. Discuss the relevance of epidemiologic and experimental evidence in recent governmental decisions to regulate exposure to asbestos in the environment.

25. Studies have shown that there are significant differences in cancer rates among different ethnic groups. For example, the Japanese have very high rates of colon cancer, but very low rates of breast cancer. It has also been demonstrated that when members of low-risk ethnic groups move to high-risk areas, their cancer risks rise to that of the high-risk area. For example, Japanese who live in the United States, where the risk of breast cancer is high, have higher rates of breast cancer than Japanese who live in Japan. What are some of the possible explanations for this interesting phenomenon? What factors may explain why the Japanese have higher rates of colon cancer than other ethnic groups?

Internet Activities are critical thinking exercises using the resources of the World Wide Web to enhance the principles and issues covered in this chapter. For a full set of links and questions investigating the topics described below, log on to **http://biology.brookscole.com/cummings7**

1. *The "Cancer Gene" TP53 (Li-Fraumeni Syndrome).* The tumor suppressor gene *TP53* (also known as *p53*) helps improve the rate of DNA repair; mutations in this gene are implicated in the onset of many cancers. At the *TP53 Cancer GeneWeb* website, you can explore links to many different sources of information on *TP53*. From this website, under the "Other Related Resources" heading, you can link to the Weizmann Institute's *p53* home page and read about the expression, cellular function, and involvement in disease of this critical gene.

2. *Family Histories of Breast Cancer.* Breast cancer screening followed by prophylactic (preventive) mastectomy has become an option for some women with strong family histories of breast cancer. At Lawrence Berkeley National Laboratory's *ELSI* website, link to the "Breast Cancer Screening" page. Within the main page, read about breast cancer in the introduction, and then scroll down to "What Would You Do?" and consider how you would answer the questions posed in the various classroom scenarios.

3. *Cancer Case Studies.* The *Genetics of Cancer* website provides access to a number of case studies including family history information and pedigrees and then poses personal and ethical questions. Select one of the cases, evaluate it, and answer the questions.

How would you vote now?

Women who carry mutations in either of two cancer genes, *BRCA1* and *BRCA2,* have a significantly higher risk of developing breast cancer. Julie, the woman discussed at the beginning of this chapter, has several female family members with one of these mutant genes. However, neither Julie nor her mother carries the mutant gene. On the basis of this information, Julie decided not to have surgery to remove her breasts. Now that you know more about cancer, and the genetics it involves, what do you think? If a female member of your family carried one of the two breast cancer genes and asked your advice, would you recommend that she have her breasts removed, even though not everyone carrying this mutation will develop breast cancer? Visit the Human Heredity Companion website to find out more on the issue, then cast your vote online at **http://biology.brookscole.com/cummings7**

 For further reading and inquiry, log on to **InfoTrac College Edition,** your world-class online library, including articles from nearly 5,000 periodicals, at **http://infotrac.thomsonlearning.com**

13

An Introduction to Cloning and Recombinant DNA

O N a farm in Wisconsin, Bob Schauf rises early every day to milk the cows in his herd and send the collected milk out for processing. After he is finished with the herd, he milks one last cow, Blackrose II. When he has collected her milk, he saves some for his personal use, and pours the rest down the drain. Why? Blackrose II is a clone, the offspring of Blackrose, a world champion Holstein. Blackrose II's DNA has not been genetically modified, only copied from her mother, and as such, there are no regulations governing the sale of animal products from clones. The United States Food and Drug Administration (FDA) has asked farmers and biotech companies to temporarily abstain from selling milk, meat, or other products from cloned cows. The FDA has asked the National Academy of Sciences to review the safety of products derived from cloned animals. As part of the evaluation, milk from uncloned and cloned cows is being chemically analyzed at the University of Wisconsin and Utah State University. The testing process is complex, because milk is a mixture with about 100,000 components. The results to date show no differences between the milk of a clone and the cow from which it was cloned. The final report from the National Academy of Sciences is expected early in 2005.

Why clone milk cows? The average cow produces about 3.9 gallons of milk a day. Blackrose II produces about 11.5 gallons of milk per day, triple the average. Herds of cloned high producers would allow farmers to significantly reduce herd size without affecting milk production. There are other possible benefits for consumers, as well. In recent years, dairy cows have been given growth hormones to stimulate milk production, a controversial practice that would be unnecessary if the milk from clones is judged to be safe.

In this chapter, we will examine how plants, animals, and DNA molecules are cloned, and in the next chapter, we discuss how this technology is being used.

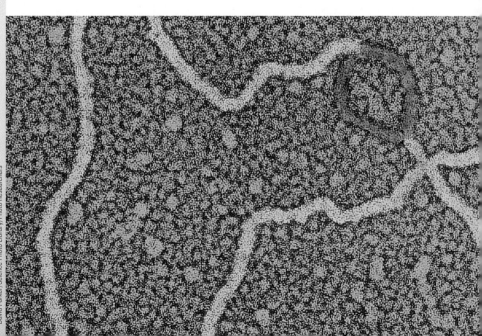

How would you vote?

Cloned cows do not contain any foreign genes and are not genetically modified. Cloning proponents argue that milk from cloned cows is safe for human consumption. Watchdog groups that monitor genetically engineered foods are concerned about the safety of products from cloned animals, and want more research to prove the safety of such products before they are allowed in the marketplace. Do you think that milk and meat from cloned animals is safe and should be sold without labeling? Visit the Human Heredity Companion website to find out more on the issue, then cast your vote online at **http://biology.brookscole.com /cummings7**

Keep in mind as you read

■Cloned plants and animals are used in research, agriculture, and medicine.

■Cloning DNA uses techniques of biochemistry, genetics, and molecular biology.

■Finding a specific gene in a cloned library requires a molecular probe.

■The polymerase chain reaction copies DNA without cloning.

13.1 What Are Clones?

When a fertilized egg divides to form two cells, these cells usually stay attached to one another and become parts of one embryo. In a small number of cases, the cells separate from each other and form separate embryos that become identical twins. Because they are derived from a single ancestral cell (in this case a fertilized egg), identical twins are clones. **Clones** are defined as molecules, cells, or individuals derived from a single ancestor. Methods for producing clones are not new; the Greek philosopher Plato wrote about cloning fruit trees 3,000 years ago. On the other hand, cloning DNA molecules, cells, and animals became possible only in recent decades.

The ability to produce many copies of specific DNA molecules has had a great impact on genetic research. This impact has spread beyond the research lab and is used in many areas, including the criminal justice system, child support cases, archaeology, dog breeding, and environmental conservation. In this chapter, we will review the methods used to clone DNA molecules, identify them, and analyze them. In the following chapter, we will examine how cloning is used in research and everyday life, and in Chapter 15, we will discuss the Human Genome Project, a direct outgrowth of recombinant DNA technology. Before we consider the basic techniques of DNA cloning, let's look briefly at how plants and animals can be cloned.

■**Clones** Genetically identical molecules, cells, or organisms all derived from a single ancestor.

Plants can be cloned from single cells.

We have genetically manipulated plants and animals for thousands of years by selective breeding. Organisms with desirable characteristics were chosen for breeding, and the offspring with the best combination of characteristics were used for breeding the next generation. Although this method seems slow and unreliable, Charles Darwin observed that after only a few generations, breeders produced many varieties of pigeons and minks.

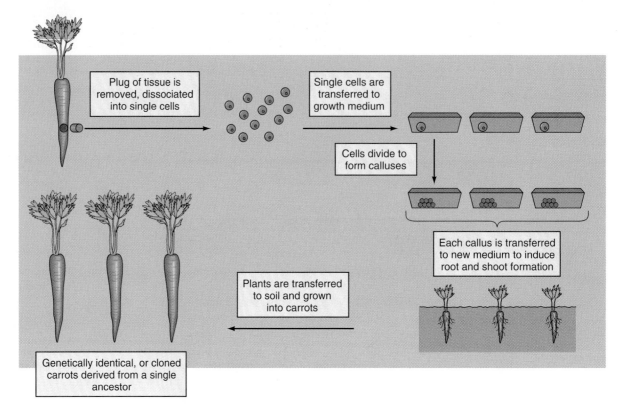

▲ FIGURE 13.1 Cloning carrots. A plug of tissue is removed and separated into single cells. Each cell is placed in a separate dish with growth medium. As single cells divide, they form a mass of cells called a callus. Calluses are transferred to a new medium to induce the formation of roots and shoots. The developing plants are then transferred to soil, where they grow into genetically identical copies, or clones, of the original carrot.

In the 1950s, Charles Steward grew individual carrot cells in the laboratory using special nutrients (▶ Figure 13.1). These single cells grew and divided to form a ball of cells known as a callus (▶ Figure 13.2). When the calluses were transferred to a different nutrient medium, they grew into full-sized carrots. Because they were all derived from a single ancestor, the carrots grown from the calluses are all clones. Variations of Steward's method have been used to clone plants of several different species. Cloning is used in the paper and lumber industry to produce trees of uniform size, growth rate, and disease resistance. For example, scientists selected a loblolly pine tree that was resistant to disease, grew rapidly, and had high wood content. A core sample was taken from this tree and separated into single cells. The cells were grown until they formed a callus and then were converted into seedlings. Thousands of these seedlings were planted and grew into a forest of genetically identical trees that mature at the same time. This allows pulpwood companies to plant and harvest trees on a predetermined schedule and is now widely used in timber farming.

Animals can be cloned by several methods.

Cloning of farm animals such as cattle and sheep moved from the research laboratory to the business world more than 20 years ago. Embryo splitting and nuclear transfer are two of the most widely used methods of cloning animals. To clone an animal by embryo splitting, an egg is collected and fertilized by *in vitro* fertilization (IVF). The embryo develops in the dish to form an embryo containing 8 to 16 cells. A technician then separates the cells from one another using a microscope and very fine needles. The individual cells are grown in the laboratory to form ge-

▲ FIGURE 13.2 This cloned plant was grown from a single cell removed from a parental plant.

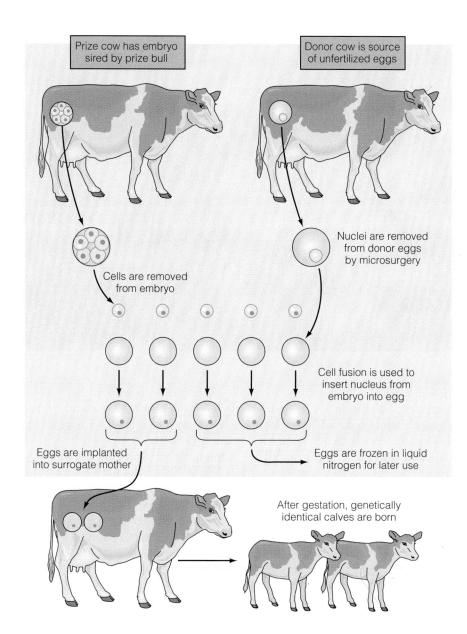

Prize cow has embryo sired by prize bull

Donor cow is source of unfertilized eggs

Nuclei are removed from donor eggs by microsurgery

Cells are removed from embryo

Cell fusion is used to insert nucleus from embryo into egg

Eggs are implanted into surrogate mother

Eggs are frozen in liquid nitrogen for later use

After gestation, genetically identical calves are born

◀ FIGURE 13.3 Mammals, including cows, have been cloned by cell fusion. This process has two stages. First, unfertilized eggs are collected from a donor cow, and the nucleus is removed from each egg by microsurgery. Second, the embryo to be cloned is recovered and separated into single cells. Embryo cells are fused with the donor eggs. Each egg then contains a genetically identical nucleus. These eggs can be frozen in liquid nitrogen for future use or implanted into the uterus of a surrogate mother to develop. All offspring of these eggs would be genetically identical copies, or clones, of the original embryo.

netically identical embryos. The embryos are then transplanted into surrogate mothers for development. This method extends nature's way of producing identical twins or triplets and can be used to clone any mammalian embryo, including human embryos.

The second method of animal cloning, nuclear transfer, is more difficult but can result in a larger number of cloned offspring. The first successful cloning of mammals by nuclear transfer was done in 1986. In this experiment, sheep eggs were collected, and the nucleus was removed from each egg by dissection under a microscope. One by one, the enucleated eggs were fused with cells from a 16- to 32-cell embryo. The fused cells were grown into embryos and transplanted into surrogate mothers for development. Ideally, if the original embryo contains 16 or 32 cells, then 16 or 32 genetically identical offspring or clones can be produced (▶ Figure 13.3). Case 1 at the end of the chapter deals with cloning animals.

With improvements in the techniques of cell fusion it became possible to use cells from older and older embryos. The cloning of Dolly the sheep, reported in 1997, was the first time that an adult cell (in this case from the udder) was successfully used to produce a cloned animal (▶ Figure 13.4). Cloning Dolly was a significant event because it showed that even nuclei from highly specialized adult cells can direct all stages of development when placed into eggs. Earlier experiments led scientists to the

> **FIGURE 13.4** Dolly the sheep, cloned by nuclear fusion of an adult cell, with her surrogate mother.

conclusion that adult cells could not be used in cloning experiments. In 1998, a research team successfully cloned more than two dozen mice. Instead of using cell fusion, they directly injected nuclei from adult cells into enucleated eggs (▶ Figure 13.5). This method has a much higher success rate than cell fusion.

Animal cloning has had a great impact on farming. Sheep, cattle, goats, and pigs have all been cloned. Farmers can now produce herds of genetically identical animals, all of which have valuable traits such as superior wool, milk, or meat production. Case 2 at the end of the chapter discusses human cloning.

Although the cloning of plants and animals has changed agriculture, the cloning of DNA molecules has revolutionized everything from laboratory research to health care to the food we eat. Using methods of DNA cloning, we can now find genes, map them, and transfer them between species. These methods are used to find carriers of genetic disorders, perform gene therapy, and create disease-resistant food plants. In the rest of this chapter, we will describe how DNA molecules are cloned and analyzed. In the following chapter, we will look at some of the ways this technology is being used in genetics, medicine, the criminal justice system, and the biotechnology industry.

Keep in mind

■ Cloned plants and animals are used in research, agriculture, and medicine.

13.2 Cloning Genes Is a Multistep Process

Cloning DNA (including DNA encoding genes) produces a large number of identical molecules, all of which are copies of a single DNA molecule. The methods for cloning DNA molecules are built on discoveries in genetics and biochemistry made in the late 1960s and early 1970s. As a group, these methods are called **recombinant DNA technology**. With recombinant DNA technology, it is possible to

1. Cut DNA at specific base sequences using enzymes called (restriction endonucleases,) producing a collection of DNA fragments;

■ **Recombinant DNA technology**
A series of techniques in which DNA fragments are linked to self-replicating vectors to create recombinant DNA molecules, which are replicated in a host cell.

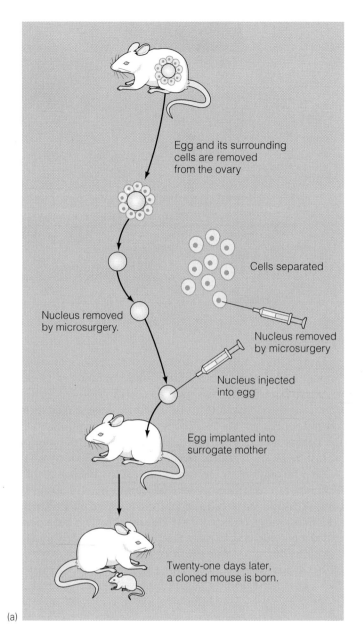

(a)

Egg and its surrounding cells are removed from the ovary

Cells separated

Nucleus removed by microsurgery.

Nucleus removed by microsurgery

Nucleus injected into egg

Egg implanted into surrogate mother

Twenty-one days later, a cloned mouse is born.

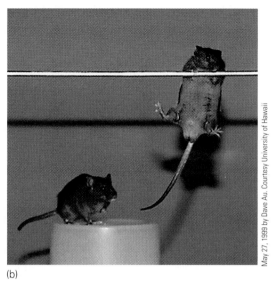

(b)

May 27, 1999 by Dave Au. Courtesy University of Hawaii

◄ FIGURE 13.5 (a) The strategy for cloning mice by injection of nuclei from adult cells. (b) A cloned mouse (on the bar) and its parent (lower left).

2. Connect these DNA fragments to carrier DNA molecules, called vectors, creating recombinant DNA molecules;
3. Transfer a recombinant DNA molecule to a host cell, where it is copied to form hundreds or thousands of exact copies (clones) of the recombinant molecule;
4. Retrieve the cloned DNA fragment from the vector in large quantities; and
5. Turn on genes in cloned DNA and purify large amounts of the encoded gene product for use in research, medicine, and industry.

Restriction enzymes cut DNA at specific sites.

The story of how recombinant DNA technology developed is an example of how basic research in one field often has an unexpected impact on other areas. It might seem odd that mapping the gene for cystic fibrosis and producing human insulin in bacteria were made possible by research into how some bacteria resist infection by viruses, but this is often how science works. The discovery that bacteria resist viral infection by producing enzymes that cut the viral DNA into pieces was one of the first steps in the development of recombinant DNA technology.

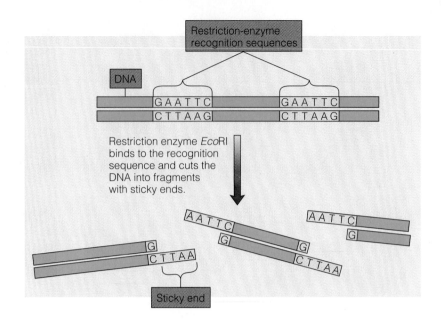

In the mid-1970s, Hamilton Smith and Daniel Nathans discovered a number of bacterial enzymes that attach to DNA molecules and cut both strands of DNA at specific nucleotide sequences. In bacteria, the enzymes cut up the DNA of invading viruses, protecting the bacteria against lethal infections. These proteins, called **restriction enzymes** or **restriction endonucleases,** are a key component in recombinant DNA technology. A restriction enzyme attaches to DNA and moves along the molecule until it finds a specific nucleotide sequence called a recognition site. Once at the recognition site, the enzyme cuts through both strands of the DNA. Each type of restriction enzyme (from different bacteria) has its own recognition and cutting site. The recognition and cutting site for an enzyme isolated from *Escherichia coli*, a bacterium that lives in the human intestine, is shown in ► Figure 13.6. This recognition and cutting sequence is called a **palindrome,** because it reads the same on either DNA

■ **Restriction enzymes** Bacterial enzymes that cut DNA at specific sites.

■ **Palindrome** A word, phrase, or sentence that reads identically in both directions. In DNA, a sequence of nucleotides that reads identically on both strands when read in the 5′ to 3′ direction.

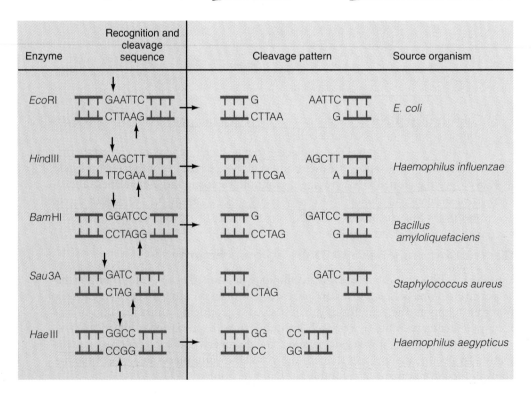

► **FIGURE 13.8** Some common restriction enzymes and their cutting sites.

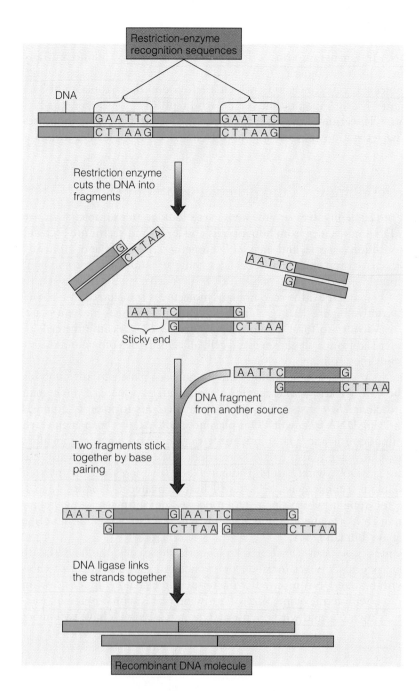

Recombinant DNA molecules can be created using DNA from two different sources, a restriction enzyme and DNA ligase. DNA from each source is cut with the restriction enzyme, and the resulting fragments are mixed. In the mixture, the complementary ends of the two types of DNA will associate by forming hydrogen bonds. These linked fragments can be covalently joined by treatment with DNA ligase, creating recombinant DNA molecules.

Restriction-enzyme recognition sequences

DNA

GAATTC   GAATTC
CTTAAG   CTTAAG

Restriction enzyme cuts the DNA into fragments

G CTTAA   AATTC
                           G

AATTC   G
      G   CTTAA
Sticky end

AATTC   G
      G   CTTAA
DNA fragment from another source

Two fragments stick together by base pairing

AATTC   GAATTC   G
G   CTTAAG   CTTAA

DNA ligase links the strands together

Recombinant DNA molecule

strand (when read in the 5′ to 3′ direction). Words like mom, pop, and radar are palindromes, as are phrases such as "Todd erases a red dot" or "live not on evil." Some restriction enzymes, like *Eco*RI, don't cut straight across the DNA. Instead they cut each strand at different places, creating single-stranded tails. These single-stranded tails are "sticky" and can reassociate with other DNA molecules with complementary (if necessary, review complementary base pairing in Chapter 8) tails to produce recombinant DNA molecules (▶ Figure 13.7). At this stage these fragments are only held together by hydrogen bonding. To chemically bond the fragments, an enzyme called DNA ligase is added. DNA ligase forms chemical bonds between the fragments, creating recombinant DNA molecules composed of human DNA and plasmid DNA. Cutting DNA from different organisms with the same restriction enzyme and then mixing them together can create new combinations, called recombinant DNA molecules. The recognition and cutting sites for several restriction enzymes are shown in ▶ Figure 13.8.

Vectors serve as carriers of DNA.

■ Vectors Self-replicating DNA molecules that are used to transfer foreign DNA segments between host cells.

Vectors carry DNA fragments into host cells where they are copied (cloned). Many of the vectors used in recombinant DNA work are derived from self-replicating, circular DNA molecules called plasmids, which are found in the cytoplasm of bacterial cells (▶ Figure 13.9). A map of one such plasmid, pBR322, is shown in ▶ Figure 13.10. The middle section of the diagram shows the location of recognition sites that can be cut by restriction enzymes and used to insert DNA molecules from another organism.

There are several steps in the process of cloning DNA.

Cloning a DNA molecule involves several steps. Let's look at an example (▶ Active Figure 13.11). If DNA molecules from human cells and a plasmid vector are each cut with the same restriction enzyme, and then mixed together, the sticky ends from any two fragments having complementary sequences can pair and form recombinant plasmids containing human DNA and plasmid DNA (▶ Active Figure 13.11a–f). To clone this piece of human DNA, the recombinant molecule is transferred into a bacterial host cell (▶ Active Figure 13.11g). The host cell is placed on a nutrient plate, where it grows and divides to form a colony (▶ Figure 13.12). Because the cells in each colony are derived from a single ancestral cell, all cells in the colony and the recombinant plasmids they contain are clones.

How do we know that a bacterial colony on the nutrient plate actually contains a plasmid, and how do we know whether this plasmid carries a piece of the human DNA we want to clone? Scientists use several clever methods to identify colonies with plasmids carrying DNA fragments. Our plasmid, pBR322, has been engineered to carry two antibiotic-resistance genes, one for tetracycline and one for ampicillin. Each of these resistance genes also carries a restriction site. If we cut the pBR322 DNA and the human DNA with *Bam*HI, human DNA fragments will be inserted into the tetracycline-resistance gene, which will be inactivated (▶ Figure 13.13). However, in the test tube in which this reaction occurs, not all plasmids will combine with a human DNA fragment. Some plasmids will re-form a circular molecule without carrying any human DNA.

After the plasmids have been transferred to bacterial host cells, the bacteria are grown on nutrient plates. Colonies formed from a bacterial cell carrying a recombinant plasmid (with human DNA inserted at the tetracycline gene) will not grow on plates supplemented with tetracycline, but will grow on plates with ampicillin. Bacterial cells that take up a plasmid with no inserted human DNA will have two intact

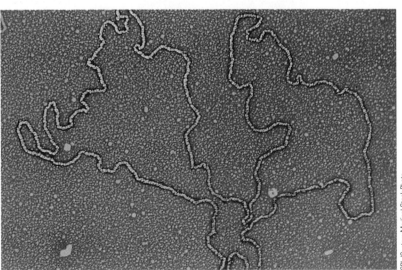

▶ FIGURE 13.9 A plasmid isolated from a bacterial cell. Plasmids are used as vectors for cloning DNA.

SPL/Custom Medical Stock Photo

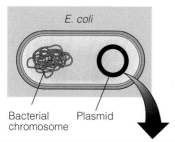

E. coli

Bacterial chromosome Plasmid

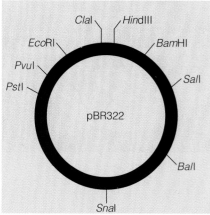

ClaI HindIII
EcoRI BamHI
PvuI
PstI SalI

pBR322

BalI

SnaI

Cut with restriction enzyme BamHI

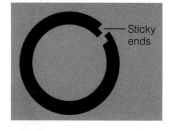

Sticky ends

▲ FIGURE 13.10 The plasmid pBR322 contains restriction sites used to carry DNA fragments for cloning.

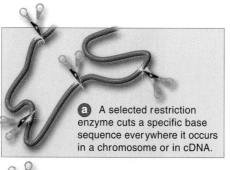

a A selected restriction enzyme cuts a specific base sequence everywhere it occurs in a chromosome or in cDNA.

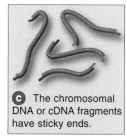

c The chromosomal DNA or cDNA fragments have sticky ends.

b The same enzyme cuts the same sequence in plasmid DNA.

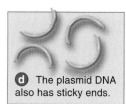

d The plasmid DNA also has sticky ends.

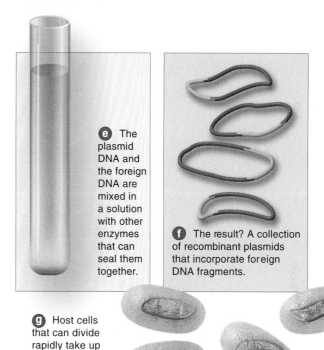

e The plasmid DNA and the foreign DNA are mixed in a solution with other enzymes that can seal them together.

f The result? A collection of recombinant plasmids that incorporate foreign DNA fragments.

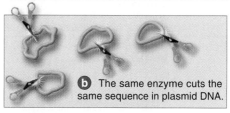

g Host cells that can divide rapidly take up the recombinant plasmids.

▲ ACTIVE FIGURE 13.11 The steps in cloning DNA. (a) The DNA to be cloned is cut with a restriction enzyme. (b) A plasmid vector is cut with the same restriction enzyme. The cut DNA with sticky ends (c–d) is mixed to allow formation of recombinant DNA molecules (e). (f) The result is a collection of recombinant plasmids carrying foreign DNA. (g) Recombinant plasmids are inserted into a bacterial host cell. At each bacterial cell division, the plasmid is replicated, producing many copies, or clones, of the foreign DNA.

HUMAN
Genetics ⊜ Now™

Learn more about DNA cloning by viewing the animation on Human GeneticsNow at **http://now.ilrn.com/cummings7**

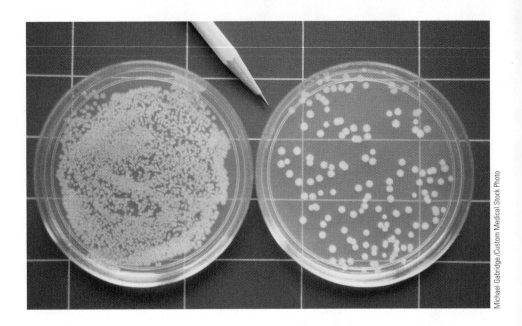

▶ FIGURE 13.12 Colonies of bacteria on Petri plates. Each colony is descended from a single cell. Therefore, each colony is a clone.

Michael Gabridge/Custom Medical Stock Photo

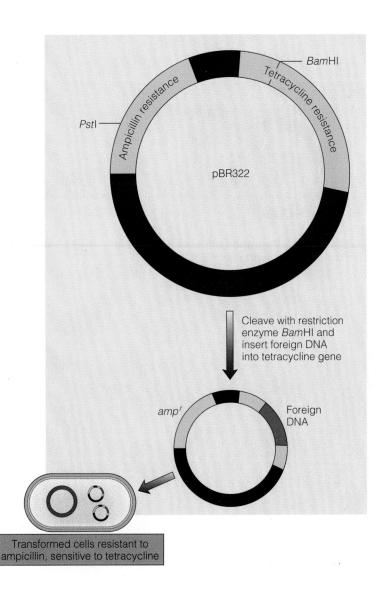

▶ FIGURE 13.13 The plasmid pBR322 carries two antibiotic-resistance genes, one for tetracycline and one for ampicillin. If DNA is inserted into the tetracycline gene, bacterial host cells that take up this plasmid will be resistant to ampicillin, but not tetracycline. Cells that have not taken up a plasmid will be killed by either antibiotic.

antibiotic resistance genes, and the cells will grow on plates with ampicillin and on plates with tetracycline. So, to find colonies that carry human DNA fragments, we look for colonies that will grow on ampicillin plates but will not grow on plates with tetracycline. If we collect all those colonies, we will have a set of cloned human DNA fragments.

The steps involved in cloning DNA can be summarized as follows:

1. DNA to be cloned is cut with restriction enzymes to produce fragments that end in specific sequences.
2. These segments are mixed with vector molecules cut with the same restriction enzyme, producing recombinant DNA molecules joined together by the enzyme DNA ligase.
3. Plasmids carrying DNA fragments are transferred into bacterial cells, where the recombinant plasmids replicate and produce many copies, or clones, of the recombinant DNA molecule.
4. Colonies carrying recombinant DNA molecules are identified, collected, and grown, the host cells are broken open, and the recombinant plasmids are extracted.

In our example, we can isolate the cloned human DNA by treating the recombinant plasmid with the same restriction enzyme used in cloning. The cloned human DNA can then be used in further experiments or transferred to new vectors and host cells to synthesize the protein encoded by the cloned human DNA.

Keep in mind

■ Cloning DNA uses techniques of biochemistry, genetics, and molecular biology.

Scientists working in the field of recombinant DNA technology were among the first to realize there might be unrecognized dangers in using and releasing recombinant organisms. Accordingly, as DNA cloning began, they called for a moratorium on all such work until safety issues were discussed and resolved (see Genetics in Society: Asilomar: Scientists Get Involved). After years of discussion and experimentation, there is now general agreement that such work poses little risk, but the episode demonstrates that scientists are concerned about the risks and the benefits of their work.

13.3 Cloned Libraries

Because each cloned fragment of human DNA is relatively small compared with the size of the genome, many clones are needed to hold all the genes carried by humans. A collection of clones containing all of the DNA sequences (and therefore, all the genes) carried by an individual or a subset of those genes is called a **genetic library**. Libraries can carry all the genes of an individual (a genomic library), the genes from a single chromosome, or only those genes expressed in a certain cell type. In a genomic library, the number of clones needed to carry all the genes depends, of course, on the size of the genome and the size and number of fragments carried by the vectors in the library.

If we make a human genomic library using DNA fragments about 1,700 bases long (1.7 kilobases), we would need about 8 million plasmids to make sure our library included all the genetic information from one human cell. Other vectors, like **yeast artificial chromosomes (YACs)**, can carry DNA fragments about 1 mil-

■ Genetic library   In recombinant DNA terminology, a collection of clones that contains all of the genetic information in an individual.

■ Yeast artificial chromosome (YAC) A cloning vector that has telomeres and a centromere that can accommodate large DNA inserts and uses the eukaryote yeast as a host cell.

lion bases long. YACs are vectors with telomeres and a centromere that replicate in yeast host cells. A human genomic library can be carried in just over 3,000 YACs. As we will see in the following sections, it is easier to search through 3,000 clones rather than 8 million clones when looking for a certain gene. The Human Genome Project used YACs and other large-capacity vectors in sequencing the human genome.

Using recombinant DNA technology, it is now possible to make genomic libraries and recover any and all genes carried by an organism. These genes can be used in research, screening for genetic disorders, gene therapy, or commercial applications in the pharmaceutical or agricultural industry. Genomic libraries from many organisms are now available, including bacteria, yeasts, crop plants, many endangered species, and humans. Genomic libraries are the basic resource used in genome projects, a topic we will discuss in the following chapter.

Genetics in Society

Asilomar: Scientists Get Involved

The first steps in creating recombinant DNA molecules were taken in 1973 and 1974. Scientists immediately realized that modifying the genetic information in *Escherichia coli,* a bacterium from the human gut, could be potentially dangerous. A group of scientists asked the National Academy of Sciences to appoint a panel to assess the risks and the need to control recombinant DNA research. A second group published a letter in *Science* and in *Nature,* two leading scientific journals, calling for a moratorium on certain kinds of experiments until the potential hazards could be assessed. Shortly afterward, an international conference was held at Asilomar, California, to consider whether recombinant DNA technology poses any dangers and whether this form of research should be regulated. In 1975, a set of guidelines resulting from this conference was published under the direction of the National Institutes of Health (NIH), a government agency that sponsors biomedical research in the United States. In 1976, NIH published a new set of guidelines prohibiting certain kinds of experiments and dictated that other types of experiments were to be conducted only with appropriate containment to prevent the release of bacterial cells carrying recombinant DNA molecules. NIH also called for research to accurately assess the risks, if any, associated with recombinant DNA techniques.

In the meantime, legislation was proposed in Congress and at the state and local levels to regulate or prohibit the use of recombinant DNA technology. The federal legislation was withdrawn after exhaustive sessions of testimony and reports. By 1978, research had demonstrated that the common K-12 laboratory strain of *E. coli* was much safer for use as a host cell than originally thought. Several projects showed that K-12 could not survive in the human gut or outside a laboratory setting. Other work showed that recombinant DNA is produced in nature, and there are no detectable serious effects. In 1982, NIH issued a new set of guidelines eliminating most of the constraints on recombinant DNA research. No experiments are currently prohibited. The most important lesson from these events is that the scientists who developed the methods were the first to call attention to the possible dangers of recombinant DNA research, and they did so based only on its potential for harm. There were no known cases of the release of recombinant DNA–carrying host cells into the environment. Scientists voluntarily shut down their research work until the situation could be properly and objectively assessed. Only when they reached a consensus that there was no danger did their work resume. Contrary to their portrayal in the popular media, scientists do care about the consequences of their work and do become involved in socially important issues.

13.4 Finding a Specific Clone in a Library

A genomic library can contain thousands or millions of different clones. Once a library is available, the difficulty is finding the clone containing a gene of interest. Humans have about 25,000 genes. To study a specific gene, the cloned DNA encoding the gene must be isolated from the thousands of clones in a human genome library. Most often, a specific cloned gene is identified using a labeled nucleic acid molecule called a **probe.** Probes are short DNA or RNA molecules with a nucleotide sequence complementary to some portion of the gene you are looking for. Probes are usually labeled with radioactivity. Probes identify genes in a library by their ability to base pair or hybridize with a complementary base sequence in the gene you are searching for (▶ Figure 13.14).

With a genomic library and a probe in hand, a specific gene can be identified and isolated. One method of gene hunting is shown in ▶ Active Figure 13.15. The bacterial cells of the library are grown on a nutrient plate at low density (▶ Figure 13.12, right side). A DNA-binding filter (nitrocellulose or nylon) is placed on top of the colonies. Some of the cells from the colonies stick to the filter. The filter is lifted off and placed in solutions to rupture the cells and convert the released DNA into single strands that stick to the filter. A radioactive probe is added and allowed to hybridize. The probe will hybridize only with the colony containing the gene with complementary base pairs. The filter is washed to remove excess probe and laid down on a piece of x-ray film. Radioactivity from the probe exposes the film, showing the location of the colony with the gene you are looking for. With this information, you

■ **Probe** A labeled nucleic acid used to identify a complementary region in a clone or genome.

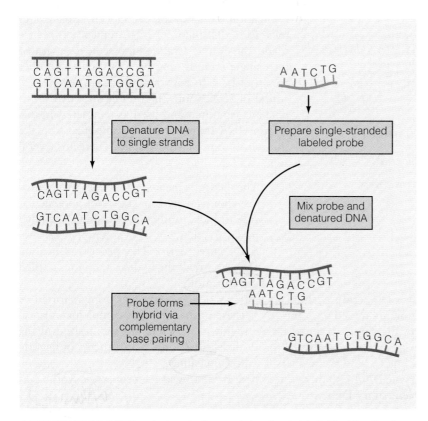

▲ FIGURE 13.14 A DNA probe is a single-stranded molecule labeled for identification in some way. Both chemical and radioactive labels are used. The single-stranded probe forms a double-stranded hybrid molecule with regions complementary to the DNA being studied.

(a) Bacterial colonies, each derived from a single cell, grow on a culture plate. Each colony is about 1 millimeter across.

(b) A nitrocellulose or nylon filter is placed on the plate. Some cells of each colony adhere to it. The filter mirrors how the colonies are distributed on the culture plate.

(c) The filter is lifted off and put into a solution. The cells stuck to it rupture; the cellular DNA sticks to the filter.

(d) Also, the DNA is denatured to single strands at each site. A radioactively labeled probe is added to the filter. It binds to DNA fragments that have a complementary base sequence.

(e) The probe's location is identified by exposing the filter to x-ray film. The image that forms on the film reveals the colony that has the gene of interest.

▲ ACTIVE FIGURE 13.15 Using a probe to identify a colony in a genome library carrying a specific gene.

HUMAN Genetics ⊜ Now™ Learn more about finding specific genes by viewing the animation on Human GeneticsNow at
http://now.ilrn.com/cummings7

■ **Polymerase chain reaction (PCR)**
A method for amplifying DNA segments using cycles of denaturation, annealing to primers, and DNA-polymerase directed DNA synthesis.

can return to the plate and grow cells only from the colony containing the gene you want to study.

Keep in mind

■ Finding a specific gene in a cloned library requires a molecular probe.

13.5 A Revolution in Cloning: The Polymerase Chain Reaction

Cloning with vectors and host cells is labor intensive and time consuming. Fortunately, it is not the only way to make many copies of a DNA molecule. A technique called the **polymerase chain reaction (PCR)**, invented in 1986, has revolutionized, and in some cases replaced, host cell cloning.

PCR uses single strands of DNA as a template to make a complementary strand with the enzyme DNA polymerase, producing a double-stranded molecule. This method parallels the way DNA replication works in the cell nucleus. (See Chapter 8 for a discussion of DNA replication.) Once made, the double-stranded molecule is separated into single strands, and each is used in another round of PCR to copy a new strand. After every round of replication, the double-stranded products are separated and used as templates. In this way, a DNA fragment can be amplified to make thousands or millions of copies. There are several steps in PCR:

1. The DNA is heated to break the hydrogen bonds between the two polynucleotide strands, producing two single-stranded molecules that will serve as templates (▶ Active Figure 13.16). This process of heating DNA to break the hydrogen bonds is called denaturation.

2. At a lower temperature, short nucleotide sequences acting as primers for DNA replication are mixed with the DNA; these primers bind to complementary regions on the single-stranded DNA fragments. Primers are synthesized in the laboratory and are usually 20 to 30 nucleotides long. This process of heating and cooling to bind the primers is called annealing.

3. The enzyme DNA polymerase is added. Beginning at the primers, it synthesizes a complementary DNA strand.

These three steps (▶ Active Figure 13.16a–e) make up one PCR cycle. The cycle is repeated (▶ Active Figure 13.16f–h) by heating the mixture and converting the double-stranded DNA into single strands, each of which will serve as a template in another round of synthesis. The amount of DNA present doubles with each PCR cycle. After n cycles, there is a 2^n increase in the amount of

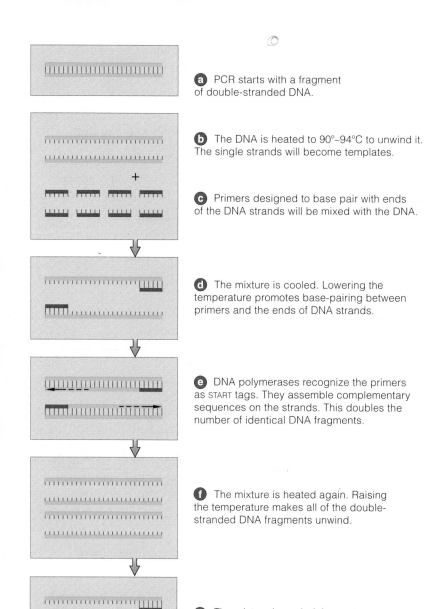

a PCR starts with a fragment of double-stranded DNA.

b The DNA is heated to 90°–94°C to unwind it. The single strands will become templates.

c Primers designed to base pair with ends of the DNA strands will be mixed with the DNA.

d The mixture is cooled. Lowering the temperature promotes base-pairing between primers and the ends of DNA strands.

e DNA polymerases recognize the primers as START tags. They assemble complementary sequences on the strands. This doubles the number of identical DNA fragments.

f The mixture is heated again. Raising the temperature makes all of the double-stranded DNA fragments unwind.

g The mixture is cooled. Lower temperature promotes base pairing between more primers added to the mixture and the single strands.

h DNA polymerase action doubles the number of identical fragments. *Repeating the sequence of reactions doubles the number of DNA fragments each time.* Billions of fragments are rapidly synthesized, as in the rows of PCR systems below that are copying human DNA.

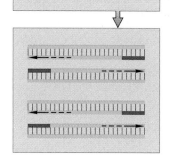

David Parker/SPL/Photo Researchers

▲ **ACTIVE FIGURE 13.16** The polymerase chain reaction. Repeated cycles can amplify the original DNA sequence by more than a million times.

 Learn more about copying DNA with the polymerase chain reaction by viewing the animation on Human GeneticsNow at
http://now.ilrn.com/cummings7

Table 13.1 DNA Sequence Amplification by PCR	
Cycle	**Number of Copies**
0	1
1	2
5	32
10	1,024
15	32,768
20	1,048,576
25	33,544,432
30	1,073,741,820

double-stranded DNA (▶ Table 13.1). The power of PCR allows DNA to be copied in a test tube instead of a host cell, and millions of copies can be made in hours rather than weeks.

The DNA to be amplified by PCR does not have to be purified and can be present in minute amounts; even a single DNA molecule can serve as a starting point. DNA from many sources has been used for PCR, including dried blood, hides from extinct animals such as the quagga (a zebralike African animal hunted to extinction in the late nineteenth century), single hairs, mummified remains, and fossils. To date, the oldest DNA used in the PCR technique has been extracted from insects preserved in amber for about 30 million years (▶ Figure 13.17). The DNA amplified from these samples is being used to study how specific genes have changed over long stretches of evolutionary time. PCR is also used in clinical diagnosis, forensic applications, and other areas, including conservation. Some of these applications are discussed in the next chapter.

Keep in mind

■ The polymerase chain reaction copies DNA without cloning.

13.6 Analyzing Cloned Sequences

Once a cloned sequence containing all or part of a gene has been identified and selected from a library, it becomes a useful tool. It can be used as a probe to find and study regulatory sequences on adjacent chromosome regions, to investigate the internal organization of the gene, and to study its expression in cells and tissues. For these studies, geneticists routinely use several methods, some of which are outlined in the following sections.

■ **Southern blot** A method for transferring DNA fragments from a gel to a membrane filter, developed by Edward Southern for use in hybridization experiments.

The Southern blot technique can be used to analyze cloned sequences.

Edward Southern discovered a way of using cloned DNA fragments that have been separated by size (using a technique called electrophoresis). Once separated, the fragments are transferred to filters and screened with probes. This procedure, known as a **Southern blot,** has many applications. It is used to find differences in normal and mutant alleles, to identify related genes in other organisms, and to study gene evolution. To make a Southern blot, DNA is extracted and cut into fragments with one or more restriction enzymes. A solution containing the fragments is placed in a slot at the top of a gel made of agarose. The DNA fragments are then separated by gel electrophoresis. In this technique, an electric current is passed through the gel. Because the DNA fragments have a negative charge, they migrate through the gel toward the positive pole. Small fragments migrate through the agarose faster than large fragments, and the DNA is separated by size. After the fragments are separated, the gel is stained to show the number and location of the fragments and photographed (▶ Figure 13.18). The DNA in the gel is converted into single strands by immersing the gel in a chemical bath. In the next step, the single-stranded fragments are transferred from the gel to a sheet of DNA-binding membrane (made from nitrocellulose or a nylon derivative).

Alfred Pasieka/SPL/Photo Researchers

▲ FIGURE 13.17 Insects embedded in amber are the oldest organisms from which DNA has been extracted and cloned.

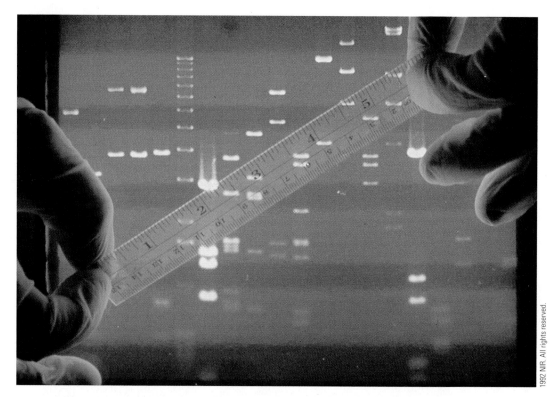

▲ FIGURE 13.18 A gel stained to show the separation of restriction fragments by electrophoresis.

To transfer the fragments, a sheet of DNA-binding membrane is placed on top of the gel, which is in a dish containing a salt solution. Paper towels and a weight are placed on top of the membrane. The salt solution flows up through the gel and the nylon membrane by capillary action, wetting the paper towels. As the solution moves from the gel into and through the membrane, the DNA fragments are carried from the gel to the membrane. The DNA fragments stick to the membrane, while the solution passes through to the paper towels. A thick sponge under the gel acts as a wick to allow the salt solution to flow through the gel (▶ Figure 13.19).

After the DNA has been transferred, the membrane is placed in a heat-sealable food storage bag along with the labeled, single-stranded probe being used in the experiment. The bag is heat-sealed, placed in a heated incubator, and then the hybridization reaction occurs. Only those single-stranded DNA fragments attached to the membrane with a base sequence complementary to the probe's sequence will form hybrids. After hybridization, the unbound probe is washed away, and the hybridized fragments are visualized. For radioactive probes, a piece of x-ray film is placed over the filter. The radioactivity in the probe exposes parts of the film. The film is developed, and a pattern of one or more bands are analyzed and compared with patterns from other experiments.

DNA sequencing can be done for an entire genome.

The ability to determine the nucleotide sequence of DNA has greatly enhanced our understanding of gene organization, the regulation of gene expression, and the evolutionary history of genes. It is also the basic method in genome projects, which seek to determine the nucleotide sequence of an entire genome. The success of the Human Genome Project and hundreds of other genome projects is derived from advances in the technology of **DNA sequencing.** There are several ways of sequencing DNA; the

▥ **DNA sequencing**   A technique for determining the nucleotide sequence of a fragment of DNA.

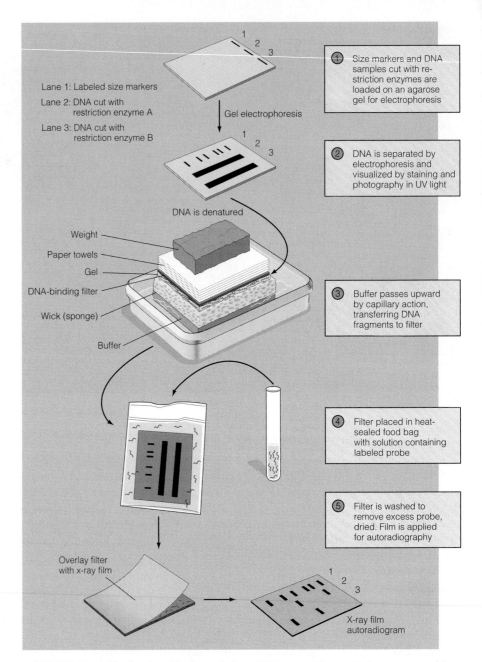

Lane 1: Labeled size markers

Lane 2: DNA cut with
restriction enzyme A

Lane 3: DNA cut with
restriction enzyme B

Gel electrophoresis

DNA is denatured

Weight
Paper towels
Gel
DNA-binding filter
Wick (sponge)
Buffer

Overlay filter
with x-ray film

X-ray film
autoradiogram

① Size markers and DNA
samples cut with re-
striction enzymes are
loaded on an agarose
gel for electrophoresis

② DNA is separated by
electrophoresis and
visualized by staining and
photography in UV light

③ Buffer passes upward
by capillary action,
transferring DNA
fragments to filter

④ Filter placed in heat-
sealed food bag
with solution containing
labeled probe

⑤ Filter is washed to
remove excess probe,
dried. Film is applied
for autoradiography

▲ FIGURE 13.19 The Southern blotting technique. DNA is cut with a restriction enzyme, and the fragments are separated by gel electrophoresis. The DNA in the gel is denatured to single strands, and the gel is placed on a sponge partially immersed in buffer. The gel is covered with a DNA-binding membrane, layers of paper towels, and a weight. Capillary action draws the buffer up through the sponge, the gel, the DNA-binding membrane, and the paper towels. This movement of buffer transfers the pattern of DNA fragments from the gel to the membrane. The membrane is placed in a heat-sealed food bag with the probe and a small amount of buffer. After hybrids have been allowed to form, the excess probe is washed away, and the regions of hybrid formation are visualized by overlaying the membrane with a piece of x-ray film. After development, regions of hybrid formation appear as bands on the film.

most widely used method is automated (▶ Active Figure 13.20). The DNA to be sequenced is separated into single strands. Each of these serves as a template for the synthesis of a complementary strand. A reaction uses DNA polymerase, a primer, and the four nucleotides (A, T, C, and G) found in DNA. In addition, a small amount of chemically altered nucleotides (A*, T*, C* and G*) is added.

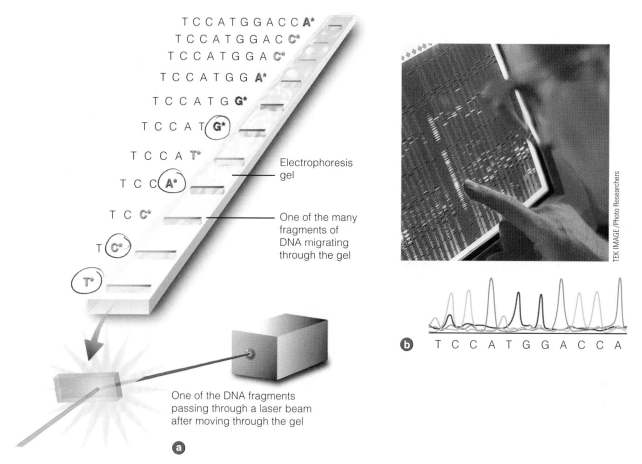

TCCATGGACC **A***
TCCATGGAC **C***
TCCATGGA **C***
TCCATGG **A***
TCCATG **G***
TCCAT **G***
TCCA **T***
TCC **A***
TCC **C***
T **C***
T*

Electrophoresis
gel

One of the many
fragments of
DNA migrating
through the gel

One of the DNA fragments
passing through a laser beam
after moving through the gel

a

b TCCATGGACCA

TEK IMAGE/Photo Researchers

▲ ACTIVE FIGURE 13.20 Automated DNA sequencing. (a) Newly synthesized DNA fragments are labeled at their ends with a modified
nucleotide that fluoresces in a specific color. The gel is scanned with a laser. The color is recorded and translated into the nucleotide sequence by software. (b) The sequence is stored and interpreted by software. Each peak in a printout shows the light fluoresced by a labeled
nucleotide. The Human Genome Project used this method to determine the sequence of the 3.2 billion nucleotides in the human genome.

HUMAN
Genetics Now™ Learn more about DNA sequencing by viewing the animation on Human GeneticsNow at
http://now.ilrn.com/cummings7

As the DNA polymerase copies the template strand, it may add a normal nucleotide, or one of the altered nucleotides. If a normal nucleotide is added, synthesis
continues. If an altered nucleotide is added, synthesis stops. Because this choice is
made at each nucleotide, and because there are thousands of template strands in the
reaction, new strands of every possible length accumulate in the reaction mixture.
Each of these strands ends in an altered nucleotide. This collection is separated by
size by gel electrophoresis, and a laser beam scans the fragments to identify the altered nucleotide by the color it fluoresces. Assembling the data on all the altered nucleotides at the end of fragments produces the DNA sequence.

Banks of hundreds of sequencing machines are used in large-scale sequencing projects such as the Human Genome Project (see Genetic Journeys: DNA Sequencing).
Each machine can sequence several hundred thousand nucleotides per day.

DNA sequencing provides information about the size and organization of genes
and the nature and function of the gene products they encode. Analysis of mutant genes provides information about how genes are altered by mutational events.
Sequencing is also used to study the evolutionary history of genes and species-to-
species variation.

DNA Sequencing

In 1977, Fred Sanger and colleagues initiated the field of genomics (the study of genomes) by applying a newly developed method of DNA sequencing to sequence the 5,400 nucleotides in the genome of a virus called ΦX174 (phi-x-174). The procedure was slow and laborious, and over the next few years, only the very small genomes of other viruses were sequenced. Fast-forward 24 years to June 2001 to the announcement in the White House Rose Garden that a draft sequence of the 3.2 billion nucleotides in the human genome had been finished. The final human genome sequence, some 600,000 times larger than ΦX174, was published in 2003. In the decades between these important milestones, the development of newer, automated sequencing methods made it possible to sequence the larger and more complex genomes of eukaryotes, including the 3.2 billion nucleotides that make up the human genome. The development of this technology was slow. The first genome of a free-living organism, a bacterium with 1,830,000 nucleotides in its genome, was not sequenced until 1995, 18 years after Sanger's work.

Progress in DNA sequencing was accelerated by the development of automated DNA sequencing machines and software to store, manipulate, and analyze DNA sequence information. Using the original Sanger method, scientists could sequence several hundred nucleotides of DNA in 3 to 5 days. Automated sequencing machines are able to sequence several hundred thousand nucleotides in a day. The Human Genome Project used banks of several hundred automated sequencing machines running day and night to sequence our genome. The results of genome projects are available on many databases, and scientists use these to study the evolution of genes, gene families, and species.

In addition to genome projects, DNA sequencing is used for other applications, including drug discovery and development, identifying disease-causing organisms, analyzing environmental contaminants, and conservation. Along with cloning and PCR, DNA sequencing is one of the basic methods in recombinant DNA technology.

Genetics in Practice

Genetics in Practice case studies are critical thinking exercises that allow you to apply your new knowledge of human genetics to real-life problems. You can find these case studies and links to relevant websites at **http://biology.brookscole.com /cummings7**

CASE 1

Cloning an animal by nuclear transfer was first done in 1952 using frogs. This and later work showed that the first few cell divisions after fertilization produce cells with nuclei that form complete embryos when transferred into unfertilized eggs. As an embryo develops further, the cells lose this property, and the success of nuclear transfer rapidly declines. Nuclear transfer in mammals proved to be more difficult. In the 1980s several groups developed transgenic livestock, and genetically modified pigs were considered as sources of organs for transplantation to human patients.

If animals can be derived from cells in culture, it should be possible to make genetic modifications, including the removal or addition of specific genes. This has been achieved in mice using embryonic stem (ES) cells, but to date no one has been successful in obtaining ES cells from cattle, sheep, or pigs. Dr. Ian Wilmut at the Roslin Institute in Edinburgh, Scotland, thought nuclear transfer might provide an alternative to ES cells. A major breakthrough came in 1995 when Wilmut and colleagues produced lambs by nuclear transfer from cells of early embryos.

So, how was Dolly produced? Dolly was the first mammal cloned using a nucleus from an adult animal. Cells were collected from the udder of a 6-year-old ewe and cultured for several weeks in the laboratory. Individual cells were then fused with unfertilized eggs from which the nuclei had been removed. Two hundred seventy-seven of these reconstructed eggs—each now with a diploid nucleus from the adult animal—were cultured for six days in temporary recipients. Twenty-nine of the eggs appeared to have developed to the blastocyst stage and were implanted into surrogate Scottish Blackface ewes. One gave rise to a live lamb, Dolly, some 148 days later. Dolly was born on July 5, 1996.

1. Why did the birth of Dolly stun the scientific world, considering that cloning of animals was already taking place?

2. What animal cloning has taken place since Dolly? Have any advancements been made in the cloning technique?

3. Because adult cloning can be done in sheep, does that mean it can be done in humans?

4. What policies or laws need to be in place to regulate animal cloning?

CASE 2

Success in cloning animals has generated controversy over whether or not to clone humans, in the form of embryos or adults. Researchers at the Andrology Institute of America plan to clone humans to help infertile couples have children. Another organization, Clonaid, is working to clone a human using the nucleus of a cell from a baby that died after having surgery. Others have proposed research programs in what is being called therapeutic cloning. The goal is to create stem cells that can be directed to differentiate into functional tissues for use in transplantation. Reprogramming a differentiated adult nucleus will be accomplished by transferring it to an unfertilized egg from which the nucleus has been removed. The nuclear-transplanted eggs are grown to the blastocyst stage. However, instead of transferring the embryos for development, the embryo's stem cells are collected. The stem cells will be grown in the lab and, in the presence of growth factors, will differentiate into specialized cells, such as liver cells, kidney cells, or insulin-producing cells of the pancreas. A South Korean research team produced a stem cell line from a human embryo in 2004, and the lab that cloned Dolly has received clearance to begin work on making human stem cell lines for use in medicine.

Adding to the ongoing debate about human cloning and embryonic stem cell research, therapeutic cloning will not use embryos left over from IVF procedures, but requires the creation of an embryo that will be destroyed.

1. What are the ethical concerns in cloning human beings? In therapeutic cloning?

2. What diseases could be treated with therapeutic cloning?

3. If therapeutic cloning is allowed for the treatment of disease, will we begin a slide down the "slippery slope" toward cloning human beings?

4. What laws must be put in place to regulate human cloning? Who should decide what laws are adopted?

Summary

13.1 What Are Clones?

■ Cloning is the production of identical copies of molecules, cells, or organisms from a single ancestor. The development and refinement of methods for cloning higher plants and animals represents a significant advance in genetic technology that will speed up the process of improving crops and the production of domestic animals.

13.2 Cloning Genes Is a Multistep Process

■ These developments have been paralleled by the discovery of methods to clone segments of DNA molecules. This technology is founded on the discovery that a series of enzymes, known as restriction endonucleases, recognize and cut DNA at specific nucleotide sequences. Linking DNA segments produced by restriction-enzyme treatment with vectors, such as plasmids or engineered viral chromosomes, produces recombinant DNA molecules.

■ Recombinant DNA molecules are transferred into host cells, and cloned copies are produced as the host cells grow and divide. A variety of host cells can be used, but the most common is the bacterium E. coli. The cloned DNA molecules can be recovered from the host cells and purified for further use.

13.3 Cloned Libraries

■ A collection of cloned DNA sequences from one source is a library. The clones in the library are a resource for work on specific genes.

13.4 Finding a Specific Clone in a Library

■ Clones for specific genes can be recovered from a library by using probes to screen the library.

13.5 A Revolution in Cloning: The Polymerase Chain Reaction

■ PCR is used to make many copies of a DNA molecule without using restriction enzymes, vectors, or host cells. It is faster and easier than conventional cloning.

13.6 Analyzing Cloned Sequences

■ Cloned sequences are characterized in several ways, including Southern blotting and DNA sequencing.

Questions and Problems

HUMAN
Genetics ⊘ Now™ Preparing for an exam? Assess your understanding of this chapter's topics with a pre-test, a personalized learning plan, and a post-test at **http://now.ilrn.com/cummings7**

What Are Clones?

1. Cloning is a general term used for whole organisms and single genes. Define what we mean when we say we have a clone.

2. Nuclear transfer to clone cattle is done by which of the following techniques?
   a. An 8-cell embryo is divided into two 4-cell embryos and implanted into a surrogate mother.
   b. A 16-cell embryo is divided into 16 separate cells, and these cells are allowed to form new 16-cell embryos that are directly implanted into surrogate mothers.
   c. A 2-cell embryo is divided into 2 separate cells and implanted into a surrogate mother.
   d. A 16-cell embryo is divided into 16 separate cells and fused with enucleated eggs. The fused eggs are then implanted into surrogate mothers.
   e. None of the above.

3. Dolly made headlines in 1997 when her birth was revealed. Why was her cloning so important, considering that cattle and sheep were already being cloned through embryo splitting?

Cloning Genes Is a Multistep Process

4. What is meant by the term *recombinant DNA*?
   a. DNA from bacteria and viruses
   b. DNA from different sources that are not normally found together
   c. DNA from restriction enzyme digestions
   d. DNA that can make RNA and proteins
   e. none of the above

5. Restriction enzymes:
   a. recognize specific nucleotide sequences in DNA
   b. cut both strands of DNA
   c. often produce single-stranded tails
   d. do all of the above   e. do none of the above

6. Restriction enzymes are derived from bacteria. Why don't bacterial chromosomes get cut with the restriction enzymes present in the cell? Why do bacteria have these enzymes?

7. The following DNA sequence contains a six-base palindromic sequence that is a recognition and cutting site for a restriction enzyme. What is this sequence? Which enzyme will cut this sequence?
   5′ CCGAGTAAGCTTAC 3′
   3′ GGCTCATTCGAATG 5′

8. Why is it an advantage that many restriction enzyme recognition sites are palindromes? Remember, restriction enzymes bind DNA and move along the molecule, looking for a recognition site.

9. Assume restriction enzyme sites as follows: E = *Eco*RI, H = *Hin*dIII, and P = *Pst*I. What size bands (in kilobases) would be present when the DNA shown below is cut with
   a. *Eco*RI   b. *Hin*dIII and *Pst*I
   c. all three enzymes

   2 kb   5 kb   3 kb   3.5 kb   4 kb   6 kb
   ——E——H——P——E——H——

10. Which enzyme is responsible for covalently linking DNA strands together?
    a. DNA polymerase   b. DNA ligase
    c. *Eco*RI   d. restriction enzymes
    e. RNA polymerase

11. When cloning human DNA, why is it necessary to insert the DNA into a vector such as a bacterial plasmid?

12. Briefly describe how to clone a segment of DNA. Start with cutting the DNA of interest with a restriction enzyme.

Cloned Libraries

13. A cloned library of an entire genome contains
    a. the expressed genes in an organism
    b. all the genes of an organism
    c. only a representative selection of genes
    d. a large number of alleles of each gene
    e. none of the above

14. You are given the task of preparing a cloned library from a human tissue culture cell line. What type of vector would you select for this library and why?

A Revolution in Cloning: The Polymerase Chain Reaction

15. The steps in the polymerase chain reaction (PCR) are:
    a. breaking hydrogen bonds, annealing primers, and synthesizing using DNA polymerase
    b. restriction cutting, annealing primers, synthesizing using DNA polymerase
    c. ligating, restriction cutting, and transforming into bacteria

---

d. DNA sequencing and restriction cutting

e. transcription and translation

16. You are running a PCR to generate copies of a fragment of the cystic fibrosis (CF) gene. Beginning with two copies at the start, how much of an amplification of this fragment will be present after six cycles in the PCR machine?

17. Why is PCR so revolutionary? Describe two applications of PCR.

Analyzing Cloned Sequences

18. Name three of the many applications for Southern blotting.

19. A new gene, causing an inherited form of retinal degeneration, has been cloned from mice. This disease leads to blindness in affected individuals. As a researcher in a human vision lab, you want to see if a similar gene exists in humans (genes with similar sequences usually mean similar functions). With your knowledge of the Southern blot procedure, how would you go about doing this? Start with the isolation of human DNA.

20. A base change (A to T) is the mutational event that created the mutant sickle cell anemia allele of beta globin. This mutation destroys an *Mst*II restriction site normally present in the beta-globin gene. This difference between the normal allele and the mutant allele can be detected by Southern blotting. Using a labeled beta-globin gene as a probe, what differences would you expect to see for a Southern blot of the normal beta-globin gene and the mutant sickle cell gene?

21. What exactly is a DNA sequence of a gene?

22. What kind of information can a DNA sequence provide to a researcher studying a disease-causing gene?

23. The cloning methods outlined in this chapter allow researchers to generate many copies of a gene they wish to study through the use of restriction enzymes, vectors, bacterial host cells, the creation of genetic libraries, and PCR. Once useful quantities of a disease-causing gene are available by cloning, what kind of questions do you think should be asked about the gene?

Internet Activities

Internet Activities are critical thinking exercises using the resources of the World Wide Web to enhance the principles and issues covered in this chapter. For a full set of links and questions investigating the topics described below, log on to **http://biology.brookscole.com/cummings7**

1. *Biotechnology Review.* Cold Spring Harbor's *DNA Learning Center* provides interactive animations of PCR, Southern blotting, and cycle sequencing. Review these techniques by selecting "Resources" and then "Biology Animation Library."

2. *Recombinant DNA Problem Set.* At the University of Arizona's *Biology Project: Molecular Biology,* you can work through an interactive problem set on recombinant DNA technology. Correct answers will be summarized; if you answer incorrectly, you will be provided with a tutorial to increase your understanding.

3. *Cloning in the News: A Critical Analysis.* At the *Access Excellence About Biotech* website, Tom Zinnen presents an analysis of the accuracies and inaccuracies in news reporting about Dolly, the first cloned sheep.

4. *Conceiving a Clone.* At the student-created *Conceiving a Clone* website, you can access details on cloning and biotechnology, compare different cloning techniques, and view animation of cloning processes. Follow the "Interactions" link to create a clone on the web, test your cloning knowledge, or participate in polls about cloning.

How would you vote now?

Cloned animals are created through manipulation of cells and genes; however, they do not contain any foreign genes and are not genetically modified. Livestock animals like cows have been cloned, and proponents of cloning argue that milk from cloned cows is safe for human consumption. Watchdog groups that monitor genetically engineered foods are concerned about the safety of products from cloned animals, and want more research to prove the safety of such products before they are allowed in the marketplace. Now that you know how cloned animals are made, and about recombinant DNA technology, what do you think? Is milk and meat from cloned animals safe, and should these products be sold without labeling? Visit the Human Heredity Companion website to find out more on the issue, then cast your vote online at **http://biology.brookscole.com/cummings7**

For further reading and inquiry, log on to **InfoTrac College Edition,** your world-class online library, including articles from nearly 5,000 periodicals, at **http://infotrac.thomsonlearning.com**

14

Biotechnology and Society

THE story begins simply enough: a boy was born in England to parents from Ghana. Because he was born in England, the boy was automatically a British citizen. As a youngster, he returned to Ghana to live with his father, leaving behind his mother, two sisters, and a brother. Some years later he returned to Britain, intending to live with his mother and siblings. At this point, the story gets complicated. Immigration authorities suspected that the boy was an impostor and thought he was either an unrelated child or a nephew of the boy's mother. Based on their suspicions, the boy was denied residence in Britain. The boy's family fought to help establish his identity so he could live in the country of his birth. The first round of medical tests used blood types, as well as genetic markers normally used to match organ donors and recipients. The results confirmed that the boy was closely related to the woman he claimed was his mother, but the tests could not tell whether she was his mother or an aunt.

The family turned to Alec Jeffreys, a scientist at the University of Leicester, for help. They asked if DNA fingerprinting, a technique developed in Jeffreys's research laboratory, could establish the boy's identity. To complicate the situation, the mother's sisters and the boy's father were not available for testing, and the mother was not sure about the boy's paternity. Despite these problems, Jeffreys agreed to take on the case. He took blood samples from the boy, the children he believed were his brother and sisters, and the woman who claimed to be his mother. DNA was extracted from the white blood cells in each sample and treated with restriction enzymes to cut the DNA at specific nucleotide sequences. The resulting fragments were separated by size and visualized by Southern blotting. The resulting pattern of bands, known as a DNA fingerprint, was analyzed to determine the boy's identity. The results showed two things. The boy has the same father as his brother and his

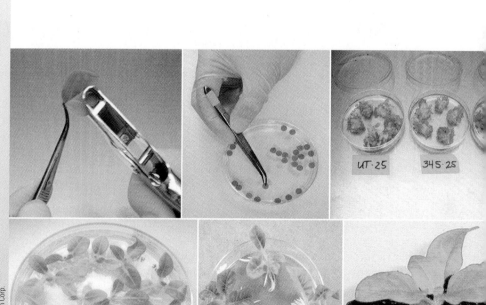

sisters because they all share DNA fragments contributed by the father. The most important question was whether the boy and his "mother" were related.

Jeffreys found that 25 fragments of the woman's DNA matched those of the boy, indicating that she was in fact the boy's mother. The chance that they were unrelated was calculated as 2×10^{-15}, or about one in a quadrillion. Faced with this evidence, immigration authorities reversed their position and allowed the boy to remain in Britain with his family.

DNA fingerprinting was one of the first of many techniques developed as part of the ongoing revolution in genetic technology. As described in the previous chapter, this revolution began in the 1970s with the discovery that DNA from different organisms could be combined to create recombinant DNA molecules. In this chapter, we examine how this technology is being used in areas of human genetics, biology, medicine, and agriculture.

How would you vote?

Identification problems like the one involving the boy from Ghana can be avoided by having a DNA identification card. These cards, now offered by several companies, contain a photo, a thumbprint or retinal scan, personal information such as height, weight, eye color, and so on, and a DNA profile that makes personal identification a certainty. Police in Britain want to expand the national DNA database to include all citizens, not just criminals, to provide a means of personal identification in the form of national identity cards. Advocates point out that such ID cards can protect children who are lost or kidnapped, help locate missing persons, prevent fraud, identify actual and potential terrorists, and be a deterrent to crime. Critics argue that DNA databases were set up to solve crimes, not to put everyone's profile into a database to deter crime, and that this technology can be used by governments to track the normal day-to-day activities of citizens instead of fighting crime. If DNA identification cards became available through a government agency, would you get one? How would you feel if such cards were required of all citizens? Visit the Human Heredity Companion website to find out more on the issue, then cast your vote online at **http://biology.brookscole.com /cummings7**

Keep in mind as you read

■ Biotechnology is an outgrowth of recombinant DNA technology.

■ Many crop plants have been genetically modified.

■ Genetic disorders can be diagnosed using biotechnology.

■ DNA profiles are based on variations in the copy number of DNA sequences.

■ The uses of biotechnology have produced unresolved ethical issues.

When developed in the late 1970s, recombinant DNA technology was used to transfer foreign genes into bacterial cells so that research scientists could clone large quantities of specific genes to study gene organization, function, and regulation. Later, it became possible to produce human proteins in host cells and to transfer genes between species of plants and animals. As a result of these and other advances, the use of recombinant DNA spread far beyond the research laboratory and is having a significant impact on our lives.

Keep in mind

■ Biotechnology is an outgrowth of recombinant DNA technology.

■ **Biotechnology** The use of recombinant DNA technology to produce commercial goods and services.

Biotechnology is the use of recombinant DNA technology to produce commodities or services. Biotechnology has altered our lives by changing the way we produce our food, diagnose and treat disease, trace our ancestry, gather evidence in criminal cases, identify human remains after disasters or war, and protect endangered species. It even helps establish whether our pets are really purebred. In this chapter, we will discuss a cross section of the ways biotechnology is being used to (1) make pharmaceutical products in genetically altered plants and animals, (2) produce new varieties of agricultural plants and animals, (3) generate animal models of human diseases, (4) diagnose and treat disease, and (5) prepare DNA profiles used in forensic applications and other fields. We will also consider the social, ethical, economic, and environmental controversies generated by the use of biotechnology.

14.1 Biopharming: Making Medical Molecules in Animals and Plants

The production of insulin to treat diabetes was one of the first commercial uses of biotechnology. Children and adults with type I diabetes cannot make insulin, a protein that controls blood sugar levels. Left untreated, diabetes is a silent, crippling disease that kills its victims. To control blood sugar levels, diabetics must inject themselves with insulin every day (▶ Figure 14.1). For decades, insulin was obtained from cows and pigs, but in 1982, human insulin produced by recombinant DNA technology became available as a safe, pure, and reliable medication that is now widely used to treat diabetes.

Before the development of biotechnology, human proteins used in the treatment of disease, such as insulin, growth hormone, and blood clotting factors, were collected from many sources, including animals in slaughterhouses, donated human blood, or even human cadavers. In some cases, proteins from these sources exposed people to serious and potentially fatal risks. For example, most hemophiliacs carry an X-linked mutation that prevents them from making a protein that is important in forming blood clots. As a result, hemophiliacs can have episodes of uncontrollable bleeding when they receive a minor injury.

Until the 1960s, there was no effective treatment for hemophilia. A process for making clotting factor by concentrating blood serum from donated blood was developed in the 1960s, and this serum-concentrated protein was given intravenously to stop episodes of bleeding. When HIV infection and AIDS first became a serious public health problem in the early1980s, donated blood was not tested for HIV (human immunodeficiency virus). As a result, about 60% of all hemophiliacs became infected with HIV from concentrated blood serum, which was prepared from pooled blood donations. Many of these infected individuals went on to develop AIDS. Hemophiliacs using the serum-concentrated clotting factor also faced a risk of hepatitis infection.

Production of clotting factors VIII (hemophilia A; OMIM 306700) and IX (hemophilia B; OMIM 306900) using recombinant DNA technology began in the early 1990s. The process does not use blood or blood cells, and since the introduction of these products, no new cases of infection with HIV or hepatitis from clotting factors have been reported. Clotting factor and insulin are only two of the dozens of human

▲ FIGURE 14.1 Some diabetics must inject themselves with insulin several times a day.

Table 14.1 Some Products Made by Recombinant DNA Technology

Product	Use
Atrial natriuretic factor	Treatment for hypertension, heart failure
Bovine growth hormone	Improve milk production in dairy cows
Cellulase	Breakdown cellulose in animal feed
Colony stimulating factor	Treatment for leukemia
Epidermal growth factor	Treatment of burns, improve survival of skin grafts
Erythropoietin	Treatment for anemia
Hepatitis B vaccine	Prevent infection by hepatitis B virus
Human insulin	Treatment for diabetes
Human growth hormone	Treatment for some forms of dwarfism, other growth defects
Interferons (alpha, gamma)	Treatment for cancer, viral infections
Interleukin-2	Treatment for cancer
Superoxide dismutase	Improve survival of tissue transplants
Tissue plasminogen activator	Treatment of heart attacks

proteins made by biotechnology. Some others are listed in ▶ Table 14.1. In the next section, we will take a more detailed look at how human proteins are made by biotechnology and how they are used in treating genetic disorders.

Human proteins can be made in animal hosts.

Infants born with a severe form of the autosomal recessive disorder Pompe disease (OMIM 232300) have difficulties breathing and feeding. Affected newborns develop muscle weakness and enlargement of the heart within a few months. As infants, they never reach normal developmental milestones such as rolling over, sitting up, or standing. Untreated babies usually die in the first year of life. Because they carry two mutant copies of the gene, affected infants are unable to make AGLU (alpha glucosidase), an enzyme that enters lysosomes (see Chapter 2 for a description of lysosomes), where it breaks down glycogen, making it available as a molecular energy source for cells. As a result, glycogen cannot be broken down, and it accumulates in muscle cells of affected children, producing the weakness associated with this disease.

To produce AGLU for treating Pompe disease, the human gene for AGLU was isolated and cloned using recombinant DNA techniques. The cloned gene was inserted into a vector designed to express the gene it carries only in milk-producing cells. The recombinant vector carrying the human AGLU gene was injected into fertilized rabbit eggs, which developed into embryos and then into adult rabbits. These rabbits are called **transgenic** organisms because they have received a gene from another species by recombinant DNA technology. Female transgenic rabbits produce the human AGLU enzyme in their milk. The human enzyme is purified from collected rabbit milk (▶ Figure 14.2) and given intravenously to the affected infants in weekly doses.

In one clinical study, infants with Pompe disease developed normal levels of AGLU activity in their muscles after treatment with the enzyme. These infants gained strength, and some were able to sit and stand after receiving transgenic AGLU. A final round of clinical trials is now underway, and if successful, AGLU produced by biotechnology will be available worldwide for treating this form of Pompe disease.

A similar strategy is being used to treat other diseases caused by lysosomal enzyme defects, as well as several other human genetic disorders. In the near future, small

■ **Transgenic** Refers to the transfer of genes between species by recombinant DNA technology; transgenic organisms have received such a gene.

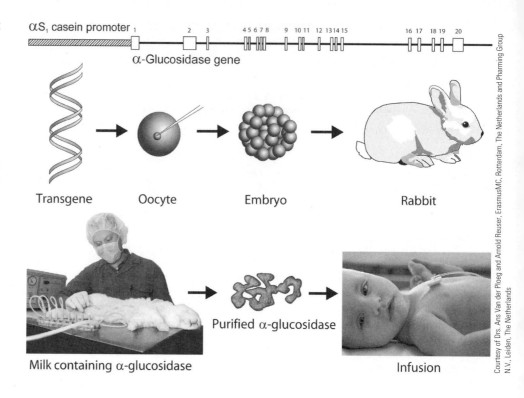

▶ **FIGURE 14.2** Making recombinant AGLU from rabbit's milk. The gene for human AGLU (alpha glucosidase) is inserted into a vector next to a control sequence (the αS₁ casein promoter) that activates the gene only in milk-producing cells. The recombinant vector is injected into a fertilized egg, which develops into an embryo and then an adult rabbit. Milk from transgenic female rabbits is collected. AGLU is purified from the milk, and then given intravenously to children with Pompe disease.

αS₁ casein promoter

α-Glucosidase gene

Transgene Oocyte Embryo Rabbit

Milk containing α-glucosidase

Purified α-glucosidase

Infusion

Courtesy of Drs. Ans Van der Ploeg and Arnold Reuser, ErasmusMC, Rotterdam, The Netherlands and Pharming Group N.V., Leiden, The Netherlands

herds of animals will be able to supply the world's need for many human proteins used in the treatment of disease.

Transgenic plants are also being used to produce human proteins for medical use. Human growth hormone has been experimentally produced in genetically modified tobacco plants. Once extracted from the plants, the hormone can be used in treating dwarfism and other disorders. Most work with using transgenic plants to make human proteins is in early stages of development, but soon, fields of transgenic plants may replace laboratories for making human proteins.

Biopharming can be used to make edible vaccines.

For hundreds of years, vaccination programs have effectively prevented many infectious diseases. The most dramatic use of vaccinations to control disease was the decades-long campaign to eliminate smallpox, a deadly viral disease that kills one in four people it infects. In the twentieth century alone, smallpox was responsible for hundreds of millions of deaths. Through an intensive vaccination program coordinated by the World Health Organization, the world was declared free of smallpox in 1977. However, vaccination programs for other diseases in developing countries are often ineffective, because many vaccines have to be refrigerated, a luxury that is often unavailable, especially in rural areas. In addition, infections from reused, nonsterile needles cause medical problems. The Children's Vaccine Initiative, a program started in 1992, aims to produce cheap, edible vaccines made in transgenic food plants for childhood vaccination programs in developing countries.

The idea is that edible plant parts, including fruit, can be used as a source of oral vaccines by transferring a gene from the disease-causing virus or bacterium into a plant. The encoded protein from the virus or bacterium would be expressed in the fruit and, when eaten, will stimulate production of antibodies that prevent infection. Such vaccines would be inexpensive and would not require investment in industrial capacity for production, purification, and distribution. In addition, these vaccines would not have to be injected under sterile conditions by trained medical staffs.

Edible vaccines are still in development, but in trials, volunteers were successfully vaccinated against bacterial diarrhea by eating uncooked transgenic potatoes. If fur-

■ **Edible vaccine** A vaccine produced in fruit or vegetables by gene transfer.

▶ **ACTIVE FIGURE 14.3** A gene incorporated into a bacterial plasmid is used to transfer a genetic trait into a plant cell. As the plant cell divides, some cells form embryos that develop into transgenic plants. Such plants are sometimes called genetically modified, or GM, organisms. Genes for herbicide resistance and resistance to insect pests were transferred to crop plants in this way.

HUMAN Genetics ⒮Now™ Learn more about making transgenic plants by viewing the animation on Human GeneticsNow at
http://now.ilrn.com/cummings7

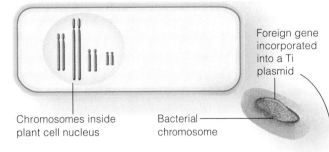

Chromosomes inside plant cell nucleus

Bacterial chromosome

Foreign gene incorporated into a Ti plasmid

ⓐ The foreign gene is transferred into a plant cell. It becomes incorporated into one of the plant's chromosomes.

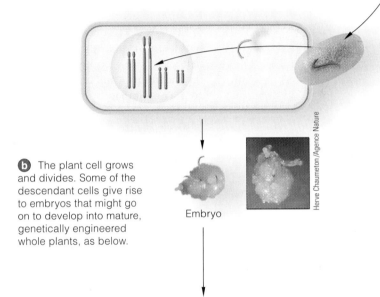

ⓑ The plant cell grows and divides. Some of the descendant cells give rise to embryos that might go on to develop into mature, genetically engineered whole plants, as below.

Embryo

Herve Chaumeton/Agence Nature

ther tests are successful, genes producing this vaccine and others, including vaccines against hepatitis B and diphtheria, will be transferred to bananas and other plants that are eaten raw and that are already grown in many parts of the world. Sometime in the near future, it seems likely that genetically engineered plants will be used to vaccinate infants, children, and adults against a variety of infectious diseases.

14.2 Genetically Modified Foods

Farmers have been genetically modifying crop plants for thousands of years by selective breeding. In the last 50 to 60 years, the use of artificial selection and hybridization has dramatically increased yield and nutritional value of crop plants. Corn yields, for example, have increased about fourfold over this time period.

About 10 years ago, scientists began using recombinant DNA methods to transfer single genes into crop plants to create transgenic crop plants with new characteristics (▶ Active Figure 14.3). The transferred gene can originate from another plant, an animal, or even fungi or bacteria. The transferred gene confers a specific new trait on the plant, such as resistance to insects or herbicides. Transgenic plants transfer only one or a few specific genes, whereas traditional breeding methods transfer hundreds or thousands of genes at a time. In the news media, transgenic plants are

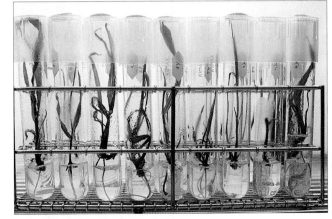

Lowell Georgia/Corbis

Genetically modified organisms (GMOs) A general term used by the media to refer to transgenic plants or animals.

often called **genetically modified organisms (GMOs)** or genetically modified (GM) plants. These terms are unfortunate, because all crop plants have been genetically modified over thousands of years by selection and crossbreeding.

Keep in mind

■ Many crop plants have been genetically modified.

Transgenic crop plants can be made resistant to herbicides and disease.

To combat weeds, farmers across the world use over 100 different herbicides at an annual cost of more than $10 billion. In spite of this effort, weeds destroy more than 10% of all crops worldwide, and pesticide runoff from treated fields is a major cause of pollution in local water supplies. The U.S. Geological Survey reports that more than 90% of all U.S. streams and fish that have been tested contain pesticides (see Spotlight on Bioremediation: Using Bugs to Clean up Waste Sites). Transgenic plants resistant to herbicides and plants resistant to insects are now available in many parts of the world. Herbicide-resistant crops carry genes for resistance to broad-spectrum herbicides such as glyphosate and glufosinate. These herbicides kill all plants in the field except the crop plant. These herbicides break down quickly in the soil, reducing the number and amount of herbicides needed and also reducing runoff and environmental impact.

Other transgenic crops including corn, cotton, and soybeans are resistant to insect pests (▶ Figure 14.4). These crops (known as Bt crops) carry a bacterial gene that produces a toxin that is released in the insect gut, causing death.

▶ FIGURE 14.4 The photograph shows the results of insect infestation on normal cotton bolls (*right*) and transgenic cotton (*left*) modified to resist attack by insects.

Agricultural Research Service/USDA

In the United States, the first transgenic crops were grown in 1996. ▶ Figure 14.5 shows the dramatic increase in the use of transgenic crops between 1996 and 2003. Similarly, the percentage of acreage planted with transgenic crops has increased dramatically since 2000. Planting transgenic crops resistant to herbicides and insect pests has reduced the number of pesticides used, as well as the amount of pesticide needed to control weeds and insects.

About 60% to 70% of the foods in your supermarket contain at least a small amount of one or more of the transgenic plants approved for commercial production in the United States (▶ Table 14.2). Products made from corn, soybeans, and cottonseed and canola oils account for almost all foods that contain transgenic ingredients. As more plants, such as nutritionally enhanced rice, are approved, this overall percentage will likely increase.

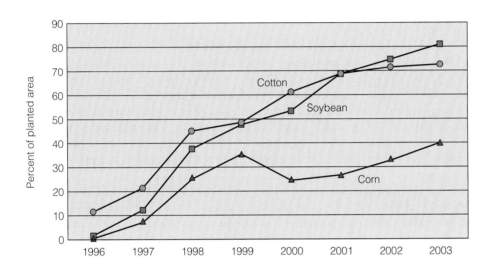

◀ FIGURE 14.5 U.S. adoption of transgenic crops from 1996 to 2003.

Table 14.2	Genetically Modified Crops Approved for Commercial Use in the United States	
Crop Plants	**New Trait**	**Source of Gene**
Corn, cotton, potato, tomato	Insect resistance (Bt crops)	Soil bacterium
Canola, corn, cotton, flax, rice, soybeans, sugar beets	Herbicide resistance	Several bacteria
Tomato	Delayed ripening	Tomato
Canola, soybeans	Altered oil content	Soybean
Corn, radicchio (chicory)	Prevention of cross-pollination with wild plants	Soil bacterium

▲ FIGURE 14.6 Normal rice grains (*left*) are white. Grains of golden rice (*right*), produced by genetic engineering, contain a precursor to vitamin A. This transgenic rice is one tool in fighting vitamin A deficiency, which causes blindness and health problems in many parts of the world.

Professor Ingo Potrykus

Transgenic crops can be used to enhance the nutritional value of plants.

Nutritional enhancement of food is nothing new. Read the labels: milk has added vitamin D, flour and products made from flour are enriched with B vitamins, and even salt has added iodine. Biotechnology is being used to modify crops to increase their nutritional value.

Most crop plants are deficient in one or more nutrients we need in our diets. For example, diets deficient in vitamin A are a serious health problem in more than 70 countries, primarily in Asia. Between 100 and 150 million children in these countries are vitamin A deficient, and 500,000 of these children become permanently blind each year, as a result. Several strategies are being used to combat vitamin A deficiency. One of these involves genetically modifying rice, a food staple in most countries with vitamin A deficiency. Golden rice (▶ Figure 14.6) is a transgenic strain constructed by transferring three genes into rice plants. Two of these genes come from daffodils and one gene is from bacteria. With these added genes, rice plants synthesize beta-carotene, a precursor to vitamin A, that turns the rice grains a golden color. When the rice is eaten, beta-carotene is converted into vitamin A.

Golden rice strains are now undergoing field trials, and once these are completed in 2006, golden rice will be available for planting in the Philippine Islands and then in other countries in Southeast Asia. This use of biotechnology along with other programs will be used to reduce vitamin A deficiency in the diet of millions of people.

Other projects to increase the nutritional value of crop plants are focusing on increasing the levels of fatty acids, antioxidants, minerals, and other vitamins to help combat dietary deficiencies. Long-chain fatty acids, known as omega fatty acids, are important in eye and brain development in newborns and in the prevention of cardiovascular disease. Diets rich in certain fish are the usual source of omega fatty acids, but overfishing has depleted wild populations, and fish and fish oils are often contaminated with mercury, pesticides, and flame retardants. Recent research has created model transgenic plants that produce these nutritionally valuable oils. Transfer of these genes to crops such as rape (from which canola oil is produced), sunflowers, and soybeans will provide a dietary source of omega fatty acids. Overall, more than 40% of the world's population suffers from a dietary lack of nutrients, and transgenic modification of crop plants is one way to combat this problem.

What are some concerns about genetically modified organisms?

The development and use of transgenic crops has generated controversy in many developed nations, including the United States and countries of Western Europe, as well as developing nations in Africa. Many questions about transgenic crops have been raised, most of which concern safety and environmental impact: Is it safe to eat food carrying part of a viral gene that switches on transgenes? Will Bt-resistant insects develop? Will disease-causing bacteria acquire the antibiotic-resistant genes used as markers in transgenic crops? We clearly need to consider health and environmental risks as transgenic crops become more common and new ones are developed.

Most food produced from transgenic crops contains one or more proteins encoded by transferred genes. Are foods containing this new protein safe to eat? In general, if the proteins are not toxic, allergenic, and do not have any other negative physiological effects, they are not considered to be a significant health hazard. In the case of herbicide-resistant food plants, the protein is readily degraded in digestive fluids, is nontoxic to mice at doses thousands of times higher than any potential human exposure, and has no similarity to known toxins or allergens. After a decade of widespread use, no human health risks have been demonstrated. To ensure the safety of GM foods, standards for evaluating proteins in genetically modified foods are already in place in European countries and are being developed in the United States.

Several potential environmental risks associated with transgenic crops have been identified. These include transfer of herbicide resistance or insect resistance from crop plants to weeds and wild plants, loss of biodiversity caused by transgenic plants hybridizing with wild varieties of the same species, and possible deleterious impact on ecosystems. As production of pharmaceuticals in transgenic plants becomes more common, it will be necessary to develop safeguards to keep such plants out of the food supply and to ensure there is no transfer of genes to wild plants.

However, most of these problems are identical to those we face in using conventional crop plants, and there is no current scientific evidence that genetically modified crops are inherently different from nonmodified crops. GM crops have a different evolutionary history and are genetically different than conventional crops, but there are no demonstrated ecological or environmental risks unique to GM crops. However, as the modification of crop plants becomes more complex, some combinations of traits may be generated and these crops may require specific management procedures.

Humans have been genetically modifying plants and animals by artificial selection for 8,000 to 10,000 years, when agriculture began. The genetic makeup of corn, sheep, cattle, tobacco, even dogs and cats has been shaped by choosing individuals with desired characteristics to be the parents of the next generation. The development of biotechnology has only changed the ways these modifications are made and the range of species that can donate genes. To continue improving crops, farm animals, pets, and garden plants by conventional breeding techniques and biotechnology, relevant scientific questions, economic issues, and public perceptions must be addressed by research, testing, and public education.

14.3 Transgenic Animals as Models of Human Diseases

Information gained from genome projects has revealed that many if not most human genes are present in other organisms such as the mouse and even the fruit fly, *Drosophila*. More than 90% of the genes involved in human diseases are also present in the mouse. This makes the mouse an excellent model to explore the genetic, biochemical, and cellular aspects of human diseases. Mouse models of human diseases can be created by transferring human disease genes into mice and using the resulting transgenic mice to achieve several goals: (1) produce an animal with symptoms that mirror those in humans, (2) use the model to study the development and progress of the disease, and (3) test treatments that hopefully will cure the model organism of a human disease.

What is the process for making transgenic animals?

Mice have been extremely useful as models of human neurodegenerative diseases. As an example, let's look at how a transgenic mouse model of Huntington disease (HD; OMIM 143100) was constructed. HD is a neurodegenerative disorder that develops in adulthood and is inherited as an autosomal dominant trait. Affected individuals gradually lose control of their arms and legs in mid-adulthood, undergo personality changes, dementia, and eventually die some 10 to 15 years after symptoms begin. The mutant gene causes the accumulation of a protein called huntingtin, which disrupts many cellular functions. Mouse models of HD are explored in more detail in Chapter 18. Here we will outline how such models are constructed and used.

To make a transgenic HD mouse, a copy of the mutant human *HD* gene was cloned into a vector, and the vector carrying the *HD* gene was microinjected into the nucleus of a fertilized mouse egg (▶ Figure 14.7). The injected eggs were implanted into foster mothers, who gave birth to transgenic mice. These newborns were examined by isolating DNA from their tail tips and analyzing for the presence and expression of the mutant *HD* gene by Southern blot analysis or by polymerase chain reaction (PCR). Adult transgenic mice expressing the human *HD* gene develop symptoms that parallel those seen in humans, including the development of protein aggregates and cell death in the brain and cells of the nervous system.

Scientists use animal models to study human diseases.

HD mice are being used to study the progressive destruction of brain structures that occur in the earliest stages of the disease process, something that is impossible to do in humans. In addition, HD mice are used to link changes in brain structure with changes in behavior.

Research on HD has identified several molecular mechanisms that play important roles in the early stages of the disease. HD mice are being used to screen drugs to identify those that improve symptoms and or reverse brain damage. Several candidate drugs have been identified, and these are being used in human clinical trials as experimental treatments for HD. Similar methods have been used to construct animal models of other human genetic disorders and infectious diseases (▶ Table 14.3).

14.4 Testing for Genetic Disorders

Before we explore the rationale, methods, and issues surrounding genetic testing, let's define the differences between genetic testing and genetic screening. **Genetic testing** determines whether someone has a certain genotype. Testing is done on individuals to identify those who have or may carry a genetic disease, those at risk of producing a genetically defective child, and those who may have a genetic susceptibility to environmental agents. **Genetic screening** is done on populations in which there is a risk for a particular genetic disorder. Genetic testing is most often a matter of choice, whereas genetic screening is often a matter of law. In this section we will concentrate

■ **Genetic testing** The use of methods to determine if someone has a genetic disorder, will develop one, or is a carrier.

■ **Genetic screening** The systematic search for individuals in a population who have certain genotypes.

▶ FIGURE 14.7 Microinjection of fertilized eggs is one way to transfer genes into mammals.

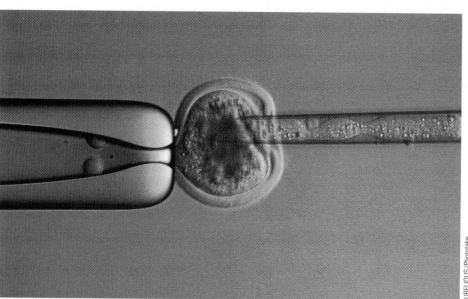

on how biotechnology is used in one type of screening and several types of genetic testing.

There are several types of testing programs:

■ Newborn screening tests infants within 48 to 72 hours after birth for a variety of genetically controlled metabolic disorders.
■ Carrier testing is done on members of families or cultural groups with a history of a genetic disorder, such as sickle cell anemia or cystic fibrosis.
■ Prenatal testing tests a fetus for a genetic disorder such as cystic fibrosis.
■ Presymptomatic testing, also called predictive testing, identifies individuals who will develop adult-onset genetic disorders such as Huntington disease or polycystic kidney disease (PCKD).

Newborn screening is universal in the United States.

All states and the District of Columbia require newborns to be tested for a range of metabolic disorders. These programs began in the 1960s with screening for phenylketonuria (PKU) and gradually expanded. Most states screen for 3 to 8 disorders, but states that use a new method, called tandem mass spectrometry, can now screen for 30 to 50 metabolic disorders.

Both carrier and prenatal testing are done to screen for genetic disorders.

Prenatal testing is used to detect genetic disorders and birth defects in the fetus. Using a combination of techniques, more than 200 single-gene disorders can be diagnosed by prenatal testing (▶ Table 14.4). In most cases, testing is done only when there is a family history or another indication that testing is recommended. If there is a family history for autosomal recessive disorders such as Tay-Sachs disease or sickle cell anemia, the parents are usually tested to determine if they are heterozygous carriers of the disease. If tests for both parents are positive, the fetus has a 25% chance of being affected. In such cases, prenatal testing can determine the genotype of the fetus.

Table 14.4	Some Metabolic Diseases and Birth Defects That Can Be Diagnosed by Prenatal Testing	
Acatalasemia	Gaucher disease	Niemann-Pick disease
Adrenogenital syndrome	G6PD deficiency	Oroticaciduria
Chédiak-Higashi syndrome	Homocystinuria	Progeria
Citrullinemia	I-Cell disease	Sandhoff disease
Cystathioninuria	Lesch-Nyhan syndrome	Spina bifida
Cystic fibrosis	Mannosidosis	Tay-Sachs disease
Fabry disease	Maple syrup urine disease	Thalassemia
Fucosidosis	Marfan syndrome	Werner syndrome
Galactosemia	Muscular dystrophy, X-linked	Xeroderma pigmentosum

For other conditions, such as Down syndrome (trisomy 21), chromosome analysis is the most direct way to detect an affected fetus. Testing for Down syndrome is usually done because of maternal age, not because there is a family history of genetic disease. Because the risk of Down syndrome increases dramatically with the age of the mother (see Chapter 6), chromosomal analysis of the fetus is recommended for all pregnancies in which the mother is age 35 or older.

Samples for prenatal testing can be obtained in several ways, including amniocentesis and chorionic villus sampling (CVS), both of which are described in Chapter 6. The fluids and cells obtained for testing can be analyzed by several techniques, including karyotyping, biochemistry, and recombinant DNA techniques.

Because recombinant DNA technology analyzes the genome directly, it is the most specific and sensitive method currently available. The accuracy, sensitivity, and ease with which recombinant DNA technology can be used to identify a genetic disease and susceptibilities carried by an individual have raised a number of legal and ethical issues that have yet to be resolved. Before we discuss these issues in a later section, we will take a closer look at how recombinant DNA methods are used in prenatal testing. Sickle cell anemia provides a good illustration.

Prenatal testing can diagnose sickle cell anemia.

As described in Chapter 4, sickle cell anemia is a recessive trait caused by a mutant form of the gene encoding the oxygen-carrying protein beta globin. This disorder is common in people with family origins in West Africa, the low-lying areas around the Mediterranean Sea, and parts of the Middle East and Asia. Affected individuals have intermittent painful episodes when red blood cells change shape, blocking small blood vessels. If not treated, sickle cell anemia can be life-threatening. The beta-globin gene is not expressed before birth, so the condition cannot be diagnosed prenatally by analyzing fetal hemoglobin, but it can be diagnosed using recombinant DNA techniques.

The mutation that causes sickle cell anemia destroys a cutting site for the enzyme *Mst*II within the beta-globin gene. If the mutation is present, *Mst*II makes one less cut in the DNA, producing fewer, longer fragments. The result is an altered pattern of bands on Southern blots (▷ Figure 14.8). The distinctive pattern of bands can be used to determine the genotypes of family members and the fetus.

For prenatal diagnosis, fetal cells are collected by amniocentesis or CVS; for family members, cells are obtained from a blood sample. DNA is extracted from the cells and cut with *Mst*II. The fragments produced by the enzyme are visualized on Southern blots. The genotypes can be read directly from the pattern of bands on the blot. The band pattern can be used to tell whether the fetus or family members have a normal homozygous, heterozygous, or a homozygous recessive genotype.

Prenatal diagnosis using amniocentesis can be done after the fifteenth week of development and by CVS as early as the eighth week of development, although it is usually done at 10 to 12 weeks. At present, several dozen genetic disorders can be detected prenatally using recombinant DNA techniques, and the list is growing steadily.

Embryos can be tested for genetic disorders.

Instead of waiting until the eighth or fifteenth week of development, biotechnology can be used to diagnose genetic disorders in the earliest stages of embryonic development. In this form of testing, eggs are fertilized in the laboratory and allowed to develop in a culture dish for several days. By this time, the egg has divided several times and formed an embryo consisting of a few cells. For testing, one of the six to eight embryonic cells (called **blastomeres**) is removed (▷ Figure 14.9). DNA is extracted from this single cell and amplified by PCR. The amplified DNA is tested to determine if the embryo is homozygous or hemizygous for a genetic disorder. Blastomere testing can be used for many common autosomal recessive disorders, such as Tay-Sachs disease, and most X-linked disorders, including muscular dystrophy, Lesch-Nyhan syndrome, and hemophilia.

■ **Blastomere** A cell produced in the early stages of embryonic development.

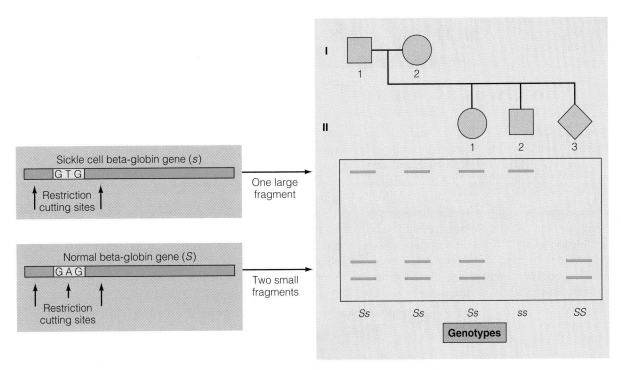

▲ FIGURE 14.8 Prenatal diagnosis for sickle cell anemia by Southern blot analysis. In sickle cell anemia, the mutation destroys a cutting site for a restriction enzyme. As a result, the number and size of restriction enzyme fragments are altered. Because the mutant allele has a distinctive band pattern, the genotypes of family members can be read directly from the blot. The parents (I-1, I-2) are heterozygotes. The first child (II-1) is a heterozygous carrier; the second child (II-2) is affected with sickle cell anemia; and the unborn child (II-3) is homozygous for the normal alleles and therefore will be unaffected and will not be a carrier.

Another method, called **polar body biopsy,** can test for genetic disorders even before fertilization takes place (see Genetics in Society: Who Owns a Genetic Test?). If a woman is heterozygous for an X-linked recessive disorder, at meiosis I, the X chromosome bearing the mutant allele can segregate into the cell destined to be the oocyte, or into the much smaller polar body attached to the larger secondary oocyte (see Chapter 7 to review gametogenesis). The polar body can be removed by micromanipulation (▶ Figure 14.10). If testing with DNA probes shows the mutant allele is in the polar body, then the oocyte must carry the normal allele. Oocytes that pass this test can be used for IVF, ensuring that her sons will not be affected with the disorder.

▓ **Polar body biopsy** Removing a polar body by micromanipulation to test whether it or the oocyte carries an allele for an X-linked recessive disorder. This test can be done prior to fertilization.

Prenatal testing is associated with some risks.

Although many genetic disorders and birth defects can be detected by prenatal testing, the technique has some limitations. With these procedures, there are measurable risks to the mother and fetus, including infection, hemorrhage, fetal injury, and spontaneous abortion. For amniocentesis, the risk of miscarriage is 0.5% to 1% and for CVS the risk of miscarriage is 1% to 3%.

Conventional strategies for prenatal testing will not always detect the majority of certain defects. For example, amniocentesis is recommended for all mothers over 34 years of age to test for Down syndrome, because maternal age is the biggest risk factor for having a child with Down syndrome. Older women have only about 5% to 7% of all children, but they have 20% of all Down syndrome children. This means that about 80% of all Down syndrome births are to mothers who are not tested by amniocentesis. This is because they are younger than 35 and are not candidates for amniocentesis. Younger mothers have 80% of the Down syndrome children, but they also have 93% to 95% of all children. Prenatal testing for Down syndrome is limited to older mothers because they have a much higher risk of having an affected child.

Genetic testing on a large scale is not always possible. For some disorders, like

▲ FIGURE 14.9 Removal of blastomere from early human embryo.

▲ FIGURE 14.10 Removal of a polar body attached to a human oocyte.

Who Owns a Genetic Test?

A few months after birth, Daniel and Debbie Greenberg's son Jonathon was diagnosed with Canavan disease (OMIM 271900), a rare and fatal genetic disorder. A second child, Amy, was also affected with this devastating disease. Canavan disease is caused by a mutation in a gene on chromosome 17 encoding the enzyme aspartoacylase. Affected children are unable to properly metabolize amino acids in their diet. Lack of this enzyme causes seizures, loss of vision, and progressive degeneration of the central nervous system. Most affected children die before the age of 18 months.

The family approached Dr. Reuben Matalon, then at the University of Illinois-Chicago, and asked for his help in identifying and cloning the gene for Canavan disease and in the development of a genetic test for this condition. Working with several nonprofit organizations, the Greenbergs started a registry for families with the disease and helped obtain tissue samples and funding for the research. After several years of work, Dr. Matalon and his colleagues identified, mapped, and cloned the gene. With the gene in hand, a genetic test was developed for use in prenatal testing and in carrier screening.

Soon after, the Canavan Foundation and other non-profit groups began offering tests for Canavan disease, a disorder that mostly affects those of Jewish descent with family origins in central and eastern Europe. Unknown to the Canavan Foundation and other organizations offering the test, Dr. Matalon's employer, Miami Children's Hospital, had patented the gene and licensed the test, charging a royalty fee of $12.50 for each test. The ironic outcome was that the people and organizations who donated money and tissue samples as part of the effort to find the gene for Canavan disease were forced to pay to use the test they helped develop.

In 2000, Daniel Greenberg, other parents of children with Canavan disease, and three non-profit organizations sued Dr. Matalon, Miami Children's Hospital, and its research institute to prevent the hospital from enforcing its patent rights and asking that testing for Canavan disease be done without royalty payments.

In 2003, a federal judge allowed the suit to proceed to trial. The case was expected to set important legal precedents on the relationship among individuals and organizations that contribute resources to genetic research, the researchers, and the use of the results. On one hand, scientists must be free to work unhampered by oversight and unreasonable restrictions imposed by contributors. On the other hand, if investigators commercialize the results of their work without the knowledge or participation of contributors, families with affected children will be less willing to participate in projects to identify genes and develop genetic tests that can help others.

In September 2003, a confidential settlement was reached in the case, allowing Miami Children's Hospital to continue to license the test and collect royalties but to allow license-free use of the Canavan gene in research.

Left unresolved is the issue of who has ownership of genetic tests developed with contributions from affected individuals. Lessons have been learned from this case, and in the last few years, families and organizations working with researchers on other genetic disorders have negotiated written agreements about gene patenting the use of genetic tests developed from samples provided by families of affected children. In commenting on the settlement, Dr. Matalon focused attention on the central issue in this and related disputes when he said "This is a disease where collaboration between investigators of the disease and families of affected children remains critical for advancing knowledge, for prevention, and hopefully, for helping affected children."

sickle cell anemia, there is a single mutation present in all cases, so testing is efficient and uncovers all cases. In cystic fibrosis (CF), however, over 1,300 different mutations have been identified (see Chapter 11), and testing for all these mutations is impractical. Many of these mutations are found only in one family, while others are found primarily in one ethnic group. Using a panel of 25 mutations to test for CF produces different levels of sensitivity (the proportion of carriers or affected fetuses correctly identified) (see ▶ Table 14.5). Thus, at the moment, CF testing is not widely performed.

Presymptomatic testing can be done for some genetic disorders.

Presymptomatic testing detects genetic disorders that develop sometime during adulthood. Those with a family history of certain dominantly inherited disorders are candidates for this type of testing. Some of these include Huntington disease (HD; OMIM 143100), amyotrophic lateral sclerosis (ALS; OMIM 105400), and polycystic kidney disease (PCKD; OMIM 173900). PCKD usually appears in adults between the ages of 35 and 50. This disease is characterized by the formation of cysts in one or both kidneys; these cysts grow and gradually destroy the kidney. Treatment options include dialysis and transplantation of a normal kidney, but many affected individuals die prematurely. Because PCKD is a dominant trait, anyone heterozygous or homozygous for the gene will be affected. The PCKD gene maps to chromosome 16, and genetic testing can determine which family members carry a mutant allele and will most likely develop this disorder. Testing can be done prenatally or at any age before (or after) the condition appears.

Presymptomatic testing, and in fact, all genetic testing, can have a serious impact on someone's life, and the life of his or her family. Some of these repercussions will be discussed after a review of some new developments in genetic testing.

Table 14.5 Cystic Fibrosis Testing for 25 Mutations: Success Rate in Different Ethnic Groups

78 percent among non-Hispanic Caucasian couples

52 percent among Hispanic Caucasian couples

42 percent among African American couples

88 percent among Ashkenazi Jewish couples

24 percent among Asian American couples

14.5 DNA Microarrays in Genetic Testing

The recombinant DNA methods described earlier can test for a single genetic disorder or, at most, several alleles of a mutant gene. A new technology using DNA microarrays can be used to screen for all known mutations in a disease gene and can even be used to screen someone's entire genome for mutations in any of the 25,000 to 30,000 genes we carry (▶ Figure 14.11).

GeneChip® Probe Arrays

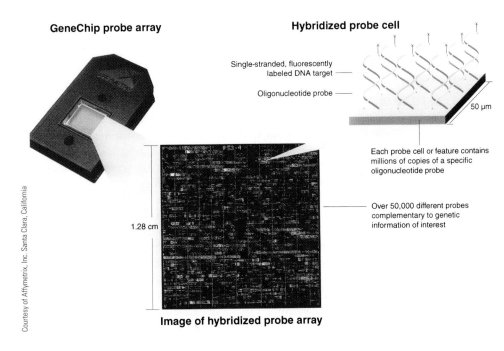

GeneChip probe array

Hybridized probe cell

Single-stranded, fluorescently labeled DNA target

Oligonucleotide probe

50 μm

Each probe cell or feature contains millions of copies of a specific oligonucleotide probe

Over 50,000 different probes complementary to genetic information of interest

1.28 cm

Image of hybridized probe array

Courtesy of Affymetrix, Inc. Santa Clara, California

◀ FIGURE 14.11 A commercial DNA array, known as a DNA chip. Each chip has between 50,000 and 500,000 fields. Each field contains many copies of a short, single-stranded DNA sequence. The sequence in each field is different from the sequence in all other fields. The DNA to be tested is converted into single strands, tagged with fluorescent dyes, and hybridized to the DNA attached to the chip. Unbound molecules are washed off the chip, and a laser scans the chip and detects the fluorescence that signals hybridization with a probe. DNA microarrays are used to detect mutations and to measure gene expression.

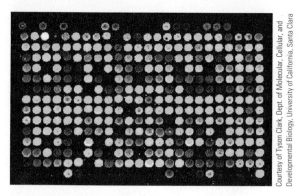

Courtesy of Tyson Clark, Dept. of Molecular, Cellular, and Developmental Biology, University of California, Santa Clara

▲ FIGURE 14.12 A DNA microarray after hybridization. In this case, DNA from two sources was used. One was tagged with a red dye, the other with a green dye. The yellow spots are produced when both of the tagged DNAs bind to the same field. Black spots mean that neither tagged DNA bound to that field.

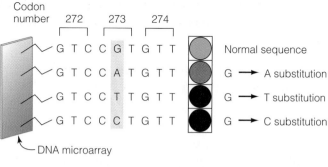

▲ FIGURE 14.13 A diagram of four of the several thousand fields on a DNA microarray carrying the sequence of the human *p53* gene. The fields shown here carry the sequence of codons 272–274 (the gene has 393 codons), and read from left to right. The sequence in the top field (the top row) is the normal sequence. The sequence in the three lower fields contains all possible sequence changes in the middle nucleotide of codon 273 (shaded box). The results of a hybridization experiment are shown. DNA from the normal *p53* gene was tagged with a green dye, and DNA from the patient being tested was tagged with a red dye. The color pattern of these four fields is shown at the right. The top field is green because only the DNA from the normal gene matches the DNA sequence in this field. The second field is red because only the DNA from the patient matches the DNA sequence in this field. The last two fields are black because neither the normal DNA nor the DNA from the patient matches the DNA in these fields. Based on this test, we can say that the patient carries a mutation in the *p53* gene. The *p53* gene is mutated in about 60% of all cancers.

■ **DNA microarray** A series of short nucleotide sequences placed on a solid support (such as glass) that have several different uses, such as detection of mutant genes or differences in the pattern of gene expression in normal and cancerous cells.

DNA microarrays are glass plates, subdivided into fields arranged to form a grid. In commercial microarrays, the fields are very small, only about as wide as a human hair, and several hundred thousand fields fit onto a one-half inch by one-half inch piece of glass. Linker molecules are used to attach short (15–25 nucleotides) single-stranded DNA molecules to each field. The DNA in each field differs from the DNA in the neighboring field by one nucleotide. To test for mutations in a single gene, the array contains DNA molecules with the normal sequence and all possible mutant combinations.

Testing uses a two-color system. Single-stranded DNA from a normal copy of the gene is labeled with a green fluorescent probe, and DNA from the gene being tested is labeled with a red fluorescent probe. The two labeled DNAs are mixed, and the solution is placed on the array. If a labeled DNA molecule is exactly complementary to DNA in a field, it will bind. After hybridization has taken place, the solution is washed off the array. Only labeled molecules base-paired with DNA on the fields will remain on the array; the rest of the labeled DNA will be washed off. A laser scans the array and detects the pattern of fluorescence as a series of dots, with each dot corresponding to a field (▶ Figure 14.12). A linked software program then analyzes the pattern of dots. The results can determine if an individual carries a mutant allele for a genetic disorder.

DNA microarrays are used to screen for mutations in many cancer-related genes, including *BRCA1* and *BRCA2*, genes in mutant form that predispose women to breast cancer. To illustrate the use of DNA microarrays in cancer screening, we will use the *p53* gene as an example. This gene is mutated in 60% of all cancers, including breast cancer. *p53* is a tumor-suppressor gene and encodes a 393-amino acid. ▶ Figure 14.13 shows the fields on a DNA microarray containing the normal nucleotide sequence of codons 272–274 and the normal sequence and all three possible mutant nucleotides for the middle nucleotide in codon 273. DNA from a normal copy of the gene is labeled with a green fluorescent dye, and DNA from the sample being tested is labeled with a red fluorescent dye. The hybridization results are shown in the array. Green dots are present in fields where the sequence of the normal gene exactly matched the

DNA on the array. Any yellow dots would represent fields where both the normal DNA and the DNA being tested are complementary to the DNA on the array (red + green = yellow). The presence of a red dot indicates that the DNA being tested carries a mutation in the *p53* gene. This dot corresponds to a G → A substitution in the middle nucleotide of codon 273. This mutation changes amino acid 273 from arginine (Arg) to histidine (His), inactivating the p53 protein, allowing cells carrying this mutation to escape cell cycle control and become cancerous.

14.6 DNA Profiles as Tools for Identification

The use of DNA samples to identify individuals was developed in the 1980s, when Alec Jeffreys and his colleagues at the University of Leicester discovered variations in the length of certain DNA sequences called **minisatellites,** located at many different chromosomal sites. Identification based on minisatellites was called **DNA fingerprinting.** Minisatellites and shorter nucleotide sequences called **short tandem repeats (STRs)** are used in numerous ways, including court cases, studies of human evolution, and tracing migratory patterns of humans across the world. Although the technology began with the use of minisatellites, STRs are now routinely used, and the term *DNA fingerprint* has been replaced by the term *DNA profile.*

■ Minisatellite Nucleotide sequences 14 to 100 base pairs long organized into clusters of varying lengths; used in construction of DNA fingerprints.

■ DNA fingerprint Detection of variations in minisatellites used to identify individuals.

■ Short tandem repeat (STR) Short nucleotide sequences 2 to 9 base pairs long organized into clusters of varying lengths; used in the construction of DNA profiles.

■ DNA profile The pattern of STR allele frequencies used to identify individuals.

Keep in mind

■ DNA profiles are based on variations in the copy number of DNA sequences.

DNA profiles can be made from short tandem repeats (STRs).

STRs are short, repeated sequences ranging from 2 to 9 base pairs in length. For example, the repeat

CCTTCCCTTCCCTTCCCTTCCCTTCCCTTC

contains six copies of the five-nucleotide sequence CCTTC. There are thousands of different STRs on different chromosomes, and because each locus can have dozens of alleles, heterozygosity is common. The number of repeats in any given STR varies so much from one person to another that by analyzing a number of different STRs in a DNA sample, a **DNA profile** that is unique to an individual (except, of course, in the case of identical twins) can be produced (▶ Figure 14.14).

To make a profile from a DNA sample, STRs are amplified by polymerase chain reaction (PCR) before Southern blotting and visualization. Because this method uses PCR, profiles can be prepared from very small DNA samples from many possible sources, including single hairs, licked envelope flaps, toothbrushes, cigarette butts, and the dried saliva on the back of a postage stamp. DNA profiles can also be obtained from very old samples of DNA, increasing its usefulness in legal cases. In fact, DNA profiles have been prepared from mummies more than 2,400 years old.

In the United States, a standardized set of 13 STRs called the CODIS panel (Combined DNA Index System) is used by law enforcement and other government agencies in preparing DNA profiles.

DNA profiles are used in the courtroom.

Forensics is the use of scientific knowledge in civil and criminal law. DNA fingerprints were first used as scientific evidence in a criminal case in 1986. In the small English village of Narborough, two girls were raped and murdered within a three-year

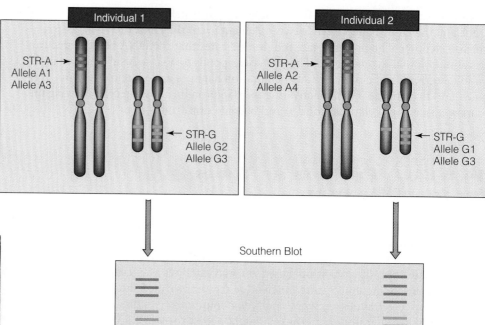

◀ FIGURE 14.14 Different alleles of a short tandem repeat (STR) on homologous chromosomes of two individuals (*above*) produce different patterns on a Southern blot. Patterns of 4 to 13 different STRs are used to produce DNA profiles for an individual.

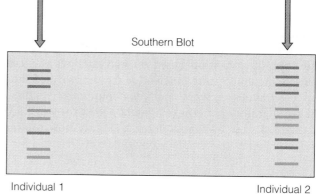

▲ FIGURE 14.15 At top, technicians analyze DNA profiles. At bottom, DNA profiles from evidence at a crime scene and profiles from seven suspects in the case (circled numbers). One suspect's profile matches that derived from blood at the crime scene; can you identify which one?

period. Faced with few leads, the police asked Alec Jeffreys at the nearby University of Leicester to analyze DNA samples recovered from the victims. His analysis showed that the same man had committed both murders, and he prepared a DNA fingerprint of the killer. Presented in court, this DNA fingerprint was admitted as evidence and freed an innocent man who had been jailed for one of the murders. To find the man who matched the killer's DNA, police asked over 4,000 men from the surrounding area to provide a DNA sample for analysis. The killer arranged for another man to give a DNA sample in his place, and would have escaped detection, but was caught when he bragged about the deception in a pub. This historic case was described in the book *The Blooding*, by Joseph Wambaugh.

In the United States, DNA evidence has been used in more than 50,000 criminal investigations and trials since 1987. DNA profiles are now used in about 10,000 criminal cases every year, 75% of which involve criminal sexual assault. They are also used in about 200,000 civil cases each year for paternity testing. DNA analysis is performed in state and local police crime labs, private labs, and the Federal Bureau of Investigation (FBI) lab in Washington, D.C. In criminal cases, DNA is often extracted from biological material left at a crime scene. This can include blood, tissue, hair, skin fragments, and semen. DNA profiles are prepared from evidence and compared with those of the victim and any suspects in the case (▶ Figure 14.15).

Usually, a DNA sample is tested using at least four different STRs from the CODIS panel to develop a profile of DNA. In a criminal case, if a suspect's profile does not match that of the evidence, then he or she can be cleared, or excluded, as the criminal. About 30% of DNA profile results clear innocent people by exclusion.

Analysis of DNA profiles combines probability theory, statistics, and population genetics to estimate how frequently a particular profile is found in an individual and in the general population. This is done in a series of steps. A DNA sample is analyzed to establish which STR alleles are present. Once the STR alleles are known, the population frequency for each STR allele is multiplied together to produce an estimate of the probability that anyone carries that combination of alleles. The frequencies are multiplied because STRs on different chromosomes are inherited independently.

Table 14.6 Calculating Probabilities in DNA Profiles

Locus	Allele	Frequency in Population	Combined Frequencies
CSF1PO	1	1 in 25 (0.040)	—
STRTPOX	2	1 in 100 (0.010)	Alleles 1 × 2 = 1 in 2,500
STRvWA	3	1 in 320 (0.0031)	Alleles 1 × 2 × 3 = 1 in 800,000
STRD5S818	4	1 in 75 (0.0133)	Alleles 1 × 2 × 3 × 4 = 1 in 60,000,000

▶ Table 14.6 shows how STR alleles provide a profile with a high degree of reliability. In the example, an individual is heterozygous for the following alleles: allele 1 of CSF1PO, allele 2 for TPOX, allele 3 for vWA, and allele 4 for D5S818. Allele 1 is found in 1 of 25 individuals in the population, and allele 2 is found in 1 of every 100 people in the population. Based on these frequencies, the probability of having a specific genotype at each locus can be calculated. For CSF1PO allele 1 *and* TPOX allele 2, this probability is 1 in 2,500 (1/25 × 1/100), meaning that this combination of alleles is found in 1 in every 2,500 individuals.

When the population frequencies of the alleles for CSF1PO, TPOX, vWA, and D5S818 are combined, the frequency of this combination is 1 in about 60 million. This means that the probability that someone has this combination of alleles is 1 person in 60 million (about 4 or 5 individuals in the United States). If an analysis includes all 13 CODIS STR markers, the chance that anyone has a particular combination is about 1 in 100 trillion, more than enough to identify someone in Earth's population of 6 billion.

The FBI began cataloging DNA profiles of convicted felons in 1998, and the database now contains more than 1,700,000 profiles. Many states are now collecting DNA samples from anyone arrested for a felony and entering their DNA profiles in a database. As they accumulate more and more profiles, DNA databases are becoming important tools in solving crimes. Over a three-year period in Virginia, for example, more than 1,600 crimes without suspects were solved by matching DNA profiles from the crime scene with profiles in the state database.

DNA profiles have many other uses.

DNA profiles have many uses outside the courtroom. Biohistorians use DNA analysis to correctly identify bodies and body parts of famous and infamous people whose graves have been moved several times, or whose graves have been newly discovered. The identification of Czar Nicholas II and many of his family members in unmarked graves in Russia is one example of how DNA profiles are used in biohistory (see Genetic Journeys: Death of a Czar). DNA analysis has been completed or is planned on the remains of Christopher Columbus, Juan Pizarro, Jesse James, and several Egyptian pharaohs.

DNA profiles are also used on less famous clients. In a case involving purebred dogs, a female Shih Tzu (▶ Figure 14.16a) was mated to a male Shih Tzu but shortly after had an unplanned encounter with a male Coton du Tulear, a breed closely resembling the Shih Tzu (▶ Figure 14.16b). Two puppies were born, and DNA analysis was used to settle the question of whether the pups were purebred Shih Tzus or a mixed breed. DNA samples were obtained from all five dogs (the female, the two males, and the two puppies) and analyzed by DNA fingerprinting. The results showed that one of the pups was a pure Shih Tzu, and the other was the offspring of the Coton du Tulear and was, therefore, a mixed breed.

(a)
Cydney Conger/Corbis

(b)
Yann Arthus-Bertrand/Corbis

▲ FIGURE 14.16 (a) A purebred Shih Tzu. (b) A purebred Coton du Tulear, a breed that closely resembles the Shih Tzu.

Genetic Journeys

Death of a Czar

Czar Nicholas Romanov II of Russia was overthrown in the Bolshevik Revolution that began in 1917. He and Empress Alexandra (granddaughter of Queen Victoria); their daughters Olga, Tatiana, Marie, and Anastasia; and their son Alexei (who had hemophilia) were taken prisoners. In July 1918, revolutionists announced that the czar had been executed, but for many years the fate of his family was unknown. In the 1920s, a Russian investigator, Nikolai Solokof, reported that the czar, his wife and children, and four others were executed at Ekaterinburg, Russia, on July 16, 1918, and their bodies were buried in a grave in the woods near the city. Other accounts indicated that at least one family member, Anastasia, escaped to live in Western Europe or the United States. Over the years, the mystery surrounding the family generated several books and movies.

In the late 1970s, two Russian amateur historians began investigating Solokof's accounts, and, after a painstaking search, nine skeletons were dug from a shallow grave at a site 20 miles from Ekaterinburg in July 1991. All of the skeletons bore marks and bullet wounds indicating violent death. Forensic experts examined the remains and, using computer-assisted facial reconstructions and other evidence, concluded that the remains included those of the czar, the czarina, and three of their five children. The remains of two children were missing: the son, Alexei, and one daughter. DNA analysis was used to confirm the findings.

The investigators used a threefold strategy in the DNA analysis. DNA was extracted from bone fragments and used for sex testing, for DNA typing to establish family relationships, and for mitochondrial DNA testing to trace maternal relationships. The sex testing used a six-nucleotide pair difference in a gene carried on both the X and Y chromosome. The results indicated that the skeletons were of four males and five females, confirming the results of physical analysis. Probes for five short tandem repeat (STR) sequences were used to test the family relationship. The results showed that skeletons 3 to 7 were a family group; 4 and 7 were parents; and 3, 5, and 6 were children. If the remains are those of the Romanovs, the combination of the sex tests and the STR tests establish that the remains of one of the princesses and the czarevitch, Alexei, are missing.

To determine whether the remains belonged to the Romanovs, investigators conducted mitochondrial DNA (mtDNA) testing. Because mtDNA is maternally inherited, living relatives of the czarina, including Prince Philip, the husband of Queen Elizabeth of England, were included in the tests. This analysis shows an exact match between the remains of the czarina, the three children, and living relatives. mtDNA from the czar matched that of two living maternal relatives, confirming that the remains are those of the czar, his wife, and three of his children.

The fate of Alexei and one daughter remains unknown, but historical accounts suggest that their bodies were burned or buried separately. This study overcame several technical challenges but clearly established the identity of the skeletons as those of the Romanovs. The results show the power that genetic technology can bring to solving a historical mystery. Robert Massie describes the intrigues, mysteries, and the science surrounding the search for the Romanovs in his book *The Romanovs: The Final Chapter.*

UPI/Corbis

14.7 Social and Ethical Questions about Biotechnology

The development of methods for cloning plants and animals by recombinant DNA technology is the foundation of the biotechnology industry, a growing and dynamic multibillion-dollar component of our economy. These methods are revolutionizing biomedical research, the diagnosis and treatment of human diseases, agriculture, the justice system, and even the production of chemicals. Unlike many other technologies, biotechnology raises serious social and ethical issues that need to be identified, discussed, and resolved.

Keep in mind

■ The uses of biotechnology have produced unresolved ethical issues.

Unfortunately, the technology and its applications are developing much faster than public policy, legislation, and social norms. Because we *can* do some things does not mean we *should* do them without a consensus from society. For example, should very short children receive recombinant DNA–produced growth hormone to make them average in height? The U.S. Food and Drug Administration approved the use of growth hormone for this purpose in 2003. If we accept that we can treat short children with growth hormone to increase their adult height, is it ethical to treat children of normal or above average height to enhance their chances of becoming professional basketball or volleyball players?

The current guidelines of the Food and Drug Administration do not require identifying labels on food produced by recombinant DNA technology. Should such food be labeled, and if so, why? It can be argued that food products have been genetically manipulated for thousands of years and that gene transfer is simply an extension of past practices. It can also be claimed that consumers have a right to know that gene transfer has altered the food products they buy and use.

Genetic testing can identify those who are carriers of genetic disorders, those who are at risk for a genetic disorder, and those who will develop a genetic disease later in life. With genome scanning, it will be possible to map out someone's lifetime risks for all the major diseases, including those that may not develop for decades. This technology raises many ethical questions. Testing often reveals information about disease genes carried by parents, siblings, and children of the one who was tested. What should be done with that information? How can we protect access to the results of genetic testing? Should we test for serious or fatal disorders for which there is no treatment?

The legal, moral, and ethical consequences of recombinant DNA technology and its use need to be considered carefully. Some of these issues are being addressed by the Human Genome Project through ELSI, the Ethical, Legal, and Social Implications program, which brings together individuals from biology, social sciences, history, law, ethics, and philosophy to explore issues and lay out public policy alternatives. In other areas, however, discussion, education, and policy formation lag far behind the technology. As citizens, it is our responsibility to become informed about these issues and to participate in formulating policy about the use of this technology.

Genetics in Practice

 Genetics in Practice case studies are critical thinking exercises that allow you to apply your new knowledge of human genetics to real-life problems. You can find these case studies and links to relevant websites at **http://biology.brookscole.com/cummings7**

CASE 1

Todd and Shelly Z. were referred for genetic counseling because of advanced maternal age (Shelley was over 40 years old) in their current pregnancy. While obtaining the family history, the counselor learned that during their first pregnancy, the couple had elected to have an amniocentesis for prenatal diagnosis of cytogenetic abnormalities because Shelly was 36 years old at the time. During that pregnancy, Shelly and Todd reported that they were also concerned that Todd's family history of cystic fibrosis (CF) increased their risk for having an affected child. Todd's only sister had CF, and she had severe respiratory complications.

The genetic counseling and testing was performed at an outside institution, and the couple had not brought copies of the report with them. They did state that they had completed studies to determine their CF carrier status and that Todd was found to be a CF carrier, but Shelly's results were negative. The couple was no longer concerned about their risk of having a child with CF based on these results. To support their belief, they had a healthy 5-year-old son who had a negative sweat test at the age of 4 months. The counselor explained the need to review the records and scheduled a follow-up appointment.

The test report from their first pregnancy only tested for the delta F508 mutation, the most common mutation in CF. The report confirmed that Shelly is not a carrier of this mutation and that Todd is. This report reduced Shelly's risk status from 1 in 25 to about 1 in 300. More information has been learned about the different mutations in the CF gene since the last time they received genetic counseling. The counselor conveyed the information about recent advances in CF testing to the couple, and Shelly decided to have her blood drawn for CF mutational analysis with an expanded panel of mutations. Her results showed that she is a carrier for the W1282X CF mutation. The family was given a 25% risk for CF for each of their pregnancies based on their combined molecular test results. They proceeded with the amniocentesis because of the risks associated with advanced maternal age and requested fetal DNA analysis for CF mutations. The fetal analysis was positive for both parental mutations, indicating that the fetus had a greater than 99% chance of being affected with CF.

1. How is it that the fetus has a greater than 99% chance of being affected with CF if each parent carries a different mutation? Is the fetus homozygous or heterozygous for these mutations?

2. If this child has CF, what are the chances that any future child will have this disease? Does the fact they have a healthy 5-year-old son affect the chances of having future children with CF?

CASE 2

Can DNA fingerprinting uniquely identify the source of a sample? Because any two human genomes differ at about three million sites, no two persons (except identical twins) have the same DNA sequence. Unique identification with DNA typing is therefore possible if enough sites of variation are examined. However, the DNA typing systems used today examine only a few sites of variation and have only limited resolution for measuring the variability at each site. There is a chance that two persons might have DNA patterns (i.e., genetic types) that match at the small number of sites examined. Nonetheless, even with today's technology, which uses three to five loci, a match between two DNA patterns can be considered strong evidence that the two samples came from the same source. How is DNA fingerprinting currently being used?

■ Paternity and maternity testing: Because a person inherits his or her STRs from his or her parents, STR patterns can be used to establish paternity and maternity. The patterns are so specific that a parental STR pattern can be reconstructed even if only the children's STR patterns are known (the more children produced, the more reliable the reconstruction). Parent-child STR pattern analysis has been used to solve standard father-identification cases and more complicated cases of confirming legal nationality and, in instances of adoption, biological parenthood.

■ Criminal identification and forensics: DNA isolated from blood, hair, skin cells, or other genetic evidence left at the scene of a crime can be compared, through STR patterns, with the DNA of a criminal suspect to determine guilt or innocence. STR patterns are also useful in establishing the identity of a homicide victim, either from DNA found as evidence or from the body itself. However, in cases of identical twins, where one twin has committed a crime, DNA evidence may not be helpful.

Personal identification: The notion of using DNA fingerprints as a sort of genetic bar code to identify individuals has been discussed, but this is not likely to happen any time in the foreseeable future. The technology required to isolate, keep on file, and analyze millions of very specific STR patterns is both expensive and impractical. Social security numbers, picture IDs, and other more mundane methods are much more likely to remain the ways to establish personal identification. However, the FBI and other police agencies and the military are enthusiastic proponents of DNA databanks. The practice has civil-liberties implications because police can go on "fishing expeditions" with DNA readouts.

1. A police chief in a large U.S. city has proposed that everyone who is arrested be DNA fingerprinted; this information would be stored in a national database, even if those arrested are later found to be innocent. The chief is actively seeking funding from his state legislature to begin such a program. Do you support such a program? Why or why not? Could there be abuses of this program?

2. Several companies advertise DNA fingerprinting services on the Internet for genealogy research. Given that most ancestors are dead and not available for such testing, how can this service be of value in genealogy? Are there privacy concerns about what the companies do with the results of such testing?

Summary

14.1 Biopharming: Making Medical Molecules in Animals and Plants

- Genetic engineering is used to manufacture proteins used in treating human diseases. This provides a constant supply that is uncontaminated by disease-causing agents. These proteins are made in bacteria, cell lines from higher organisms, animals, and even plants.

14.2 Genetically Modified Foods

- Gene transfer into crop plants has conferred resistance to herbicides, insect pests, and plant diseases. In addition, gene transfer is being used to increase the nutritional value of foods as a way of combating diseases such as vitamin A deficiency.

14.3 Transgenic Animals as Models of Human Diseases

- The transfer of disease-causing human genes creates transgenic organisms that are used to study the development of human diseases and the effects of drugs and other therapies as methods of treating these disorders.

14.4 Testing for Genetic Disorders

- Genetic testing is the search for individuals who have a particular genotype. Genetic screening tests general populations that may have a low frequency for a disorder. Carrier screening is the search for heterozygotes that may be at risk of producing a defective child. Prenatal screening can also detect chromosome abnormalities, such as Down syndrome, and birth defects, such as spina bifida, that may have a genetic component.

- Recombinant DNA–based prenatal testing can detect genetic disorders such as sickle cell anemia that cannot otherwise be detected before birth. Testing can also be done on embryos following *in vitro* fertilization and before implantation. Methods for visualizing the fetus can be used to check for visible genetic disorders and developmental problems.

14.5 DNA Microarrays in Genetic Testing

- Testing for a wide range of genetic disorders is possible using DNA chips, which can hold thousands of genes. This technology can be used to identify individuals who will develop late-onset genetic disorders such as polycystic kidney disease (PCKD) or Huntington disease, as well as those who are at risk for disorders such as diabetes.

14.6 DNA Profiles as Tools for Identification

- DNA profiles use variations in the length of short repetitive DNA sequences to identify individuals with a high degree of accuracy and reliability. This method is used in many areas including law enforcement, biohistory, conservation, and the study of human populations.

14.7 Social and Ethical Questions about Biotechnology

- Recombinant DNA technology and its many applications have developed faster than societal consensus, public policy, and laws governing its use. In addition, efforts to inform legislators, members of the legal and medical profession, and the public at large about the applications of this technology have often lagged behind its commercial use. An effort to educate and debate the risks and benefits of how we use genetic engineering will ensure a balanced approach to this technology.

Questions and Problems

HUMAN Genetics Now™ Preparing for an exam? Assess your understanding of this chapter's topics with a pre-test, a personalized learning plan, and a post-test at http://now.ilrn.com/cummings7

Biopharming: Making Medical Molecules in Animals and Plants

1. What advantages do edible vaccines made from transgenic plants have over traditional vaccines?

Genetically Modified Foods

2. The creation of transgenic crop plants using recombinant DNA methods involves the transfer of just one or a small number of genes to the plants, in contrast to classical breeding methods in which hundreds or even thousands of genes are transferred at once. Explain why this is true. If fewer genes are transferred during the creation of transgenic crops, why are they looked upon as potentially more dangerous?

3. What are "Bt crops"? What potential risks are associated with this technology?

4. "Golden" rice strains are transgenic crops that have been genetically engineered to produce elevated levels of beta-carotene. These strains were developed in order to help prevent:
   a. tuberculosis
   b. malaria
   c. vitamin A deficiency
   d. protein deficiency
   e. hypertension

Transgenic Animals as Models of Human Diseases

5. You are a neuroscientist working on a particular neurodegenerative disease, and discover that the disease is caused by the expression of a mutant gene in the brain. A colleague of yours suggests that you use this new information to develop a transgenic mouse model of the disease. How might you do this? How would you use the animal to study the disease or look for new treatments?

Testing for Genetic Disorders

6. What is the difference between genetic testing and genetic screening?

7. Cystic fibrosis is an autosomal disease that mainly affects the white population, and 1 in 20 whites are heterozygotes. Now that the gene has been mapped to chromosome 7, assume that restriction fragment length polymorphism (RFLP) markers are available to diagnose heterozygotes. Should a genetic screening program for cystic fibrosis be instituted? Should the federal government fund this? Should the program be voluntary or mandatory, and why?

8. The measurement of alpha-fetoprotein levels is used to diagnose neural tube defects. For every 1,000 such tests, approximately 50 positive cases will be detected. However, up to 20 (40%) of these cases may be false positives. In a false positive, the alpha-fetoprotein level is elevated, but the child has no neural tube defect. Your patient has undergone testing of the maternal blood for alpha-fetoprotein, and the results are positive. She wants to abort a defective child but not a normal one. What are your recommendations?

9. You are a governmental science policy advisor, and you learn about a new technique being developed that promises to accurately predict IQ based on a particular combination of genetic markers. You also learn that this technique may potentially be applied to preimplantation testing, so that parents would be able to not only select an embryo for implantation that is free of a genetic disorder but also select one that is likely to be relatively smart. What policy recommendations would you make concerning this technology? Do you think parents should have the right to choose any characteristic of their children, or should preimplantation be limited to ensuring that embryos are free of genetic disorders? Should guidelines be imposed to regulate this process, or should it even be banned?

DNA Microarrays in Genetic Testing

10. One of the major advantages of using DNA microarrays for genetic testing is that large numbers of genes can be screened for mutations at the same time. What features of microarrays allow this possibility?

11. You decide to use a DNA microarray-based assay to determine if a patient of yours carries a mutation in a particular disease-associated gene. To do this, you label DNA from the patient with a red probe, and label control DNA (which is known to not contain any mutations) with a green probe. The two labeled DNA samples are allowed to hybridize to a microarray that contains short segments of DNA corresponding to both the normal and various mutated versions of the gene. After washing off nonhybridized DNA, a laser scans the array and presents you with a pattern of colored dots. How would you know if a mutation is present in the patient's DNA? What would the pattern look like if no mutations were present?

DNA Profiles as Tools for Identification

12. STRs are
   a. used for DNA fingerprinting
   b. repeated sequences present in the human genome
   c. highly variable in copy number
   d. all of the above
   e. none of the above

13. A crime is committed and the only piece of evidence the police are able to gather is a small bloodstain. The forensic scientist at the crime lab is able to
   • extract DNA from the blood,
   • cut the DNA with a restriction enzyme,
   • separate the fragments by electrophoresis, and
   • transfer the DNA from a gel to a membrane and probe with radioactive DNA.
   Probe 1 is used to visualize the pattern of bands. The forensic scientist compares the band pattern in the evidence (E) with the patterns from the suspects (S1, S2). The first probe is removed, the membrane is hybridized using another probe (Probe 2), and the band patterns

are compared. This process is repeated for Probe 3 and Probe 4.
   a. Based on the results of this testing, can either of the suspects be excluded as the one who committed the crime?
   b. If so, which one? Why?
   c. Is the pattern from the evidence consistent with the band pattern of one of the suspects? Which one?

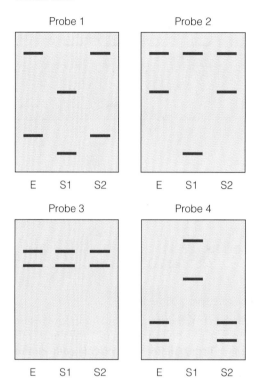

14. You are serving on a jury in a murder case in a large city. The prosecutor has just stated that DNA fingerprinting shows that the suspect must have committed the crime. He says that for cost reasons, two STR probes were used instead of four, but that the results are just as accurate. The first probe detected a locus that has a population frequency of 1 in 100 (1%), and the second probe has a population frequency of 1 in 500 (0.2%).
   a. What is the combined frequency of these alleles in the population?
   b. Does this point to the suspect as the perpetrator of the crime?
   c. Is there any other evidence you would like from the DNA lab?
   d. Do you think that most people on juries understand basic probabilities, or do you think that DNA evidence can be used to mislead jurors into reaching false conclusions?

15. DNA fingerprinting has been used in criminal identification, establishing bloodlines of purebred dogs, identifying endangered species, and studying extinct animals. Can you think of other uses for DNA fingerprinting?

16. A paternity test is conducted using PCR to analyze an RFLP that consistently produces a unique DNA fragment pattern from a single chromosome. Examining the results of the following Southern blot, which male(s) can be excluded as the father of the child? Which male(s) could be the father of the child?

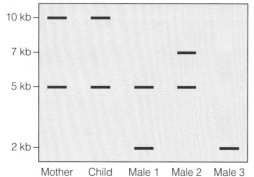

Social and Ethical Questions about Biotechnology

17. What are some of the potential dangers of recombinant DNA technology? How should these potential problems be investigated to determine whether they are in fact dangers?

18. Answer the following questions with a "yes" or a "no," and support each view with a coherent reason.
    a. Should very short children receive recombinant human growth hormone (HGH) to reach average height?
    b. Should genetically modified foods be labeled in grocery stores?
    c. Is it acceptable for physicians to treat children of normal height with HGH to make them taller if they and their parents want this?

Internet Activities

Internet Activities are critical thinking exercises using the resources of the World Wide Web to enhance the principles and issues covered in this chapter. For a full set of links and questions investigating the topics described below, log on to **http://biology.brookscole.com/cummings7**

1. *DNA Analysis.* This technique can be used to identify and compare human DNA samples on the basis of DNA fingerprints. Conclusions from these comparisons can be used in paternity cases, criminal trials, and many other applications. University of Arizona's *Biology Project: Human Biology* web page provides several problem sets and activities to enhance your understanding of DNA analysis. The two DNA Forensics problem sets allow you to try your hand at answering questions about DNA data and also provide tutorials to explain difficult concepts. The two Blackett Family DNA activities provide opportunities to work with DNA fingerprints and the newer STR (short tandem repeat) technique.

2. *Discussing Biotechnology and Genetically Engineered Products.* The *Union of Concerned Scientists* website provides information and a forum to discuss scientific aspects of biotechnology, as well as the marketing of genetically engineered products. At the website, click on the "Genetically Engineered Food" link to find the answers to frequently asked questions about biotechnology, as well as fact sheets on current biotechnology issues. Read the position statement on the "Genetically Engineered Food" web page. What are the positions—and potential biases—of the scientists in this organization? Do you share them, or do you disagree? And don't forget to check out the "Transgenic Cafe"!

How would you vote now?

DNA identification cards offered by several companies contain a photo, a thumbprint or retinal scan, personal information such as height, weight, eye color, and so on, and a CODIS DNA profile that makes personal identification a certainty. Police in Britain want to expand the national DNA database to include all citizens, not just criminals, to provide a means of personal identification. Advocates point out that such ID cards can protect children who are lost or kidnapped, help locate missing persons, prevent fraud, identify actual and potential terrorists, and be a deterrent to crime. Critics argue that DNA databases were set up to solve crimes, not to put everyone's profile in a database to deter crime, and that this technology can be used by governments to track the normal day-to-day activities of citizens instead of fighting crime. Now that you know more about the power of DNA identification, what do you think? Would you get a government DNA identification card if one was offered, and would you support making such cards mandatory? Visit the Human Heredity Companion website to find out more on the issue, then cast your vote online at **http://biology.brookscole.com /cummings7**

 For further reading and inquiry, log on to **InfoTrac College Edition,** your world-class online library, including articles from nearly 5,000 periodicals, at **http://infotrac.thomsonlearning.com**

15

The Human Genome Project and Genomics

THE Human Genome Project (HGP) took 13 years and cost $3 billion to complete. Although overall costs were about a dollar per nucleotide, most of this cost went into technology development, and by the time sequencing started, the real sequencing costs were about $0.05 per nucleotide. The sequencing method used in the HGP was developed in 1977, and efforts to develop newer, faster, and cheaper methods of DNA sequencing have received widespread attention with the completion of the HGP.

To jump-start efforts, the National Institutes of Health has initiated a grant program to develop technologies to lower the cost of human genome sequencing to $1,000. In addition, the J. Craig Venter Foundation has offered a $500,000 prize to the individual or organization that can reliably sequence a human genome for $1,000 or less.

At a cost of $1,000, genome sequencing would be within the reach of many individuals in the United States. At that price, it is likely that sequencing will be routinely offered at major medical centers and possibly even advertised directly to consumers over the Internet. Low cost and widespread availability of genome sequencing raise ethical and legal questions that are now being considered even as the race to develop the technology is accelerating.

Knowing you are at risk for developing heart disease, breast cancer, or diabetes may influence you to develop and deploy lifestyle strategies that can lower such risks. On the other side, once your genome has been sequenced, who owns and has access to such data? Should this information become part

Celera, Inc.

of your medical record, accessible by physicians, hospitals, employers, and insurance companies? If sequencing shows you will develop a devastating and fatal disease in middle age, should you inform your siblings, who may also be at risk for this disease? Should you have your children tested to see if they will develop this disorder as adults?

Many of these same issues have been considered as part of the HGP, and this chapter examines the methods, results, and impact of sequencing the human genome, including the ethical, legal, and social implications of genomics, the study of genomes.

How would you vote?

As a gift for your twenty-first birthday, your parents announce that they will pay to have your genome sequenced. This gift will allow you to have first-hand information about any genetic disorders you carry in the heterozygous condition and may pass on to your children, as well as defining any genetic risk factors for adult onset disorders. You can use this information to make reproductive decisions and to adjust your lifestyle early enough to maximally reduce your risk of developing adult onset disorders. The sequence will not be stored in any database and will be given to you on compact discs for you to have read and interpreted in the health care setting of your choice. Would you accept your parents' gift and have your genome sequenced? Visit the Human Heredity Companion website to find out more on the issue, then cast your vote online at **http://biology.brookscole.com/cummings7**

Keep in mind as you read

■ The Human Genome Project grew out of methods originally developed for basic research: recombinant DNA technology and DNA sequencing.

■ Genomics relies on interconnected databases and software to analyze sequenced genomes and to identify genes.

■ The human genome has a surprisingly small number of genes and produces a surprisingly large number of proteins using a number of different mechanisms.

■ Genomics is impacting basic research in biology and generating new methods of diagnosis and treatment of disease.

15.1 Genomic Sequencing Is an Extension of Genetic Mapping

If we want to know how many genetic disorders humans can have and how to develop treatments for these diseases, we need to map all of the genes in the human genome and find out what these genes do. Genetic mapping is therefore one of the basic activities of human genetics, as it is in other areas of genetics. There are several ways of mapping genes. The most basic method, developed in the 1930s, involved searching for genetic evidence that two genes are near each other on the same chromosome (called **linkage** in the language of geneticists). Genes on the same chromosome are said to show linkage because they tend to be inherited together.

In 1936, Julia Bell and J. B. S. Haldane discovered the first example of linkage in humans. Using pedigree analysis, they showed that the genes for hemophilia (OMIM 306700) and color blindness (OMIM 303800) are both on the X chromosome and are therefore linked.

■ **Linkage** A condition in which two or more genes do not show independent assortment. Rather, they tend to be inherited together. Such genes are located on the same chromosome. By measuring the degree of recombination between linked genes, the distance between them can be determined.

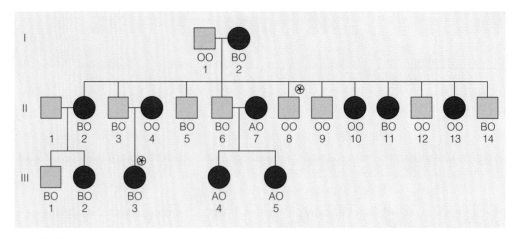

▲ FIGURE 15.1 Linkage between nail-patella syndrome and the ABO blood type locus. Darkly shaded symbols in this pedigree represent those who have nail-patella syndrome, an autosomal dominant trait. Genotypes for the ABO locus are shown below each symbol. Nail-patella syndrome and the *B* allele are present in I-2; they tend to be inherited together in this family, and they are identified as linked genes. Individuals marked with an asterisk (II-8 and III-3) inherited the nail-patella allele or the *B* allele alone. This separation of the two alleles occurred by recombination.

Although genes can be assigned directly to the X chromosome by their unique pattern of inheritance, it is more difficult to map genes to individual autosomes. This requires a large family that carries two genetic disorders with accurate pedigrees that span several generations. Because most genetic disorders are rare, it is difficult to find such families. To get around this problem, geneticists hoped to use one common trait such as blood types and find a genetic disorder that is inherited with a specific blood type. Because everyone has a blood type, it was thought that this would make the search for linkage easier. The search for linkage between blood type and a genetic disorder began in the 1930s, but the first case of linkage between a genetic trait and a blood type was not discovered until 1955, when researchers found linkage between the ABO blood type (the *I* locus; OMIM 110300) and an autosomal dominant condition called nail-patella syndrome (OMIM 161200). Nail-patella syndrome causes deformities in the nails and kneecaps.

In this pedigree (▶ Figure 15.1), type B blood (I^B) and nail-patella syndrome occur together in individual I-2. Examination of the pedigree through the next two generations shows the two traits are inherited together (linked). However, the pedigree shows that two individuals inherited only *one* of these two alleles, either the nail-patella allele or the B blood type allele, but not both (individuals II-8 and III-3). The separation of the two alleles is the result of crossing-over between the two genes. The rate of crossing-over between two linked genes (discussed in Chapter 2) can be used to construct a genetic map listing the order and distance between genes on a chromosome.

Recombination frequencies are used to make genetic maps.

A genetic map shows the order of genes on a chromosome, and the distance between any two genes. A genetic map is made in two steps: (1) finding linkage between two genes establishes they are on the same chromosome, (2) measuring how frequently crossing-over takes place between them establishes the distance between them (▶ Figure 15.2). The units of distance are expressed as a percentage of recombination, where 1 map unit is equal to a frequency of 1% recombination. (This unit is also known as a **centimorgan,** or **cM.**)

■ **Centimorgan (cM)** A unit of distance between genes on chromosomes. One centimorgan equals a value of 1% crossing-over between two genes.

▶ FIGURE 15.2 Crossing-over between homologous chromosomes during meiosis involves the exchange of chromosomal parts. In this case, crossing-over between the genes for blood type (alleles *B, O,* of gene *I*) and nail-patella syndrome (*N*) produces new combinations of alleles that show up in pedigrees. The frequency of crossing-over is proportional to the distance between the genes, allowing construction of a genetic map for this chromosomal region.

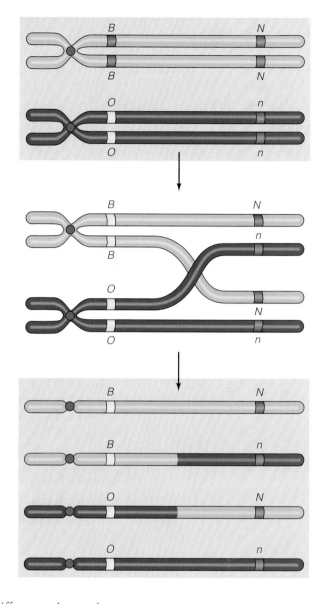

Let's use this method to calculate the distance between the gene for nail-patella syndrome and the gene for the ABO blood group by measuring how often they recombine by crossing-over. In the pedigree (▶ Figure 15.1), the nail-patella gene and the gene for the B blood type have separated in 2 of the 16 individuals (II-8 and III-3). From the frequency of recombination (2/16 = 0.125, or 12.5% recombination), the distance between the gene for the ABO locus and the gene controlling nail-patella syndrome can be calculated as 12.5 map units (▶ Figure 15.3).

For accuracy, either a much larger pedigree or many more pedigrees need to be examined to determine the extent of recombination between these two genes. When a large series of families is combined in an analysis, the map distance between the ABO locus and the nail-patella locus is about 10 map units.

It took about 20 years to find the first example of linkage between autosomal genes, and by 1969, only five cases of linkage had been discovered. It was apparent that the effort to map all human genes by linkage analysis using pedigrees was going nowhere.

Recombinant DNA technology radically changed gene-mapping efforts.

Beginning in 1980, geneticists began mapping cloned DNA sequences to specific human chromosomes. Most often, these sequences weren't genes, but were simply markers that detected differences in restriction enzyme cutting sites or differences in the number of repeated DNA sequences in a cluster. Once these markers were assigned to chromosomes, they could be used to follow the inheritance of these molecular markers and a genetic disorder in pedigrees in the same way that pedigrees were used to link ABO blood types and nail-patella syndrome. This method, called **positional cloning,** was used to map the genes for cystic fibrosis, neurofibromatosis, Huntington disease, and dozens of other genetic

■ **Positional cloning** A recombinant DNA-based method of mapping and cloning genes with no prior information about the gene product or its function.

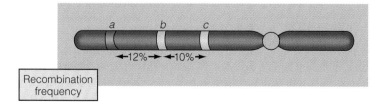

Recombination frequency

▲ FIGURE 15.3 Linked genes are carried on the same chromosome. Crossover frequencies are used to construct genetic maps, giving the order and distance between genes on a chromosome.

Table 15.1 Some of the Genes Identified by Positional Cloning

	OMIM		OMIM
Chromosome 4		Chromosome 17	
Huntington disease	143100	Breast cancer (*BRCA1*)	113705
		Neurofibromatosis (*NF1*)	162200
Chromosome 5			
Familial polyposis (APC)	175100	Chromosome 19	
		Myotonic dystrophy	160900
Chromosome 7			
Cystic fibrosis	219700	Chromosome 21	
		Amyotrophic lateral sclerosis	105400
Chromosome 11			
Wilms tumor	194070	X Chromosome	
Ataxia-telangiectasia	208900	Duchenne muscular dystrophy	310200
		Fragile-X syndrome	309550
Chromosome 13		Adrenoleukodystrophy	300100
Retinoblastoma	180200		
Chromosome 16			
Polycystic kidney disease	173900		

conditions (▶ Table 15.1) and to produce genetic maps of most human chromosomes (▶ Figure 15.4). Although positional cloning still identified one gene at a time, by the late 1980s, more than 3,500 genes and markers had been assigned to human chromosomes, putting geneticists closer to the goal of producing a high-resolution map of all human genes.

Keep in mind

■ The Human Genome Project grew out of methods originally developed for basic research: recombinant DNA technology and DNA sequencing.

15.2 Origins of the Human Genome Project

The development of recombinant DNA technology and its use in gene mapping was a springboard for discussions on how to map all the genes in the human genome (recall that a genome is the set of genes carried by an individual). After several years of planning, in 1990 the Human Genome Project (HGP) began in the United States. This program was coordinated by two agencies, the National Institutes of Health and the Department of Energy, under the direction of James Watson, one of the discoverers of DNA structure.

Instead of finding and mapping markers and disease genes one at a time, the HGP set out to sequence all the DNA in the human genome, identify and map the thousands of genes we carry to the 24 chromosomes, and establish the function of all these genes. To deal with the impact that genetic information would have on society, the HGP set up the ELSI program (Ethical, Legal, and Social Implications) to ensure that genetic information would be safeguarded and not used in discriminatory ways. The ethical implications of the HGP are discussed later in this chapter. Once established in the United States, the HGP quickly grew into an international effort involving programs in other countries, especially Japan, Germany, and England, all coordinated by the Human Genome Organization (HUGO).

Although not reflected in the name, the Human Genome Project also included projects to sequence the genomes of organisms used in experimental genetics (▶ Table 15.2). These include bacteria, yeast, a roundworm, the fruit fly, and the mouse.

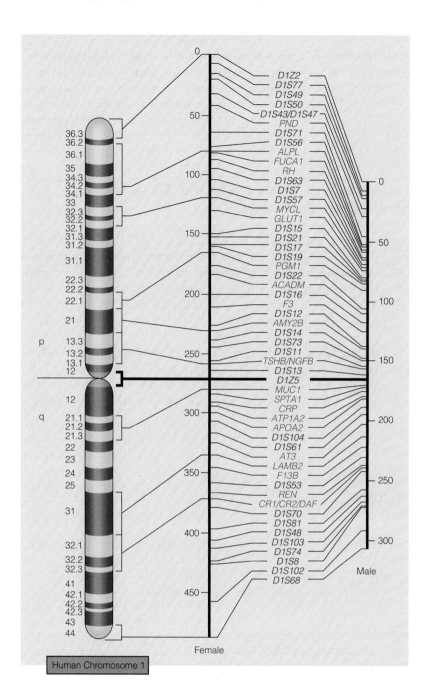

◀ FIGURE 15.4 Genetic map of human chromosome 1. At the left is a drawing of the chromosome. The two vertical lines at the right represent genetic maps derived from studies of recombination in males and females. Between the genetic maps are the order and location of 58 loci, some of which are genes (*red*) and others (*blue*) that are genetic markers detected using recombinant DNA techniques. The map in females is about 500 cM long, and in males it is just over 300 cM. This is a result of differences in the frequency of crossing-over in males and females. This map provides a framework for locating genes on the chromosome as part of the Human Genome Project.

Table 15.2 Organisms Included in the Human Genome Project

Organism	Genome Size Million base pairs (Mb)	Estimated Number of Genes
Escherichia coli (bacterium)	4.6	4,300
Saccharomyces cerevisiae (yeast)	12	6,000
Caenorhabditis elegans (roundworm)	97	19,000
Arabidopsis thaliana (plant)	120	20,000
Drosophila melanogaster (fruit fly)	165	13,300
Mus musculus (mouse)	3,000	30,000
Homo sapiens (human)	3,200	25,000–30,000

Table 15.3 Goals of Genomics

Create genetic and physical maps of genomes.

Find location of all genes in a genome and annotate each gene.

Compile lists of expressed genes and nonexpressed sequences.

Elucidate gene function and gene regulation.

Identify all proteins encoded by a genome and their functions.

Compare gene proteins between species.

Characterize DNA variations with and between genomes.

Implement and manage web-based databases.

The history and timelines for these projects are shown in ▶ Figure 15.5. In addition, genome-sequencing projects are underway for literally hundreds of other bacterial species (in addition to those in the Human Genome Project), many of which cause disease. Genome projects have been completed or are underway for dozens of higher organisms. To date, the genomes for over 125 species have been completely sequenced, and hundreds more are being sequenced.

15.3 Genome Projects Have Created New Scientific Fields

The human genome contains about 3.2 billion nucleotides of DNA. The sheer size of our genome required the development of new technologies, including automated methods of DNA sequencing; advances in software to collect, analyze, and store the information derived from genome sequencing; and the creation of web-based databases and research tools to access genome sequence information. The study of genomes by these methods is called **genomics.** The goals of genomics are outlined in ▶ Table 15.3.

Methods for sequencing DNA were developed in the 1970s. One method uses four separate chemical reactions using radioactive isotopes (one for each base, A, T, C, and G), and the fragments generated by these reactions are separated by size in four adjacent lanes in a gel (▶ Figure 15.6). The sequence can be read from the bottom

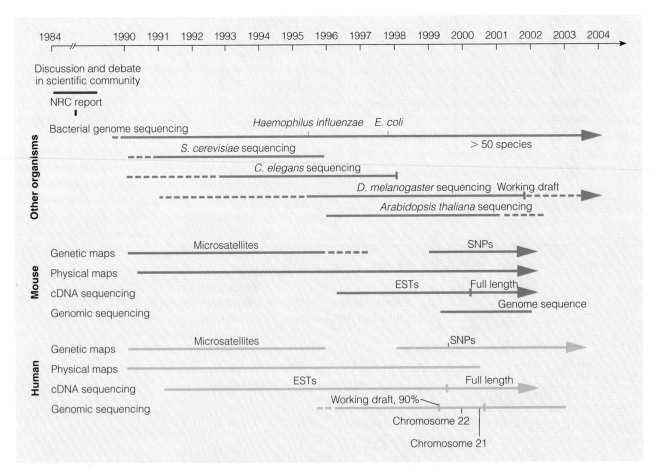

▲ **FIGURE 15.5** The history and timelines for the Human Genome Project and some organisms used in genetic research. (SNPs are single nucleotide polymorphisms; ESTs are expressed sequence tags. Both are DNA markers that are used in constructing maps.)

up. In ▶ Figure 15.6, the lowest fragment is in the A lane, so that is the first base in the sequence. Reading from this base upwards, the sequence is AATCGGCCGCT. This method is useful for sequencing small amounts of DNA but is somewhat labor-intensive, requiring several days to produce a few hundred bases of sequence. To sequence whole genomes, new technology was needed to speed up and automate the process. Genome projects refined DNA sequencing so that only a single reaction is needed. Each base is labeled with a different fluorescent dye, and the resulting sequence can be read from a single lane on a gel by a scanner (to review this process, see ▶ Active Figure 13.20). Machines called DNA sequencers were developed to automatically sequence several hundred thousand bases per day. The HGP used banks of hundreds of DNA sequencing machines linked to computers that controlled the machines and stored the resulting data (▶ Figure 15.7).

Once a genome sequence has been deciphered, it must be stored in a database so it can be assembled and analyzed to identify genes. In the case of the human genome, this is no trivial task. Our genome contains over 3 billion nucleotides. If our genome was organized into books, it would require 200 volumes, each containing 1,000 pages (▶ Figure 15.8). A new field called **bioinformatics** was developed to use computer hardware and software to store, analyze, and visualize genomic information. Bioinformatics is a growing field that encompasses more than just genome sequence data. It plays important roles in other fields including:

- **Comparative genomics**—comparing genomes of different species, looking for similarities and differences in genes and clues to the evolutionary history of a species.
- **Structural genomics**—deriving three-dimensional structures for proteins.
- **Pharmacogenomics**—analyzing genes and proteins to identify targets for therapeutic drugs.

In the following sections, we will see how genomics and bioinformatics are used to collect, analyze, and store information about the human genome.

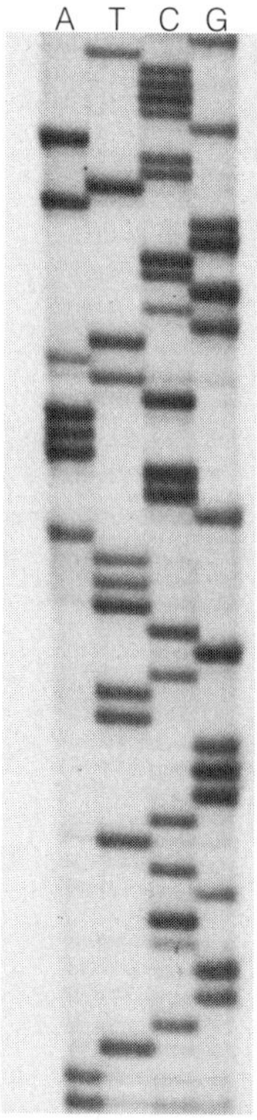

15.4 Genomics: Sequencing, Identifying, and Mapping Genes

Genomics begins by sequencing an organism's genome. Geneticists use two strategies for this task. The first, called the **clone-by-clone method,** was used by the publicly funded Human Genome Project (▶ Figure 15.9). In this method, genomic DNA is cut into pieces with restriction enzymes, and a collection of clones covering the entire genome (the collection is called a genomic library) is made. Next, the clones are

Courtesy the author

▲ **FIGURE 15.6** A DNA sequencing gel showing the separation of fragments in the four sequencing reactions. The sequence is read from the bottom, starting with the lowest band in any lane, then the next lowest, and so on. In this gel, the sequence begins with AATCGGCCGCT.

▨ **Genomics** The study of the organization, function, and evolution of genomes.

▨ **Bioinformatics** The use of computers and software to acquire, store, analyze, and visualize the information from genomics.

▨ **Comparative genomics** Compares genomes of different species in order to look for clues to the evolutionary history of genes or a species.

▨ **Structural genomics** Derives three-dimensional structures for proteins.

▨ **Pharmacogenomics** Analyzes genes and proteins to identify targets for therapeutic drugs.

▨ **Clone-by-clone method** A method of genome sequencing that begins with genetic and physical maps and sequences clones after they have been placed in order.

Volker Steger/SPL /Photo Researchers

◀ **FIGURE 15.7** Gene sequencing computers used in human genome research at Celera Corporation in Maryland.

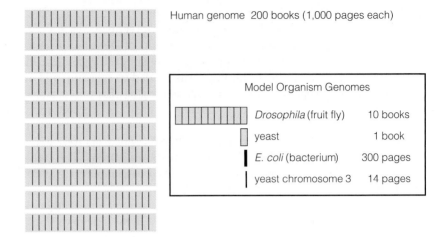

▶ **FIGURE 15.8** If the human genome was assembled in book form, it would require 200 volumes, each 1,000 pages in length. By comparison, the genome of the bacterium *E. coli* would occupy 300 pages.

■ **Shotgun cloning** A method of genome sequencing that selects clones at random from a genomic library, and after sequencing them, assembles the genome sequence using software analysis.

pieced together into genetic and physical maps that cover each chromosome (▶ Figure 15.9a). By using different restriction enzymes to create several libraries, overlapping fragments are created, helping organize the clones into genetic and physical maps. In a third step, each clone is sequenced, and software assembles the sequence from each library into the genomic sequence.

The second strategy for genome sequencing is called **shotgun cloning.** In this method (▶ Figure 15.9b), a genomic library is prepared, but no genetic or physical maps are created. Instead, by using restriction enzymes that cut at different sites, libraries containing overlapping fragments are created. Clones are selected at random from each library and sequenced. Software, called assembler programs, detects overlaps in the sequences obtained from each library and organizes the information into a genome sequence.

The government-sponsored genome project, headed by Francis Collins, used the clone-by-clone method in 1990 and by 1998 had assembled most of the maps and clones required, but the actual sequencing had not started on a large scale. In 1999, a privately funded human genome project, coordinated by Celera Corporation, led by J. Craig Venter, was announced. This project successfully used shotgun cloning to sequence several genomes, including that of the bacterium *Haemophilus influenzae,* the first organism to have its genome sequenced. Celera intended to use this method to sequence the human genome in 18 months instead of 10 years. This announcement set off an intense competition between the two projects to be first with the sequence. Venter's sequence was completed in early 2000, and Collins's project finished a few months later.

At a press conference at the White House in June 2000, the public and private human genome projects jointly announced they had completed a draft of the human genome sequence, and, in February 2001, they each published their results. The sequence published in 2001 was incomplete, and much work remained unfinished. There were gaps in the sequence to be closed and discrepancies to be resolved. In 2003, an additional sequence from the gene-coding portion of the human genome was published. However, neither project sequenced the 15% of the genome found in heterochromatic regions of chromosomes. These regions were omitted because it is very difficult to clone DNA sequences from heterochromatin, and it was thought that heterochromatin, which surrounds centromeres and is found at telomeres, did not contain any genes. New advances in cloning and sequencing technology are being used to sequence the DNA from heterochromatic regions, so it will be several years before the entire human genome is sequenced.

Scientists can analyze genomic information with bioinformatics.

In addition to the traditional laboratory notebooks, geneticists are now using large-scale databases to store, analyze, visualize, and share the results of their work on ge-

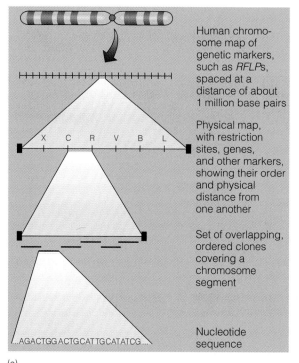

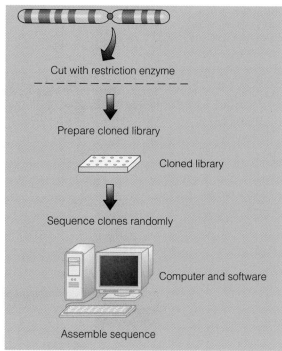

(a)

(b)

▲ **FIGURE 15.9** Cloning approaches. (a) The clone-by-clone method of genome sequencing. First, a genomic library is used to make genetic maps and physical maps of each chromosome. Clones are organized into overlapping sets, and each clone is sequenced. The genome sequence is assembled as the clones are sequenced. (b) In the shotgun method, a genomic library is prepared, and clones are chosen randomly for sequencing. Software programs assemble the sequence. The Human Genome Project sponsored by the National Institutes of Health and the Department of Energy used the clone-by-clone method of sequencing. The genome project organized by Celera Corporation used the shotgun method.

nomes. One of the main tasks in bioinformatics is managing the large amount of information generated by genome projects and storing this information in a way that it can be easily accessed.

One of the first steps after a genome is sequenced is organizing the sequence (called compiling the sequence) and checking it for accuracy. To ensure that the sequence is error free, genomes are sequenced more than once. The publicly funded genome project sequenced the human genome 12 times, and the privately funded project sequenced the genome approximately 35 times. Using bioinformatics, each team compiled and checked its sequence, and each team derived an accurate sequence, called a consensus sequence.

Keep in mind

■ Genomics relies on interconnected databases and software to analyze sequenced genomes and to identify genes.

Annotation is used to find where the genes are.

Once a genomic sequence has been compiled, the next task is to find all the genes that encode gene products (both proteins and RNA products). This process is called **annotation.** Looking at a DNA sequence (▶ Figure 15.10), it is not obvious whether it contains genes (remember, only 5% of human DNA encodes genes). If this sequence contains one or more genes, it isn't clear where they begin and end. How then, are genes identified in a DNA sequence?

Genes leave certain identifiable footprints that clue us in to their location in a

■ **Annotation** The analysis of genomic nucleotide sequence data to identify the protein-coding genes, the non-protein-coding genes, their regulatory sequences, and their function(s).

► FIGURE 15.10 DNA sequence recovered from a database. Analysis of the sequence using gene-searching software shows several ORFs (exons) bordered by the splice junctions between introns and exons (blue). A sequence just before the coding region marks the site at which transcription begins (green). Analysis of this sequence shows it encodes the human beta-globin gene. This protein is part of hemoglobin, the oxygen-carrying protein found in red blood cells.

```
gagccacacc  ctagggttgg  ccaatctact  cccaggagca  gggagggcag  gagccagggc
tgggcataaa  agtcagggca  gagccatcta  ttgcttacat  ttgcttctga  cacaactgtg
ttcactagca  acctcaaaca  gacaccatgg  tgcacctgac  tcctgaggag  aagtctgccg
ttactgccct  gtggggcaag  gtgaacgtgg  atgaagttgg  tggtgaggcc  ctgggcaggt
tggtatcaag  gttacaagac  aggtttaagg  agaccaatag  aaactgggca  tgtggagaca
gagaagactc  ttgggtttct  gataggcact  gactctctct  gcctattggt  ctattttccc
acccttaggc  tgctggtggt  ctacccttgg  acccagaggt  tctttgagtc  ctttgggggat
ctgtccactc  ctgatgctgt  tatgggcaac  cctaaggtga  aggctcatgg  caagaaagtg
ctcggtgcct  ttagtgatgg  cctggctcac  ctggacaacc  tcaagggcac  ctttgccaca
ctgagtgagc  tgcactgtga  caagctgcac  gtggatcctg  agaacttcag  ggtgagtcta
tgggacccct  gatgttttct  ttccccttct  tttctatggt  taagttcatg  tcataggaag
gggagaagta  acagggtaca  gtttagaatg  ggaaacagac  gaatgattgc  atcagtgtgg
aagtctcagg  atcgttttag  tttcttttat  ttgctgttca  taacaattgt  tttcttttgt
ttaattcttg  ctttcttttt  ttttcttctc  cgcaattttt  actattatac  ttaatgcctt
aacattgtgt  ataacaaaag  gaaatatctc  tgagatacat  taagtaactt  aaaaaaaaac
tttacacagt  ctgcctagta  cattactatt  tggaatatat  gtgtgcttat  ttgcatattc
ataatctccc  tactttattt  tctttttattt  ttaattgata  cataatcatt  atacatattt
atgggtaaa  gtgtaatgtt  ttaatatgtg  tacacatatt  gaccaaatca  gggtaatttt
gcatttgtaa  ttttaaaaaa  tgctttcttc  ttttaatata  cttttttgtt  tatcttattt
ctaatacttt  ccctaatctc  tttctttcag  ggcaataatg  atacaatgta  tcatgcctct
ttgcaccatt  ctaaagaata  acagtgataa  tttctgggtt  aaggcaatag  caatatttct
gcatataaat  atttctgcat  ataaattgta  actgatgtaa  gaggtttcat  attgctaata
gcagctacaa  tccagctacc  attctgcttt  tattttatgg  ttgggataag  gctggattat
tctgagtcca  agctaggccc  ttttgctaat  catgttcata  cctcttatct  tcctcccaca
gctcctgggc  aacgtgctgg  tctgtgtgct  ggcccatcac  tttggcaaag  aattcaccccc
accagtgcag  gctgcctatc  agaaagtggt  ggctggtgtg  gctaatgccc  tggcccacaa
gtatcactaa  gctcgctttc  ttgctgtcca  atttctatta  aaggttcctt  tgttccctaa
gtccaactac  taaactgggg  gatattatga  agggccttga  gcatctggat  tctgcctaat
aaaaaacatt  tattttcatt  gcaatgatgt  atttaaatta  tttctgaata  ttttactaaa
```

DNA sequence. If a DNA sequence encodes a protein, its nucleotide sequence is an **open reading frame**, or **ORF**, that encodes amino acids. Control regions at the beginning of genes are marked by identifiable sequences (CAAT or CCAAT). Splice sites between exons and introns have a predictable sequence, as do the sites at the ends of genes where a polyA tail is added. Computer software scans sequences, searching for ORFs and other features of genes. Analysis of the sequence in ► Figure 15.10 shows a control region and three ORFs (exons) of a gene. The two unshaded regions between the exons represent introns, regions removed as mRNA is processed. Using this sequence to search databases containing sequence information from other species, we discover that this is the sequence of the human beta-globin gene. Annotation of the human genome is an ongoing process; when it is completed, we will have a final count of how many genes are in our genome.

Geneticists work to discover gene functions and gene products.

After this gene has been identified by annotation, its amino acid sequence is derived (► Figure 15.11) and compared with sequences already in protein databases. This sequence matches that for the human beta-globin, confirming the identity of the DNA sequence. A match to a known protein provides information about the function of the gene product encoded by the gene (see Spotlight on Our Genetic Relative). So far, functions have been assigned to about 60% of the genes identified in the human genome (► Figure 15.12). Geneticists are working to identify functions for the rest of the genes.

■ **Open reading frame (ORF)**   The codons in a gene that encode the amino acids of the gene product.

► FIGURE 15.11 The amino acid sequence derived from the DNA coding sequences in Figure 15.10. The amino acids are represented by one-letter symbols. The sequence is 146 amino acids long, and protein databases confirm this is the sequence of human beta globin.

```
         10          20          30          40          50          60
          |           |           |           |           |           |
VHLTPEEKSA VTALWGKVNV DEVGGEALGR LLVVYPWTQR FFESFGDLST PDAVMGNPKV

         70          80          90         100         110         120
          |           |           |           |           |           |
KAHGKKVLGA FSDGLAHLDN LKGTFATLSE LHCDKLHVDP ENFRLLGNVL VCVLAHHFGK

        130         140
          |           |
EFTPPVQAAY QKVVAGVANA LAHKYH
```

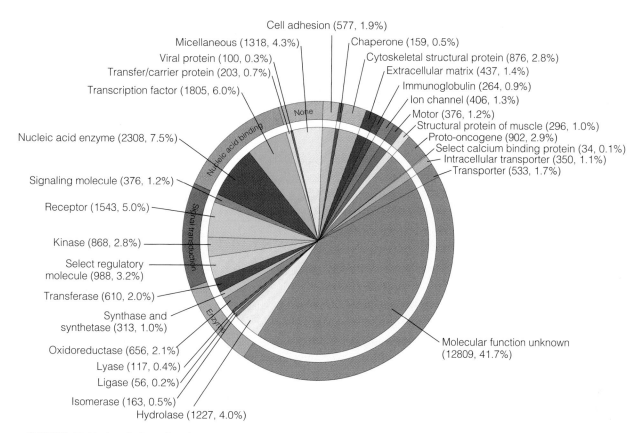

Cell adhesion (577, 1.9%)
Chaperone (159, 0.5%)
Micellaneous (1318, 4.3%)
Cytoskeletal structural protein (876, 2.8%)
Viral protein (100, 0.3%)
Extracellular matrix (437, 1.4%)
Transfer/carrier protein (203, 0.7%)
Immunoglobulin (264, 0.9%)
Transcription factor (1805, 6.0%)
Ion channel (406, 1.3%)
Motor (376, 1.2%)
Structural protein of muscle (296, 1.0%)
Nucleic acid enzyme (2308, 7.5%)
Proto-oncogene (902, 2.9%)
Select calcium binding protein (34, 0.1%)
Intracellular transporter (350, 1.1%)
Signaling molecule (376, 1.2%)
Transporter (533, 1.7%)
Receptor (1543, 5.0%)
Kinase (868, 2.8%)
Select regulatory molecule (988, 3.2%)
Transferase (610, 2.0%)
Synthase and synthetase (313, 1.0%)
Molecular function unknown (12809, 41.7%)
Oxidoreductase (656, 2.1%)
Lyase (117, 0.4%)
Ligase (56, 0.2%)
Isomerase (163, 0.5%)
Hydrolase (1227, 4.0%)

▲ FIGURE 15.12 A preliminary functional assignment for over 26,000 genes in the human genome. Just over 12,000 of these genes (41%) have no known function and represent the largest class of genes identified to date, emphasizing the work that needs to be completed before we fully understand our genome. The other classes of genes were assigned because they were previously known from other experiments or are similar to proteins of known function from other organisms. Among the most common genes are those involved in nucleic acid (DNA and RNA) metabolism (7.5% of all the identified genes).

15.5 What Have We Learned So Far about the Human Genome?

Several general features of the human genome have been identified in analyzing the sequence information.

- Although the genome is large and contains over 3 billion nucleotides of DNA, only approximately 5% of this DNA encodes genetic information. The rest does not code for anything and is spacer DNA. About half of our DNA is composed of non-gene sequences that are repeated thousands of times.
- Genes are not equally distributed along chromosomes. Gene-rich clusters on all 24 chromosomes are separated from each other by long stretches of gene-poor regions. The gene-poor regions correspond to the banded regions of stained chromosomes (see Chapter 6 for a discussion of banding and karyotypes).
- Not all the genes in the human genome have been identified, but humans have 25,000 to 30,000 genes, far fewer than the predicted number of 80,000 to 100,000.
- There are many more different proteins in the human body than there are genes. This discrepancy is explained by the fact that the mRNAs from many genes are processed in several different ways so that 30,000 to 40,000 genes can produce up to 300,000 different proteins.

Table 15.4   Comparison of Selected Genomes

Organism	Approximate Size of Genome (Date Completed)	Number of Genes	Approximate Percentage of Genes Shared with Humans	Web Access to Genome Databases
Bacterium (*Escherichia coli*)	4.6 million bp (1997)	4,300	Not determined	http://www.genome.wisc.edu/
Fruit fly (*Drosophila melanogaster*)	165 million bp (2000)	~13,300	50%	http://www.fruitfly.org/sequence/index.html http://Flybase.bio.indiana.edu/
Humans (*Homo sapiens*)	3,200 million bp (February 2001)	25,000–30,000	100%	http://www.ornl.gov/hgmis/
Mouse (*Mus musculus*)	3,000 million bp (2001)	~30,000	~90%	http://www.informatics.jax.org/
Plant (*Arabidopsis thaliana*)	120 million bp (2000)	~20,000	Not determined	http://www.arabidopsis.org/
Roundworm (*Caenorhabditis elegans*)	97 million bp (1998)	19,000	40%	http://www.genome.wustl.edu/projects/celegens
Yeast (*Saccharomyces cerevisiae*)	12 million bp (1996)	~6,000	31%	http://www.yeastgenomes.org

Source: Howard Hughes Medical Institute (2001), *The Genes We Share with Yeast, Flies, Worms, and Mice: New Clues to Human Health and Disease.*

■ The human genome is very similar to those of other higher organisms. We share about half our genes with the fruit fly *Drosophila* and more than 90% of our genes are shared with mice (▶ Table 15.4).

Spotlight on...

Our Genetic Relative

Analysis of the mouse genome sequence has shown it is very similar to the human genome. More than 90% of the protein-coding genes identified in the mouse genome have a match in the human genome. Although last linked by a common ancestor some 65 million years ago, this close genetic relationship means that many human genetic disorders can be studied using mice.

Keep in mind

■ The human genome has a surprisingly small number of genes and produces a surprisingly large number of proteins using a number of different mechanisms.

Findings from genome projects are changing our concepts of several genetic processes, including mutation and gene function. A new mechanism of mutation, trinucleotide expansion (discussed in Chapter 11), has been identified. This mechanism is important in disorders of the nervous system, and understanding why this is so will play an important role in treating these disorders. Another unexpected finding is that mutations in a single gene can give rise to different genetic disorders, depending on how the gene is affected. For example, the *RET* gene (OMIM 164761) encodes a cell-surface receptor protein that transfers signals across the plasma membrane. Depending on the type and location of mutations within this gene, four distinct genetic disorders can result: two types of cancers called multiple endocrine neoplasias (OMIM 171400 and OMIM 162300), another cancer called familial medullary thyroid carcinoma (OMIM 155240), and Hirschsprung disease (OMIM 142623), a disorder in which parts of the large intestine are unconnected to the nervous system.

Another significant discovery is that some mutations in DNA repair genes can destabilize distant regions of the genome and make them susceptible to more mutations, often resulting in cancer (these mutations were discussed in Chapter 12). These and similar discoveries have already had an impact on the diagnosis, treatment, and genetic counseling for several groups of genetic disorders.

15.6 Using Genomics and Bioinformatics to Study a Human Genetic Disorder

In studying any genetic disorder, several important questions must be answered:

- Where is the gene located?
- What is the normal function of the protein encoded by this gene?
- How does the mutant gene or protein produce the disease phenotype?

For some genes, like the CF (cystic fibrosis) gene, these questions were easy to answer. First, the CF gene was isolated by positional cloning, and in the process, was mapped to a region on the long arm of chromosome 7 (▶ Figure 15.13). Second, soon after the gene was isolated and cloned, it was sequenced. The nucleotide sequence was converted into an amino acid sequence, which was compared with protein sequences already in databases. Searching protein databases revealed that the CF protein has an amino acid sequence that closely resembles the sequence of proteins that reside in the cell's membrane and which control the flow of ions into and out of the cell. Because CF patients have problems with chloride ion flow, the protein function in normal individuals and the problem with chloride ions in CF individuals could be pinpointed.

In other cases, these questions are not so easy to answer. Suppose a gene for a genetic disorder is cloned and sequenced, but the amino acid sequence derived from the nucleotide sequence does not match the sequence of any known protein in any database. Since no similar protein is known, how can scientists understand what the normal protein does, and how a mutant form of the protein causes disease? This is not a hypothetical question; more than half the genes identified by the Human Genome Project have no known function. Let's look at how this problem was solved in the case of one genetic disorder.

Friedreich ataxia (FRDA; OMIM 229300) is a progressive and fatal neurodegenerative disorder inherited as an autosomal recessive trait. FRDA occurs in approximately 1 in 50,000 individuals and is a disease of young people. Symptoms appear between puberty and the age of 25 and include the progressive loss of muscle coordination (ataxia) and the formation of skeletal deformities. There is a gradual loss of cells in the nervous system and brain, and early death results.

Using a strategy of positional cloning, geneticists mapped the gene to chromosome 9 and then isolated, cloned, and sequenced the FRDA gene. Database scans with the amino acid sequence quickly reached a dead end. The only matches to the FRDA protein, named frataxin, were to proteins of unknown function in yeast and a nematode.

To match frataxin with proteins already in databases, researchers used short stretches of the frataxin amino acid sequence and searched protein databases hoping to find matches to proteins that carry these shorter sequences. One search matched part of the frataxin protein with a protein present in several bacterial species, all of which were related to purple bacteria. As it turns out, purple bacteria are the closest living relatives to ancient bacterial species that evolved into mitochondria, so it seemed possible that frataxin might be found in the cell's mitochondria and play some role in mitochondrial energy production.

Using these clues, another research group found that frataxin *is* located in mitochondria, and they worked out the three-dimensional structure of the protein (▶ Figure 15.14). However, because the protein has an unusual three-dimensional structure and because its amino acid sequence is different than any

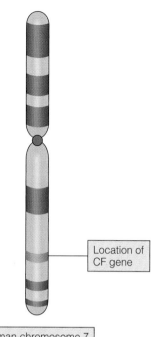

Human chromosome 7

▲ FIGURE 15.13 The gene for cystic fibrosis was mapped to the long arm of chromosome 7 by its position relative to known DNA markers.

■ **Friedreich ataxia** A progressive and fatal neurodegenerative disorder inherited as an autosomal recessive trait with symptoms appearing between puberty and the age of 25.

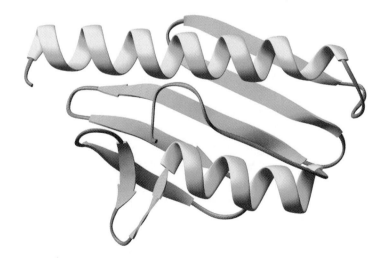

▲ FIGURE 15.14 A three-dimensional model showing the molecular structure of the human frataxin protein. This protein is necessary for normal mitochondrial function. A decreased production or function of this protein causes the autosomal recessive disorder Friedreich ataxia. This disorder is one of the first whose molecular basis was uncovered using analysis of genomic information.

protein domains with a known function, the function of frataxin is still unknown. Work with yeast and mouse models of FRDA have shown that frataxin is somehow involved in mitochondrial function, and may play a role in iron metabolism.

Analysis of frataxin genes isolated from several hundred people with FRDA has shown that the mutant gene has a large number of GAA triplet repeats near the beginning of the gene, which reduces expression by stopping transcription.

With this information in hand, efforts are now focused on developing therapies to treat FRDA. This example illustrates the power of genomics and bioinformatics to help unravel the mysteries of human genetic disorders.

15.7 Proteomics Is an Extension of Genomics

Now that we are identifying all the genes in our genome using genomics, the next step is to understand their function. As genes are identified, they are classified into functional groups (see the listing in ▶ Figure 15.12, for example). However, understanding gene function involves more than just identifying the protein it encodes. Proteins work in a cellular environment with many other proteins, and protein-protein interactions, modifications, and chemical linkage of proteins to other molecules such as lipids or carbohydrates are important parts of protein function. **Proteomics** is the study of all the proteins present in a cell at a given time.

Gene expression profiles are gathered by separating all the proteins in a cell type by their mass and electric charge (some amino acids carry an electrical charge) (▶ Figure 15.15). Although such profiles are very complex, they can play important roles in disease diagnosis and treatment. For example, some forms of breast cancer have a protein profile that indicates they will spread rapidly and have a high fatality rate, while other forms have profiles associated with slow growth and high survival rates. With this molecular information in hand, specific therapies can be designed for each form of this cancer.

Proteomics is a rapidly developing field that has several crucial roles:

- Understanding gene function and its changing role in development and aging.
- Identifying proteins that are markers (called biomarkers) for diseases. These proteins can be used in developing diagnostic tests.
- Finding proteins that are targets for the development of drugs to treat diseases and genetic disorders.

Several recent breakthroughs in proteomic techniques coupled with new developments in information science make this new field an essential partner with genomics in medicine.

15.8 Ethical Concerns about Human Genomics

The original planners of the HGP realized that the information generated by a genome project would have many uses, and thus they set up the ELSI program (Ethical, Legal, and Social Implications) as part of the genome project. As the project progressed, scientists and the public became concerned about the lack of coordination between the potential uses of genomic information and the development of policy and laws safeguarding genetic information of all kinds.

ELSI has focused on the areas listed in ▶ Table 15.5. The ELSI program uses research grants, meetings, symposia, workshops, and public forums to discuss and debate these issues and to formulate policy options and help frame legislation to ensure that members of the public are protected against the misuse of genetic information. Some protections are in place, but many gaps remain.

■ **Proteomics** The study of the expressed proteins present in a cell at a given time under a given set of circumstances.

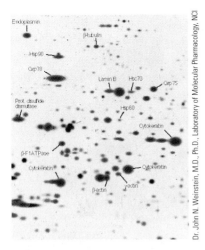

▲ **FIGURE 15.15** The proteins expressed in a cell are separated by size and electric charge and displayed on a gel. The spots on this gel represent the proteins in a gene expression profile.

Dr. John N. Weinstein, M.D., Ph.D., Laboratory of Molecular Pharmacology, NCI

Table 15.5 New Developments Needed In Genomics

Biology Research	Health Care	Society
1. Identify the structural and functional components of the human genome.	1. Identify genes that contribute to disease and determine how they interact with environmental factors.	1. Establish policies that effectively use and protect personal genetic information.
2. Find how gene and protein networks contribute to the phenotype.	2. Look for genes that contribute to health and resistance to disease.	2. Identify the genetic components of ethnicity and consider the consequences of this knowledge.
3. Measure how much variation is present in human DNA.	3. Use genomics to predict disease susceptibility and drug responses.	3. Assess the impact of learning genomic contributions to behavior.
4. Understand how evolution has shaped genomes.	4. Develop new drugs based on new understanding of gene action.	4. Define the ethical boundaries for the use of genomics and genetic information.
5. Provide free and widespread access to genetic information and databases.	5. Investigate how genetic information is given to patients and how that information is used.	
	6. Explore how to use genomics to improve health care for all.	

Even as ELSI works to resolve these issues, new areas of concern are appearing. Advertising campaigns on television and in newspapers and magazines are marketing genetic tests directly to consumers. Tests for breast cancer, cystic fibrosis, hemochromatosis, and other genetic disorders are offered. While this advertising raises public awareness and has increased requests for testing, research has shown that the ads do not accurately convey the tests' ability to detect or predict a genetic disorder, nor do these ads recommend that these tests be done in consultation with a health care provider or a genetic counselor (see Genetics in Society: Who Owns Your Genome?). Internet ads offer consumers face creams or nutritional supplements especially formulated for their genetic background. These developments indicate the need for policy and regulatory oversight of the marketing of genetic products and services.

15.9 Looking Beyond the Genome Project: What the Future Holds

In 2003, Francis Collins and others involved in the Human Genome Project published a paper that described the impact of genomics on science and society and their vision for the future of genomics research. This paper was the product of more than two years of discussion and debate involving hundreds of scientists and nonscientists. The paper points out the challenges involved in converting genome information into improved health and is organized around three major themes: research in biology, health care, and soci-

▲ FIGURE 15.16 The Human Genome Project forms the foundation for future developments in biology, health care, and social issues. Cutting across these areas are six fields that require new initiatives and new developments to fulfill the promise of genomics.

ety. The themes or areas are depicted as floors in a building, all resting on the foundation of the human genome project (▶ Figure 15.16).

The themes or "floors" are interconnected, and developments in one area impact the others. For example, analysis of the human genome sequence can identify a disease-causing gene, such as Freidreich ataxia (FRDA). Using this knowledge, physicians can now diagnose this heritable disorder. However, the ability to diagnose a fatal disease for which there is no treatment raises ethical and social questions that need to be considered and resolved.

For each theme, scientists have posed several challenges. These challenges represent goals for the research community and are listed in ▶ Table 15.5. To help achieve these goals, six areas cutting across all three themes need to be developed:

1. Resources, such as genome sequences and libraries of cloned DNA sequences;
2. Technology, including new sequencing methods and ways of monitoring gene expression and linking it to disease;
3. Software for computational biology to reveal protein structure, protein-protein interactions in disease, and to evaluate environmental factors on health;

Genetics in Society

Who Owns Your Genome?

John Moore, an engineer working on the Alaska oil pipeline, was diagnosed with a rare and fatal form of cancer known as hairy cell leukemia. This disease causes overproduction of one type of white blood cell known as a T-lymphocyte. These cells secrete a growth factor that activates other blood cells that kill cancer cells. Moore went to the UCLA Medical Center for treatment and was examined by Dr. Golde, who recommended that Moore's spleen be removed in an attempt to slow down or stop the cancer. For the next 8 years, John Moore returned to UCLA for checkups. Unknown to Moore, Dr. Golde and his research assistant applied for and received a patent on a cell line and products of the cell line derived from Moore's spleen. The cell line, named Mo, produced a protein that stimulates the growth of two types of blood cells that are important in identifying and killing cancer cells. Arrangements were made with Genetics Institute, a small start-up company, and then Sandoz Pharmaceuticals to develop the cell line and produce the growth-stimulating protein. Moore filed suit to claim ownership of his cells and asked for a share of the profits derived from the sale of the cells or products from the cells. Eventually, this case went through three courts, and in July 1990, 11 years after the case began, the California Supreme Court ruled that patients such as John Moore do not have property rights over any cells or tissues removed from their bodies that are later used to develop drugs or other commercial products.

This case was the first in the nation to establish a legal precedent over the commercial development and use of human tissue. The National Organ Transplant Act of 1984 prevents the sale of human organs. Current laws allow the sale of human tissues and cells but do not define ownership interests of donors. Questions originally raised in the Moore case remain largely unresolved in laws and public policy.

These questions are now being raised in many other cases as well. Who owns fetal and adult stem cell lines established from donors, and who has ownership and a commercial interest in diagnostic tests developed through cell and tissue donations by affected individuals? Who benefits from new genetic technologies based on molecules, cells, or tissues contributed by patients? Are these benefits, financial, medical, and ethical, being fairly distributed? What can be done to ensure that risks and benefits are distributed in an equitable manner?

Gaps between technology, laws, and public policy developed with the advent of recombinant DNA technology in the 1970s, and in the intervening decades, these gaps have not been closed. These controversies are likely to continue as new developments in technology continue to outpace social consensus about their use.

4. Training of scientists, physicians, and scholars in interdisciplinary skills to tie together biology, computer science, and the social sciences;
5. Ethical, legal, and social implications, including the protection of human subjects and genomic information; and
6. Education of health professionals and the public about genomics and the development of reliable resources that everyone seeking information about genomics can consult.

Keep in mind

■ Genomics is impacting basic research in biology and generating new methods of diagnosis and treatment of disease.

The shape of genomics in the coming years and its impact on our everyday lives is visible in the ideas outlined by these scientists. Genomics already presents many opportunities and challenges, and these will grow larger and more numerous in the near future. We can already choose genetic testing for ourselves, our children, and unborn children. We can take drugs and medication produced by gene transfer. Soon we will be able to have our genomes scanned to provide an inventory of all the genetic disorders we carry but will not develop, as well as any genetic diseases we may develop later in life or be susceptible to. This information can be used to help us decide whether to test embryos and fetuses to ensure they are free of genetic disorders and even whether to have children. Education, discussion, and access to information are vital parts of genomics.

Genetics in Practice

 Genetics in Practice case studies are critical thinking exercises that allow you to apply your new knowledge of human genetics to real-life problems. You can find these case studies and links to relevant websites at **http://biology.brookscole.com /cummings7**

CASE 1

James sees an advertisement in a magazine for an at-home genetic test that promises to deliver personalized nutritional advice based on an individual's genetic profile. The company can test for genetic variations, the advertisement states, that predispose individuals to developing health conditions, such as heart disease and bone loss, or that affect how they metabolize certain foods. If such variations are detected, the company can provide specific nutritional advice that would help counteract their effects. Always keen to take any steps available to ensure the best possible health for their family, James and his wife, Sally, decide that they should both be tested, as should their 11-year-old-daughter, Patty. They order three kits and wait for them to arrive.

Once the kits arrive, the family members use cotton swabs to take cell samples from each of their cheeks and place the swabs in three individually labeled envelopes.

They mail the envelopes back to the company, along with questionnaires regarding their diets. Four weeks later, they receive three individual reports detailing the results of the tests and providing extensive guidelines about what foods they should eat. Among the results is the finding that James has a particular allele in a gene that may make him vulnerable to the presence of free radicals in his cells. The report suggests that he increase his intake of antioxidants, such as vitamins C and E, and highlights a number of foods that are rich in these vitamins. The tests also show that Sally has several genetic variations that indicate that she may be at risk for elevated bone loss. The report recommends that she try to minimize this possibility by increasing her intake of calcium and vitamin D and lists a number of foods that she could emphasize in her diet. Finally, the report shows that Patty has a genetic variation that may mean that she has a lowered ability to metabolize saturated fats, putting her at risk for developing heart disease. The report points to ways in which she can lower her

(continued)

intake of saturated fats and lists various types of foods that would be beneficial for her.

A number of companies now offer genetic testing services, promising to deliver personalized nutritional or other advice based on people's genetic profiles. Generally, these tests fall into two different categories, with individual companies offering unique combinations of the two types. The first type of test detects alleles of known genes that encode proteins playing an established role in, for example, counteracting free radicals in cells or in building up bone. In such cases, it is easy to see why individuals carrying alleles that may encode proteins with lower levels of activity may be more vulnerable to free radicals or more susceptible to bone loss.

A second type of test examines genetic variations that may have no clear biological significance (that is, they may not occur within a gene, or they may not have a detectable effect on gene activity), but which have been shown to have a statistically significant correlation with a disease or a particular physiological condition. For example, a variation may be frequently detected in individuals with heart disease, even though the reason for the correlation between the variation and the disease may be entirely mysterious.

1. Of the two types of genetic tests described in the example, which do you think is more reliable?

2. Do you think that companies should be allowed to market such tests directly to the public, or do you believe that only a physician should be able to order them?

3. What kind of regulations, if any, should be in place to ensure that the results of these tests are not abused?

4. Do you think parents should be able to order such a test for their children? What if the test indicates that a child is at risk for a disease for which there is no known cure?

CASE 2

Both the Human Genome Project (HGP) and Celera's genome sequencing project were faced with an interesting dilemma: Whose DNA should be sequenced? Whose genome should serve as a "representative" human genome? The HGP and Celera answered the question in different ways. The HGP started with a collection of DNA samples from a large number of individuals and then randomly selected a small number of samples that were used for sequencing. Celera, on the other hand, used a mixture of DNA samples from several individuals of diverse ethnicity (and including, as later revealed, the DNA of Celera's founder, Craig Venter). In both cases, because only a

handful of genomes was used in the sequencing, the final sequences only represented a small fraction of the total variation present in humanity.

In 1991, as the genome project got underway, a group of anthropologists and geneticists proposed carrying out an ambitious project to address this issue, namely, to study the genetic variation of all human populations. The Human Genome Diversity Project, as it came to be called, proposed to obtain blood samples from a wide variety of peoples throughout the world and to sequence their genomes in conjunction with the official human genome project. According to the organizers of the project, the information obtained in the project would shed light on the history of different human populations, would illuminate the biological relationships between populations, and would likely be useful for understanding the causes and genetic features of various diseases.

Although the proposal won some initial support, it soon ran into several major, and ultimately insurmountable, obstacles. Certain scientists questioned its scientific rationale, casting doubt on whether the project would actually yield important information about human history and disease. But the greatest difficulties came from many of the indigenous populations that the organizers hoped would be participating in the project. They viewed the project suspiciously, raising questions about its true purpose and value: Who would "own" the genetic information that was generated during the project? Would it be patented? Who would benefit from drugs or other products developed using the information? What other purposes could the genetic information be used for, either good or bad? Wouldn't the millions of dollars that would be spent in the project be better used trying to directly help indigenous populations?

Despite the attempts of the project organizers to address these questions, the project's critics never relented, and the project was essentially abandoned in the late 1990s. Since then, other alternative approaches have been initiated to try to address issues related to human genetic diversity, such as the "HapMap" project that officially got underway in 2002.

1. Is it ethical for western scientists to involve indigenous populations in their research studies, or should they limit their studies to populations living in their own countries?

2. Do you think such a project would be likely to help indigenous populations? Do you think the objections to the project were reasonable?

3. If a scientist makes a medically important discovery using DNA obtained from an indigenous group, should the discovery be patentable? How should any benefits arising from such a discovery be shared?

Summary

15.1 Genomic Sequencing Is an Extension of Genetic Mapping

- Because mutant genes are the basis of genetic disorders, mapping helps us identify genes that cause disease and is the first step in developing diagnostic tests and treatments for these disorders.

15.2 Origins of the Human Genome Project

- Instead of finding and mapping markers and disease genes one by one, scientists organized the Human Genome Project (HGP) to sequence all the DNA in the human genome, identify and map the thousands of genes to the 24 chromosomes we carry, and assign a function to all the genes in our genome.

15.3 Genome Projects Have Created New Scientific Fields

- The sheer size of the human genome required the development of new technologies, including automated methods of DNA sequencing and advances in software to collect, analyze, and store the information derived from genome sequencing. The study of genomes by these methods is called genomics. Bioinformatics is the use of software, computational tools, and databases to store, organize, analyze, and visualize genomic information.

15.4 Genomics: Sequencing, Identifying, and Mapping Genes

- Geneticists developed two strategies for genome sequencing. One method, called the clone-by-clone method, uses clones from a genomic library that have been arranged to cover an entire chromosome. After the order of the clones is known, they are sequenced. The second method, called shotgun cloning, randomly selects clones from a genomic library and sequences them. Once the sequence of the clones is known, assembly software organizes them into the genomic sequence.

15.5 What Have We Learned So Far about the Human Genome?

- The Human Genome Project has provided several unexpected findings about our genome. There are about 25,000 to 30,000 genes, far fewer than the 80,000 to 100,000 genes that were predicted.

- Only about 5% of the genome actually encodes proteins, and about half the genome is non-gene-containing DNA sequences that are present in hundreds of thousands of copies. What this DNA does is still a mystery.

- Genes are organized in clusters, separated by long stretches of DNA that does not code for any genes.

15.6 Using Genomics and Bioinformatics to Study a Human Genetic Disorder

- Analysis of genome sequence information has led to the identification of disease-causing genes, including one for Freidreich ataxia, a progressive and fatal neurodegenerative disorder.

15.7 Proteomics Is an Extension of Genomics

- Proteomics is the study of the structure and function of proteins. This new field is playing an important role in analyzing proteins to develop new diagnostic tests for genetic disorders and to identify targets for drug development.

15.8 Ethical Concerns about Human Genomics

- To deal with the impact that genomic information would have on society, the HGP set up the ELSI program (Ethical, Legal, and Social Implications) to ensure that genetic information would be safeguarded and not used in discriminatory ways. ELSI works to develop policy guidelines for the use of genomic information.

15.9 Looking Beyond the Genome Project: What the Future Holds

- Information from the Human Genome Project will have a significant impact on research, health care, and society. Information from any one of these areas will effect the others. Six fields have been targeted for development as genomic and genetic information grows in its importance to and impact on our daily lives.

HUMAN
Genetics ⓔ Now™ Preparing for an exam? Assess your understanding of this chapter's topics with a pre-test, a personalized learning plan, and a post-test at **http://now.ilrn.com/cummings7**

Genomic Sequencing Is an Extension of Genetic Mapping

1. The gene controlling ABO blood type and the gene underlying nail-patella syndrome are said to show *linkage*. What does that mean in terms of their relative locations in the genome? What does it mean in terms of how the two traits are inherited with respect to each other?

2. Hemophilia and color blindness are both recessive conditions caused by genes on the X chromosome. To calculate the recombination frequency between the two genes, you draw a large number of pedigrees that include grandfathers with both hemophilia and color blindness, their daughters (who presumably have one chromosome with two normal alleles and one chromosome with two mutant alleles), and their sons. Analyzing all of the pedigrees together shows that 25 grandsons have both color blindness and hemophilia, 24 have neither of the traits, 1 has color blindness only, and 1 has hemophilia only. How many centimorgans separate the hemophilia locus from the locus for color blindness?

3. Prior to the advent of recombinant DNA technology, why was it so difficult for geneticists to map human genes using pedigrees? How did recombinant DNA technology help move things forward?

Origins of the Human Genome Project

4. In what years did the publicly funded Human Genome Project begin and end? Who participated in the project?

5. What were the scientific goals of the Human Genome Project?

Genome Projects Have Created New Scientific Fields

6. How many nucleotides does the human genome contain?

7. Which of the following best describes the process of DNA sequencing?
   a. DNA is separated on a gel, and the different bands are then labeled with fluorescent nucleotides and scanned with a laser.
   b. A laser is used to fluorescently label the nucleotides present within DNA, the DNA is run on a gel, and then the DNA is broken into fragments.
   c. Nucleotides are scanned with a laser and incorporated into the DNA that has been separated on a gel, and then the DNA is amplified using PCR.
   d. Fragments of DNA are produced in a reaction that labels them with any of four different fluorescent dyes, and the fragments are then run on a gel and scanned with a laser.
   e. DNA is broken down into its constituent nucleotides, and the nucleotides are then run on a gel and purified using a laser.

8. Which of the following is NOT an activity carried out in the field of **bioinformatics?**
   a. collecting and storing DNA sequence information produced by various genome sequencing projects
   b. analyzing genomic sequences to determine the location of genes
   c. determining the three-dimensional structure of proteins
   d. comparing genomes of different species
   e. none of the above

Genomics: Sequencing, Identifying, and Mapping Genes

9. How does the sequencing strategy used by the Human Genome Project differ from that used by Celera Corporation?

10. What was the first organism to have its genome sequenced? Which sequencing strategy was used?

11. Although both the Human Genome Project and Celera's human genome sequencing project are considered to be complete, in reality, neither of the groups actually sequenced 100% of the genome. Why not? What regions were left unsequenced?

12. Once an organism's genome has been sequenced, how do geneticists usually go about trying to pinpoint the location of the genes?

What Have We Learned So Far about the Human Genome?

13. What percentage of the DNA in the genome actually corresponds to genes? What makes up the rest?

14. When the human genome sequence was finally completed, scientists were surprised to discover that the genome contains far fewer genes than expected. How many genes are present in the human genome? Scientists have also found that there are many more different kinds of proteins in human cells than there are different genes in the genome. How can this be explained?

15. One unexpected result of the sequencing of the human genome was the finding that mutations in a single gene can be responsible for multiple distinct disorders. For example, mutations in the *RET* gene can cause two different types of multiple endocrine neoplasias, familial medullary thyroid carcinoma, and Hirschsprung disease. How do you think mutations in a single gene can have such diverse effects?

Using Genomics and Bioinformatics to Study a Human Genetic Disorder

16. Sequence comparison studies revealed that the product of the CF (cystic fibrosis) gene shares strong similarity to proteins known to be involved in
   a. transcription.   b. translation.

c. transport of ions across the cell membrane.

d. mRNA splicing.

e. movement of proteins across the Golgi membrane.

17. You join a lab that studies a rare genetic disorder that causes affected individuals to have unusually fast-growing, bright green hair. You are joining the lab at a fortuitous moment, as the gene causing the disorder has just been cloned. Despite this breakthrough, however, it is still unclear what the function of the gene is, and the lab director asks you for suggestions about how to go about trying to determine this. What do you recommend?

Proteomics Is an Extension of Genomics

18. How does proteomics differ from genomics? What kind of information can proteomics provide that is not available from genomics studies?

Ethical Concerns about Human Genomics

19. How did the Human Genome Project attempt to deal with the social and ethical issues that were bound to arise from the sequencing of the human genome?

15.9 Looking Beyond the Genome Project: What the Future Holds

20. In 2003, Francis Collins and his colleagues published a paper addressing the impact of genomics research and looking toward the future, focusing on three major themes: research in biology, health care, and society. Can you predict at least one future genomics-related development in each of these three areas? What potential benefits and dangers do you see for each of them?

Internet Activities

Internet Activities are critical thinking exercises using the resources of the World Wide Web to enhance the principles and issues covered in this chapter. For a full set of links and questions investigating the topics described below, log on to **http://biology.brookscole.com/cummings7**

1. *Comparative Genomics.* Access the Department of Energy's *Genome* site. Click on the link titled "Beyond HGP." Scroll down and click on the "Functional and Comparative Genomics Fact Sheet." Read and answer the questions about functional genomics and comparative genomics. Pay close attention to the questions about model organisms and the comparison between humans and mice.

2. *Genomic Analysis.* There are many sites covering the analysis of genomic sequence data. One of the best introductions to this topic is the *2Can Bioinformatics Educational Resources* page. Access this site and select the tutorial on "Nucleotide Analysis." Follow the directions for checking two unknown sequences to see if they contain vector sequences mixed in with the genomic sequence. On subsequent pages, you can see how well vector sequences detected in your search match known vector sequences already in databases, and you can read about the sequence information stored in genomic databases.

How would you vote now?

As a gift for your twenty-first birthday, your parents announce that they will pay to have your genome sequenced. This gift will allow you to have first-hand information about any genetic disorders you carry in the heterozygous condition and may pass on to your children, as well as defining any genetic risk factors for adult onset disorders. You can use this information to make reproductive decisions and to adjust your lifestyle early enough to maximally reduce your risk of developing adult onset disorders. The sequence will not be stored in any database and will be given to you on compact discs for you to have read and interpreted in the health care setting of your choice. Now that you know more about the impact of genomics, would you accept your parents' gift and have your genome sequenced? Visit the Human Heredity Companion website to find out more on the issue, then cast your vote online at **http://biology.brookscole.com/cummings7**

For further reading and inquiry, log on to **InfoTrac College Edition,** your world-class online library, including articles from nearly 5,000 periodicals, at **http://infotrac.thomsonlearning.com**

Reproductive Technology, Gene Therapy, and Genetic Counseling

16

A T 11:47 p.m. on July 25, 1978, a 5 pound, 12 ounce baby girl named Louise Brown was born in Oldham, England. Unlike the millions of other children born that year, her birth was an international media event, and she was called the "baby of the century" and a "miracle baby." Although some praised her birth as a medical miracle, others talked of her birth as the beginning of an era of genetically engineered children and even human-animal hybrids raised in artificial wombs. Why all the fuss? Louise was the first human born after *in vitro* fertilization (IVF), a procedure in which an egg is fertilized outside the body (*in vitro* literally means "in glass"), and the developing embryo is implanted into the uterus to complete its growth.

The development of IVF by Patrick Steptoe and Robert Edwards was a long, slow process. After 9 years, and more than 80 failed attempts, Steptoe repeated his procedure once more in November 1977 by recovering an egg from Lesley Brown. He first made a small incision (about ½ inch long) in her abdomen. Then he inserted a tube-like laparoscope to examine the ovary and made another small incision to remove an egg. He gave it to Edwards, who put it in a sterile glass dish and mixed it with semen from Lesley's husband. The dish containing the fertilized egg in a solution of nutrients was placed in an incubator for 2½ days, after which the developing embryo was implanted

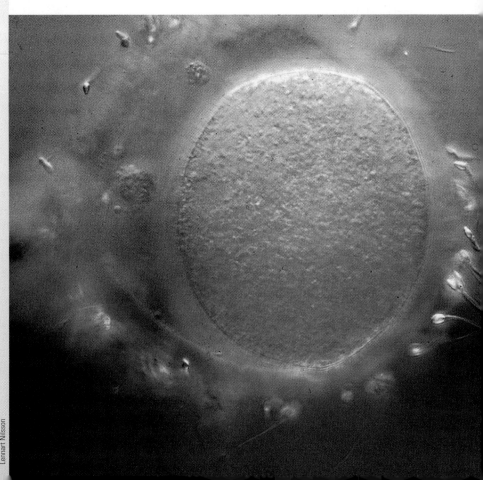

into Lesley's uterus through a tube inserted into the vagina. For the first time, the procedure was a success, and Louise was born on July 25 of the following year. In the United States, IVF is now a routine procedure available at over 350 clinics, and to date more than 45,000 babies have started life in a glass dish.

This procedure, which is now so commonplace, was a remarkable achievement for its time and was the first step in the development of methods, collectively called assisted reproductive technologies (ART), used to help infertile couples. Some of these methods are outlined in this chapter. The use of ART along with recombinant DNA techniques has created a powerful new technology for sex selection, the diagnosis of genetic disorders, and gene therapy.

How would you vote?

IVF produces more embryos than are implanted. Surplus embryos are routinely stored in liquid nitrogen. Some are used in subsequent attempts at pregnancy if IVF implantation fails; other are stored in case the couple wants another child at a future date. These embryos have several possible fates. They can be stored indefinitely, thawed and discarded, donated to researchers for use in stem cell research, or donated to other couples. Some nations, like Sweden and Great Britain, limit the time unused embryos can be stored before destruction. If you were having IVF, what would you do with extra embryos? Visit the Human Heredity Companion website to find out more on the issue, then cast your vote online at **http://biology.brookscole.com/cummings7**

16.1 Gaining Control over Reproduction

Over the past 50 years, steady advances in genetics, physiology, and molecular biology have allowed us to control many aspects of human reproduction. The techniques developed can be used to reduce or eliminate the chances of conception (contraception) or to enhance the chances of conception (**assisted reproductive technologies,** or **ART**). Reproductive technologies can also correct defective functions of the reproductive system and manipulate the physiology of reproduction.

Contraception uncouples sex from pregnancy.

To uncouple sexual intercourse from fertilization and pregnancy, one of the three stages in reproduction must be blocked: (1) release and transport of gametes, (2) fertilization, or (3) implantation. Aside from complete abstinence from sexual intercourse (▶ Figure 16.1), no method of birth control is completely successful in preventing pregnancy or sexually transmitted diseases (STDs).

■ **Assisted reproductive technologies (ART)** The collection of techniques used to help infertile couples have children.

EXTREMELY EFFECTIVE

Total abstinence	100%
Tubal ligation or vasectomy	99.6%
Hormonal implant (Norplant)	99%

HIGHLY EFFECTIVE

IUD + slow-release hormones	98%
IUD + spermicide	98%
Depo-Provera injection	96%
IUD alone	95%
High-quality latex condom + spermicide with nonoxynol-9	95%
Oral contraceptive	94%

EFFECTIVE

Cervical cap	89%
Latex condom alone	86%
Diaphragm + spermicide	84%
Sympto-thermal rhythm method	84%
Vaginal sponge + spermicide	83%
Foam spermicide	82%

MODERATELY EFFECTIVE

Spermicide cream, jelly, suppository	75%
Rhythm method (daily temperature)	74%
Withdrawal	74%
Condom (cheap brand)	70%

UNRELIABLE

Douching	40%
Chance (no method)	10%

▲ FIGURE 16.1 Effectiveness of various methods of birth control. The percent effectiveness is a measure of how many women in a group of 100 will not become pregnant in a year when using a given method of birth control. For example, foam spermicides are rated as 82% effective. This means that for every 100 women who use this as their only method of contraception for one year, 18 will have unplanned pregnancies.

■ **Tubal ligation** A contraceptive procedure for women in which the oviducts are cut, preventing eggs from reaching the uterus.

■ **Vasectomy** A contraceptive procedure for men in which each vas deferens is cut and sealed to prevent the transport of sperm.

Surgery can prevent gamete transport.

Surgical procedures that physically prevent the release and transport of gametes (such as tubal ligation and vasectomy) are very effective in controlling pregnancy. In **tubal ligation,** the woman's oviducts are cut, and the ends are tied off or cauterized. This prevents sperm from reaching oocytes released from the ovaries. In males, sperm release is prevented by **vasectomy.** In this procedure, small incisions are made on each side of the scrotum and a section of each vas deferens is withdrawn and cut. The cut ends are sealed to prevent transport and release of sperm.

Blocking egg production controls fertility.

Preventing release of eggs by manipulating the hormonal cycle controlling maturation and release of the oocyte from the ovary is also an effective method of preventing pregnancy. Most birth control pills contain a combination of hormones including estrogens and progesterones. These hormones prevent ovulation (the release of an egg from the ovary) by inhibiting the normal release of hormones required for egg maturation. These pills must be taken daily for a certain number of days each month. Instead of pills, injection of Depo-Provera (a progesterone) inhibits ovulation for several months. Hormone-containing time-release capsules (such as Norplant) can be implanted under the skin. These implants slowly release a steady supply of hormones that stop ovulation for several years before the capsule needs replacement.

Several progesterone-estrogen combination pills prevent ovulation or implantation when prescribed up to 72 hours after unprotected intercourse as emergency contraception pills (also known as morning-after pills). Another type of drug, RU-486, is an anti-progesterone abortifacent that can be used up to seven weeks after the beginning of the last menstrual period.

Physical and chemical barriers block fertilization.

To prevent fertilization, both physical and chemical barriers can be used. Condoms are latex or gut sheaths and pouches worn over the penis or inserted into the vagina before intercourse. These membranes prevent sperm from entering the vagina and uterus. Only latex condoms are highly effective in preventing STDs, including AIDS. Diaphragms are flexible caps that fit over the opening of the cervix and are inserted before intercourse. They prevent sperm from entering the cervix and swimming up the uterus to the oviducts. Intrauterine devices (IUDs) are made of metal or plastic and are inserted into the uterus where they interfere with the attachment of the embryo to the uterine wall. They can be left in place for up to two years. Chemical barriers, such as spermicidal jelly or foam placed

Sovereign/Phototake

▲ FIGURE 16.3 One method of assisted reproduction is intracytoplasmic sperm injection (ISCI), in which a single sperm (in needle) is injected into the egg.

IVF, however, is not the only option. Gametes can be collected and placed into the woman's oviduct, a procedure known as **GIFT, gamete intrafallopian transfer.**

For cases of male infertility, many couples, however, now choose another option, called **intracytoplasmic sperm injection (ICSI).** In this procedure, an egg is collected and injected with a carefully selected single sperm from the male partner (▶ Figure 16.3). The zygote is cultured in the laboratory for a few days and transferred to the uterus of the female partner for development.

Reproductive technology has altered traditional and accepted patterns of reproduction and redefined the meaning of parenthood. For example, in the United States, surrogate motherhood and its variations are a reproductive option. Surrogate parenthood can take many forms. In one version, a woman is artificially inseminated by sperm and carries the child to term. After the child is born, she surrenders the child to the father and his mate. In this case, the surrogate is both the genetic and gestational mother of the child. In another version, a couple provides both the oocyte and the sperm for IVF. A surrogate is implanted with the developing embryo and serves as the gestational mother but is genetically unrelated to the child she bears.

Following the discovery that it is the age of the oocyte, not the age of the reproductive system, that is responsible for infertility as women age, women are now becoming mothers in their late 50s and early 60s. After hormonal treatment to prepare the uterus, they are implanted with zygotes produced by IVF using eggs donated by younger women. In addition, fertilized eggs can now be collected and frozen for later use, separating fertilization from development. This allows younger women to collect oocytes while they are young, when the risks for chromosome abnormalities in the offspring are low. The oocytes can be fertilized by IVF and the resulting embryos, frozen in liquid nitrogen, can be stored for years. The embryos can be thawed and implanted over a period of years, including menopause, allowing women to extend their childbearing years.

■ **Gamete intrafallopian transfer (GIFT)** A procedure in which gametes are collected and placed into a woman's oviduct.

■ **Intracytoplasmic sperm injection (ICSI)** A treatment to overcome defects in sperm count or motility; an egg is fertilized by microinjection of a single sperm.

Although the most rapid advances in assisted reproductive technologies (ART) began after the 1978 birth of Louise Brown by IVF, one of the first recorded uses of ART occurred in the early 1770s. An Englishman with a malformation of the urethra collected his semen in a syringe and injected it into his wife's vagina; she became pregnant and gave birth to a child. In 1866, Dr. J. Sims reported the first intrauterine insemination in the United States; he later used this procedure over 50 times to aid infertile couples. Dr. William Panacost performed the first artificial insemination using donor sperm in 1884. Oklahoma was the first state to legalize artificial insemination in 1967.

These and other unconventional means of generating a pregnancy have developed more rapidly than the social conventions and laws governing their use (see Spotlight on Reproductive Technologies from the Past). In the process, controversy about the moral, ethical, and legal grounds for using these techniques has arisen but is not yet resolved.

16.4 Ethical Issues in Reproductive Technology

Although ART have been responsible for thousands of births, there are also significant risks to both parents and children. While the benefits have been significant, several unexpected risks have emerged from using these alternative methods of reproduction. Some of the risks have been well documented, whereas others are still matters for debate and more study. In other cases, the use of ART raises ethical questions (see Genetics in Society: The Business of Making Babies). We'll discuss some of these risks and questions in the following sections.

Reproductive technology carries risks to parents and children.

These risks include a threefold increase in ectopic pregnancies (a situation where the fertilized egg implants outside the uterus, and the placenta and embryo begin to develop there) and multiple births (35% of IVF couples have twins or triplets).

Infants born by means of ART have an increased risk of low birth weight and often require prolonged hospital care. When ICSI is used in ART, there is an increased risk of transmitting genetic defects to males. About 13% of infertile males with low sperm counts carry a small deletion on the Y chromosome. With ICSI, this infertility is passed on to their sons. The same is true for some chromosomal abnormalities such as Klinefelter syndrome. Questions arise whether it is ethical to use ICSI to produce sons who will be infertile.

There has been a long-standing debate about whether children conceived by means of ART have increased risks for birth defects. While this issue has not been resolved, it is important that couples considering ART be informed of this and other potential risks.

Selecting siblings to be donors is a possibility.

Screening embryos by preimplantation genetic diagnosis (PGD) is used by parents who are carriers of genetic disorders that would be fatal to any children born with the disorder (such as Tay-Sachs disease or cystic fibrosis). This allows implantation of embryos known to be free of the disease.

In the early 1990s, Jack and Linda Nash used PGD to screen embryos after their daughter, Molly, was born with Fanconi anemia (OMIM 227650), a usually fatal disorder of the bone marrow. In this case, PGD allowed them to have a healthy child, but they also had the embryos screened to find one that would be a suitable stem cell donor for Molly. Umbilical cord blood from their son, Adam, was transfused into Molly, who is now healthy (▶ Figure 16.4). (See Genetic Journeys: Saving Cord Blood.) At the time, bioethicists debated whether it was ethical to have a child destined to be a donor for a sibling. This case was complicated by the fact that the parents planned to have other children and used PGD to screen out embryos with Fanconi anemia.

In 2004, physicians reported that they helped four couples use IVF and PGD to have babies that are tissue-matched to siblings with leukemia. In these cases, the embryos were not screened for genetic disorders, but only for alleles that would allow the chil-

The Business of Making Babies

New technology has made the business of human fertilization a part of private enterprise. One in eight couples in the United States (over 5 million couples of childbearing age) is classified as infertile, and most of these couples want to have children. The first successful *in vitro* fertilization (IVF) in the United States was done in 1981 at the Medical College of Virginia at Norfolk. Since then, more than 400 hospitals and clinics that use IVF and ART have opened. Many of these clinics are associated with university medical centers, but others are operated as freestanding businesses. Most operate only at a single location, but national chains are becoming part of the business. One of the largest, the Sher Institute for Reproductive Medicine, has eight locations nationwide, with plans for more. Some clinics are public companies that have sold stock to raise start-up money or to cover operating costs. Typically, clinics have revenues of $2 million/year or less.

Charges for IVF range from $7,500 to $15,000, and because success rates are less than 50% for each IVF, several attempts (four to six) are usually required. Because costs are not usually covered by insurance, IVF is a major expense for couples who want children. In fact, spending on IVF reached the $1 billion mark in 2002. Because of the high start-up costs and expertise required, it is possible that the field will undergo significant consolidation and eventually be dominated by a small number of companies through franchising agreements. Some investment analysts predict that IVF will grow into a $6 billion annual business.

In IVF, egg maturation is induced with drugs, and the mature eggs are recovered by laparoscopy (a small incision is made in the abdominal wall, and a fiber-optic device is used to recover the eggs). In an alternative procedure, an ultrasonically guided needle is inserted into the vagina and moved up to the ovary to remove the eggs. The eggs are fertilized in a dish (*in vitro* literally means "in glass"). After the fertilized egg begins to develop, it is implanted into the woman's uterus.

If extra eggs are recovered, they are fertilized, and the resulting embryos are frozen in liquid nitrogen. This eliminates the need to retrieve eggs every month for fertilization. If fertilization is successful, the extra embryos can remain in storage for implantation at a later time, or they can be donated to another couple.

dren produced from these embryos to serve as transplant donors for their siblings. These cases have reignited the debate on whether it is ethical to select for genotypes that have nothing to do with a genetic disorder, and whether screening to benefit someone else is acceptable. Some countries, including Great Britain, have laws against using PGD for sex selection or for screening embryos to be donors unless they are also screened to avoid a genetic disorder, but the United States has no such restrictions.

Advocates of embryo screening to match donor and recipient say there are no associated ethical issues, but critics wonder if embryo screening for transplant compatibility will eventually lead to screening for sex of the embryo or for traits such as eye color. A survey by the Genetics and Public Policy Center at The Johns Hopkins University shows that 61% of the Americans surveyed approved of using PGD to select an embryo for the benefit of a sibling, but it also revealed that 80% of those surveyed were concerned that reproductive genetic technologies could get "out of control."

▲ **FIGURE 16.4** Molly Nash and her brother Adam. Molly's parents used *in vitro* fertilization and prenatal genetic diagnosis to avoid having another child with Fanconi anemia and to select a compatible stem cell donor for Molly.

Saving Cord Blood

Two California boys, Blayke and Garrett LaRue, are alive today thanks to umbilical cord blood donated to cord blood banks by two anonymous mothers, one in New York and another in Germany. Both Blayke and Garrett were diagnosed with a rare and fatal genetic disorder called X-linked lymphoproliferative disorder (XLP; OMIM 308240) after their brother Layne died of liver failure. Layne had been ill with mononucleosis, an infection caused by the Epstein-Barr virus. Mononucleosis is usually not fatal, but XLP destroys the ability of the immune system to respond to infection, and extreme sensitivity to the Epstein-Barr virus is one of the hallmarks of this disorder.

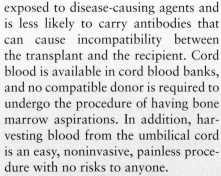

After Blayke and Garrett were diagnosed with XLP, tissue matches for both boys were made through the National Marrow Donor Program's cord blood bank. The boys' immune systems were destroyed by chemotherapy and replaced with the immune system generated by the transplanted stem cells.

Cord blood transplants have several advantages over bone marrow transplants for treating XLP and other immune disorders. Cord blood has not been exposed to disease-causing agents and is less likely to carry antibodies that can cause incompatibility between the transplant and the recipient. Cord blood is available in cord blood banks, and no compatible donor is required to undergo the procedure of having bone marrow aspirations. In addition, harvesting blood from the umbilical cord is an easy, noninvasive, painless procedure with no risks to anyone.

There are numerous reports of cord blood being used to cure diseases associated with blood cells, but unfortunately, most cord blood is discarded along with the umbilical cord and placenta after birth. Doctors at UCLA, where the LaRue boys were treated, are encouraging mothers to consider donating the cord blood of their babies to a cord blood bank so that other lives can be saved.

16.5 Gene Therapy Promises to Correct Many Disorders

Although preimplantation genetic diagnosis and other methods of genetic testing allow couples to have children free of genetic disorders, about 5% of all newborns have a genetic or chromosomal disorder. A recombinant DNA–based method, called gene therapy, has been developed to treat disorders caused by mutations in single genes. As discussed in Chapter 4, disorders such as cystic fibrosis and hemophilia are caused by mutations in single genes. Gene therapy puts cloned copies of normal genes into cells that carry defective copies. These normal genes make functional proteins that result in a normal phenotype.

What are the strategies for gene transfer?

There are several methods for transferring cloned genes into human cells, including viral vectors, chemical methods to transfer DNA across the cell membrane, and physical methods such as microinjection or fusion of cells with vesicles that carry cloned DNA sequences.

Viral vectors, especially retroviruses, are the most commonly used method for gene therapy. Retroviruses are used because they readily infect human cells. These vectors are genetically modified by removing some viral genes. This prevents the virus from causing disease and makes room for a human gene to be inserted (▶ Figure 16.5). Once the recombinant virus carrying a human gene is inside the cell, the viral DNA inserts itself into a human chromosome, where it becomes part of the genome.

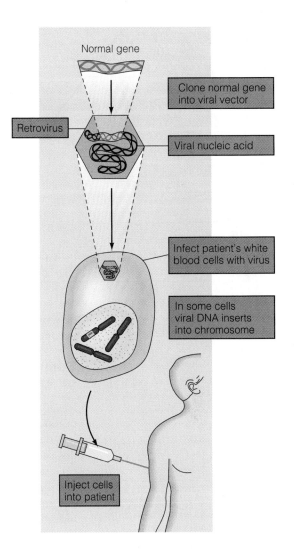

FIGURE 16.5 The most widely used method of gene therapy uses a virus as a vector to insert a normal copy of a gene into the white blood cells of a patient who has a genetic disorder. The normal gene becomes active, and the cells are reinserted into the affected individual, curing the genetic disorder. Because white blood cells die after a few months, the procedure has to be repeated regularly. In the future, it is hoped that transferring a normal gene into the mitotically active cells of the bone marrow will make gene therapy a one-time procedure.

Labels in figure:
- Normal gene
- Clone normal gene into viral vector
- Retrovirus
- Viral nucleic acid
- Infect patient's white blood cells with virus
- In some cells viral DNA inserts into chromosome
- Inject cells into patient

Gene therapy showed early promise.

Gene therapy began in 1990, when a human gene for the enzyme adenosine deaminase (ADA) was inserted into a retrovirus and then transferred into the white blood cells of a young girl, Ashanti De Silva (▶ Figure 16.6), who had a form of severe combined immunodeficiency disease (SCID; OMIM 102700). She had no functional immune system and was prone to infections, many of which can be fatal. The normal *ADA* gene, inserted into her white blood cells, encodes an enzyme that allows cells of the immune system to mature properly. As a result, she now has a functional immune system and is leading a normal life. Unfortunately, gene therapy for other children with SCID was unsuccessful, and Ashanti remains the only success story for SCID gene therapy.

Keep in mind

■ Gene therapy has not yet fulfilled its promises of treating genetic disorders.

Gene therapy has also experienced setbacks and restarts.

In the early to mid 1990s, gene therapy trials were started for several genetic disorders, including cystic fibrosis and familial hypercholesterolemia. Over a 10-year period, more than 4,000 people underwent gene transfer. Unfortunately, these trials were largely failures and led to a loss of confidence in gene therapy. Hopes for gene therapy plummeted even further in September 1999, when Jesse Gelsinger died during gene therapy. His death was triggered by a massive immune response to the vector, a modified adenovirus (adenoviruses cause colds and respiratory infections).

In 2000, two French children who underwent successful gene therapy for an X-linked form of SCID developed leukemia. In these children, the recombinant virus inserted itself into a gene that controls cell division, activating the gene, and causing uncontrolled production of white blood cells and the symptoms of leukemia.

Gene therapy has not been a total failure, however. It is used successfully to treat cancer, cardiovascular disease, and HIV infection (▶ Figure 16.7). In fact, gene therapy is more often used to treat cancer than any other condition. In spite of these limited success stories, gene therapy is still an experimental procedure performed on only a few carefully selected patients, under strict regulation by government agencies.

Most of the problems with gene therapy have been traced to the vectors. Efforts are now directed at developing new vectors that are less visible to the immune system and that transfer genes to target cells with a higher efficiency. Some successes in animal models and human transfers are encouraging researchers and clinicians to continue work on gene therapy. As

▲ **FIGURE 16.6** Ashanti De Silva was the first human to undergo gene therapy.

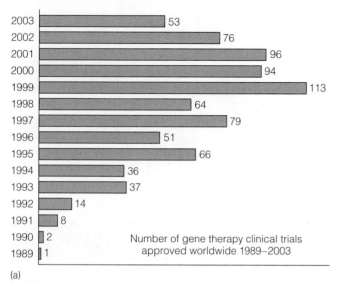

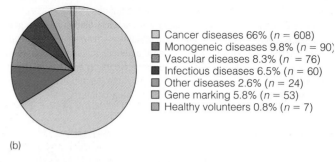

Cancer diseases 66% ($n = 608$)
Monogeneic diseases 9.8% ($n = 90$)
Vascular diseases 8.3% ($n = 76$)
Infectious diseases 6.5% ($n = 60$)
Other diseases 2.6% ($n = 24$)
Gene marking 5.8% ($n = 53$)
Healthy volunteers 0.8% ($n = 7$)

(b)

◀ **FIGURE 16.7** An overview of gene therapy trials. (a) Worldwide gene therapy trials by year from 1989 to 2003. (b) Target disorders for gene therapy. Most gene therapy trials are for cancer (66%) and not single gene (monogenic) disorders (9.8%).

new vectors are developed, gene therapy will undoubtedly fulfill its early promise and become a commonplace method of treating disease.

Some gene therapy involves stem cells, gene targeting, and therapeutic cloning.

In some genetic disorders, gene transfer into adult stem cells followed by transplantation has the potential to avoid some of the problems associated with vectors. First, let's define what stem cells are and then discuss how they can be used to treat genetic disorders. Embryonic stem (ES) cells are derived from a small cluster of about 100 cells (the inner cell mass) in early mammalian embryos (▶ Figure 16.8). They can also be created by transferring the nucleus from a somatic cell into an egg that has had its nucleus removed. The egg divides, forming an inner cell mass, and cells removed and grown in culture dishes form ES cells. This second method is called **somatic cell nuclear transfer,** or therapeutic cloning, since the ES cells contain the genome of the cell used as a source of the nucleus. Stem cells can also be found in adult tissues. Our ability to heal wounds depends on the existence of stem cells that can divide and form new blood vessels, connective tissue, muscle, and so on. All stem cells are classified by their potential. **Totipotent** cells from early embryos can form every cell type in the body; adult stem cells are **pluripotent,** able to form a smaller number of cell types, or **multipotent,** able to form only a few cell types.

For gene therapy, attention is focused on using stem cells from someone with a genetic disorder, whether they are adult cells or ones created by therapeutic cloning. These cells can be used as recipients in gene transfer or can be genetically reprogrammed and used to treat neurodegenerative disorders, diabetes, or leukemia.

Adult stem cells modified by gene transfer are being developed to treat cases of osteogenesis imperfecta (OI; OMIM 166200 and others), an autosomal dominant disorder of collagen genes that causes life-threatening bone malformations. Adult stem cells were isolated from bones removed from OI patients during surgery. A recombinant viral vector carrying a normal collagen gene was designed to insert into the mutant collagen genes in these stem cells. After gene transfer, the stem cells were analyzed and found to produce normal collagen and bone. Clinical trials with these genetically targeted stem cells are being proposed. Because these adult stem cells can form cartilage, fat, and muscle in addition to bone, they may be useful in treating several types of genetic disorders.

There are ethical issues related to gene therapy.

At present, gene therapy is done using an established set of ethical and medical guidelines. All patients are volunteers, gene transfer is started after the case has un-

■ **Somatic cell nuclear transfer**
A cloning technique that transfers a somatic cell nucleus to an enucleated egg, which is stimulated to develop into an embryo. Inner cell mass cells are collected from the embryo and grown to form a population of stem cells. Also called therapeutic cloning.

■ **Totipotent** The ability of a stem cell to form every cell type in the body; characteristic of embryonic stem cells.

■ **Pluripotent** The ability of a stem cell to form most of the cell types in the body.

■ **Multipotent** The restricted ability of a stem cell to form only one or a few different cell types.

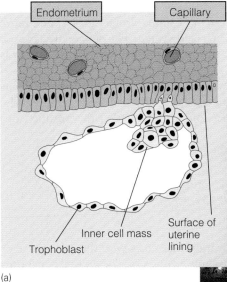

The process of implantation. (a) A blastocyst as it begins to implant in the uterine wall. The inner cell mass will form the embryo and is the source of embryonic stem cells. (b) Embryonic stem cells growing in culture. From here, cells can be transplanted to form a variety of tissues and organs.

Endometrium

Capillary

Inner cell mass

Surface of uterine lining

Trophoblast

(a)

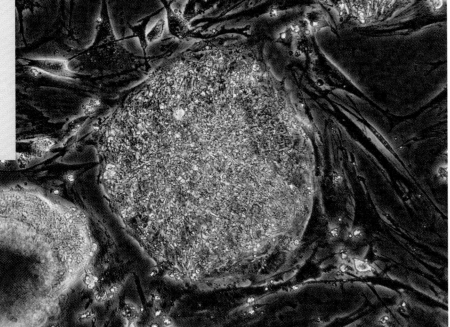

James Thomson Research Laboratory, University of Wisconsin–Madison

(b)

dergone several reviews, and the trials are monitored to protect the patients' interests. Newer guidelines instituted after Gelsinger's death strengthen these protections and coordinate the role of government agencies that regulate gene therapy. Other ethical concerns have not yet been resolved, as described next.

At present, gene therapy uses somatic cells as targets for transferred genes. This form of gene therapy is called **somatic gene therapy.** In somatic therapy, genes are transferred into somatic cells of the body; the procedure involves only a single target tissue, and only one person is treated (only after obtaining informed consent and permission for the treatment).

Two other forms of gene therapy are not yet used, mainly because ethical issues surrounding them have not been resolved. One of these is **germ-line therapy,** in which germ cells (cells that produce eggs and sperm) are targets for gene transfer. In germ-line therapy, the transferred gene would be present in all cells of the individual produced from the genetically altered gamete, including his or her germ cells. As a result, members of future generations are affected by this gene transfer, without their consent. Do we have the right to genetically modify others without their consent? Can we make this decision for members of future generations? These and other ethical concerns are not resolved, and germ-line therapy is currently prohibited.

The second form of gene therapy, **enhancement gene therapy,** raises even more ethical concerns. If we discover genes that control a desirable trait, such as intelligence or athletic ability, should we use them to enhance someone's intellectual ability or athletic skills? For now, the consensus is that we should not use gene transfer for such purposes. However, a recent decision by the U.S. Food and Drug Administration allows the use of growth hormone produced by recombinant DNA technology to enhance the growth of children who have no genetic disorder or disease but are likely to be shorter than average adults. Critics point out that approving gene transfer for

■ **Somatic gene therapy** Gene transfer to somatic target cells to correct a genetic disorder.

■ **Germ-line therapy** Gene transfer to gametes or the cells that produce them. Transfers a gene to all cells in the next generation, including germ cells.

■ **Enhancement gene therapy** Gene transfer to enhance traits such as intelligence or athletic ability rather than to treat a genetic disorder.

enhancement is only a short step from approving a gene product for enhancement. As the number of genes identified in the Human Genome Project grows, and more cloned genes are available, this issue is likely to be revisited. The debate over the use of germ-line therapy and enhancement therapy has long-term consequences for us as a species.

There is also an ethical debate about the use of embryonic stem cells in the treatment of disease. Questions about whether the use or creation of human embryos destroys nascent life are unresolved, as are questions about whether therapeutic cloning is a step toward cloning humans.

16.6 Genetic Counseling Assesses Reproductive Risks

Genetic counseling A process of communication that deals with the occurrence or risk that a genetic disorder will occur in a family.

Genetic counseling is a process of communication that deals with the occurrence or risk for a genetic disorder in a family. Counseling involves one or more trained professionals who help an individual or family to understand each of the following:

- The medical facts, including the diagnosis, progression, management, and any available treatment for a genetic disorder
- The way heredity contributes to the disorder and the risk of having children with this disorder
- The alternatives for dealing with the risk of recurrence
- How to adjust to the disorder in an affected family member or to the risk of recurrence

Genetic counselors achieve these goals in a nondirective way. They provide all the information available to individuals or family members, so the person or family can make the decisions best suited to them based on their own cultural, religious, and moral beliefs.

Keep in mind

- Genetic counseling educates individuals and families about genetic disorders and helps them make decisions about reproductive choices.

Who are genetic counselors?

Genetic counselors are health care professionals with specialized graduate training and experience in the areas of medical genetics, psychology, and counseling. They usually work as members of a multidisciplinary health care team and offer information and support to families that have relatives with genetic conditions or that may be at risk for a variety of inherited conditions. Genetic counselors identify families at risk, investigate the problem in the family, interpret information about the disorder, analyze inheritance patterns and risk of recurrence, and review available options with the family (▶ Figure 16.9).

Why do people seek genetic counseling?

People seek genetic counseling for many reasons. Typically, cases involve an individual or family with a history of a genetic disorder, cancer, birth defect, or developmental disability. Women older than 35 years and individuals from specific ethnic groups in which particular genetic conditions occur more frequently are counseled to teach them about their increased risk for genetic or chromosomal disorders and the diag-

◀ FIGURE 16.9 In a genetic counseling session, the counselor uses the information from pedigree construction, medical records, and genetic testing to educate and inform a couple about their risks for genetic disorders.

Martha Cooper/Peter Arnold, Inc.

nostic testing that is available. Counseling is especially recommended for the following individuals or families:

- Women who are pregnant, or are planning to become pregnant, after age 35
- Couples who already have a child with mental retardation, an inherited disorder, or a birth defect
- Couples who would like testing or more information about genetic defects that occur more frequently in their ethnic group
- Couples who are first cousins or other close blood relatives
- Individuals who are concerned that their jobs, lifestyle, or medical history may pose a risk to a pregnancy, including exposure to radiation, medications, chemicals, infection, or drugs
- Women who have had two or more miscarriages or babies who died in infancy
- Couples whose infant has a genetic disease diagnosed by routine newborn screening
- Those that have, or are concerned that they might have, an inherited disorder or birth defect
- Pregnant women who have been told that, based on ultrasound tests or blood tests for alpha fetoprotein, their pregnancy may be at increased risk for complications or birth defects

How does genetic counseling work?

Most people go for counseling after a prenatal test or after the birth of a child with a genetic condition. The counselor usually begins by constructing a detailed family and medical history, or pedigree.

Prenatal screening and cytogenetic or biochemical tests can be used along with pedigree analysis to help determine what, if any, risks are present. The counselor uses as much information as possible to establish whether the trait in question is genetically determined.

If the trait is genetically determined, the counselor constructs a risk assessment profile for the couple. In this process, the counselor uses all of the information available to explain the risk of having another child affected with the condition or to explain the risk that the individual who is being counseled will be affected with the condition. High-risk conditions include dominantly inherited disorders (50% risk if one parent is heterozygous), simple autosomal recessively inherited conditions (25%

when both parents are heterozygotes), and certain chromosomal translocations. Often conditions are difficult to assess because they involve polygenic traits or conditions that have high mutation rates (like neurofibromatosis).

Genetic counselors explain basic concepts of biology and inheritance to all couples. This helps them understand how genes, proteins, or cell-surface antigens are related to the defects seen in their child or family. The counselor provides information that allows informed decision making about future reproductive choices. Reproductive alternatives, such as adoption, artificial insemination, *in vitro* fertilization, egg donation, and surrogate motherhood are options that the counselor presents to the couple.

What are some future directions in genetic counseling?

As the Human Genome Project accelerates the number of genetic disorders that can be detected by heterozygote and prenatal screening, and as these techniques become more available, the role of the genetic counselor will become more important. The Human Genome Project is changing the focus of genetic counseling from reproductive risks to adult-onset conditions, such as cancer and Huntington disease. Although counseling sessions address reproductive risks for these conditions, the primary focus in these conditions is the individual being counseled. Areas addressed in these sessions include the risk of inheriting the gene, the potential severity of the condition, and the age of onset. This information allows individuals to develop lifestyles that may reduce the impact of the disorder, make decisions about having children, and plan for medical care they may require later in life.

Genetics in Practice

Genetics in Practice case studies are critical thinking exercises that allow you to apply your new knowledge of human genetics to real-life problems. You can find these case studies and links to relevant websites at **http://biology.brookscole.com/cummings7**

CASE 1

Jan, a 32-year-old woman, and her husband, Darryl, have been married for 7 years. They attempted to have a baby on several occasions. Five years ago they had a first-trimester miscarriage followed by an ectopic pregnancy later that same year. Jan continued to see her OB/GYN physician for infertility problems but was very unsatisfied with the response. After four miscarriages, she went to see a fertility specialist who diagnosed her with severe endometriosis and polycystic ovarian disease (detected by hormone studies). The infertility physician explained that these two conditions were hampering her ability to become pregnant and thus making her infertile. She referred Jan to a genetic counselor.

At the appointment, the counselor explained to Jan that one form of endometriosis is a genetic disorder, inherited as an autosomal dominant trait, and that polycystic ovarian disease can also be a genetic disorder and is the most common reproductive disorder among women. The counselor recommended that a detailed family history of both Jan and Darryl would help establish whether Jan's problems have a genetic component, and whether any of her potential daughters would be at risk for one or both of these disorders. In the meantime, Jan is in the process of taking hormones, and she and Darryl are considering alternative modes of reproduction.

1. Using the information in ▶ Figure 16.2, explain the reproductive options that are open to Jan and Darryl.

2. Would ISCI be an option? Why or why not?

3. Jan is concerned about using ART. She wants to be the genetic mother and have Darryl be the genetic father of any children they have. What methods of ART would you recommend to this couple?

CASE 2

Trudy was a 33-year-old woman who went with her husband, Jeremy, for genetic counseling. Trudy has had three miscarriages. The couple has a 2-year-old daughter who is in good health and is developing normally. Chromosomal analysis was done on the products of conception of the last miscarriage and was found to be 46,XY. The last miscarriage occurred in January 1999. Peripheral blood samples for both parents were taken at the time and sent to the laboratory. Trudy's chromosomes were 46,XX and Jeremy's were 46,XY,t(6;18)(q21;q23). Jeremy appears to have a balanced translocation between chromosome number 6 and 18. There is no family history of stillbirths, neonatal death, infertility, mental retardation, or birth defects. Jeremy's parents both died in their 70s from heart disease, and he is unaware of any pregnancy losses experienced by his parents or siblings.

The recurrence risks associated with a balanced translocation between chromosome 6 and 18 were discussed in detail. The counselor used illustrations to demonstrate the approximately 50% risk of unbalanced gametes; the other 50% of the gametes result in either normal or balanced karyotypes. The family was informed that empirical risk for unbalanced conceptions is significantly less than the 50% relative risk. Prenatal diagnostic procedures were described, including amniocentesis and chorionic villus sampling. The benefits, risks, and limitations of each were described.

The couple indicated a desire to pursue another pregnancy and were interested in proceeding with an amniocentesis.

1. Draw each of the possible combinations of chromosomes 6 and 18 that could be present in Jeremy's gametes, showing how there is an approximately 50% chance that they are either normal, or balanced, and a 50% chance that they are unbalanced.

2. Trudy became pregnant again, and an amniocentesis showed that the fetus received the balanced translocation from her father. Is she likely to have any health problems because of this translocation? Will it affect her in any way?

Summary

16.1 Gaining Control over Reproduction

- Many aspects of human reproduction can be controlled by contraception to reduce or eliminate the chances of pregnancy by manipulating one or more stages of reproduction: gamete production and/or transport, fertilization, and implantation.

16.2 Infertility Is a Common Problem

- In the United States, about 13% of all couples are infertile. Infertility has many causes, including genetics, hormone imbalance, and sexually transmitted diseases.

16.3 Assisted Reproductive Technologies (ART) Expand Childbearing Options

- ART is a collection of techniques used to help infertile couples have children. These techniques have developed ahead of legal and social consensus about their use.

16.4 Ethical Issues in Reproductive Technology

- The use of ART raises several unresolved ethical issues. These include health risks to both parents and their offspring resulting from ART and the use of preimplantation genetic diagnosis to select sibs who are suitable tissue or organ donors for other members of the family.

16.5 Gene Therapy Promises to Correct Many Disorders

- Gene therapy transfers a normal copy of a gene into target cells of individuals carrying a mutant allele. After initial successes, gene therapy suffered several setbacks, including the death of a participant. Ethical issues over the use of germ-line therapy and enhancement therapy are unresolved, and these therapies are not used.

16.6 Genetic Counseling Assesses Reproductive Risks

- Genetic counseling involves developing an accurate assessment of a family history to determine the risk of genetic disease. In many cases, this is done after the birth of a child affected with a genetic disorder, but in other cases counseling is entirely retrospective. Decisions about whether to have additional children, to undergo abortion, or even to marry are always left to those being counseled.

HUMAN
Genetics ⓔ Now™ Preparing for an exam? Assess your understanding of this chapter's topics with a pre-test, a personalized learning plan, and a post-test at **http://now.ilrn.com/cummings7**

Gaining Control over Reproduction

1. Explain how the following methods prevent conception.
   a. vasectomy
   b. tubal ligation
   c. birth control pills
2. RU-486 is a controversial drug. What makes it different from birth control methods?

Assisted Reproductive Technologies (ART) Expand Childbearing Options

3. How does IVF differ from artificial fertilization?
4. What is the difference between gamete intrafallopian transfer (GIFT) and intracytoplasmic sperm injection?
5. Why should women consider collecting and freezing oocytes for use later in life when they want to have children? What are the risks associated with older women having children?

Ethical Issues in Reproductive Technology

6. What do you think are the legal and ethical issues surrounding the use of IVF? How can these issues be resolved? What should be done with the extra gametes that are removed from the woman's body but never implanted in her uterus?
7. Researchers are currently learning how to transfer sperm-making cells from fertile male mice into infertile male mice in the hopes of learning more about reproductive abnormalities. These donor spermatogonia cells have developed into mature spermatozoa in 70% of the cases, and some recipients have gone on to father pups (as baby mice are called). This new advance opens the way for a host of experimental genetic manipulations. It also offers enormous potential for correcting human genetic disease. One potentially useful human application of this procedure is treating infertile males who wish to be fathers.
   a. Do you foresee any ethical or legal problems with the implementation of this technique? If so, elaborate on the concerns.
   b. Could this procedure have the potential for misuse? If so, explain how.

Gene Therapy Promises to Correct Many Disorders

8. Gene therapy involves:
   a. the introduction of recombinant proteins into individuals
   b. cloning human genes into plants
   c. the introduction of a normal gene into an individual carrying a mutant copy
   d. DNA fingerprinting
   e. none of the above

9. In selecting target cells to receive a transferred gene in gene therapy, what factors do you think would have to be taken into account?
10. The prospect of using gene therapy to alleviate genetic conditions is still a vision of the future. Gene therapy for adenosine deaminase deficiency has proven to be quite promising, but many obstacles remain to be overcome. Currently, the correction of human genetic defects is done using retroviruses as vectors. For this purpose, viral genes are removed from the retroviral genome, creating a vector capable of transferring human structural genes into sites on human chromosomes within target tissue cells. Do you see any potential problems with inserting pieces of a retroviral genome into humans? If so, are there ways to combat or prevent these problems?
11. Is gene transfer a form of eugenics? Is it advantageous to use gene transfer to eliminate some genetic disorders? Can this and other technology be used to influence the evolution of our species? Should there be guidelines for the use of genetic technology to control its application to human evolution? Who should create and enforce these guidelines?

Genetic Counseling Assesses Reproductive Risks

12. A couple who wishes to have children visits you, a genetic counselor. There is a history of a deleterious, recessive trait in males in the woman's family but not in the man's family. The couple is convinced that because his family shows no history of this genetic disease, they are at no risk of having affected children. What steps would you take to assess this situation and educate this couple?
13. A couple has had a child born with neurofibromatosis. They come to your genetic counseling office for help. After taking an extensive family history, you determine that there is no history of this disease on either side of the family. The couple wants to have another child and wants to be advised about risks of having another child with neurofibromatosis. What advice do you give them?
14. You are a genetic counselor and your patient has asked to be tested to determine if she carries a gene that predisposes her to early-onset cancer. If your patient has this gene, there is a 50/50 chance that all of her siblings inherited this gene and there is also a 50/50 chance that it will be passed on to their offspring. Your patient is very concerned about confidentiality and does not want anyone in her family to know she is being tested, including her identical twin sister. Your patient is tested and found to carry an altered gene that gives her an 85% lifetime risk of developing breast cancer and a 60% lifetime risk of developing ovarian cancer. At the result-

disclosure session, she once again reiterates that she does not want anyone in her family to know her test results.

a. Knowing that a familial mutation is occurring in this family, what would be your next course of action in this case?

b. Is it your duty to contact members of this family despite the request of your patient? Where do your obligations lie—with your patient or with the patient's family? Would it be inappropriate to try to convince the patient to share her results with her family members?

15. A young woman (proband) and her partner are referred for prenatal genetic counseling because the woman has a family history of sickle cell anemia. The proband has sickle cell trait (*Ss*), and her partner is not a carrier nor does he have sickle cell anemia (*SS*). Prenatal testing indicates that the fetus is affected with sickle cell anemia (*ss*). The results of this and other tests indicate that the only way the fetus could have sickle cell disease is if the woman's partner is not the father of the fetus. The couple is at the appointment seeking their test results.

a. How would you handle this scenario? Should you have contacted the proband beforehand to explain the results and the implications of these results?

b. Is it appropriate to keep this information from the partner because he believes he is the father of the baby? What other problems do you see with this case?

Internet Activities

 Internet Activities are critical thinking exercises using the resources of the World Wide Web to enhance the principles and issues covered in this chapter. For a full set of links and questions investigating the topics described below, log on to **http://biology.brookscole.com/cummings7**

1. *Overview and History of Genetic Counseling.* At the *Access Excellence: Classics Collection* site, click on the link to the article "Genetic Counseling: Coping with the Impact of Human Disease." This article gives an overview of the history of genetic counseling and how genetic counseling is used today. How does the use of genetic information by eugenicists early in the twentieth century compare to the use of genetic information by genetic counselors today? (For review, you may want to refer to Chapter 1, where eugenics was first discussed.) What kinds of ethical questions and issues may arise as a result of genetic counseling?

2. *Genetic Counseling Resources.* The *New York Online Access to Health* program has an excellent home page on genetic disorders and genetic counseling. This site is a good place to start if you or someone in your family has any concerns about genetic disorders.

How would you vote now?

The surplus embryos created in the process of IVF are routinely stored in liquid nitrogen. Some may be used in subsequent attempts at pregnancy, but many remain in storage. These embryos have several possible fates: they can be stored indefinitely, thawed and discarded, donated to researchers for use in stem cell research, or donated to other couples. Some nations, like Sweden and Great Britain, limit the time unused embryos can be stored before destruction. Now that you know more about IVF, ART, and the issues surrounding reproductive technology, what do you think? If you were having IVF, what would you want done with the extra embryos? Visit the Human Heredity Companion website to find out more on the issue, then cast your vote online at **http://biology.brookscole.com/cummings7**

 For further reading and inquiry, log on to **InfoTrac College Edition,** your world-class online library, including articles from nearly 5,000 periodicals, at **http://infotrac.thomsonlearning.com**

17

Genes and the Immune System

ABOUT every two hours, someone in the United States dies while waiting for an organ transplant. At any given time, about 50,000 people are waiting for organ transplants. Although more Americans are signing pledge cards to become organ donors at death, the demand for organs far outstrips the supply. To address this shortage, scientists and biotechnology companies are developing an alternative source of organs: animals. Nonhuman primates such as baboons or chimpanzees are poor candidates as organ donors; these are endangered species and harbor viruses (HIV originated in nonhuman primates) that may cause disease in humans. Most attention is focused on using a strain of mini-pigs developed over 30 years ago as organ donors. These pigs have major organs (hearts, livers, kidneys, etc.) that are about the same size as those of adult humans and have similar physiology.

The major stumbling block to xenotransplants (transplants across species) using pigs as organ donors, or any other animal, for that matter, is rejection by the immune system of the recipient. To overcome this problem, researchers have transferred human genes to pigs so that their organs carry molecular markers found on human organs. Other workers have deleted specific pig genes to make their organs look more like human organs to the immune system. More radical approaches to making pigs and humans compatible for transplants involves altering the immune system of the human recipient so that a transplanted pig organ will be tolerated. To do this, purified bone marrow cells from the donor pig are infused into the recipient. In this way, the recipient's immune system accepts the donor pig organ. Trials across species in animal–animal transplants have been successful.

Proponents of xenotransplantation point to the lives that will be saved if

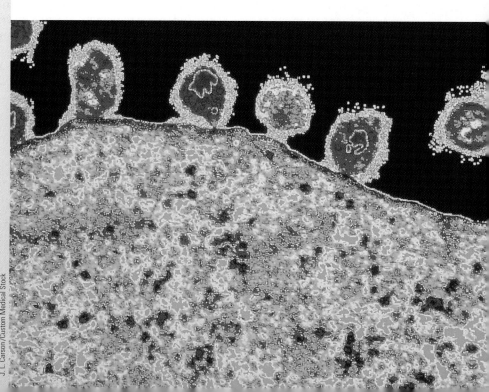

J. L. Carson/Custom Medical Stock

pig organs can be used for organ transplants. Opponents point out that there is no evidence that pig organs will work properly in humans and that pig organs may harbor harmful viruses that will be transferred to the human recipient. Others question the ethics of genetically modifying animals with human genes or modifying humans by transplanting parts of the pig's immune system.

How would you vote?

Organ donations are unable to keep up with the demand, and thousands of people die each year while awaiting transplants. Using pigs that have been genetically modified to carry human genes that prevent transplant rejection and modifying the immune system of human recipients by injection of pig bone marrow cells are two methods of overcoming the inherent problems of organ transplantation between species. Do you think it is ethical to genetically modify pigs with human genes, or to modify humans by giving them a pig immune system to accept transplanted organs? Visit the Human Heredity Companion website to find out more on the issue, then cast your vote online at **http://biology.brookscole.com /cummings7**

17.1 The Immune System Defends the Body Against Infection

In the course of an average day, we encounter **pathogens** (disease-causing agents) of many kinds: viruses, bacteria, fungi, and parasites. Fortunately, we possess several levels of defense against infection. Each level brings an increasingly aggressive response to attempts to invade and cause damage. Humans have three levels of defense: (1) the skin and the organisms that inhabit it, (2) nonspecific responses such as inflammation, and (3) specific responses in the form of an immune reaction.

The skin is a barrier to infectious agents, such as viruses and bacteria, and prevents them from entering the body. The skin's outer surface is home to bacteria, fungi, and even mites, but they cannot penetrate the protective layers of dead skin cells to cause infection. In fact, these organisms help defend the body against infection by inhabiting the skin and body linings, setting up conditions that are unfavorable for pathogens.

Nonspecific responses, such as inflammation, are designed to (1) block entry of disease-causing agents into the body and (2) block the spread of infectious agents if they get into the body. If these defenses do not stop the disease-causing agents, the immune system can make use of two types of specific responses: antibody-mediated immunity and cell-mediated immunity. In addition, the immune system is responsible for the success or failure of blood transfusions and organ transplants.

Keep in mind as you read

■ Humans have three defenses against infection: the skin, inflammation, and the immune response.

■ The immune system has two components: antibody-mediated immunity and cell-mediated immunity.

■ The A and O blood types are the most common, and B and AB are the rarest.

■ Disorders of the immune system can be inherited or acquired by infection.

■ **Pathogens**   Disease-causing agents.

In this chapter, we examine the cells of the immune system and their mobilization in an immune response. We also consider how the immune system determines blood groups and affects mother–fetus compatibility. The immune system also plays a role in organ transplants and in determining risk factors for a wide range of diseases. Finally, we describe a number of disorders of the immune system, including how AIDS acts to cripple the immune response of infected individuals.

Keep in mind

■ Humans have three defenses against infection: the skin, inflammation, and the immune response.

17.2 The Complement System Kills Microorganisms

The **complement system** is a chemical defense system that works in both nonspecific responses (inflammation) and specific responses (immune response). Its name derives from the way it complements the action of the immune system. About 20 different complement proteins are synthesized in the liver and circulate in the bloodstream in the form of inactive precursors. Complement proteins can mount several different reactions to infection. In one response, complement proteins at the site of infection bind to bacterial cells, activating a second protein, which in turn activates a third, and so forth, in a cascade of activation. Several components in this pathway form a large, cylindrical multiprotein, called the **membrane-attack complex (MAC)**. The MAC embeds itself in the plasma membrane of an invading microorganism, creating a pore (▶ Figure 17.1). Fluid flows into the cell in response to an osmotic gradient, eventually bursting the cell (▶ Active Figure 17.2).

In addition to directly destroying microorganisms, some complement proteins guide phagocytes to the site of infection. Other components aid the immune response by binding to the outer surface of microorganisms and marking them for destruction.

17.3 The Inflammatory Response Is a General Reaction

If microorganisms penetrate the skin or the cells lining the respiratory, digestive, or urinary systems, a nonspecific response, called the inflammatory reaction, develops (▶ Active Figure 17.3). Cells infected by microorganisms release chemical signals, including **histamine.** These signals increase blood flow in the affected area (that is why the area around a cut or scrape gets red and warm). The increased heat creates an unfavorable environment for microorganism growth, mobilizes white blood cells, and raises the metabolic rate in nearby cells. These reactions promote healing. Additional white blood cells migrate to the area in response to the chemical signals, to engulf and destroy the invading microorganisms.

If infection persists, capillaries in the infected area become leaky and plasma flows into the injured tissue, causing it to swell. Complement proteins become activated and attack the invading bacteria. Clotting factors in the plasma trigger a cascade of small blood clots that seal off the injured area, preventing the escape of invading organisms. Several types of white blood cells, including macrophages, are recruited to destroy the invading bacteria. Finally, the area becomes targeted by white blood cells that clean up dead viruses, bacteria, or fungi and dispose of dead cells and debris.

This chain of events, beginning with the release of chemical signals and ending with cleanup, is the **inflammatory response.** This response is an active defense mechanism

■ **Complement system** A chemical defense system that kills microorganisms directly, supplements the inflammatory response, and works with (complements) the immune system.

■ **Membrane-attack complex (MAC)** A large, cylindrical multiprotein that embeds itself in the plasma membrane of an invading microorganism and creates a pore through which fluids can flow, eventually bursting the microorganism.

■ **Histamine** A chemical signal produced by mast cells that triggers dilation of blood vessels.

■ **Inflammatory response** The body's reaction to invading microorganisms, a nonspecific active defense mechanism that the body employs to resist infection.

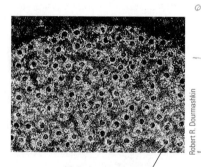

Hole in membrane

Robert R. Dourmashkin

▲ FIGURE 17.1 The complement system forms membrane attack complexes (MACs) in response to infection. The MACs insert themselves into the plasma membrane of the invading cell, forming a pore. This causes water to flow into the cell by osmosis, bursting the cell.

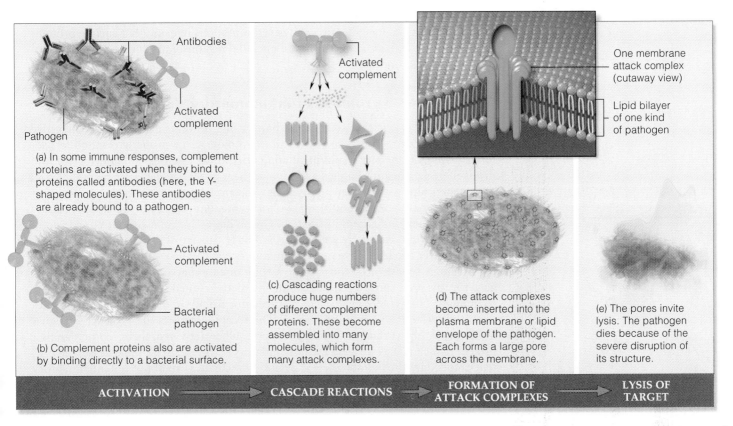

(a) In some immune responses, complement proteins are activated when they bind to proteins called antibodies (here, the Y-shaped molecules). These antibodies are already bound to a pathogen.

(b) Complement proteins also are activated by binding directly to a bacterial surface.

(c) Cascading reactions produce huge numbers of different complement proteins. These become assembled into many molecules, which form many attack complexes.

(d) The attack complexes become inserted into the plasma membrane or lipid envelope of the pathogen. Each forms a large pore across the membrane.

(e) The pores invite lysis. The pathogen dies because of the severe disruption of its structure.

ACTIVATION → CASCADE REACTIONS → FORMATION OF ATTACK COMPLEXES → LYSIS OF TARGET

▲ ACTIVE FIGURE 17.2 The complement system responds in several ways to infection. (a) One way begins with binding to antibodies. In this example, the antibodies are bound to the surface of bacterial cells. (b) The complement system can also be activated by binding directly to the surface of an invading bacterial cell. Both pathways lead to the formation of membrane attack complexes (MACs) and the destruction of the invading cell.

HUMAN Genetics ℰ Now™ Learn more about the complement system by viewing the animation on Human GeneticsNow at **http://now.ilrn.com/cummings7**

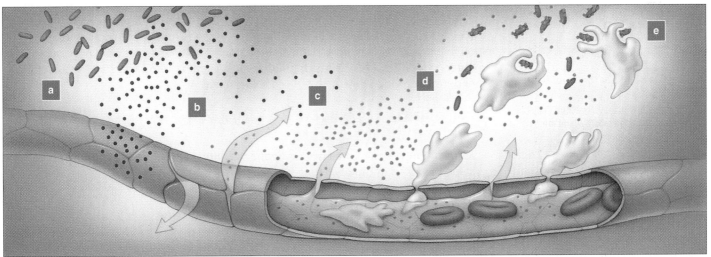

(a) Bacteria invade a tissue and directly kill cells or release metabolic products that damage tissue.

(b) Mast cells in tissue release histamine, which then triggers arteriole vasodilation (hence redness and warmth) as well as increased capillary permeability.

(c) Fluid and plasma proteins leak out of capillaries; localized edema (tissue swelling) and pain result.

(d) Complement proteins attack bacteria. Clotting factors wall off inflamed area.

(e) Neutrophils, macrophages, and other phagocytes engulf invaders and debris. Macrophage secretions attract even more phagocytes, directly kill invaders, and call for fever and for T and B cell proliferation.

▲ ACTIVE FIGURE 17.3 The acute inflammatory response following a bacterial infection.

HUMAN Genetics ℰ Now™ Learn more about the inflammatory response by viewing the animation on Human GeneticsNow at **http://now.ilrn.com/cummings7**

that the body employs to resist infection. This reaction is usually enough to stop the spread of infection. In some cases, however, mutations in genes that encode proteins involved in the inflammatory response alter the response, producing clinical symptoms of an inflammatory disease.

Genetics can be related to inflammatory diseases.

The inner cell layer of the intestine is a barrier that prevents bacteria in the digestive system from crossing into the body. Failure to monitor or properly respond to bacteria crossing this barrier results in inflammatory bowel diseases. Inflammatory bowel diseases are genetically complex and involve the interaction of environmental factors with genetically predisposed individuals. Ulcerative colitis (OMIM 191390) and Crohn disease (OMIM 266600) are two forms of inflammatory bowel disease caused by malfunctions in the immune system. Crohn disease occurs with a frequency of 1 in 1,000 individuals, mostly young adults. The frequency of this disorder has greatly increased over the last 50 years, presumably as the result of unknown environmental factors. A genetic predisposition to Crohn disease maps to chromosome 16. The gene for this predisposition has been identified and cloned using recombinant DNA techniques. The *NOD2* gene encodes a receptor found on the surface of monocytes and other cells of the immune system. The receptor detects the presence of signal molecules on the surface of invading bacteria. Once activated, the receptor signals a protein in the nucleus to begin the inflammatory response. In Crohn disease, the protein encoded by the mutant allele is defective and causes an abnormal inflammatory response that damages the intestinal wall. Carrying the mutant allele of *NOD2* confers only a predisposition; unknown environmental factors and other genes are probably involved in this disorder.

17.4 The Immune Response Is a Specific Defense Against Infection

If the nonspecific inflammatory response fails to stop an infection, another, more powerful system—the immune response—is called into action. The immune system generates a chemical and cellular response that neutralizes and/or destroys viruses, bacteria, fungi, and cancer cells. The immune system is more effective than the nonspecific defense system and has a memory component that remembers previous encounters with infectious agents. Immunological memory allows a rapid, massive response to a second exposure to a foreign substance.

Keep in mind

■ The immune system has two components: antibody-mediated immunity and cell-mediated immunity.

How does the immune response function?

The immune response is mediated by white blood cells called **lymphocytes.** The two main cell types in the immune system are called **B cells** and **T cells.** Both cell types are formed by mitotic division from **stem cells** located in bone marrow. Both play a mediating role in the immune response.

Once produced, B cells remain in the bone marrow until they mature. As they mature, each B cell becomes genetically programmed to produce large quantities of a unique protein called an **antibody.** Antibodies are displayed on the surface of the B cell and bind to foreign molecules and microorganisms such as bacterial or fungal

■ Lymphocytes White blood cells that originate in bone marrow and mediate the immune response.

■ B cells A type of lymphocyte that matures in the bone marrow and mediates antibody-directed immunity.

■ T cells A type of lymphocyte that undergoes maturation in the thymus and mediates cellular immunity.

■ Stem cells Cells in bone marrow that produce lymphocytes by mitotic division.

■ Antibody A class of proteins produced by B cells that bind to foreign molecules (antigens) and inactivate them.

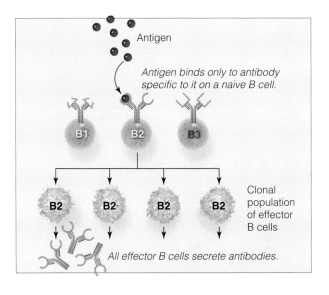

◀ ACTIVE FIGURE 17.4 Clonal selection. An antigen binds to a specific antibody on a B cell called a naïve B cell because it has not encountered an antigen before. This encounter triggers mitosis and the buildup of a large population of cells derived from the activated cell. Since all cells in the population are derived from a single ancestor, they are clones.

HUMAN Genetics ⒺNow™ Learn more about clonal selection by viewing the animation on Human GeneticsNow at http://now.ilrn.com/cummings7

cells or toxins in order to inactivate them. Molecules that bind to antibodies are called **antigens** (*anti*body *gen*erators) because they trigger, or generate, an antibody response. Most antigens are proteins or proteins combined with polysaccharides, but *any* molecule, regardless of its source, that can bind to an antibody is an antigen.

T cells are formed in the bone marrow. While still immature, these cells migrate from the bone marrow to the thymus gland and become programmed to produce unique cell-surface proteins called **T-cell receptors (TCRs).** These receptors bind to proteins on the surface of cells infected with viruses, bacteria, or intracellular parasites. Mature T cells circulate in the blood and concentrate in lymph nodes and the spleen.

It is important to remember that each B cell makes only one type of antibody and each T cell makes only one type of receptor. Since there are literally billions of possible antigens, there are billions of possible combinations of antibodies and TCRs. When an antigen binds to a TCR or antibody on the surface of a T or B cell, it stimulates that cell to divide, producing a large population of genetically identical descendants, or clones, all with the same TCR or antibody. This process is called clonal selection (▶ Active Figure 17.4).

Specific molecular markers on cell surfaces also play a role in the immune response. Each cell of our body carries recognition molecules that prevent the immune system from attacking our organs and tissues. These markers are encoded by a set of genes on chromosome 6 called the **major histocompatibility complex (MHC).** The MHC proteins can bind to antigens, stimulating the immune response. MHC proteins also play a major role in successful organ transplants, as will be described in a later section.

The immune system has two interconnected parts, **antibody-mediated immunity,** regulated by B cell antibody production, and **cell-mediated immunity,** controlled by T cells (▶ Active Figure 17.5). The two systems are connected by T cells called **helper T cells.** Antibody-mediated reactions detect antigens circulating in the blood or body fluids and interact with helper T cells, which in turn signal the B cell with antibodies against that antigen to divide.

Cell-mediated immunity attacks cells of the body infected by viruses or bacteria. T cells also protect against infection by parasites, fungi, and protozoans. One group of T cells can also kill cells of the body if they become cancerous.

▶ Table 17.1 compares the antibody-mediated and cell-mediated immune reactions.

The antibody-mediated immune response involves several stages.

The antibody-mediated immune response has several stages: antigen detection, activation of helper T cells, and antibody production by B cells. A specific cell type of

■ **Antigens** Molecules carried or produced by microorganisms that initiate antibody production.

■ **T-cell receptors (TCRs)** Unique proteins on the surface of T-cells that bind to specific proteins on the surface of cells infected with viruses, bacteria, or intracellular parasites.

■ **Major histocompatibility complex (MHC)** A set of genes on chromosome 6 that encode recognition molecules that prevent the immune system from attacking a body's own organs and tissues.

■ **Antibody-mediated immunity** Immune reaction that protects primarily against invading viruses and bacteria using antibodies produced by plasma cells.

■ **Cell-mediated immunity** Immune reaction mediated by T cells directed against body cells that have been infected by viruses or bacteria.

■ **Helper T cell** A lymphocyte that stimulates production of antibodies by B cells when an antigen is present.

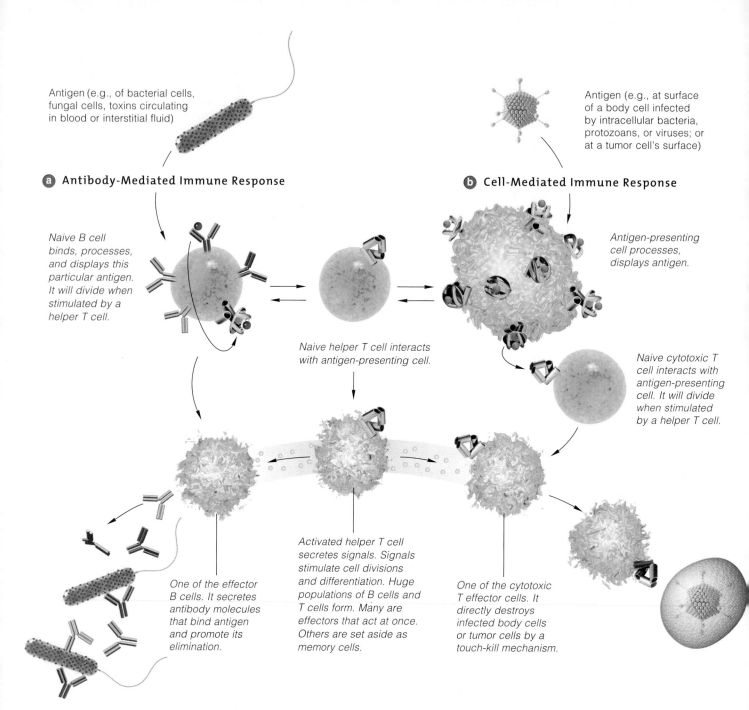

a Antibody-Mediated Immune Response

Antigen (e.g., of bacterial cells, fungal cells, toxins circulating in blood or interstitial fluid)

Naive B cell binds, processes, and displays this particular antigen. It will divide when stimulated by a helper T cell.

Naive helper T cell interacts with antigen-presenting cell.

One of the effector B cells. It secretes antibody molecules that bind antigen and promote its elimination.

Activated helper T cell secretes signals. Signals stimulate cell divisions and differentiation. Huge populations of B cells and T cells form. Many are effectors that act at once. Others are set aside as memory cells.

b Cell-Mediated Immune Response

Antigen (e.g., at surface of a body cell infected by intracellular bacteria, protozoans, or viruses; or at a tumor cell's surface)

Antigen-presenting cell processes, displays antigen.

Naive cytotoxic T cell interacts with antigen-presenting cell. It will divide when stimulated by a helper T cell.

One of the cytotoxic T effector cells. It directly destroys infected body cells or tumor cells by a touch-kill mechanism.

○ ▲ **ACTIVE FIGURE 17.5** Overview of the cell-cell interactions in the antibody-mediated and cell-mediated immune responses. Memory cells produced in the first encounter with an antigen (cells that first encounter an antigen are called naïve cells) are activated in subsequent infections with this antigen and mount a rapid, massive response to the antigen.

HUMAN Genetics ⒺNow™ Learn more about the antibody-mediated and cell-mediated immune responses by viewing the animation on Human GeneticsNow at **http://now.ilrn.com/cummings7**

the immune system controls each of these steps. Let's start with a B cell as it encounters an antigen and follow the stages of antibody production and immune response (▶ Active Figure 17.6). In this example, a B cell with antibodies displayed on its surface wanders through the circulatory system and the spaces between cells searching for foreign (nonself) antigenic molecules, viruses, or microorganisms. When it encounters an antigen, the antigen binds to a surface antibody, and the antigen molecule is internalized and partially destroyed with enzymes. Small fragments

Table 17.1 Comparison of Antibody-Mediated and Cell-Mediated Immunity

Antibody-Mediated	Cell-Mediated
Principal cellular agent is the B cell. B cell responds to bacteria, bacterial toxins, and some viruses.	Principal cellular agent is the T cell. T cells respond to cancer cells, virally infected cells, single-celled fungi, parasites, and foreign cells in an organ transplant.
When activated, B cells form memory cells and plasma cells, which produce antibodies to these antigens.	When activated, T cells differentiate into memory cells, cytotoxic cells, suppressor cells, and helper cells; cytotoxic T cells attack the antigen directly.

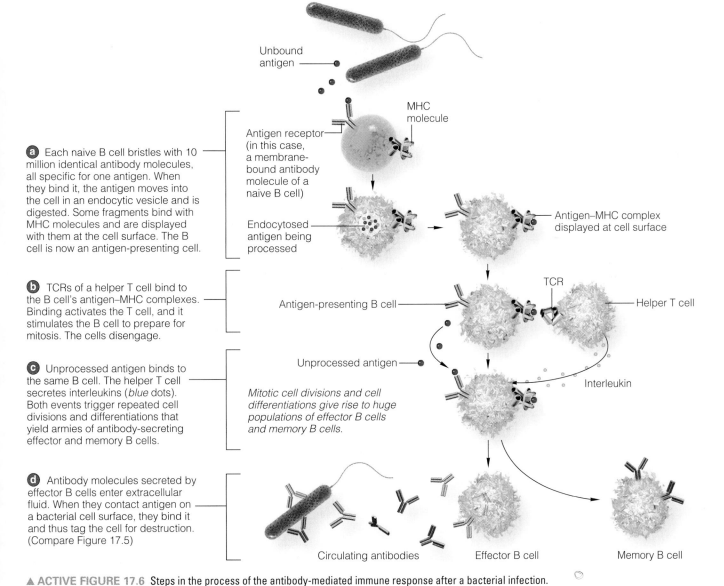

a Each naive B cell bristles with 10 million identical antibody molecules, all specific for one antigen. When they bind it, the antigen moves into the cell in an endocytic vesicle and is digested. Some fragments bind with MHC molecules and are displayed with them at the cell surface. The B cell is now an antigen-presenting cell.

b TCRs of a helper T cell bind to the B cell's antigen–MHC complexes. Binding activates the T cell, and it stimulates the B cell to prepare for mitosis. The cells disengage.

c Unprocessed antigen binds to the same B cell. The helper T cell secretes interleukins (*blue* dots). Both events trigger repeated cell divisions and differentiations that yield armies of antibody-secreting effector and memory B cells.

d Antibody molecules secreted by effector B cells enter extracellular fluid. When they contact antigen on a bacterial cell surface, they bind it and thus tag the cell for destruction. (Compare Figure 17.5)

Unbound antigen

MHC molecule

Antigen receptor (in this case, a membrane-bound antibody molecule of a naive B cell)

Endocytosed antigen being processed

Antigen–MHC complex displayed at cell surface

TCR

Antigen-presenting B cell

Helper T cell

Unprocessed antigen

Interleukin

Mitotic cell divisions and cell differentiations give rise to huge populations of effector B cells and memory B cells.

Circulating antibodies

Effector B cell

Memory B cell

▲ ACTIVE FIGURE 17.6 Steps in the process of the antibody-mediated immune response after a bacterial infection.

HUMAN Genetics ⊘ Now™   Learn more about the antibody-mediated immune response by viewing the animation on Human GeneticsNow at **http://now.ilrn.com/cummings7**

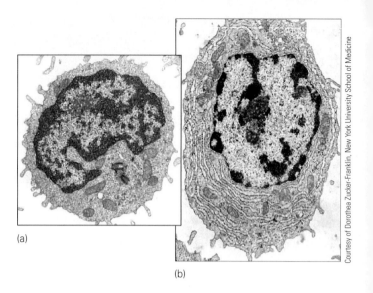

(a)

(b)

Courtesy of Dorothea Zucker-Franklin, New York University School of Medicine

■ **Effector cells** Daughter cells of B cells, which synthesize and secrete 2,000 to 20,000 antibody molecules per second into the bloodstream.

■ **Memory B cell** A long-lived B cell produced after exposure to an antigen that plays an important role in secondary immunity.

■ **Immunoglobulins (Ig)** The five classes of proteins to which antibodies belong.

of the antigen move to the outer surface of the B cell membrane, and the B cell becomes an antigen-presenting cell.

This antigen-presenting B cell encounters a lymphocyte called a helper T cell. Surface TCRs on the T cell make contact with the antigen fragment on the B cell, activating the T cell. The activated T cell in turn identifies and activates B cells that synthesize an antibody against the antigen encountered by the T cell. The activated B cells divide and form two types of daughter cells. The first is **effector cells,** which synthesize and secrete 2,000 to 20,000 antibody molecules *per second* into the bloodstream (▶ Figure 17.7). A second cell type, a **memory B cell,** also forms at this time. Effector cells live only a few days, but memory cells have a life span of months or even years. Memory cells are part of the immune memory system and are described in a later section.

Antibodies are molecular weapons against antigens.

Antibodies are Y-shaped protein molecules that bind to specific antigens in a lock-and-key manner to form an antigen–antibody complex (▶ Active Figure 17.8). Antibodies are secreted by effector cells and circulate in the blood and lymph, while others remain attached to the surface of B cells. Antibodies belong to a class of proteins known as **immunoglobulin (Ig)** molecules.

There are five classes of Igs, abbreviated as IgD, IgM, IgG, IgA, and IgE. Each class has a unique structure, size, and function (▶ Table 17.2). Antibody molecules have two identical long polypeptides (H chains) and two identical short polypeptides (L chains). The chains are held together by chemical bonds (▶ Active Figure 17.8).

Antibody structure is related to its functions: (1) recognize and bind antigens and (2) inactivate the antigen. At one end of the antibody is an antigen-binding site formed by the ends of the H and L chains. This site recognizes and binds to a specific antigen. Formation of an antigen–antibody complex leads to the destruction of an antigen.

Humans can produce billions of different antibody molecules, each of which can bind to a different antigen. Because there are billions of such combinations, it is impossible that each antibody molecule is encoded directly in the genome; there simply is not enough DNA in the human genome to encode hundreds of millions or billions of antibodies.

The vast number of different antibodies is produced as a result of genetic recombination in three clusters of antibody genes: the heavy-chain genes (H genes) on chromosome 14, and two clusters of light-chain genes, the L genes on chromosome 2 and the L genes on chromosome 22. These recombination events take place during B-cell maturation, producing a unique gene that produces one type of antibody. This re-

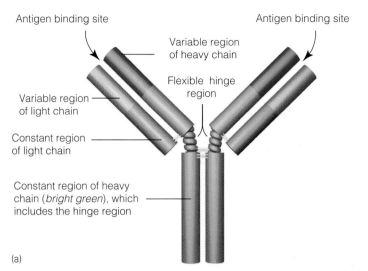

Antigen binding site
Antigen binding site
Variable region of heavy chain
Flexible hinge region
Variable region of light chain
Constant region of light chain
Constant region of heavy chain (*bright green*), which includes the hinge region

(a)

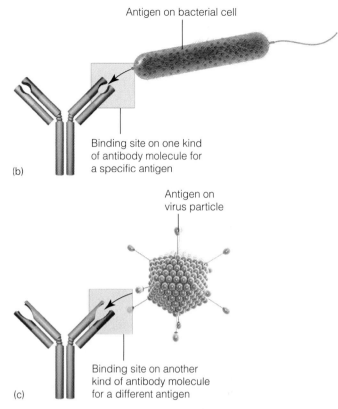

Antigen on bacterial cell

Binding site on one kind of antibody molecule for a specific antigen

(b)

Antigen on virus particle

Binding site on another kind of antibody molecule for a different antigen

(c)

▲ ACTIVE FIGURE 17.8 (a) Antibody molecules are made up of two different proteins (an H chain and an L chain). The molecule is Y-shaped and forms a specific antigen-binding site at the ends. (b, c) At the antigen-binding site, only antigens that match the antigen-binding site will fit into the site.

HUMAN Genetics ⊘ Now™ Learn more about antibodies and antigens by viewing the animation on Human GeneticsNow at http://now.ilrn.com/cummings7

Table 17.2 Types and Functions of the Immunoglobulins

Class	Location and Function
IgD	Present on surface of many B cells, but function uncertain; may be a surface receptor for B cells; plays a role in activating B cells.
IgM	Found on surface of B cells and in plasma; acts as a B-cell surface receptor for antigens secreted early in primary response; powerful agglutinating agent.
IgG	Most abundant immunoglobulin in the blood plasma; produced during primary and secondary response; can pass through the placenta, entering fetal bloodstream, thus providing protection to fetus.
IgA	Produced by plasma cells in the digestive, respiratory, and urinary systems, where it protects the surface linings by preventing attachment of bacteria to surfaces of epithelial cells; also present in tears and breast milk; protects lining of digestive, respiratory, and urinary systems.
IgE	Produced by plasma cells in skin, tonsils, and the digestive and respiratory systems, overproduction is responsible for allergic reactions, including hay fever and asthma.

arranged gene is stable and is passed on to all daughter B cells. This process of recombination makes it possible to produce billions of possible antibody combinations from only three gene sets.

T cells mediate the cellular immune response.

There are several types of T cells in the immune system (▶ Table 17.3). Helper T cells, described earlier, activate B cells to produce antibodies. **Suppressor T cells** slow down and stop the immune response and act as an "off" switch for the immune sys-

■ **Suppressor T cells** T cells that slow or stop the immune response of B cells and other T cells.

Killer T cells T cells that destroy body cells infected by viruses or bacteria. These cells can also directly attack viruses, bacteria, cancer cells, and cells of transplanted organs.

tem. A third type, the cytotoxic or **killer T cells,** also known as NK (natural killer) cells, finds and destroys cells of the body that are infected with a virus, bacteria, or other infectious agents (▶ Active Figure 17.9).

If a cell becomes infected with a virus, viral proteins will appear on its surface. These foreign antigens are recognized by receptors on the surface of a killer T cell. The T cell attaches to the infected cell and secretes a protein, perforin, which punches holes in the plasma membrane of the infected cell. The cytoplasmic contents of the infected cell leak out through these holes, and the infected cell dies and is removed by phagocytes.

Killer T cells also kill cancer cells (▶ Figure 17.10) and transplanted organs if they

a A virus particle infects a macrophage. The host cell's pirated metabolic machinery makes viral proteins, which are antigens. Macrophage enzymes cleave antigen into fragments.

b Some antigen fragments bind to class I MHC molecules, which occur on all nucleated cells. On the infected cells, antigen–MHC complexes form and move to the cell surface, where they are displayed.

e Meanwhile, antigen–MHC complexes on the first macrophage bind to TCRs of a cytotoxic T cell. Binding, in combination with interleukin signals from the helper T cell, stimulate the cytotoxic T cell to divide and differentiate, forming huge populations of effector and memory T cells.

g An effector cytotoxic T cell touch-kills the infected cell by releasing perforins and toxic chemicals (*green* dots) onto it.

h The effector disengages and reconnoiters for more targets. Its perforins punch holes in the infected cell's plasma membrane; its toxins disrupt the target's organelles and DNA, so the infected cell dies.

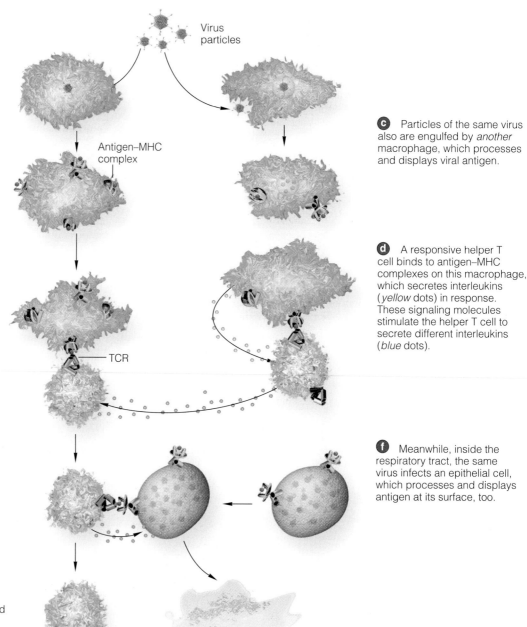

Virus particles

Antigen–MHC complex

TCR

c Particles of the same virus also are engulfed by *another* macrophage, which processes and displays viral antigen.

d A responsive helper T cell binds to antigen–MHC complexes on this macrophage, which secretes interleukins (*yellow* dots) in response. These signaling molecules stimulate the helper T cell to secrete different interleukins (*blue* dots).

f Meanwhile, inside the respiratory tract, the same virus infects an epithelial cell, which processes and displays antigen at its surface, too.

▲ **ACTIVE FIGURE 17.9** A diagram of the steps in the T-cell–mediated immune response.

HUMAN Genetics ⊘ Now™ Learn more about the T-cell–mediated immune response by viewing the animation on Human GeneticsNow at **http://now.ilrn.com/cummings7**

Table 17.3 Summary of T Cells

Cell Type	Action
Killer T cells	Destroy body cells infected by viruses and attack and kill bacteria, fungi, parasites, and cancer cells.
Helper T cells	Produce a growth factor that stimulates B-cell proliferation and differentiation and also stimulates antibody production by plasma cells; enhance activity of cytotoxic T cells.
Suppressor T cells	May inhibit immune reaction by decreasing B- and T-cell activity and B- and T-cell division.
Memory T cells	Remain in body awaiting reintroduction of antigen, when they proliferate and differentiate into cytotoxic T cells, helper T cells, suppressor T cells, and additional memory cells.

recognize them as foreign. ▶ Table 17.4 summarizes the nonspecific and specific reactions of the immune system.

The immune system has a memory function.

Ancient writers observed that exposure to certain diseases made people resistant to second infections by the same disease. B and T memory cells, produced as a result of the first infection, control this resistance. With memory cells present, a second exposure to the same antigen results in an immediate, large-scale production of antibodies and killer T cells. Because of the presence of the memory cells, the second reaction is faster, more massive, and lasts longer than the primary immune response.

The immune response controlled by memory cells is the reason that we can be vaccinated against infectious diseases. A **vaccine** stimulates the production of memory cells against a disease-causing agent. A vaccine is really a weakened, disease-causing antigen, given orally or by injection, that provokes a primary immune response and the production of memory cells. Often, a second dose is administered to elicit a secondary response that raises, or "boosts," the number of memory cells (which is why such shots are called booster shots).

Vaccines are made from killed or weakened strains (called attenuated strains) of disease-causing agents that stimulate the immune system but do not produce life-threatening symptoms of the disease. Recombinant DNA methods are now used to prepare vaccines against a number of diseases that affect humans and farm animals. The irony is that although vaccination is a proven and effective means of controlling a wide range of infectious diseases, fewer people in the United States are choosing to be vaccinated.

■ **Vaccine** A preparation containing dead or weakened pathogens that elicits an immune response when injected into the body.

■ **Blood type** One of the classes into which blood can be separated based on the presence or absence of certain antigens.

17.5 Blood Types Are Determined by Cell-Surface Antigens

Antigens on the surface of blood cells determine compatibility in blood transfusions. There are about 30 known antigens on blood cells; each of these constitutes a blood group or **blood type**.

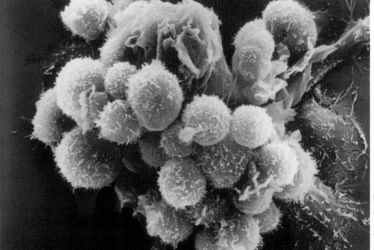

▲ FIGURE 17.10 Killer T cells (*yellow*) attacking a cancer cell (*red*).

Table 17.4 Nonspecific and Specific Immune Responses to Bacterial Invasion

Nonspecific Immune Mechanisms	Specific Immune Mechanisms
INFLAMMATION	Processing and presenting of bacterial antigen by macrophages
Engulfment of invading bacteria by resident tissue macrophages	Proliferation and differentiation of activated B-cell clone into plasma cells and memory cells
Histamine-induced vascular responses to increase blood flow to area, bringing in additional immune cells	Secretion by plasma cells of customized antibodies, which specifically bind to invading bacteria
Walling off of invaded area by fibrin clot	Enhancement by helper T cells, which have been activated by the same bacterial antigen processed and presented to them by macrophages
Migration of neutrophils and monocytes/macrophages to the area to engulf and destroy foreign invaders and to remove cellular debris	Binding of antibodies to invading bacteria and activation of mechanisms that lead to their destruction
Secretion by phagocytic cells of chemical mediators, which enhance both nonspecific and specific immune responses	Activation of lethal complement system
NONSPECIFIC ACTIVATION OF THE COMPLEMENT SYSTEM	Stimulation of killer cells, which directly lyse bacteria
Formation of hole-punching membrane attack complex that lyses bacterial cells	Persistence of memory cells capable of responding more rapidly and more forcefully should the same bacterial strain be encountered again
Enhancement of many steps of inflammation	

For successful transfusions, certain critical antigens of the donor and recipient must be identical. If transfused red blood cells do not have matching surface antigens, the recipient's immune system will produce antibodies against this antigen, clumping the transfused cells. The clumped blood cells block circulation in capillaries and other small blood vessels, with severe and often fatal results. In transfusions, two blood groups are of major significance: the ABO system and the Rh blood group.

ABO blood typing allows safe blood transfusions.

ABO blood types are determined by a gene I (for isoagglutinin) encoding an enzyme that alters a cell-surface protein. This gene has three alleles, I^A, I^B, and I^O, often written as A, B, and O. The A and B alleles each produce a slightly different version of the enzyme, and the O allele produces no gene product. Individuals with type A blood have A antigen on their red blood cells, so they do not produce antibodies against this cell-surface marker. However, people with type A blood have antibodies against the antigen encoded by the B allele (▶ Table 17.5). Those with type B blood carry the B antigen on their red cells and have antibodies against the A antigen. If you have type AB blood, both antigens are present on red blood cells and no anti-

Table 17.5 Summary of ABO Blood Types

Blood Type	Antigens on Plasma Membranes of RBCs	Antibodies in Blood	Safe to Transfuse To	Safe to Transfuse From
A	A	Anti-B	A, AB	A, O
B	B	Anti-A	B, AB	B, O
AB	A + B	None	AB	A, B, AB, O
O	—	Anti-A Anti-B	A, B, AB, O	O

bodies against A and B are made. Those with type O blood have neither antigen but do have antibodies against both the A and B antigen.

Because AB individuals have no antibodies against A or B, they can receive a transfusion of blood of any type. Type O individuals have neither antigen and can donate blood to anyone, even though their plasma contains antibodies against A and B; after transfusion, the concentration of these antibodies is too low to cause problems.

Keep in mind

■ The A and O blood types are the most common, and B and AB are the rarest.

When transfusions are made between incompatible blood types, several problems arise (see Spotlight on Genetically Engineered Blood). ▶ Figure 17.11 shows the cascade of reactions that follows transfusion of someone who is type A with type B blood. Antibodies to the B antigen are in the blood of the recipient. These bind to the transfused red blood cells, causing them to clump. The clumped cells restrict blood flow in capillaries, reducing oxygen delivery. Lysis of red blood cells releases large amounts of hemoglobin into the blood. The hemoglobin forms deposits in the kidneys that block the tubules of the kidney and often cause kidney failure.

Rh blood types can cause immune reactions between mother and fetus.

The Rh blood group (named for the rhesus monkey, in which it was discovered) includes those who can make the Rh antigen (Rh-positive, Rh$^+$) and those who cannot make the antigen (Rh-negative, Rh$^-$).

The Rh blood group is a major concern when there is incompatibility between mother and fetus, a condition known as **hemolytic disease of the newborn (HDN)**. This occurs most often when the mother is Rh$^-$ and the fetus is Rh$^+$ (▶ Active Figure 17.12). If Rh$^+$ blood from the fetus enters the Rh$^-$ maternal circulation, the mother's immune system makes antibodies against the Rh antigen. Mingling of fetal blood with that of the mother commonly occurs during birth, so the first Rh$^+$ child is often unaffected. However, in response to the presence of the Rh antigen, the mother makes antibodies against it, and any subsequent child that is Rh$^+$ evokes an immune response from the maternal immune system. Massive amounts of antibodies cross the placenta in late stages of pregnancy and destroy the fetus's red blood cells, resulting in HDN.

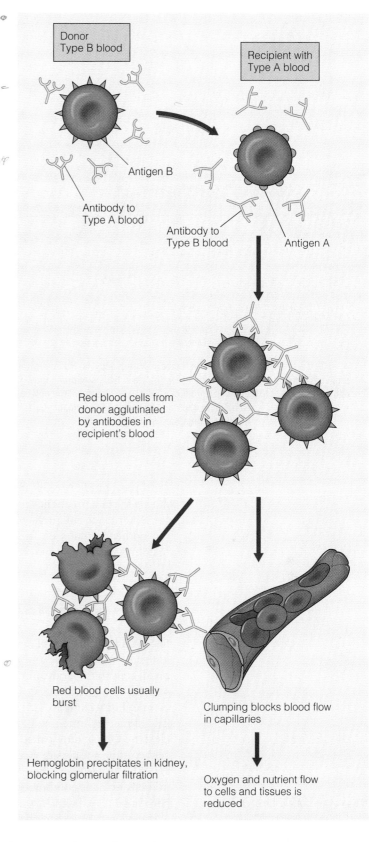

Donor
Type B blood

Recipient with
Type A blood

Antigen B

Antibody to
Type A blood

Antibody to
Type B blood

Antigen A

Red blood cells from donor agglutinated by antibodies in recipient's blood

Red blood cells usually burst

Clumping blocks blood flow in capillaries

Hemoglobin precipitates in kidney, blocking glomerular filtration

Oxygen and nutrient flow to cells and tissues is reduced

■ **Hemolytic disease of the newborn (HDN)** A condition of immunological incompatibility between mother and fetus that occurs when the mother is Rh$^-$ and the fetus is Rh$^+$.

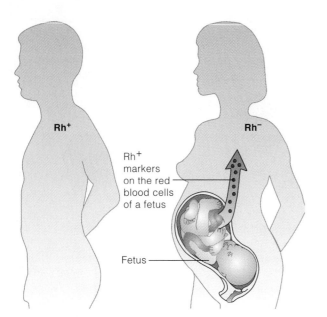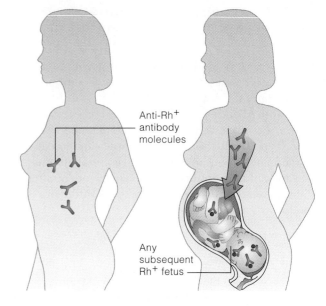

(a) A forthcoming child of an Rh⁻ woman and Rh⁺ man inherits the gene for the Rh⁺ marker. During pregnancy or childbirth, some of its cells bearing the marker may leak into the maternal bloodstream.

(b) The foreign marker stimulates her body to make antibodies. If she gets pregnant again and if this second fetus (or any other) inherits the gene for the marker, the circulating anti-Rh⁺ antibodies will act against it.

▲ ACTIVE FIGURE 17.12 The Rh factor and pregnancy. (a) Rh⁺ cells from the fetus can enter the maternal circulation at birth. The Rh⁻ mother produces antibodies against the Rh factor. (b) In a subsequent pregnancy, if the fetus is Rh⁺, the maternal antibodies cross into the fetal circulation and destroy fetal red blood cells, producing hemolytic disease of the newborn (HDN).

HUMAN
Genetics ⊗ Now™   Learn more about the Rh factor and pregnancy by viewing the animation on Human GeneticsNow at **http://now.ilrn.com/cummings7**

To prevent HDN, Rh⁻ mothers are given an Rh-antibody preparation (RhoGam) during the first pregnancy if the child is Rh⁺. The injected Rh antibodies move through the maternal circulatory system and destroy any Rh fetal cells that may have entered the mother's circulation. To be effective, this antibody must be administered before the maternal immune system can make its own antibodies against the Rh antigen.

17.6 Organ Transplants Must Be Immunologically Matched

Successful organ transplants and skin grafts depend on matches between cell-surface antigens of the donor and recipient. These antigens are cell-surface proteins found on all cells in the body and serve as identification tags, distinguishing self from nonself. In humans, a cluster of genes on chromosome 6, known as the major histocompatibility complex (MHC), produces these antigens. The *HLA* genes play a critical role in the outcome of transplants. The *HLA* complex consists of several gene clusters. One group, called class I, consists of *HLA-A*, *HLA-B*, and *HLA-C*. Adjacent to this is a cluster called class II, which consists of *HLA-DR*, *HLA-DQ*, and *HLA-DP*. A large number of alleles have been identified for each *HLA* gene, making millions of allele combinations possible. The array of *HLA* alleles on a given copy of chromosome 6 is known as a **haplotype**. Because each of us carries two copies of chromosome 6, we each have two *HLA* haplotypes (▶ Figure 17.13).

Because so many allele combinations are possible, it is rare that any two individuals have a perfect *HLA* match. The exceptions are identical twins, who have identical *HLA* haplotypes, and siblings, who have a 25% chance of being matched. In the example shown in ▶ Figure 17.13, each child receives one haplotype from each parent. As a result, four new haplotype combinations are represented in the children. (Thus siblings have a one in four chance of having the same haplotypes.)

■ **Haplotype** A cluster of closely linked genes or markers that are inherited together. In the immune system, the *HLA* alleles on chromosome 6 are a haplotype.

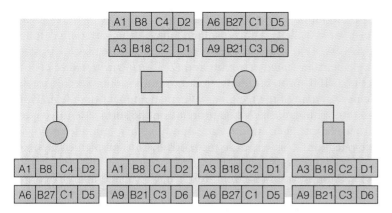

▲ FIGURE 17.13 The transmission of *HLA* haplotypes. In this simplified diagram, each haplotype contains four genes, each of which encodes a different antigen.

Successful transplants depend on *HLA* matching.

Successful organ transplants depend largely on matching *HLA* haplotypes between donor and recipient. Because there are so many *HLA* alleles, the best chance for a match is usually between related individuals, with identical twins having a 100% match. The order of preference for organ and tissue donors among relatives is identical twin, sibling, parent, and unrelated donor. Among unrelated donors and recipients, the chances for a successful match are only 1 in 100,000 to 1 in 200,000. Because the frequency of *HLA* alleles differs widely across ethnic groups, matches across groups are often more difficult. When *HLA* types are matched, the survival of transplanted organs is dramatically improved. Survival rates for matched and unmatched kidney transplants over a 4-year period are shown in ▶ Figure 17.14.

Genetic engineering makes animal–human organ transplants possible.

In the United States, about 18,000 organs are transplanted each year, but about 50,000 qualified patients are on waiting lists. Each year, almost 4,000 people on the waiting list die before receiving a transplant and another 100,000 die even *before* they are placed on the waiting list. Although the demand for organ transplants is rising, the number of donated organs is growing very slowly. Experts estimate that more than 50,000 lives would be saved each year if enough organs were available.

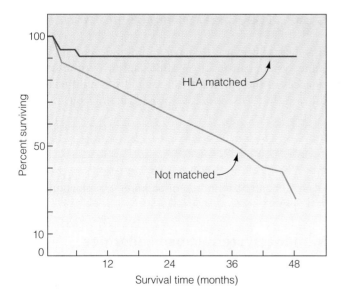

◀ FIGURE 17.14 The outcome of kidney transplants with (upper curve) and without (lower curve) *HLA* matching.

One way to increase the supply of organs is to use animal donors for transplants. Animal–human transplants (called **xenotransplants**) have been attempted many times, but with little success. Two important biological problems are related to xenotransplants: (1) complement-mediated rejection and (2) T-cell–mediated rejection. In complement rejection, MHC proteins on the organ from the donor species are detected by the complement system of the recipient. When an animal organ (e.g., from a pig) is transplanted into a human, the MHC proteins on the surface of the pig organ are so different that they trigger an immediate and massive immune response, known as hyperacute rejection. This reaction, mediated by the complement system, usually destroys the transplanted organ within hours.

To overcome this rejection, several research groups have isolated and cloned human genes that block the complement reaction. These genes were injected into fertilized pig eggs, and the resulting transgenic pigs carry human-recognition antigens on all their cells. Organs from these transgenic pigs should appear as human organs to the recipient's immune system, preventing a hyperacute rejection. Transplants from genetically engineered pigs to monkey hosts have been successful, but the ultimate step will be an organ transplant from a transgenic pig to a human.

Even if hyperacute rejection can be suppressed, however, transplanting pig organs will still cause problems with T-cell–mediated rejection of the transplant. Because transplants from pig donors to humans occur across species, the tendency toward rejection may be stronger and require the lifelong use of immunosuppressive drugs. These powerful drugs may be toxic when taken over a period of years or may weaken the immune system, paving the way for continuing rounds of infections. To deal with this problem, it may be necessary to transplant bone marrow from the donor pig to the human recipient. The resulting pig–human immune system (called a chimeric immune system) would recognize the pig organ as "self" and still retain normal human immunity. As far-fetched as this may sound, animal experiments using this approach have been successful in preventing rejection more than 2 years after transplantation without the use of immunosuppressive drugs. This same method has been used in human-to-human heart transplants to increase chances of success.

As recently as 10 years ago, the possibility of animal–human organ transplants seemed remote, more suited to science fiction than to medical fact. There are now more than 200 people in the United States who have received xenografts of animal cells or tissues. The advances described here make it likely that xenotransplants of major organs to humans will be attempted in the next few years. Although animal organ donors will probably become common in the near future, guidelines for transgenic donors and problems with immunosuppressive drugs and immune tolerance remain to be solved.

17.7 Disorders of the Immune System

We are able to resist infectious disease because we have an immune system. Unfortunately, failures in the immune system can result in abnormal or even absent immune responses. The consequences of these failures can range from mild inconvenience to systemic failure and death. In this section, we briefly catalog some ways in which the immune system can fail.

Keep in mind

■ Disorders of the immune system can be inherited or acquired by infection.

Overreaction in the immune system causes allergies.

Allergies result when the immune system overreacts to weak antigens that do not provoke an immune response in most people (▶ Figure 17.15). These weak antigens,

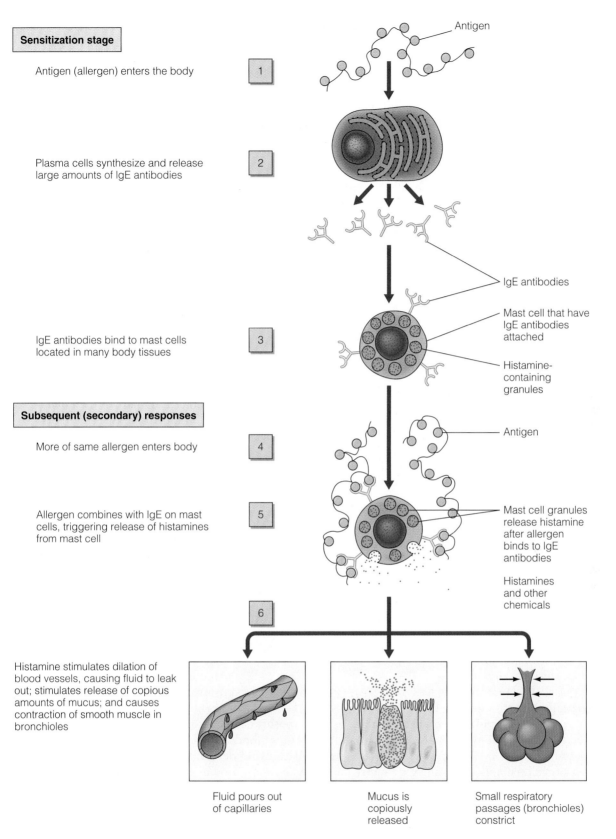

Antigen (allergen) enters the body

1

Antigen

2

Plasma cells synthesize and release large amounts of IgE antibodies

IgE antibodies

IgE antibodies bind to mast cells located in many body tissues

3

Mast cell that have IgE antibodies attached

Histamine-containing granules

Subsequent (secondary) responses

More of same allergen enters body

4

Antigen

Allergen combines with IgE on mast cells, triggering release of histamines from mast cell

5

Mast cell granules release histamine after allergen binds to IgE antibodies

Histamines and other chemicals

6

Histamine stimulates dilation of blood vessels, causing fluid to leak out; stimulates release of copious amounts of mucus; and causes contraction of smooth muscle in bronchioles

Fluid pours out of capillaries

Mucus is copiously released

Small respiratory passages (bronchioles) constrict

▲ FIGURE 17.15 The steps in an allergic reaction.

called **allergens,** include a wide range of substances: house dust, pollen, cat dander, certain foods, and even medicines such as penicillin. It is estimated that up to 10% of the U.S. population suffers from at least one allergy (see Genetic Journeys: Peanut Allergies Are Increasing). Typically, allergic reactions develop after a first exposure to an allergen. The allergen causes B cells to make IgE antibodies, instead of IgG antibodies. The IgE antibodies attach to mast cells in tissues, including those of the nose and respiratory system.

In a second exposure, the allergen binds to the IgE and the mast cells release histamine, triggering an inflammatory response that results in fluid accumulation, tissue swelling, and mucus secretion. This reaction is severe in some individuals, and

Genetic Journeys

Peanut Allergies Are Increasing

Allergy to peanuts is one of the most serious food sensitivities and is a growing health concern in the United States. Hypersensitivity to peanuts can provoke a systemic anaphylactic reaction, in which the bronchial tubes constrict, closing the airways. Fluids pass from the tissues into the lungs, making breathing difficult. Blood vessels dilate, causing blood pressure to drop, and plasma escapes into the tissues, causing shock. Heart arrhythmias and cardiac shock can develop and cause death within 1 to 2 minutes after the onset of symptoms. About 30,000 cases of food-induced anaphylactic reactions are seen in emergency rooms each year, with 200 fatalities. About 80% of all cases are caused by allergies to peanuts or other nuts. Peanut-sensitive individuals must avoid ingesting peanuts, and recognize symptoms of anaphylactic reactions. In spite of precautions, accidental exposures, caused by cooking pans previously used to cook food with peanuts, or inhaling peanut dust on airplanes, have been reported to cause anaphylactic reactions. Many peanut-sensitive people carry doses of self-injectable epinephrine (EpiPen Autoinjectors and similar products) to stop anaphylaxis in case they are exposed to peanuts.

The number of children and adults allergic to peanuts appears to be increasing. In a 1988–1994 survey of American children, allergic reactions to peanuts were twice as high as a group surveyed from 1980 to 1984. A national survey indicates that about 3 million people in the United States (about 1.1% of the population) are allergic to peanuts, tree nuts, or both. The hypersensitive reaction to one of three allergenic peanut proteins is mediated by IgE antibodies. Within 1 to 15 minutes of exposure, the IgE antibodies activate mast cells. The stimulated mast cells release large amounts of histamines and chemotactic factors, which attract other white blood cells as part of the inflammatory response. In addition, the mast cells release prostaglandins and other chemicals that trigger an anaphylactic reaction.

What is causing the increase in peanut allergies is unclear. Genetics obviously plays some part, but environmental factors appear to play a major role. For example, peanut allergies are extremely rare in China, but the children of Chinese immigrants have about the same frequency of peanut allergies as the children of native-born Americans, pointing up the involvement of environmental factors.

One proposal is that as peanuts have become a major part of the diet in the United States, especially in foods advertised to provide quick energy, that exposure of newborns and young children (1–2 years of age) to peanuts is more common. This exposure occurs through breast milk, peanut butter, or other foods. The immune system in newborns is immature and develops over the first few years of life. As a result, food allergies are more likely to develop during these first few years. In the absence of conclusive information, it is recommended that mothers avoid eating peanuts and peanut products during pregnancy or while they are nursing, and that children not be exposed to peanuts or other nuts for the first three years of life.

histamine is released into the circulatory system, causing a life-threatening decrease in blood pressure and constriction of airways in the lungs. This reaction, called **anaphylaxis** or anaphylactic shock, most often occurs after exposure to antibiotics, to the venom in bee or wasp stings, or to some foods. Prompt treatment of anaphylaxis with antihistamines, epinephrine, and steroids can reverse the reaction. As the name suggests, antihistamines block the action of histamine. Epinephrine opens the airways and constricts blood vessels, raising blood pressure. Steroids, such as prednisone, inhibit the inflammatory response. Some people who have a history of severe reaction to insect stings or foods carry these drugs with them in a kit.

Autoimmune reactions cause the immune system to attack the body.

One of the most elegant properties of the immune system is its capacity to distinguish self from nonself and destroy what it perceives as nonself. During development, the immune system "learns" not to react against cells of the body. In some disorders, this immune tolerance breaks down, and the immune system attacks and kills cells and tissues in the body. Juvenile diabetes, also known as insulin-dependent diabetes (IDDM; OMIM 222100), is an autoimmune disease. Clusters of cells in the pancreas produce insulin, a hormone that lowers blood sugar levels. In IDDM, the immune system attacks and kills the insulin-producing cells, causing life-long diabetes and the need for insulin injections to control blood sugar levels.

Other forms of autoimmunity, such as systemic lupus erythematosus (SLE; OMIM 152700), attack and slowly destroy major organ systems. ▶ Table 17.6 lists some autoimmune disorders.

Genetic disorders can impair the immune system.

Ogden Bruton, a physician at Walter Reed Army Hospital, described the first disease of the immune system in 1952. He examined a young boy who had suffered at least 20 serious infections in the preceding 5 years. Blood tests showed that the child had no antibodies. Other patients with similar problems were soon discovered. Affected individuals were boys who were highly susceptible to bacterial infections. In all cases, either B cells were completely absent, or the B cells were immature and unable to produce antibodies. Without functional B cells, no antibodies can be produced, but these individuals usually have nearly normal levels of T cells. In other words, antibody-mediated immunity is absent, but cellular immunity is normal. This heritable disorder, called **X-linked agammaglobulinemia** (**XLA**; OMIM 300300), usually appears 5 to 6 months after birth when maternal antibodies disappear and the infant's B-cell population normally begins to produce antibodies. Patients with this disorder are highly susceptible to pneumonia and streptococcal infections and pass from one life-threatening infection to another.

Individuals with XLA lack mature B cells but have normal populations of immature B-cells, indicating that the defective gene controls some stage of maturation. The *XLA* gene was mapped to Xq21.3–Xq22 and encodes an enzyme that transmits signals from the cell's environment into the cytoplasm. Chemical signals from outside the cell help trigger B-cell maturation, and the gene product that is defective in XLA plays a critical role in this signaling process. Understanding the role of this protein in B-cell development may permit the use of gene therapy to treat this disorder.

A rare genetic disorder causes a complete absence of *both* antibody-mediated and cell-mediated immune responses. This condition is called **severe combined immunodeficiency** (**SCID**; OMIM 102700, 600802, and others). Affected individuals have recurring and severe infections and usually die at an early age from seemingly minor infections. One of the longest known survivors of this condition was David, the "boy in the bubble," who died at 12 years of age after being isolated in a sterile plastic bubble for all but the last 15 days of his life (▶ Figure 17.16).

One form of SCID is caused by a deficiency of the enzyme adenosine deaminase

■ **Anaphylaxis** A severe allergic response in which histamine is released into the circulatory system.

■ **X-linked agammaglobulinemia (XLA)** A rare, sex-linked, recessive trait characterized by the total absence of immunoglobulins and B cells.

■ **Severe combined immunodeficiency disease (SCID)** A genetic disorder in which affected individuals have no immune response; both the cell-mediated and antibody-mediated responses are missing.

Table 17.6 Some Autoimmune Diseases

Addison's disease

Autoimmune hemolytic anemia

Diabetes mellitus— insulin-dependent

Graves' disease

Membranous glomerulonephritis

Multiple sclerosis

Myasthenia gravis

Polymyositis

Rheumatoid arthritis

Scleroderma

Sjögren's syndrome

Systemic lupus erythematosus

FIGURE 17.16 David, the "boy in the bubble," had severe combined immunodeficiency and lived in isolation for 12 years. He died of complications following a bone marrow transplant.

■ **Acquired immunodeficiency syndrome (AIDS)** A collection of disorders that develop as a result of infection with the human immunodeficiency virus (HIV).

(ADA). A small group of children affected with ADA-deficient SCID (OMIM 102700) is currently undergoing gene therapy to give them a normal copy of the gene. Their genetically modified white blood cells are returned to their circulatory systems by transfusion. Expression of the normal *ADA* gene in these cells stimulates development of functional T and B cells and at least partially restores a functional immune system. The recombinant DNA techniques used in gene therapy are reviewed in Chapter 13, and gene therapy is described in Chapter 16.

HIV attacks the immune system.

The immunodeficiency disorder currently receiving the most attention is **acquired immunodeficiency syndrome (AIDS)**. AIDS is a collection of disorders that develop following infection with the human immunodeficiency virus (HIV) (▶ Active Figure 17.17). Worldwide, about 40 million people are infected with HIV (▶ Figure 17.18).

HIV is a retrovirus with three components: (1) a protein coat (which encloses the other two components), (2) RNA molecules (the genetic material), and (3) an enzyme called reverse transcriptase. The viral particle is enclosed in a coat derived from a T-cell plasma membrane. The virus selectively infects and kills T4 helper cells, which act as the master "on" switch for the immune system. Inside a cell, reverse transcriptase transcribes the RNA into a DNA molecule, and the viral DNA is inserted into a human chromosome, where it can remain for months or years.

Later, when the infected T cell is called upon to participate in an immune response, the viral genes are activated. Viral RNA and proteins are made, and new viral particles are formed. These bud off the surface of the T cell, rupturing and killing the cell and setting off a new round of T-cell infection. Over the course of HIV infection, the number of helper T4 cells gradually decreases. As the T4 cell population falls, the ability to mount an immune response decreases. The result is increased susceptibility to infection and increased risk of certain forms of cancer. The eventual outcome is premature death brought about by any of a number of diseases that overwhelm the body and its compromised immune system.

HIV is transmitted from infected to uninfected individuals through body fluids, including blood, semen, vaginal secretions, and breast milk. The virus cannot live for more than 1 to 2 hours outside the body and cannot be transmitted by food, water, or casual contact.

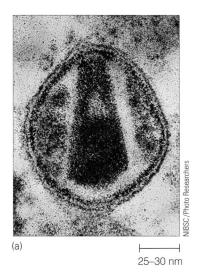

(a)

25–30 nm

NIBSC/Photo Researchers

◀ **ACTIVE FIGURE 17.17** Steps in HIV replication. (a) Electron micrograph of an HIV particle budding from the surface of an infected T cell. (b) The HIV life cycle. A CD4 lymphocyte is a helper T cell.

HUMAN
Genetics ⒺNow™ Learn more about HIV replication by viewing the animation on Human GeneticsNow at **http://now.ilrn.com/cummings7**

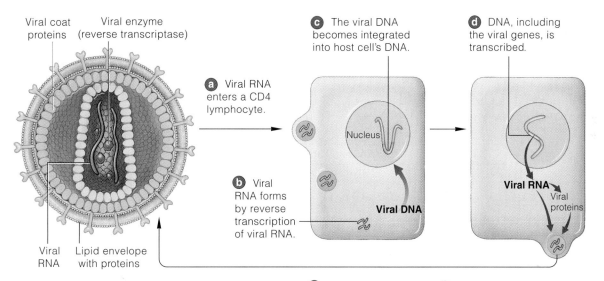

Viral coat proteins

Viral enzyme (reverse transcriptase)

c The viral DNA becomes integrated into host cell's DNA.

d DNA, including the viral genes, is transcribed.

a Viral RNA enters a CD4 lymphocyte.

Nucleus

b Viral RNA forms by reverse transcription of viral RNA.

Viral DNA

Viral RNA

Viral proteins

Viral RNA

Lipid envelope with proteins

f Virus particles that bud from the infected cell may attack a new one.

e Some transcripts are new viral RNA, others are translated into proteins. Both self-assemble as new virus particles.

(b)

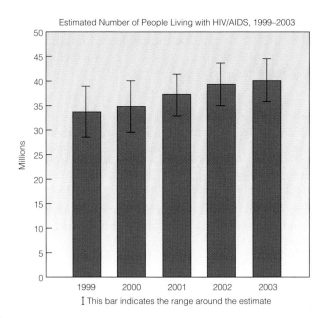

Estimated Number of People Living with HIV/AIDS, 1999–2003

Millions

I This bar indicates the range around the estimate

◀ **FIGURE 17.18** The number of people living with HIV/AIDS has increased steadily every year from 1998 to 2003.

Genetics in Practice

Genetics in Practice case studies are critical thinking exercises that allow you to apply your new knowledge of human genetics to real-life problems. You can find these case studies and links to relevant websites at **http://biology.brookscole.com /cummings7**

CASE 1

Mary and John Smith went for genetic counseling because John did not believe that their newborn son was his. They both wanted blood tests to help rule out the possibility that someone other than John was the father of this baby. The counselor explained that ABO blood typing could give some preliminary indications about possible paternity. Its use is limited, however, because there are only four possible ABO blood types, and the vast majority of people in any population have only two of those types (A and O). This means that a man may have a blood type consistent with paternity and still not be the father of the tested child. Because the allele for type O can be masked by the genes for A or B, inheritance of blood type can be unclear. More modern (and more expensive) genetic tests, such as DNA typing, would lead to a more reliable conclusion regarding paternity.

Mary's blood was tested and identified as type O. John's blood was tested and identified as type O. Based on these two parental combinations of blood types, the only possible blood type that their son could be is type O. The baby was tested and his blood type was, indeed, type O.

1. Can any absolute conclusions be drawn based upon the results of these blood tests? Why or why not?

2. If not, was it worth doing the test in the first place?

3. Why do you think DNA testing would be more reliable than blood testing for this purpose?

		And the Father is				
		A	B	AB	O	
If the Mother is	A	A or O	A, B, AB or O	A, B, or AB	A or O	The Child must be
	B	A, B, AB, or O	B or O	A, B, or AB	B or O	
	AB	A, B, or AB	A, B, or AB	A, B, or AB	A or B	
	O	A or O	B or O	A or B	O	

CASE 2

The Joneses were referred to a clinical geneticist because their 6-month-old daughter was failing to grow adequately and was having recurrent infections. The geneticist took a detailed family history (which was uninformative) and medical history of their daughter. He discovered that their daughter had a history of several ear infections against which antibiotics had no effect, she had difficulty gaining weight (failure to thrive), and had an extensive history of yeast infection (thrush) in her mouth. The geneticist did a simple blood test to check their daughter's white blood count and determined that she had severe combined immunodeficiency (SCID).

The geneticist explained that SCID is an immune deficiency that causes a marked susceptibility to infections. The defining characteristic is usually a severe defect in both the T- and B-lymphocyte systems. This usually results in one or more infections within the first few months of life.

These infections are usually serious and may even be life-threatening. They may include pneumonia, meningitis, or bloodstream infections. Based on the family history, it was possible that their daughter inherited an altered gene from each of them and therefore was homozygous for the gene that causes SCID. Each time the Joneses had a child there would be a 25% chance that the child would have SCID. Prenatal testing is available to determine whether the developing fetus has SCID.

1. Genetic testing showed that both parents were heterozygous carriers of a mutant allele of adenosine deaminase (ADA) and that the daughter is homozygous for this mutation. Are there any treatment options available for ADA-deficient SCID?

2. If the Joneses wanted to be certain that their next child would not have SCID, what types of reproductive options do you think they might have?

17.1 The Immune System Defends the Body Against Infection

- The immune system protects the body against infection by a graded series of responses that attack and inactivate foreign molecules and organisms.

17.2 The Complement System Kills Microorganisms

- The complement system participates in both the non-specific and specific immune responses in a number of ways, all of which enable it to kill invading cells.

17.3 The Inflammatory Response Is a General Reaction

- The lowest level of response to infection involves a non-specific, local, inflammatory response. This response is mediated by cells of the immune system and isolates and kills invading microorganisms. Genetic control of this response is abnormal in inflammatory diseases, including ulcerative colitis and Crohn's disease.

17.4 The Immune Response Is a Specific Defense Against Infection

- The immune system has two components: antibody-mediated immunity, which is regulated by B cells and antibody production, and cell-mediated immunity, which is controlled by T cells. The primary function of antibody-mediated reactions is to defend the body against invading viruses and bacteria. Cell-mediated immunity is directed against cells of the body that have been infected by agents such as viruses and bacteria.

17.5 Blood Types Are Determined by Cell-Surface Antigens

- The presence or absence of certain antigens on the surface of blood cells is the basis of blood transfusions and blood types. Two blood groups are of major significance: the ABO system and the Rh blood group. Matching ABO blood types is important in blood transfusions. In some cases, mother–fetus incompatibility in the Rh system can cause maternal antigens to destroy red blood cells of the fetus, resulting in hemolytic disease in the newborn.

17.6 Organ Transplants Must Be Immunologically Matched

- The success of organ transplants and skin grafts depends on matching histocompatibility antigens found on the surface of all cells in the body. In humans, the antigens produced by a group of genes on chromosome 6 (known as the MHC complex) play a critical role in the outcome of transplants.

17.7 Disorders of the Immune System

- Allergies are the result of immunological hypersensitivity to weak antigens that do not provoke an immune response in most people. These weak antigens, known as allergens, include a wide range of substances: house dust, pollen, cat hair, certain foods, and even medicines, such as penicillin. Acquired immunodeficiency syndrome (AIDS) is a collection of disorders that develop as a result of infection with a retrovirus known as the human immunodeficiency virus (HIV). The virus selectively infects and kills the T4 helper cells of the immune system.

Questions and Problems

HUMAN
Genetics Now™ Preparing for an exam? Assess your understanding of this chapter's topics with a pre-test, a personalized learning plan, and a post-test at **http://now.ilrn.com/cummings7**

The Inflammatory Response Is a General Reaction

1. What causes the area around a cut or a scrape to become warm? What is the role of this heat in the inflammatory response?

The Complement System Directly Kills Microorganisms

2. The complement system supplements the inflammatory response by directly killing microorganisms. Describe the life cycle of the complement proteins, from their synthesis in the liver to their activity at the site of an infection.

The Immune Response Is a Specific Defense Against Infection

3. Name the class of molecules that includes antibodies, and name the five groups that make up this class.
4. Discuss the roles of the three types of T cells: helper cells, suppressor cells, and killer cells.

5. Compare the general inflammatory response, the complement system, and the specific immune response.

6. Distinguish between antibody-mediated and cell-mediated immunity. What components are involved in each?

7. The molecular weight of IgG is 150,000 kd. Assuming that the two heavy chains are equivalent, the two light chains are equivalent, and the molecular weights of the light chains are half the molecular weight of the heavy chains, what are the molecular weights of each individual subunit?

8. Identify the components of cellular immunity, and define their roles in the immune response.

9. Describe the rationale for vaccines as a form of preventive medicine.

10. Researchers have been having a difficult time developing a vaccine against a certain pathogenic virus due to the lack of a weakened strain. They turn to you because of your wide knowledge of recombinant DNA technology and the immune system. How could you vaccinate someone against the virus, using a cloned gene from the virus that encodes a cell-surface protein?

11. It is often helpful to draw a complicated pathway in the form of a flow chart to visualize the multiple steps and the ways in which the steps are connected to each other. Draw the antibody-mediated immune response pathway that acts in response to an invading virus.

12. Describe the genetic basis of antibody diversity.

13. In cystic fibrosis gene therapy, scientists propose the use of viral vectors to deliver normal genes to cells in the lungs. What immunological risks are involved in this procedure?

Blood Types Are Determined by Cell-Surface Antigens

14. A man has the genotype $I^A I^A$, and his wife is $I^B I^B$. If their son needed an emergency blood transfusion, would either parent be able to be a donor? Why or why not?

15. Why can someone with blood type AB receive blood of any type? Why can a blood type O individual donate blood to anyone?

16. What is more important to match during blood transfusions: the antibodies of the donor or the antigens of the donor/recipient?

17. The following data were presented to a court during a paternity suit: (1) the infant is a universal donor for blood transfusions, (2) the mother bears antibodies against the B antigen only, and (3) the alleged father is a universal recipient in blood transfusions.
   a. Can you identify the ABO genotypes of the three individuals?
   b. Can the court draw any conclusions?

18. A patient of yours has just undergone shoulder surgery and is now experiencing kidney failure for no apparent reason. You check his chart and find that his blood is type B, but he has been mistakenly transfused with type A. Explain why he is experiencing kidney failure.

19. Assume that a single gene having alleles that show complete-dominance relationships at the phenotypic

level controls the Rh character. An Rh$^+$ father and an Rh$^-$ mother have eight boys and eight girls, all Rh$^+$.
   a. What are the Rh genotypes of the parents?
   b. Should they have been concerned about hemolytic disease of the newborn?

20. How is Rh incompatibility involved in hemolytic disease of the newborn? Is the mother Rh$^+$ or Rh$^-$? Is the fetus Rh$^+$ or Rh$^-$? Why is a second child that is Rh$^+$ more susceptible to attack from the mother's immune system?

Organ Transplants Must Be Immunologically Matched

21. What mode of inheritance has been observed for the *HLA* system in humans?

22. A burn victim receives a skin graft from her brother; however, her body rejects the graft a few weeks later. The procedure is attempted again, but this time the graft is rejected in a few days. Explain why the graft was rejected the first time, and why it was rejected faster the second time.

23. In the human *HLA* system there are 23 *HLA-A* alleles, 47 for *HLA-B*, 8 for *HLA-C*, 14 for *HLA-DR*, 3 for *HLA-DQ*, and 6 for *HLA-DP*. How many different human *HLA* genotypes are possible?

24. In the near future, pig organs may be used for organ transplants. How are researchers attempting to prevent rejection of the pig organs by human recipients?

25. A couple has a young child who needs a bone marrow transplant. They propose that preimplantation screening be done on several embryos fertilized *in vitro* to find a match for their child.
   a. What do they need to match in this transplant procedure?
   b. The couple proposes that the matching embryo be transplanted to the mother's uterus and serve as a bone marrow donor when old enough. What are the ethical issues involved in this proposal?

26. A couple have the following *HLA* genotypes: Male A12, B36, C7, DR14, DQ2, DP5/A9, B27, C3, DR14, DQ3, and DP1; Female A8, B15, C2, DR5, DQ2, DP3/A2, B20, C3, DR8, DQ1, and DP3. What is the probability that an offspring will suffer from the autoimmune disorder ankylosing spondylitis if:
   a. the child is male and has *HLA-A9*?
   b. the child is female and has *HLA-A9*?
   c. the child is male and has *HLA-DP1*?
   d. the child is male and has *HLA-C7*?

Disorders of the Immune System

27. Why are allergens called "weak" antigens?

28. Antihistamines are used as anti-allergy drugs. How do these drugs work to relieve allergy symptoms?

29. Autoimmune disorders involve a breakdown of an essential property of the immune system. What is it? How does this breakdown cause juvenile diabetes?

30. A young boy, who has suffered from over a dozen viral and bacterial infections in the past 2 years, comes to your office for an examination. You determine by test-

ing that he has no circulating antibodies. What syndrome does he have, and what are its characteristics? What component of the two-part immune system is nonfunctional?

31. AIDS is an immunodeficiency syndrome. In the flow chart you drew for Question 11, describe where AIDS sufferers are deficient. Why can't our immune systems fight off this disease?

32. An individual has an immunodeficiency that prevents helper T cells from recognizing the surface antigens presented by macrophages. As a result, the helper T cells are not activated, and they, in turn, fail to activate the appropriate B cells. At this point, is it certain that the viral infection will continue unchecked?

Internet Activities

Internet Activities are critical thinking exercises using the resources of the World Wide Web to enhance the principles and issues covered in this chapter. For a full set of links and questions investigating the topics described below, log on to **http://biology.brookscole.com/cummings7**

1. *Immune System Function.* At the *CellsAlive!* website, access the "Antibody" link for a beautiful overview of antibody structure, production, and function.

2. *HIV/AIDS and the Immune System.* The University of Arizona's *Biology Project HIV 2001* allows you to run a simulation of the spread of HIV through a population or work through a tutorial on HIV/AIDS and the immune system. The tutorial includes an overview of immune system function.

3. *Autoimmune Diseases.* The National Library of Medicine maintains the *Medline Plus Autoimmune Diseases* web page. Here you can find links to various resources on autoimmune diseases such as lupus, multiple sclerosis (MS), and rheumatoid arthritis. Scroll down to the "Anatomy/Physiology" link for a good immune system

tutorial from the National Cancer Institute or down to the "Diagnosis/Symptoms" link to find out how the standard antinuclear antibody (ANA) test used in the diagnosis of many autoimmune diseases works.

4. *Which Immune Disorders Have Genetic Bases?* Because the normal functioning of the immune response in humans requires the delicate interplay of B cells, T cells, and phagocytic cells, as well as the actions of several types of immunoglobulins and cytokines, it is easy to see that some genetic disorders are likely to be recognized as related to absent or abnormal immune function. The National Institute of Allergy and Infectious Disease maintains a fact sheet on "Primary Immune Deficiencies."

How would you vote now?

Organ donations are unable to keep up with the demand, and thousands of people die each year while awaiting transplants. Xenotransplantation techniques may increase the supply of organs by using those from other animals, such as pigs, but these techniques must address the inherent problem of immune system rejection. Using pigs that have been genetically modified to carry human genes that prevent transplant rejection and modifying the immune system of human recipients by injection of pig bone marrow cells are two methods of overcoming the problems of organ transplantation between species. Now that you know more about the immune system and its role in organ transplants, what do you think? Is it ethical to genetically modify pigs with human genes, or to modify humans by giving them a pig immune system to accept transplanted organs? Visit the Human Heredity Companion website to find out more on the issue, then cast your vote online at **http://biology.brookscole.com/cummings7**

For further reading and inquiry, log on to **InfoTrac College Edition,** your world-class online library, including articles from nearly 5,000 periodicals, at **http://infotrac.thomsonlearning.com**

18

Genetics of Behavior

ANCIENT Greeks were among the first cultures to observe that creativity and madness are linked. The Greek philosopher Socrates wrote:

> If a man comes to the door of poetry untouched by the madness of the Muses, believing that technique alone will make him a good poet, he and his sane compositions never reach perfection, but are utterly eclipsed by the performances of the inspired madman.

By the second century, it was recognized that mania and depression are opposite poles of a cycle, and their familial patterns of inheritance have been known for almost a thousand years. In the last quarter of the twentieth century, advances in genetics established that manic depression, or bipolar illness, as it is now called, is probably a polygenic trait with environmental influences. Clearly, not all poets and authors suffer from bipolar illness, in fact, most do not, and creativity should not be viewed through the filter of genetics, but evidence from authors and poets themselves and advances in medicine and genetics has established that artists, writers, and poets have a much higher rate of depression or bipolar illness than the general population.

Vincent Van Gogh's family had an extensive history of psychiatric problems, mood disorders, and suicide. His brothers, his sister, two of his uncles, and Vincent himself suffered with mental illness, most probably manic depression. All of Vincent's brilliant work as a painter was produced in a 10-year period and conveys to the viewer the intense anguish of his mental illness, which resulted in his suicide at the age of 37. In writing about his family's illness, he said:

> The root of the evil lies in the constitution itself, in the fatal weakening of families from generation to generation. . . . The root of the evil certainly lies there, and there's no cure for it.

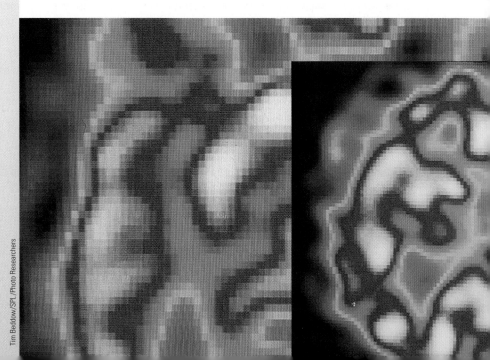

Tim Beddow/SPL/Photo Researchers

The idea that creativity and mental illness are linked is still controversial, illustrating many of the problems geneticists encounter in dissecting the genetic basis of human behavior. This chapter discusses the genetic models and methods used in studying human behavior, and the state of our knowledge about the genetic control of behavior.

How would you vote?

Biographic and scientific evidence strongly suggests that in many people, creative abilities in art and literature are linked in a complex way to mood disorders such as bouts of depression or to the onset of manic states. Because of this proposed linkage, it is possible, though not proved, that medicating mood disorders may reduce people's creativity. If you were a successful artist, author, or poet who suffers from depression, or bipolar illness, and a cure for your illness were discovered, would you elect to have the treatment, knowing your creative abilities may be diminished or even disappear, but also knowing that your risk of suicide would be reduced or eliminated? Visit the Human Heredity Companion website to find out more on the issue, and then cast your vote online at **http://biology.brookscole.com /cummings7**

18.1 Models, Methods, and Phenotypes in Studying Behavior

Pedigree analysis, family studies, adoption, and twin studies suggest that many parts of our behavior are genetically influenced. However, behaviors with a genetic component are likely to be controlled by several genes, to interact with other genes, and to be influenced by environmental components. In fact, most behaviors are *not* inherited as single-gene traits, demonstrating the need for genetic models that can explain observed patterns of inheritance. To a large extent, the model proposed to explain the inheritance of a trait determines the methods used to analyze its pattern of inheritance and the techniques to be used in mapping and isolating the gene or genes responsible for the trait's characteristic phenotype (discussed later).

Keep in mind

■ Most human behaviors are polygenic and have environmental influences.

Keep in mind as you read

■Most human behaviors are polygenic and have environmental influences.

■Transgenic animals carrying human genes are used to develop drugs and treatment strategies for behavioral disorders.

■Indirect evidence from family studies indicates that mood disorders and schizophrenia have genetic components, but no genes have yet been identified.

■Human behavior in social settings is complex and often difficult to define.

There are several genetic models of inheritance and behavior.

Several models for genetic effects on behavior have been proposed (▶ Table 18.1). The simplest model is a single gene, with a dominant or recessive pattern of inheritance that affects a well-defined behavior. Several behavioral traits—including Huntington disease, Lesch-Nyhan syndrome, fragile-X syndrome, and others—are described by such a model. Multiple-gene models are also possible. The simplest of these is a polygenic additive model in which two or more genes contribute equally in an additive manner to the phenotype. This model has been proposed (along with others) to explain schizophrenia (the inheritance of additive polygenic traits was considered in Chapter 5). Multigene models can also include situations in which one or more genes have a major effect, while other genes make smaller contributions to the phenotype. Still another multigene model involves epistasis, a form of gene interaction in which an allele of one gene masks the expression of a second gene. This form of gene interaction has been well documented in experimental genetics, although it has not yet been shown to operate in human behavioral traits.

In each of these models, the environment can significantly affect the phenotype, and the study of behavior must take this into account (see Genetic Journeys: Is Going to Medical School a Genetic Trait?). To assess the role of the environment in the phenotype, geneticists must use methods that measure the genetic and environmental contributions to a trait.

Methods of studying behavior genetics often involve twin studies.

For the most part, methods for studying behavior genetics follow the pattern used for study of other human traits. If a single gene is proposed, pedigree analysis and linkage studies, including the use of DNA markers and other methods of recombinant DNA technology, are the most appropriate methods. However, because many behaviors are polygenic, twin studies play a prominent role in human behavior genetics. Concordance and heritability values based on twin studies have established a genetic link to mental illnesses (manic depression and schizophrenia) and to behavioral traits (sexual preference and alcoholism). The results of such studies must be interpreted with caution because there are limitations inherent in interpreting heritability (see Chapter 5), and these studies often use small sample sizes, in which minor variations can have a disproportionately large effect on the outcome.

To overcome these problems, geneticists are adapting twin studies as a genetic tool to study behavior. One innovation studies the children of twins to confirm the existence of genes predisposing to a certain behavior. Twin studies are also being coupled to recombinant DNA techniques to search for behavior genes, and this combination may prove to be a powerful method for identifying such genes.

Phenotypes: How is behavior defined?

A second restriction on progress in human behavior genetics is the choice of a consistent phenotype as the basis for study. The phenotypic definition of a behavior must

Table 18.1	Models for Genetic Analysis of Behavior
Model	**Description**
Single gene	One gene controls a defined behavior
Polygenic trait	Additive model that has two or more genes
	One or more major genes with other genes contributing to phenotype
Multiple genes	Interaction of alleles at different loci generates a unique phenotype

Is Going to Medical School a Genetic Trait?

Many behavioral traits follow a familial, if not Mendelian pattern of inheritance. This observation, along with twin studies and adoption studies, reveals a genetic component in complex behavioral disorders, such as bipolar illness and schizophrenia. In most cases, these phenotypes are not inherited as simple Mendelian traits. Researchers are then faced with selecting a model to describe how a behavioral trait is inherited. Using this model, further choices select the methods used in genetic analysis of the trait. A common strategy is to find a family in which the behavior appears to be inherited as a recessive or an incompletely penetrant dominant trait controlled by a single gene. One or more molecular markers (such as RFLPs) are used in linkage analysis to identify the chromosome that carries the gene that controls the trait.

If researchers are looking for a single gene when the trait is controlled by two or more genes or by genes that interact with environmental factors, then the work may produce negative results, even though preliminary findings can be encouraging. Reports of loci for bipolar illness on chromosome 11 and the X chromosome were based on single-gene models, but after initial successes, it was found that these reports were flawed. Overall, regions on 14 chromosomes have been proposed as candidates for genes that control bipolar illness, but none have been substantiated.

To illustrate some of the pitfalls associated with model selection in behavior genetics, one study selected attendance at medical school as a behavioral phenotype and attempted to determine if the distribution of this trait in families is consistent with a genetic model. This study surveyed 249 first- and second-year medical students. Thirteen percent of these students had first-degree relatives who had also attended medical school, compared with 0.22% of individuals selected from the general population with such relatives. Thus the overall risk factor among first-degree relatives for medical school attendance was 61 times higher for medical students than for the general population, indicating a strong familial pattern. To determine whether this trait was inherited in a Mendelian manner, researchers used standard statistical analysis, which supported familial inheritance and rejected the model of no inheritance. Analysis of the pedigrees most strongly supported a simple recessive mode of inheritance, although other models, including polygenic inheritance, were not excluded. Using a further set of statistical tests, the researchers concluded that the recessive mode of inheritance was just at the border of statistical acceptance.

Similar results are often found in studies of other behavioral traits, and it is usually argued that another, larger study would confirm the results, in this case that attendance at medical school is a recessive Mendelian trait. Although it is true that genetic factors may partly determine whether one will attend medical school, it is unlikely that a single recessive gene controls this decision, regardless of the support of such a conclusion by this family study and segregation analysis of the results.

The authors of this study were not serious in their claims that a decision to attend medical school is a genetic trait, nor was it their intention to cast doubts on the methods used in the genetic analysis of behavior. Rather, it was intended to point out the folly of accepting simple explanations for complex behavioral traits.

be precise enough to distinguish it from all other behaviors and from the behavior of the control group. However, the definition cannot be so narrow that it excludes some variations of the behavior. Recall that gene mapping uses the phenotype as a guide, and so starting with an accurate description of the phenotype is very important.

For some mental illnesses, clinical definitions are provided in the *Diagnostic and Statistical Manual of Mental Disorders* of the American Psychiatric Association. For other behaviors, phenotypes are poorly defined and may not provide clues to the underlying biochemical and molecular basis of the behavior. For example, alcoholism can be defined as the development of characteristic deviant behaviors associated with excessive consumption of alcohol. Is this definition explicit enough to be useful as a phenotype in genetic analysis? Is there too much room for interpreting what is deviant behavior or what is excessive consumption? As we will see, whether the behavioral phenotype is narrowly or broadly defined can affect the outcome of the genetic analysis and even the model of inheritance for the trait.

The nervous system is the focus of behavior genetics.

In Chapter 10, we discussed the role of genes in metabolism. Mutations that disrupt metabolic pathways or interfere with the synthesis of essential gene products can influence the function of cells and in turn produce an altered phenotype. If the affected cells are part of the nervous system, the phenotype may include altered behavior. In fact, some, and perhaps many, genetic disorders affect cells in the nervous system that in turn affect behavior. In phenylketonuria (PKU), for example, brain cells are damaged by excess levels of phenylalanine, causing mental retardation and other behavioral deficits.

For much of behavior genetics, the focus is on the structure and function of the nervous system. This emphasis is reinforced by the finding that many disorders with a behavioral phenotype, including Huntington disease, Alzheimer disease, and Charcot-Marie-Tooth disease, alter the structure of the brain and nervous system. Other behavior disorders such as bipolar illness and schizophrenia may be disorders of brain function rather than structure.

18.2 Animal Models: The Search for Behavior Genes

One way to study the genetics of human behavior is to ask whether behavior genetics can be studied in model systems. Results from experimental organisms can then be used to study human behavior. Several approaches are used to study behavior in animals. One method uses two closely related species or two strains of the same species to detect variations in behavioral phenotypes. Genetic crosses are used to establish whether these variations are inherited and, if so, to determine the pattern of inheritance. More recently, the effects of single genes on animal behavior have been studied. In some cases, these studies have led to isolating and cloning genes that affect behavior.

Some behavioral geneticists study open-field behavior in mice.

Beginning in the mid-1930s, the emotional and exploratory behaviors of mice were tested by studying open-field behavior (▶ Figure 18.1). When mice are placed in a new, brightly lit environment, some mice actively explore the new area, whereas others are apprehensive and do not move about. Their nervous behavior is reflected in their elevated rates of urination and defecation. This behavior pattern is under genetic control, because strains exhibiting both types of behavior have been established.

To test the genetic components of this behavior, mice are placed in an enclosed, illuminated box whose floor is marked into squares. Counting the mouse's movements in different squares tests exploration, and emotion is quantified by counting the number of defecations. Inbred strains of mice show significant dif-

Columbus Instruments/Visuals Unlimited

▲ FIGURE 18.1 An open-field trial. Movements are automatically recorded as the mouse moves across the open field.

ferences in behavior. The BALB/cJ strain, homozygous for a recessive albino allele, shows low exploratory behavior and is highly emotional. The C57BL/6j strain has normal pigmentation, is active in exploration, and shows low levels of emotional behavior.

If these two strains are crossed and the offspring are interbred for several generations, each generation beyond the F_1 includes both albino and normally pigmented mice. When tested for open-field behavior, pigmented mice behaved like the C57 parental line, showing active exploration and low levels of emotional behavior. The albino mice behaved like the BALB parental line and showed low exploratory activity and high levels of emotional behavior, indicating that the albino gene affects behavior and pigmentation. The overall results show that open-field behavior is a polygenic trait.

Human polygenic traits, including behavioral traits, are difficult to analyze genetically, as discussed in Chapter 5. Animal models of polygenic traits offer several advantages. In mice, it is possible to control population size, genetic heterogeneity, matings, and the environment. The use of highly inbred strains and crosses between inbred strains limits the number of genetic differences between strains, and makes analysis easier.

Mapping of polygenic traits has been successfully used to identify regions of the mouse genome involved in fearfulness. Heterozygotes from crosses between inbred strains were subjected to several different behavioral tests for anxiety (▶ Figure 18.2). Over 1,600 mice from 8 different crosses were tested, and regions on three chromosomes (on mouse chromosomes 1, 4, and 15) were shown to affect behavior on all tests. High-resolution genetic mapping is being used to identify genes in these regions that control fear. Finding these genes in mice will set the stage for finding the equivalent genes in humans.

Transgenic animals are used as models of human neurodegenerative disorders.

In addition to studying behavior in animal models using mutants and inbred strains, researchers are creating animal models of nervous system disorders by constructing transgenic animals. Recall from Chapter 13 that transgenic animals are produced by transferring genes from one species to another.

Let's briefly look at how transgenic animals are used in one group of genetic disorders of the nervous system. Neurodegenerative disorders are a group of progressive and fatal diseases. Some of these disorders, such as Alzheimer disease (AD), amyotrophic lateral sclerosis (ALS), and Parkinson disease (PD), occur sporadically or from inherited mutations. Others, such as Huntington disease (HD) and

◀ FIGURE 18.2 A mouse explores a new environment from the arm of an elevated maze, a standard device for measuring fear in mice. This experiment was one in a series designed to identify genes that control fear and anxiety in mice.

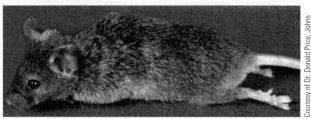

FIGURE 18.3 A transgenic mouse carrying a mutation in the *SOD1* gene, which causes paralysis of the limbs. In humans, this mutation causes amyotrophic lateral sclerosis (ALS), a neurodegenerative disease. The mutant mouse serves as a model for this disease, allowing researchers to explore the mechanism of the disease and to design therapies to treat humans affected with ALS.

Courtesy of Dr. Donald Price, Johns Hopkins University

spinocerebellar ataxias, have only a genetic cause. Transgenic animal models can be constructed only after a specific gene for a disorder has been identified and isolated. The use of transgenic models allows research on the molecular and cellular mechanisms of the disorder and on the development and testing of drugs for the treatment of these disorders.

For example, about 10% of all ALS cases are inherited as an autosomal dominant trait (OMIM 105400). Affected individuals have progressive weakness and muscle atrophy with occasional paralysis caused by degeneration of nerve cells that connect with muscles. Some of these affected individuals have a mutation in the *SOD1* gene on chromosome 21. Mutations cause the SOD1 protein to become toxic. Transgenic mice carrying a mutant human *SOD1* gene develop muscle weakness and atrophy similar to that seen in affected humans (▶ Figure 18.3). These mice are used to study how the mutant SOD1 protein selectively damages some nerve cells, but leaves others untouched, and to study the effects of drugs designed to treat this disorder.

Although mice are widely used in transgenic research, human genes transferred to *Drosophila* are also used as models of human neurodegenerative diseases. Flies carrying mutant human genes for HD and spinocerebellar ataxia 3 have been constructed and are used to study how the mutant proteins kill nerve cells and to identify genes or chemicals that can slow or prevent the loss of cells.

18.3 Single Genes Affect the Nervous System and Behavior

In this section we discuss several single-gene defects that have specific effects on the development, structure, and/or function of the nervous system and that consequently affect behavior. Following this we discuss more complex interactions between the genotype and behavior, in which the number and functional roles of genes are not well understood and in which effects on the nervous system may be subtler.

Huntington disease is a model for neurodegenerative disorders.

■ **Huntington disease** An autosomal dominant disorder associated with progressive neural degeneration and dementia. Adult onset is followed by death 10 to 15 years after symptoms appear.

Huntington disease (HD; OMIM 143100) is a useful model for single-gene disorders that affect the nervous system and have a behavioral phenotype. HD is an adult-onset neurodegenerative disorder inherited as an autosomal dominant trait. It affects about 1 in 10,000 individuals in Europe and the United States. HD was one of the first disorders to be mapped using recombinant DNA techniques (see Chapters 13 and 15). The mutation causing the disorder involves expansion of a trinucleotide repeat (this topic is covered in Chapter 11), and the disorder shows anticipation (also covered in Chapter 11). Predictive genetic testing and transgenic animal models are available, and the condition is being treated by transplantation of fetal nerve cells.

The phenotype of HD usually begins in mid-adult life as involuntary muscular movements and jerky motions of the arms, legs, and torso. As it progresses, personality changes, agitated behavior, and dementia occur. Most affected individuals die within 10 to 15 years after the onset of symptoms.

The gene for HD is located on the short arm of chromosome 4 (4p16.3). Mutant alleles carry more copies of CAG triplet repeats than the normal allele. HD is one of eight known neurodegenerative disorders caused by the expansion of a CAG trinucleotide repeat (see Chapter 11). In all cases, the mutation leads to an increase in the number of copies of the amino acid glutamine inserted into the gene product. This increase, called a polyglutamine expansion, causes the protein to become toxic and kill nerve cells. Individuals with less than 35 copies of the CAG repeat in the *HD* gene do not develop the disease, and those with 35 to 39 repeats may or may not develop HD. However, those with 40 to 60 repeats will develop HD as adults. People with more than 60 repeats will develop HD before age 20.

Anticipation is a pattern of successive earlier age of disease onset within a pedigree (see Chapter 11). In HD, anticipation is associated with expansion of the number of CAG repeats as the *HD* gene is passed from parent to offspring. Expansion of paternal copies of CAG is more likely to produce an earlier age of onset, and juvenile cases of HD are almost always related to paternal transmission. The reasons for this are unclear.

Brain autopsies of HD victims show damage to several specific brain regions, including the striatum and the cerebral cortex. In affected regions, cells fill with cytoplasmic and nuclear clusters of the mutant protein and degenerate and die (Figure 18.4). Involuntary movements and progressive personality changes accompany the degeneration and death of these neurons (neurons are cells of the nervous system). The *HD* gene encodes a large protein, huntingtin (Htt). In adult brains, Htt stimulates the production of a protein, BDNF, necessary for the survival of cells in the striatum. Mutant Htt causes a decrease in BDNF production, and cells of the striatum degenerate and die. Because the HD mutation causes the gene product to lose its function, HD can be regarded as a loss-of-function mutation. However, mutation in the *HD* gene also causes the altered Htt protein to become toxic. Whether its toxic effects are caused by intracellular aggregation, cleavage fragments of the protein, or the accumulation of other proteins in the Htt aggregates is not clear. This effect of the mutant HD protein can be regarded as a gain-of-function mutation.

Strains of transgenic mice that carry and express the mutant human *HD* gene have been developed. In these mice, the mutant *HD* gene is active in the brain and other organs. Phenotypically, these mice show progressive behavioral changes and an increasing loss of muscular control. The brains of affected transgenic mice show the same changes as affected humans: accumulation of Htt clusters leading to degeneration and loss of cells in the striatum and cerebrum (Figure 18.5). These mice are being used to study the early events in Htt accumulation and to develop treatments that work in presymptomatic stages to prevent cell death.

HD is caused by a mutation resulting in degeneration and death of cells in the striatum. Animal experiments showed that transplantation of fetal nerve cells into the striatum partially restores nerve connections, muscle control, and behavior. Investigators have started treating HD patients with transplants of fetal striatal cells to determine whether the transplanted cells would survive, make connections to other cells, and lead to improvements in muscle control and intellectual functions. Results from the first round of transplants are encouraging, adding HD to the list of disorders

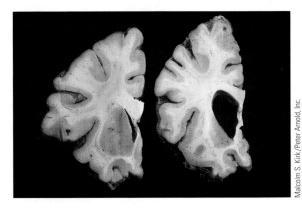

▲ **FIGURE 18.4** Section of brain with Huntington disease (*left*) and normal brain (*right*).

Malcolm S. Kirk/Peter Arnold, Inc.

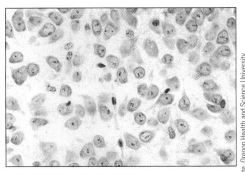

(a)

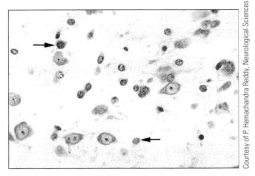

(b)

Courtesy of P. Hemachandra Reddy, Neurological Sciences Institute, Oregon Health and Science University

▲ **FIGURE 18.5** Loss of brain cells in a transgenic Huntington disease mouse. (a) Section of normal mouse brain striatum showing densely packed neurons. (b) Section of the striatum from an HD89 mouse showing extensive loss of neurons that accompany this disease.

that can be treated with such transplants. This success, however, adds to the debate about fetal stem cells, and the direction of stem cell research in this country.

Keep in mind

■ Transgenic animals carrying human genes are used to develop drugs and treatment strategies for behavioral disorders.

The link between language and brain development is still being explored.

For over forty years, linguists, psychologists, and geneticists have unsuccessfully argued over the relationship between language and genetics. The linguist Noam Chomsky observed that all children learn to speak without much instruction. Because language seems to be an inherent trait, he proposed that it might have a genetic component. The finding that specific language deficits are a familial trait and can be separated from other conditions such as autism, deafness, and mental retardation supported Chomsky's idea. However, because there was no clear-cut pattern of inheritance, any genetic link to language seemed weak.

About 10 years ago, a large, multigenerational family, the KE family, came to the attention of researchers. In this family, a very specific speech and language disorder is inherited as an autosomal dominant trait (▶ Figure 18.6). Affected members cannot correctly identify language sounds and have difficulty understanding sentences. They also have problems in making language sounds, and it is almost impossible to understand their speech.

With the cooperation of the family, investigators were able to map the disorder to a small region on the long arm of chromosome 7 and named the unknown gene in this region *SPCH1* (*SPEECH1;* OMIM 602081). Recently, a child, CS, not related

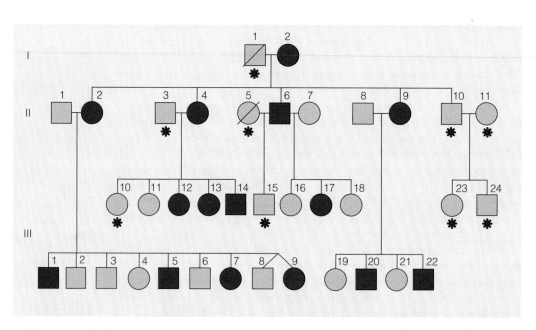

▲ FIGURE 18.6 Pedigree of a family in which some members are affected with a severe speech and language disorder (solid symbols represent affected members). Asterisks mark individuals who were not analyzed. The pattern of inheritance is consistent with an autosomal dominant trait. The gene for this disorder maps to the long arm of chromosome 7.

to the KE family, was found to have the same speech deficit as the KE family. CS carries a translocation involving the long arm of chromosome 7 in the *SPCH1* region. This allowed researchers to identify the gene, now called *FOXP2* (OMIM 605317), because it is a member of a previously identified gene family.

Affected members of the KE family have a single nucleotide change in the *FOXP2* gene, in which a G is replaced with an A. This change results in an amino acid substitution in the protein, presumably altering the protein's function and resulting in the language deficit.

FOXP2 is a member of a family of genes that encode transcription factors. Transcription factors are proteins that switch on genes or gene sets, often at specific stages of development. The FOXP2 protein is very active in fetal brains. It may be that affected members have a 50% reduction in the amount of this protein at a critical stage of brain development, leading to an abnormality of language development.

Future work on *FOXP2* may help us understand how the brain understands and processes language and to develop therapies to treat language disorders. In addition, comparing the action of *FOXP2* in the developing brains of chimpanzees and other primates may help us understand how language evolved and what separates us from our fellow primates.

18.4 Single Genes Control Aggressive Behavior and Brain Metabolism

In 1993 a new form of X-linked mild mental retardation was identified in a large European family. All affected males showed forms of aggressive and often violent behavior. Gene mapping and biochemical studies indicate that this condition exhibits a direct link between a single-gene defect and a phenotype characterized by aggressive and/or violent behavior (▶ Figure 18.7). In particular, eight males with mild or borderline mental retardation showed a characteristic pattern of aggressive and often violent behavior triggered by anger, fear, or frustration. The behavioral pheno-

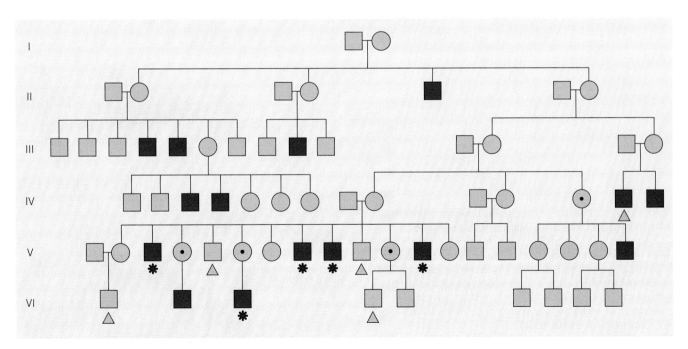

▲ FIGURE 18.7 Cosegregation of mental retardation, aggressive behavior, and a mutation in the monoamine oxidase type A (*MAOA*) gene. Affected males are indicated by the filled symbols. Symbols marked with an asterisk represent males known to carry a mutation of the *MAOA* gene; those marked with a triangle are known to carry the normal allele. Symbols marked with a dot inside represent females known to be heterozygous carriers.

type varied widely in levels of violence and time but included acts of attempted rape, arson, stabbings, and exhibitionism.

Geneticists have mapped a gene for aggression.

Using molecular markers, the gene for this behavior was mapped to the short arm of the X chromosome in region Xp11.23–11.4. A gene in this region encodes an enzyme called monoamine oxidase type A (MAOA) that breaks down a neurotransmitter (▶ Table 18.2). Neurotransmitters are chemical signals that carry nerve impulses across synapses in the brain and nervous system (▶ Active Figure 18.8). Failure to break down these chemical signals rapidly can disrupt the normal function of the nervous system.

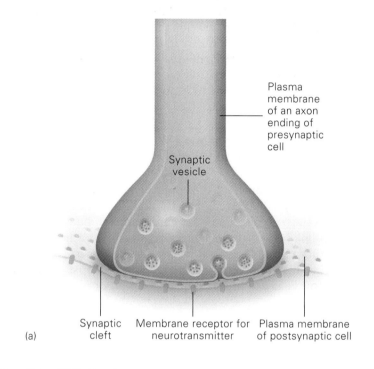

Plasma membrane of an axon ending of presynaptic cell

Synaptic vesicle

(a)

Synaptic cleft   Membrane receptor for neurotransmitter   Plasma membrane of postsynaptic cell

Presynaptic cell
Synaptic cleft
Postsynaptic cell
(b)

Dr. Constantino Sotelo

Ions (*red*) that affect membrane excitability

Molecule of neurotransmitter—synaptic cleft

Receptor for the neurotransmitter on gated channel protein in plasma membrane of postsynaptic cell

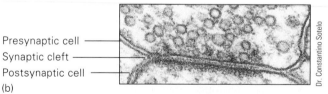

(c)

▶ ACTIVE FIGURE 18.8 The synapse and synaptic transmission. (a) A thin cleft, called the synapse, separates one cell from another. (b) An electron micrograph of a synapse between two nerve cells. (c) As a nerve impulse arrives at the synapse, it triggers the release of a chemical neurotransmitter from storage in synaptic vesicles. The neurotransmitters diffuse across the synapse and bind to receptors on the membrane of the cell on the other side of the synapse (post-synaptic neuron), where they trigger another nerve impulse by allowing ions into the postsynaptic cell.

HUMAN
Genetics ⒺNow™   Learn more about synaptic transmission by viewing the animation on Human GeneticsNow
at **http://now.ilrn.com/cummings7**

The urine of the eight affected individuals contains abnormal levels of chemicals produced by MAOA. The researchers concluded that the eight affected males carried a mutation in the gene that encodes this enzyme and that lack of MAOA activity is associated with their behavior pattern.

A follow-up study analyzed the *MAOA* gene (OMIM 309850) in five of the eight affected individuals and showed that all five carry a mutation that encodes a non-functional gene product. This mutation was also found in two female heterozygotes and is not found in any unaffected males in this pedigree. Loss of MAOA activity in affected males prevents normal breakdown of certain neurotransmitters, reflected in elevated levels of toxic compounds in the urine.

Because it is difficult to relate a phenotype such as aggression (for example, what exactly constitutes aggression?) to a specific genotype, further work is needed to determine whether mutations of *MAOA* are associated with altered behavior in other families and in animal model systems. In addition, the interaction of this disorder with external factors such as diet, drugs, and environmental stress remains to be established. However, the identification of a specific mutation associated with this behavior pattern is an important discovery and suggests that biochemical or pharmacological treatment for this disorder may be possible.

There are problems with single-gene models for behavioral traits.

Although recombinant DNA markers have been successful in identifying, isolating, and cloning single genes that affect behavior, in other cases it has produced erroneous results. In 1987, a DNA linkage study mapped a gene for bipolar illness to a region of chromosome 11. Later, individuals from the study group who did not carry the suspect copy of chromosome 11 developed the illness, indicating that the gene was not on this chromosome. Another study reported linkage between DNA markers on chromosome 5 and schizophrenia; however, the linkage was later found to be coincidental, or, at best, could only explain the disorder in a small, isolated population. These early failures to find single genes that control these disorders led to the reevaluation of single-gene models for many behavior traits and to the development of alternative models for complex traits, as described in the next section.

18.5 The Genetics of Mood Disorders and Schizophrenia

Mood disorders, also known as affective disorders, are conditions in which affected individuals have profound emotional disturbances. **Moods** are sustained emotions; **affects** are short-term expressions of emotion. Individuals with affective disorders experience periods of prolonged depression (**unipolar disorder**) or cycles of depression alternating with periods of elation (**bipolar disorder**).

Schizophrenia is a collection of mental disorders characterized by psychotic symptoms, delusions, thought disorders, and hallucinations, often called the schizoid spectrum. Schizoid individuals experience disordered thinking, inappropriate emotional responses, and social deterioration. Mood disorders and schizophrenia are complex, often are difficult to diagnose, and have genetic components and environmental triggers. There are genetic components to mood disorders and schizophrenia, but there is no clear-cut pattern of inheritance, and the roles of social and environmental factors are unclear. Nonetheless, some information about the genetics of these conditions is emerging, and, despite setbacks in identifying single genes, progress is being made in generating genetic models of these disorders. Sylvia Nasar's book, *A Beautiful Mind*, tells the story of John Nash, a mathematician, his decades-

Mood disorders A group of behavior disorders associated with manic and/or depressive syndromes.

Mood A sustained emotion that influences perception of the world.

Affect A short-term expression of feelings or emotion.

Unipolar disorder An emotional disorder characterized by prolonged periods of deep depression.

Bipolar disorder An emotional disorder characterized by mood swings that vary between manic activity and depression.

Schizophrenia A behavioral disorder characterized by disordered thought processes and withdrawal from reality. Genetic and environmental factors are involved in this disease.

long struggle with schizophrenia, and his work on game theory, for which he won a Nobel Prize.

Keep in mind

■ Indirect evidence from family studies indicates that mood disorders and schizophrenia have genetic components, but no genes have yet been identified.

Mood disorders include unipolar and bipolar illnesses.

The lifetime risk for a clinically identifiable mood disorder in the U.S. population is 8% to 9%. Depression (unipolar illness) is the most common mood disorder. It is more common in females (about a 2:1 ratio), usually begins in the fourth or fifth decade of life, and is often long lasting or a recurring condition. Depression is associated with weight loss, insomnia, poor concentration, irritability and anxiety, and lack of interest in surrounding events.

About 1% of the U.S. population suffers from bipolar illness. The age of onset occurs during adolescence or the second and third decades of life, and males and females are at equal risk for this condition. In bipolar illness, periods of manic activity alternate with depression. Manic phases are characterized by hyperactivity, acceleration of thought processes, low attention span, creativity, feelings of elation or power, and risk-taking behavior.

Family, twin, and adoption studies have linked bipolar illness to genetics. These studies show concordance of 60% for monozygotic (MZ) twins and 14% for dizygotic (DZ) twins. Adoption studies also indicate that genetic factors are involved in this disorder. Studies have also documented the risk to first-degree relatives of those with bipolar illness (▷ Figure 18.9). The fact that concordance in MZ twins is not 100% suggests that environmental factors (such as stress) interact with genetic risk. As discussed earlier, several attempts have been made to map genes for bipolar illnesses by using a single-gene model but were unsuccessful. The failure to find link-

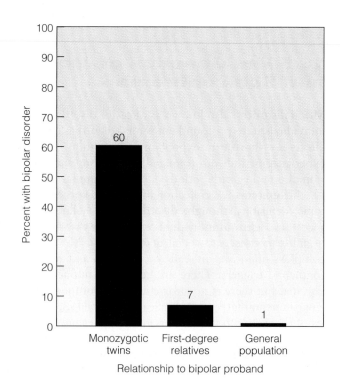

▷ FIGURE 18.9 The frequency of bipolar illness in members of monozygotic twin pairs and in first-degree relatives of affected individuals indicates that genetic factors are involved in this disease.

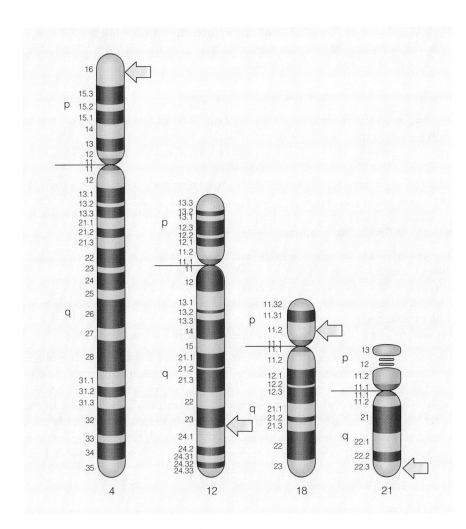

age between genetic markers and single genes for bipolar illness does not undermine the role of genes in this condition but means that new strategies of linkage analysis are required to identify the genes involved.

New strategies, using genomics, have centered on several approaches used in a worldwide effort to screen all human chromosomes for genes that control bipolar illness. Association studies are one such approach. If there is an association between a disease gene and nearby markers, this combination should occur more often in people with bipolar illness than in control populations. Association studies use DNA markers but follow the inheritance of the marker and the disorder (bipolar illness in this case) in unrelated individuals affected with the disorder. The idea is to identify portions of the genome that are more common in affected individuals than in those without the trait. Association studies have identified regions on chromosomes 4, 12, 18, and 21 as candidates that may contain genes for bipolar illness (▶ Figure 18.10). Other chromosomes, including 7, 10, 13 and 15, may also carry genes associated with this disorder. Now that these chromosome regions have been identified, they are being closely studied to search for specific genes responsible for this disorder.

In addition to its elusive genetic nature, bipolar illness remains a fascinating behavioral disturbance because of its close association with creativity. Many great artists, authors, and poets have been afflicted with bipolar illness (▶ Figure 18.11). The thought patterns of the creative mind parallel those of the manic stage of bipolar illness.

▲ FIGURE 18.11 Virginia Woolf, the author and poet, was affected by manic depression. Like other affected people, she often commented on the relationship between creativity and her illness.

In her book *Touched with Fire*, Kay Jamison explores the relationship among genetics, neuroscience, and the lives and temperaments of creative individuals, including Byron, Van Gogh, Poe, and Virginia Woolf.

Schizophrenia has a complex phenotype.

Schizophrenia is a mental illness affecting about 1% of the population (about 2.5 million people in the United States are affected). The disorder usually appears in late adolescence or early adult life. It has been estimated that half of all hospitalized mentally ill and mentally retarded individuals are schizophrenic.

Schizophrenia is a disorder of thought rather than of mood. Diagnosis is often difficult, and there is notable disagreement on the definition of schizophrenia because it has no single distinguishing feature and causes no characteristic brain pathology. Some features of the disorder include:

- Psychotic symptoms, including delusions of persecution;
- Disorders of thought; loss of the ability to use logic in reasoning;
- Perceptual disorders, including auditory hallucinations (hearing voices);
- Behavioral changes, ranging from mannerisms of gait and movement to violent attacks on others; and
- Withdrawal from reality and inability to participate in normal activities.

Models for schizophrenia generally fall into two groups: those in which biological factors (including genetics) play a major role and environmental factors are secondary and, conversely, models in which environmental factors are primary and biological factors are secondary. Some evidence points to metabolic differences in the brains of schizophrenics compared with those of normal individuals (▶ Figure 18.12), but it is unclear whether these differences are genetic. Overall, however, the best evidence supports the role of genetics as a primary factor in schizophrenia, but full expression is dependent on environmental factors.

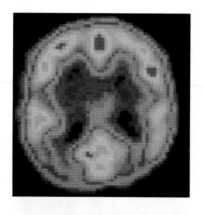
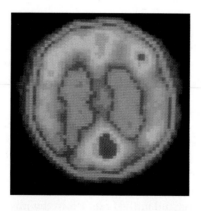
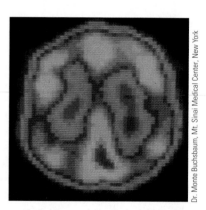
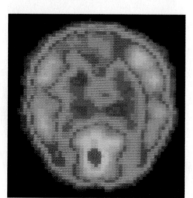
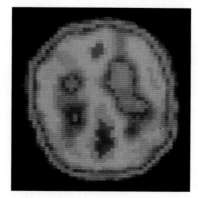

Dr. Monte Buchsbaum, Mt. Sinai Medical Center, New York

▲ **FIGURE 18.12** Brain metabolism in a set of monozygotic quadruplets, all of whom suffer from varying degrees of schizophrenia. These scans of glucose utilization by brain cells are visualized by positron emission tomography (PET scan). They show low metabolic rates in the frontal lobe (*top of each scan*) compared with nonschizophrenics (*top left image*). The frontal lobe is where cognitive ability resides.

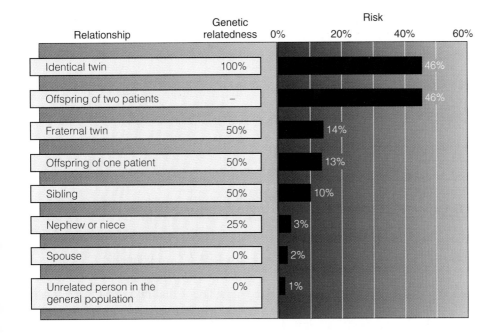

Relationship	Genetic relatedness	Risk
Identical twin	100%	46%
Offspring of two patients	–	46%
Fraternal twin	50%	14%
Offspring of one patient	50%	13%
Sibling	50%	10%
Nephew or niece	25%	3%
Spouse	0%	2%
Unrelated person in the general population	0%	1%

◀ FIGURE 18.13 The lifetime risk for schizophrenia varies with the degree of the relationship to an affected individual. The observed risks are more compatible with a multifactorial mode of transmission than a single-gene or polygenic mode of inheritance.

Risk factors for relatives of schizophrenics are high (▶ Figure 18.13), revealing the influence of genotype on schizophrenia. Overall, relatives of affected individuals have a 15% chance of developing the disorder (as opposed to 1% among unrelated individuals). Using a narrow definition of schizophrenia, the concordance value for MZ twins is 46%, versus 14% for DZ twins. MZ twins raised apart show the same level of concordance as MZ twins raised together. If a broader definition of the phenotype is used, one which combines schizophrenia and borderline or schizoid personalities, the concordance for MZ twins approaches 100%, and the risk for siblings, parents, and offspring of schizophrenics is about 45%. This strongly supports the role of genes in this disorder.

How schizophrenia (or susceptibility to this illness) is inherited is unknown. In the past, pedigree and linkage studies suggested genes on the X chromosome and a number of autosomes as sites contributing to this condition. Although all the linkage studies have been contested and contradicted by other studies, several tentative conclusions can be drawn about the genetics of schizophrenia.

A polygenic model, in which a single gene makes a major contribution, is consistent with results from family studies, concordance in MZ twins, and incidence of the disorder in the general population. Linkage studies seem more valuable than association studies for identifying genes related to schizophrenia. A group of research laboratories has formed the Schizophrenia Collaborative Linkage Group to conduct genome scans for major genes that contribute to schizophrenia and bipolar illness. In one study, 452 microsatellite markers covering all regions of the genome were used to study individuals in 54 pedigrees, each of which has multiple cases of schizophrenia. To date, linkages between schizophrenia and loci on chromosomes 3, 8, 13, and 22 have been identified.

Polygenic models suggest that several genes contribute to the phenotype of schizophrenia, but a molecular understanding of the disease has not been achieved. Instead of studying genes one at a time, researchers are using DNA microarrays to study changes in mRNA expression in several thousand genes at a time. Using microarrays, one research group has identified altered patterns of expression of genes involved in nerve cell myelination in schizophrenia (▶ Figure 18.14). Myelin is a multilayered coating around nerve cells, much like insulation around electrical wires. Myelin enhances long-distance transmission of nerve impulses, and the lack of myelin around nerve cells in schizophrenia may contribute to the abnormal behavior characteristic of this disease.

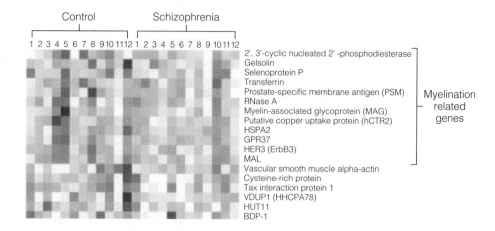

FIGURE 18.14 Relative levels of gene expression in normal brains and brains of schizophrenics. Each column shows the level of expression (*blue* is low, *red* is high), and each row is the expression level for a specific gene. There is a great reduction in expression of genes involved in myelin formation in the brains of schizophrenics compared with normal brains. Although thousands of genes were studied in this experiment, only the results for some of the genes are shown here.

Control — 1 2 3 4 5 6 7 8 9 10 11 12
Schizophrenia — 1 2 3 4 5 6 7 8 9 10 11 12

2′, 3′-cyclic nucleated 2′ -phosphodiesterase
Gelsolin
Selenoprotein P
Transferrin
Prostate-specific membrane antigen (PSM)
RNase A
Myelin-associated glycoprotein (MAG)
Putative copper uptake protein (hCTR2)
HSPA2
GPR37
HER3 (ErbB3)
MAL
Vascular smooth muscle alpha-actin
Cysteine-rich protein
Tax interaction protein 1
VDUP1 (HHCPA78)
HUT11
BDP-1

Myelination related genes

18.6 Genetics and Social Behavior

Human geneticists have long been interested in behavior that takes place in a social context, that is, behavior resulting from interactions between and among individuals. This behavior is often complex, and evidence indicates that such behavior involves multifactorial inheritance. Several traits that affect different aspects of social behavior are discussed in the following sections.

Keep in mind

■ Human behavior in social settings is complex and often difficult to define.

Tourette syndrome affects speech and behavior.

Tourette syndrome (GTS) A behavioral disorder characterized by motor and vocal tics and inappropriate language. Genetic components are suggested by family studies that show increased risk for relatives of affected individuals.

Alzheimer disease (AD) A heterogeneous condition associated with the development of brain lesions, personality changes, and degeneration of intellect. Genetic forms are associated with loci on chromosomes 14, 19, and 21.

Tourette syndrome (GTS; OMIM 137580) is characterized by motor and behavioral disorders. About 10% of affected individuals have a family history of the condition. Males are affected more frequently than females (3:1), and onset is usually between 2 and 14 years of age. GTS is characterized by episodes of motor and vocal tics that can progress to more complex behaviors involving a series of grunts and barking noises. Vocal tics include outbursts of profane and vulgar language and parrot-like repetition of words spoken by others. Because there is so much variation in expression, the incidence of the condition is unknown, but some researchers suggest that the disorder may be very common.

Family studies reveal that relatives of affected individuals are at significantly greater risk for GTS than relatives of unaffected controls. The inheritance of GTS is complex and involves a major gene and a number of minor genes, as well as environmental contributions.

Identifying genes associated with GTS is proving difficult. Recent efforts have focused on genetically isolated populations, such as the Afrikaners of South Africa. Genome scans using DNA markers on samples from affected and unaffected members of the Afrikaner population have identified regions on five autosomes that may contain genes associated with GTS. These regions are on chromosomes 2p, 8q, 11q, 20q, and 21q. Further work using data from the Human Genome Project will identify specific genes in these regions that are candidates for GTS and hopefully lead to new methods of diagnosis and treatment for this disorder.

Alzheimer disease has genetic and nongenetic components.

Alzheimer disease (AD) is a progressive and fatal neurodegenerative disease that affects almost 2% of the population of developed countries. Age is a major risk factor for AD, and as populations in these countries age, the incidence of this disease is expected to increase threefold by 2055. Ten percent of the U.S. population older than 65 years has AD, and the disorder affects 50% of those older than 80 years. The cost of treatment and care for AD is more than $80 billion.

The symptoms of Alzheimer disease begin with loss of memory; progressive dementia; and disturbances of speech, motor activity, and recognition. There is an ongoing degeneration of personality and intellect, and, eventually, affected individuals are unable to care for themselves. Brain lesions accompany the progression of AD (▷ Figure 18.15). The lesions are formed by a protein fragment, amyloid beta-protein, which accumulates outside cells in aggregates known as senile plaques. These plaques cause the degeneration and death of nearby neurons, affecting selected regions of the brain (▷ Figure 18.16). Formation of senile plaques is not specific to AD; almost everyone who lives beyond the age of 80 will have such lesions. The difference between normal aging of the brain and AD appears to be the number of such plaques (greatly increased in AD) and the rate of accumulation (earlier and faster in AD).

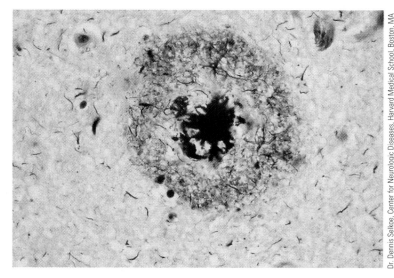

▲ FIGURE 18.15 A lesion called a plaque in the brain of an individual with Alzheimer disease. A ring of degenerating nerve cells surrounds the deposit of protein.

The genetics of AD is complex. Less than 50% of all cases can be traced to genetic causes, indicating that the environment plays a significant role in this disorder. We first examine the genetic evidence and then discuss some of the proposed environmental factors associated with AD.

The gene that encodes the amyloid beta-protein, *AD1* (OMIM 104300), is located on the long arm of chromosome 21. Mutations of this gene are responsible for an early-onset form of AD, inherited as an autosomal dominant trait. Other inherited forms of AD are caused by a mutation in a membrane protein encoded by a gene on chromosome 14 (OMIM 104311), and a gene on chromosome 1 that encodes a

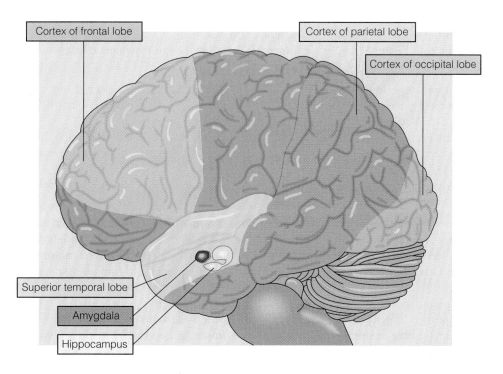

Cortex of frontal lobe

Cortex of parietal lobe

Cortex of occipital lobe

Superior temporal lobe

Amygdala

Hippocampus

◀ FIGURE 18.16 Location of brain lesions in Alzheimer disease. Plaques are most heavily concentrated in the amygdala and hippocampus. These brain regions are part of the limbic system.

membrane protein (OMIM 600759). There is evidence for AD genes on other chromosomes, and mitochondrial DNA polymorphisms may also play a role in susceptibility to this disease.

The overwhelming majority of AD cases are sporadic—not inherited. At least one gene conferring an enhanced risk for AD has been identified. This gene, *apolipoprotein E* (*APOE*; OMIM 107741), encodes a protein involved with cholesterol metabolism, transport, and storage. The *APOE* gene has three alleles (*E*2*, *E*3*, and *E*4*). Those who carry one or two copies of the *APOE*4* allele are at increased risk of AD, but the mechanism of how this happens is unclear.

Once the disease begins, and beta amyloid accumulates, several factors can influence the rate at which the disease progresses. These include:

- Free radical production stimulated by beta-amyloid accumulation;
- Calcium uptake into nerve cells caused by beta amyloid; and
- beta-amyloid toxicity to nearby nerve cells.

In summary, we know that AD has several genetic causes, and mutations in any of several genes can produce the AD phenotype. Moreover, the fact that many cases of AD cannot be traced to a genetic source may indicate that there is more than one cause for AD. In addition, several factors influence the rate at which the disease progresses. Scientists continue to investigate the role of nongenetic influences and their mechanisms, attempting to define the risk factors and rate of progression for this debilitating condition.

Research has shown that several environmental factors can reduce the risk of AD, and even reverse the disease in early stages. Intellectually stimulating jobs or environments, moderate exercise done on a regular basis, and diets low in cholesterol and saturated fats all have been shown to help stave off this condition.

Alcoholism has several components.

As a behavioral disorder, excessive alcohol consumption (OMIM 103780) has two important components. First, drinking in excess over a long time causes damage to the nervous system and other organ systems. Over time, the accumulation of damage results in altered behavior, hallucinations, and loss of memory. The second component involves the behavior patterns that lead to alcohol abuse and a loss of the ability to function in social settings, the workplace, and the home.

It is estimated that 75% of the adult U.S. population consumes alcohol. About 10% of these adults are classified as alcoholics, and the male to female ratio is about 4:1. From the genetic standpoint, alcoholism is most likely a genetically influenced, multifactorial (genetic and environmental) disorder. The role of genetic factors in alcoholism is indicated by a number of findings:

- In mice, experiments indicate that alcohol preference can be selected for; some strains of mice choose a solution of 75% alcohol over water, whereas other strains shun all alcohol.
- There is a 25% to 50% risk of alcoholism in sons and brothers of alcoholic men.
- There is a 55% concordance for alcoholism in MZ twins, and a 28% rate in same-sex DZ twins.
- Sons adopted by alcoholic men show a rate of alcoholism closer to their biological fathers than their adoptive fathers.

Genes that influence alcoholism have not been identified. Segregation analysis in families with alcoholic members has produced evidence against the Mendelian inheritance of a single major gene and for multifactorial inheritance involving several genes. Other researchers have adopted a single-gene model and find an association between an allele of a gene that encodes a neurotransmitter receptor protein (called D2) and alcoholism. This evidence is based on the finding that in brain tissue, the *A1* allele of the *D2* gene was found in 69% of the samples from severe alcoholics but in only 20% of the samples from nonalcoholics, implying that the *A1* allele is involved in alcoholism.

Subsequent linkage and association studies on the *A1* allele have not supported the idea that this allele is involved in alcoholism. Taken together, the available studies have failed to show any relationship between abnormal neurotransmitter metabolism or receptor function and alcoholic behavior.

The search for genetic factors in alcoholism illustrates the problem of selecting the proper genetic model to analyze behavioral traits. Segregation and linkage studies indicate that there is no single gene for alcoholism. If a multifactorial model involving a number of genes, each with a small additive effect, is invoked, the problem becomes more complicated. How do you prove or disprove that a given gene contributes, say, 10% to the behavioral phenotype? At present, the only method would involve studying thousands of individuals to find such effects.

Is sexual orientation a multifactorial trait?

Most people are heterosexual and prefer sexual activity with the opposite sex, but a fraction of the population is homosexual and prefers sexual activity with members of the same sex. These variations in sexual behavior have been recorded since ancient times, but biological models for these behaviors have been proposed only recently.

Twin studies and adoption studies have investigated the role of genetics in sexual orientation. One twin study involved 56 MZ twins, 54 DZ twins, and 57 genetically unrelated adopted brothers. The concordance for homosexuality was 52% for MZ twins, 22% for DZ twins, and 11% for the unrelated adopted siblings. Overall heritability ranged from 31% to 74%. Another study investigating homosexual behavior in women employed 115 twin pairs and 32 genetically unrelated adopted sisters. In this study, heritability ranged from 27% to 76%.

The results from these and other studies indicate that homosexual behavior has a strong genetic component. These studies have been challenged on the grounds that the results can be affected by the phrasing of the interview questions and by the methods used to recruit participants and that the phenotype is self-described. But the average heritability estimates for sexual preference from these studies parallel those from the Minnesota Twin Project, a long-term study of MZ twins separated at birth and reared apart. Further twin studies are needed to determine whether the heritability values are accurate. If confirmed, the studies to date indicate that homosexual behavior is a multifactorial trait that involves several genes and unidentified environmental components.

Several years ago, a group of U.S. researchers used RFLP markers to study male homosexual behavior and found linkage between one subtype of homosexuality and markers on the long arm of the X chromosome (▶ Figure 18.17). This study employed a two-step approach. First, family histories were collected from 114 homosexual males. Pedigree analysis was performed on 76 randomly selected individuals from this group, using interviews with male relatives to ascertain sexual preference. The results indicated the possibility of X-linked inheritance (OMIM 306995).

A further pedigree analysis was conducted using 38 families in which there were two homosexual brothers, based on the idea that this might show a stronger trend for X-linked inheritance because there were two homosexual siblings in the same family. The results show a stronger trend toward X-linked inheritance and an absence of paternal transmission.

Using information about traits and relatedness from the pedigree analysis, the second part of the study employed DNA markers to determine whether an X-linked locus or loci were associated with maternally transmitted male homosexual behavior in the 38 families that had two homosexual brothers. Linkage was detected to RFLP markers from the distal region of the long arm in the Xq28 region. These markers were present in two-thirds of the 32 pairs of homosexual brothers and in about one-fourth of the heterosexual brothers.

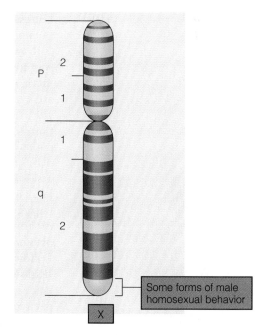

▲ FIGURE 18.17 Region of the X chromosome found by linkage analysis to be associated with one form of male homosexual behavior.

In a subsequent study of 52 homosexual brothers, a Canadian research team failed to find linkage between the Xq28 markers and homosexuality. Their analysis could not rule out the possibility of a gene with a minor contribution to this behavior but could exclude a gene with a major contribution. More work is needed to confirm the linkage relationship and to search the region for a locus or loci that affects sexual orientation.

Two important factors related to the U.S. study should be mentioned. First, seven sets of homosexual brothers did not share all of the markers in the Xq28 region, and about one-fourth of all heterosexual brothers inherited the markers but did not display homosexual behavior. This indicates that genetic heterogeneity or nongenetic factors are significant in this behavior. Second, the study cannot estimate what fraction of homosexual behavior might be related to the Xq28 region or whether this region influences lesbian sexual behavior. In spite of its preliminary nature and (to some) its controversial conclusions, this work applies genome-wide screening with molecular markers to study the role of genes in one type of male sexual behavior and represents a model for future studies in this area of behavior genetics.

18.7 Summing Up: The Current Status of Human Behavior Genetics

In reviewing the current state of human behavior genetics, several elements are apparent. Almost all studies of complex human behavior have provided only indirect and correlative evidence for the role of specific genes. Segregation studies and heritability estimates indicate that most behaviors are complex traits involving several genes. Searches for single-gene effects have proven unsuccessful to date, and initial reports of single genes that control bipolar illness, schizophrenia, and alcoholism have been retracted or remain unconfirmed.

The multifactorial nature of behavioral traits means that methods for identifying genes with small, incremental effects must be developed. Although twin and adoption studies have been valuable in behavior genetics, these studies typically involve small numbers of individuals. For example, fewer than 300 pairs of MZ twins raised apart have been identified worldwide. When traits involve multiple genes, confirmation results can require detailed examination of thousands of individuals in hundreds of families. This process is necessarily slow and labor-intensive.

Recent successes in finding genes involved in other complex traits may point the way to identifying genes associated with behavior. Chromosome regions shared by those with a genetic disorder are larger than originally thought, and fewer markers are needed to trace the coinheritance of a trait and its associated markers, making work easier and faster. Using a small number of markers, along with information from the Human Genome Project and newer methods of data analysis, researchers have identified genes for susceptibility to inflammatory bowel disease (Crohn disease) and insulin-dependent diabetes. Previously, the linkage results for these diseases showed the same inconsistency as the results for behavior disorders such as bipolar illness and schizophrenia. Putting this set of techniques to work in identifying genes for behavior along with more refined definitions of phenotype may be successful in dissecting the genetic components of mental illness and other behavioral phenotypes.

However, success in identifying susceptibility genes for behavior should not overshadow the fact that the environment plays a significant role in behavior. As confirmation of the role of genes in behavior becomes available, investigations into the role of environmental factors cannot be neglected. The history of human behavior genetics in the eugenics movement of the early part of this century provides a lesson in the consequences of overemphasizing the role of genetics in behavior. Attempts to provide single-gene explanations for complex behavior inhibited the growth of human genetics as a discipline.

The identification of genes affecting behavior may lead to improvements in diagnosis and treatment of behavior disorders but also has implications for society at large. As discussed in Chapter 15, the Human Genome Project has raised questions about the way genetic information will be disseminated and used and who will have access to this information and under what conditions. These same concerns need to be addressed for genes that affect behavior. If genes for alcoholism or homosexuality can be identified, will this information be used to predict an individual's future behavioral patterns? Who should have access to this information, and what can be done to prevent health care discrimination in employment or insurance?

Many behavioral phenotypes, such as Huntington or Alzheimer disease, are clearly regarded as abnormal. Few would argue against the development of treatments for intervening in and perhaps preventing these conditions. When do behavior phenotypes move from being abnormal to being variants? If there is a connection between bipolar illness and creativity, to what extent should this condition be treated? If genes that influence sexual orientation are identified, will this behavior be regarded as a variant or as a condition that should be treated and/or prevented?

Although research can provide information about the biological factors that play a role in determining human behavior, it cannot provide answers to questions of social policy. Social policy and laws have to be formulated by using information from research.

Genetics in Practice

Genetics in Practice case studies are critical thinking exercises that allow you to apply your new knowledge of human genetics to real-life problems. You can find these case studies and links to relevant websites at **http://biology.brookscole.com/cummings7**

CASE 1

Rachel asked to see a genetic counselor because she was concerned about developing schizophrenia. Her mother and maternal grandmother both had schizophrenia and were institutionalized for most of their adult lives. Rachel's three maternal aunts are all in their 60s and have not shown any signs of this disease. Rachel's father is alive and healthy, and his family history does not suggest any behavioral or genetic conditions. The genetic counselor discussed the multifactorial nature of schizophrenia and explained that there have been many candidate genes identified that may be mutated in individuals with this condition. However, a genetic test is not available for presymptomatic testing. The counselor explained that based on Rachel's family history and her relatedness to the individuals who have schizophrenia, her risk of developing it is approximately 13%. If an altered gene is in the family and her mother carries this gene, Rachel would have a 50% chance of inheriting it.

1. Why do you think it has been so difficult to identify genes underlying schizophrenia?

2. If a test were available that could tell you whether you were likely to develop a disorder such as schizophrenia later in life, would you submit yourself to the test? Why or why not?

CASE 2

A genetic counselor was called to the pediatric ward of the hospital for a consultation. Her patient, an 8-year-old boy, was having a "temper tantrum" and was biting his own fingers and toes. The nurse called after she noticed that he was a patient of a clinical geneticist at another institution. The counselor reviewed the boy's chart and noted a history of growth retardation and self-mutilation since the age of 3. His movements were very "jerky," and he was banging his head against the bedpost. The nurses were having a very difficult time controlling him. The counselor immediately recognized these symptoms as part of a genetic disorder known as Lesch-Nyhan syndrome.

Lesch-Nyhan syndrome is an X-linked recessive condition (Xq26) caused by mutation in the *hypoxanthine*
(continued)

phosphoribosyltransferase gene, and it affects about 1 in 100,000 males. Symptoms usually begin between the ages of 3 and 6 months. Prenatal testing is available, but there is no treatment for Lesch-Nyhan, and most affected individuals die by the second decade of life.

1. If your child had Lesch-Nyhan syndrome and you heard about an experimental gene therapy technique that had shown some promise in treating the disease but which also had significant associated risks, would you attempt to enroll your child in a clinical trial of the technique? Explain.

2. Lesch-Nyhan syndrome is quite rare (1 in 100,000 males), but its effects are devastating. Would you support an effort to screen every developing fetus for this disorder? Explain.

Summary

18.1 Models, Methods, and Phenotypes in Studying Behavior

■ Many forms of behavior represent complex phenotypes. The methods used to study inheritance of behavior encompass classical methods of linkage and pedigree analysis, newer methods of recombinant DNA analysis, and new combinations of techniques, such as twin studies combined with molecular methods. Refined definitions of behavior phenotypes are also being used in the genetic analysis of behavior.

18.2 Animal Models: The Search for Behavior Genes

■ Results from work on experimental animals indicate that behavior is under genetic control and have provided estimates of heritability. The molecular basis of single-gene effects in some forms of behavior has been identified and provides useful models to study gene action and behavior. Transgenic animals carry mutant copies of human genes and are studied to understand the action of the mutant alleles and to develop drugs for treatment of these conditions.

18.3 Single Genes Affect the Nervous System and Behavior

■ Several single-gene effects on human behavior are known. Most of these affect the development, structure, or function of the nervous system and consequently affect behavior. Huntington disease serves as a model for neurodegenerative disorders. Language and brain development are linked by genes that encode transcription factors.

18.4 Single Genes Control Aggressive Behavior and Brain Metabolism

■ Most forms of mental retardation are genetically complex multifactorial disorders. One form of X-linked retardation, associated with aggressive behavior, is linked to abnormal metabolism of a neurotransmitter (a chemical that transfers nerve impulses from cell to cell).

18.5 The Genetics of Mood Disorders and Schizophrenia

■ Bipolar illness and schizophrenia are common behavior disorders, each affecting about 1% of the population. Simple models of single-gene inheritance for these disorders have not been supported by extensive studies of affected families, and more complex forms of multifactorial inheritance seem likely.

18.6 Genetics and Social Behavior

■ Multifactorial traits that affect behavior include Tourette syndrome, Alzheimer disease, alcoholism, and sexual orientation. Twin studies combined with molecular markers have identified a region of the X chromosome that may affect one form of homosexual behavior. Others have been unable to confirm this link, leaving open the question of genetic control of sexual choice.

18.7 Summing Up: The Current Status of Human Behavior Genetics

■ The evidence for genetic control of complex behaviors is indirect. Although some progress has been made in showing linkage between certain chromosome regions and disorders including bipolar illness and schizophrenia, no genes contributing to these or other behaviors such as alcohol abuse or sexual preference have been discovered. New ways of studying linkage, coupled with new methods of analysis of linkage data and information from the Human Genome Project may lead to rapid identification of genes involved in behavior.

Preparing for an exam? Assess your understanding of this chapter's topics with a pre-test, a personalized learning plan, and a post-test at **http://now.ilrn.com/cummings7**

Models, Methods, and Phenotypes in Studying Behavior

1. What are the major differences in the methods used to study the behavior genetics of single-gene traits versus polygenic traits?

2. In human behavior genetics, why is it important for the trait under study to be defined accurately?

3. One of the models for behavioral traits in humans involves a form of interaction known as epistasis. In a simplified example involving two genes, the expression of one gene affects the expression of the other. How might this interaction work, and what patterns of inheritance might be shown?

Animal Models: The Search for Behavior Genes

4. What are the advantages of using *Drosophila* for the study of behavior genetics? Can this organism serve as a model for human behavior genetics? Why or why not?

5. You are a researcher studying an autosomal dominant neurodegenerative disorder. You have cloned the gene underlying the disorder and have found that it encodes an enzyme that is overexpressed in neurons of individuals suffering from the disorder. To better understand how this enzyme causes neurodegeneration in humans, you make a strain of transgenic *Drosophila* whose nerve cells overexpress the enzyme.

   a. How might you use these transgenic flies to try to gain insight into the disease or to identify drugs that might be useful in the treatment of the disease?

   b. Can you think of any potential limitations of this approach?

Single Genes Affect the Nervous System and Behavior

6. What type of mutation causes Huntington disease? How does this mutation result in neurodegeneration?

7. Perfect pitch is the ability to identify a note when it is sounded. In a study of this behavior, perfect pitch was found to predominate in females (24 out of 35 in one group). In one group of seven families, individuals in each family had perfect pitch. In two of these families, the affected individuals included a parent and a child. In another group of three families, three or more members (up to five) of each had perfect pitch, and in all three families, two generations were involved. Given this information, what, if any, conclusions can you draw as to whether this behavioral trait might be genetic? How would you test your conclusion? What further evidence would be needed to confirm your conclusion?

8. Opposite to perfect pitch is tone deafness, the inability to identify musical notes. In one study, a bimodal distribution in populations was found, with frequent segregation in families and sibling pairs. The author of the study concluded that the trait might be dominant. In a family study, segregation analysis suggested an autosomal dominant inheritance of tone deafness with imperfect penetrance. One of the pedigrees is presented here. On the basis of the results, do you agree with this conclusion? Could perfect pitch and tone deafness be alleles of a gene for musical ability?

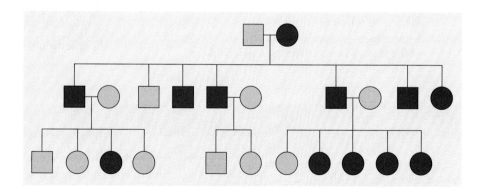

Single Genes Control Aggressive Behavior and Brain Metabolism

9. Name three genes whose mutation leads to an altered behavioral phenotype. Briefly describe the normal function of the mutated gene as well as the altered phenotype.

10. Mutations in the gene encoding monoamine oxidase type A (MAOA) have been linked to aggressive and sometimes violent behavior. Based on this finding, it is conceivable that a genetic test could be developed that could potentially identify individuals likely to exhibit such behaviors. Do you think such a test would be a good idea? What would some of the ethical and societal implications of such a test be?

The Genetics of Mood Disorders and Schizophrenia

11. The two main affective disorders are bipolar illness and schizophrenia. What are the essential differences and similarities between these disorders?

12. A pedigree analysis was performed on the family of a man with schizophrenia. Based on the known concordance statistics, would his MZ twin be at high risk for the disease? Would the twin's risk decrease if he were raised in a different environment than his schizophrenic brother?

13. You are a researcher studying bipolar disorder. Your RFLP data shows linkage between a marker on chromosome 7 and bipolar illness. Later in the study, you find that a number of individuals lack this RFLP marker but still develop the disease. Does this mean that bipolar disorder is not genetic?

14. A region on chromosome 6 has been linked to schizophrenia, but researchers have not yet found a gene. Explain this linkage and why linkage does not necessarily locate a gene.

Genetics and Social Behavior

15. Of the following findings, which does *not* support the idea that alcoholism is genetic?
    a. Some strains of mice select alcohol over water 75% of the time, whereas others shun alcohol.
    b. The concordance value is 55% for MZ twins and 28% for DZ twins.
    c. Biological sons of alcoholic men who have been adopted have a rate of alcoholism more like their adoptive fathers.
    d. There is a 20% to 25% risk of alcoholism in sons of alcoholic men.
    e. None of the above.

16. A woman diagnosed with Alzheimer disease wants to know the probability of her children inheriting this disorder. Explain to her the complications of determining heritability for this disease.

17. What types of studies have been used to suggest that sexual orientation has a genetic component?

18. In July 1996, *The Independent*, a popular newspaper published in London, England, reported a study conducted by Dr. Aikarakudy Alias, a psychiatrist who had been working on the relationship between body hair and intelligence for 22 years. Dr. Alias told the 8th Congress of the Association of European Psychiatrists that hairy chests are more likely to be found among the most intelligent and highly educated than in the general population. According to this new research, excessive body hair could also mean higher intelligence. Is correlating body hair with intelligence a valid method for studying the genetics of intelligence? Why or why not? What other factors contribute to intelligence? Is it logical to assume that individuals with little or no body hair are consistently less intelligent than their hairy counterparts? What type of study could be done to prove or disprove this idea?

19. In a long-term study of over 100 pairs of MZ and DZ twins separated shortly after birth and reared apart, one of the conclusions was that "general intelligence or IQ is strongly affected by genetic factors." The study concluded that about 70% of the variation in IQ is due to genetic variability (review the concept of heritability in Chapter 5). Discuss this conclusion, and include in your answer the relationship between IQ and intelligence and to what extent these conclusions can be generalized. In evaluating the study's conclusion, what would you like to know about the twins?

Internet Activities

Internet Activities are critical thinking exercises using the resources of the World Wide Web to enhance the principles and issues covered in this chapter. For a full set of links and questions investigating the topics described below, log on to **http://biology.brookscole.com/cummings7**

1. *The Genetics of Personality.* The *Personality Research Site* presents an overview of scientific research programs in personality psychology. Follow the "Behavior Genetics" link. The study of human behavior and behavioral disorders is complex and must account for both environmental and genetic influences. What types of studies do researchers use to attempt to tease out these differences?

Biographic and scientific evidence strongly suggests that in many people, creative abilities in art and literature are linked in a complex way to bouts of depression or to the onset of manic states. Because of this proposed linkage, it is possible that medicating mood disorders may reduce people's creativity. Now that you know more about the genetic and environmental factors that affect behavior in general, and mood disorders in particular, what do you think? If you were a successful artist, author, or poet who suffered from depression, or bipolar illness, and a cure for your illness were discovered, would you elect to have the treatment, knowing your creative abilities may be diminished or even disappear, but also knowing that your risk of suicide would be reduced or eliminated? Visit the Human Heredity Companion website to find out more on the issue, and then cast your vote online at **http://biology.brookscole.com/cummings7**

For further reading and inquiry, log on to **InfoTrac College Edition,** your world-class online library, including articles from nearly 5,000 periodicals, at **http://infotrac.thomsonlearning.com**

19

Population Genetics and Human Evolution

Chapter Outline

THE DNA sequence generated by the Human Genome Project (HGP) was assembled from the genomes of about 10 individuals and is a valuable guide to the location and identity of the 25,000 to 30,000 genes in our genome. Shortly after the HGP got underway in 1990, Luigi Cavalli-Sforza and a number of his colleagues called for the project to broaden its scope to include a study of the genetic diversity of the entire human population.

The proposed Human Genome Diversity Project (HGDP) would gather cells and blood samples from members of hundreds of indigenous groups, with emphasis on isolated populations that are in danger of being merged into larger, neighboring populations. DNA from these samples would be available to scientists to trace the ancestry and relationships of modern populations to provide insight into the origin, evolution, and diversification of our species. The DNA samples would also be used to identify genes conferring susceptibility or resistance to disease for the development of diagnostic tests and drugs. In spite of its commendable goals, representatives of many indigenous groups vehemently opposed the HGDP proposal. Their concerns centered on whether it was possible to obtain informed consent from people in such cultures, whether indigenous populations would share in ownership and patents of any cell lines or genes discovered during the project, and possible stigmatization of groups found to harbor disease-susceptibility genes.

In the face of this opposition and growing controversy, the National Research Council reviewed the proposed project and issued a report in 1997 endorsing the HGDP but setting guidelines for informed consent, sample collection, and distribution of financial interests in the outcome of the research. The HGDP reformulated its approach, recognizing the ethical and legal challenges of such work. Samples are being collected, but progress has been

Tim Davis/Stone/Getty Images

slow. About 1,000 samples from 51 populations have been collected and are available to researchers worldwide.

In this chapter we consider the population as a genetic unit and examine its organization, methods of measuring allele frequencies, and the ways in which evidence from several disciplines is providing insight into the origin of our species and its dispersal over the Earth.

19.1 The Population as a Genetic Reservoir

Humans are distributed over most of the land surfaces of the Earth (▶ Figure 19.1). As such, our species is subdivided into locally interbreeding units known as **populations.** Like individuals, populations are dynamic: they have life histories that include birth, growth, and response to the environment. Like individuals, populations can age and eventually die. Populations can be described by age structure (▶ Active Figure 19.2), geography, birth and death rates, and allele frequencies.

Populations are more genetically diverse than individuals. For example, no individual can have blood types A, B, AB, and O. Only a group of individuals can carry all three blood types. The set of genetic information carried by a population is known as its **gene pool.** For a given gene, such as *I*, the gene for ABO blood type, the pool includes all the *A, B,* and *O* alleles in the population. Zygotes produced by one

■ **Populations** Local groups of organisms belonging to a single species, sharing a common gene pool.

■ **Gene pool** The set of genetic information carried by the members of a sexually reproducing population.

▲ **FIGURE 19.1** Human population density. As this satellite view of the world at night shows, humans are not randomly distributed across the land areas of the world but are clustered into discrete populations.

C. Mayhew and R. Simmon /(NASA./GSFC) NOAA/NGDC, DMSP Digital Archive

generation represent samples selected from the gene pool to form the next generation. The gene pool of a new generation is descended from the parental generation, but for several reasons, including chance, the gene pool of the new generation may have different allele frequencies than the parental pool. Over time, changes in allele frequency can cause changes in phenotype frequency. The long-term effect of changes in allele frequency is evolutionary change.

19.2 How Can We Measure Allele Frequencies in Populations?

Many factors influence population size, including birth rate, disease, migration, and adaptation to environmental factors such as climate. As these factors change, the genetic structure of a population can also change.

In studying the genetic structure of populations, geneticists measure allele frequencies over several generations to determine whether they are stable. In our discussion, the term **allele frequency** means the frequency with which alleles of a given gene are present in a population. As several following examples will show, allele frequencies are not the same as genotype frequencies.

We cannot always directly determine allele frequencies, because we see phenotypes, not genotypes. However, in codominant alleles, phenotypes are equivalent to genotypes, and we can determine allele frequencies directly. The MN blood group is an example of a codominant system. The gene L (located on chromosome 4) has two alleles, L^M and L^N, responsible for the M and N blood types, respectively. Each allele controls the synthesis of a gene product on the surface of red blood cells independently of the other allele. Thus individuals may be type M (L^M/L^M), type N (L^N/L^N), or type MN (L^M/L^N). ▶ Table 19.1 shows the genotypes, blood types, and immunological reactions of the MN blood groups.

■ **Allele frequency** The frequency with which alleles of a given gene are present in a population.

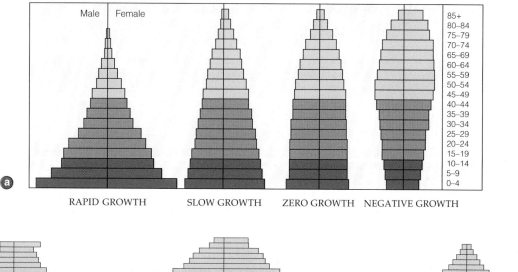

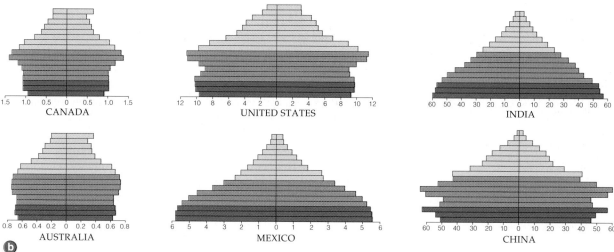

▲ ACTIVE FIGURE 19.2 (a) Age structure diagrams for countries with rapid, slow, zero, and negative rates of population growth. *Green:* population members in prereproductive years, *purple:* reproductive portion of the population, *light blue:* postreproductive members of the population. Males are to the left of the vertical bar, females to the right. Width of bar indicates proportion of population in each age group. (b) Age structures for representative countries. Scale is population size in millions.

HUMAN
Genetics Now™ Learn more about age structure diagrams by viewing the animation on Human GeneticsNow at
http://now.ilrn.com/cummings7

Codominant allele frequencies can be measured directly.

The frequency of codominant alleles can be determined simply by counting how many copies of each allele are present in a population. For example, suppose that blood typing reveals that a population of 100 people contains 54 with type M (*MM* homozygotes), 26 with type MN (*MN* heterozygotes), and 20 with type N (*NN* ho-

Table 19.1 MN Blood Groups

Genotype	Blood Type	Antigens Present	Antibody Reactions
$L^M L^M$	M	M	Anti-M
$L^M L^N$	MN	M, N	Anti-M, Anti-N
$L^N L^N$	N	N	Anti-N

■ **Hardy-Weinberg Law** The statement that allele and genotype frequencies remain constant from generation to generation when the population meets certain assumptions.

Table 19.2 Determining Allele Frequencies for Codominant Genes by Counting Alleles

Genotype	MM	MN	NN	Total
Number of individuals	54	26	20	100
Number of L^M alleles	108	26	0	134
Number of L^N alleles	0	26	40	66
Total	108	52	40	200

Frequency of L^M in population: $134/200 = 0.67 = 67\%$

Frequency of L^N in population: $66/200 = 0.33 = 33\%$

mozygotes). The 54 people with type M carry 108 copies of the *M* allele (54 individuals, each of whom carries 2 *M* alleles). The 26 MN individuals carry an additional 26 *M* alleles, so the population has a total of 134 *M* alleles ($108 + 26 = 134$). Each member of the population carries two copies of the *L* gene, for a total of 200 alleles (100 individuals, each of whom has 2 alleles = 200). The frequency of the *M* allele is $134/200$, or 0.67 (67%). The frequency of the *N* allele can be calculated by counting 40 *N* alleles in the *NN* homozygotes (20 individuals, each of whom has 2 *N* alleles) and an additional 26 *N* alleles in the *MN* heterozygotes, a total of 66 copies of the *N* allele. The frequency of the *N* allele in the population is $66/200$, or 0.33 (33%). ▶ Table 19.2 summarizes this method of calculating gene frequencies in codominant populations.

Recessive allele frequencies cannot be measured directly.

In genes with codominant alleles, there is a direct relationship between phenotype and genotype. However, if one allele is recessive, there is no direct relationship between phenotype and genotype because heterozygotes and homozygotes for the dominant allele have identical phenotypes. In this situation, we cannot measure allele frequencies by counting, because we don't know how many people in the population have a homozygous dominant genotype and how many are heterozygotes. Godfrey Hardy and Wilhelm Weinberg independently developed a mathematical formula that can be used to determine allele frequencies when one or more alleles are recessive and a number of conditions (described later) are met. This method is known as the **Hardy-Weinberg Law.**

Keep in mind

■ The frequency of recessive alleles in a population cannot be measured directly.

19.3 The Hardy-Weinberg Law Measures Allele and Genotype Frequencies

After Mendel's work became widely known, there was intense debate about whether the principles of Mendelian inheritance applied to humans. One of the first genetic traits identified in humans was a dominant allele, brachydactyly (OMIM 112500) (▶ Figure 19.3). Because the phenotypic ratio of dominant traits is 3:1 in children of heterozygotes (1 *AA*, 2 *Aa*, 1 *aa*), it was thought that over time, the phenotypic ratio for this dominant allele would become 3:1 in the human population.

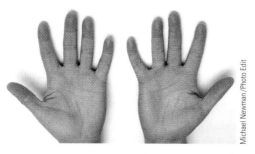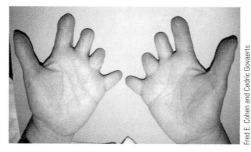

▲ FIGURE 19.3 Brachydactyly is a dominantly inherited trait that causes shortening of the fingers.

An English mathematician, Godfrey Hardy, and a German physician, Wilhelm Weinberg, independently recognized that this reasoning was false. The argument failed to recognize that there is a difference between *how a trait is inherited* (in this case a dominant trait with a 3:1 ratio) and the *frequency* of the dominant and recessive alleles in the population. Hardy and Weinberg each developed a simple mathematical model to estimate allele frequency in a population and to describe how alleles combine to form genotypes.

What are the assumptions for the Hardy-Weinberg Law?

The model developed by Hardy and Weinberg is based on a number of assumptions:
- The population is large. In practical terms, this means that the population is large enough that there are no errors in measuring allele frequencies.
- No genotype is better than any other; that is, all genotypes have equal ability to survive and reproduce.
- Mating in the population is random (see Spotlight on Selective Breeding Gone Bad).
- Other factors that change allele frequency, such as mutation and migration, are absent or rare events and can be ignored.

How can we calculate allele and genotype frequencies?

Let's see how the model works in a population carrying an autosomal gene with two alleles, *A* and *a*. In this population, the frequency of the dominant allele *A* in gametes is represented by p, and the frequency of the recessive allele *a* in gametes is represented by q. Because the sum of p and q represents 100% of the alleles for that gene in the population, $p + q = 1$. A Punnett square can be used to predict the genotypes produced by the random combination of these gametes (▶ Figure 19.4).

In combining gametes, the chance that both the egg and sperm will carry the *A* allele is $p \times p = p^2$. The chance that the gametes will carry different alleles is $(p \times q) + (p \times q) = 2pq$. The chance that both gametes carry recessive alleles is $q \times q = q^2$. Although p^2 represents the chance that both gametes will carry an *A* allele, p^2 also represents the frequency of the homozygous *AA* genotype in the new generation. In the same way, $2pq$ is a measure of heterozygote (*Aa*) frequency, and q^2 represents the frequency of homozygous recessive (*aa*) individuals. In other words, the distribution of genotypes in the next generation can be expressed as $p^2 + 2pq + q^2 = 1$, where 1 represents 100% of the genotypes in the new generation. This equation formulates the Hardy-Weinberg Law, which states that both allele and genotype frequencies will remain constant from generation to generation in a large, interbreeding population in which mating is random and there is no selection, migration, or mutation.

The formula can be used to calculate allele frequencies (*A* and *a* in our example) and the frequency of the various genotypes in a population. To show how the model works, let's begin with a population for which we already know the frequency of the alleles. Suppose we have a large, randomly mating population in which the frequency of an autosomal dominant allele *A* is 60% and the frequency of the recessive allele *a*

	Sperm	
	$A(p)$	$a(q)$
Eggs $A(p)$	AA (p^2)	Aa (pq)
$a(q)$	Aa (pq)	aa (q^2)

▲ FIGURE 19.4 The frequency of the dominant and recessive alleles in the gametes of the parental generation determines the frequency of the alleles and the genotypes of the next generation.

Sperm

	$A (p = 0.6)$	$a (q = 0.4)$
A ($p = 0.6$)	AA ($p^2 = 0.36$)	Aa ($pq = 0.24$)
a ($q = 0.4$)	Aa ($pq = 0.24$)	aa ($q^2 = 0.16$)

Eggs

▲ FIGURE 19.5 The frequency of alleles and genotypes in a new generation in which the alleles in the parental generation are present at a frequency of 0.6 for the dominant allele (A) and 0.4 for the recessive allele (a).

is 40%. This means that $p = 0.6$ and $q = 0.4$. Because A and a are the only two alleles, the sum of $p + q$ equals 100% of the alleles:

$$p(0.6) + q(0.4) = 1$$

In this population, 60% of the gametes carry the dominant allele A, and 40% carry the recessive allele a. ▶ Figure 19.5 shows the distribution of genotypes in the next generation.

In the new generation, 36% ($p^2 = 0.6 \times 0.6$) of the offspring will have the genotype AA, 48% ($2pq = 2[0.6 \times 0.4]$) will be Aa, and 16% ($q^2 = 0.4 \times 0.4$) will have the genotype aa.

We can also use the Hardy-Weinberg equation to calculate the frequency of the A and a alleles in the new generation. The frequency of A is

$$p^2 + 1/2(2pq)$$
$$0.36 + 1/2(0.48)$$
$$0.36 + 0.24 = 0.6 = 60\%$$

For the recessive allele a, the frequency is

$$q^2 + 1/2(2pq)$$
$$0.16 + 1/2(0.48)$$
$$0.16 + 0.24 = 0.40 = 40\%$$

Because $p + q = 1$, we could have calculated the value for the recessive allele by subtraction:

$$p + q = 1$$
$$q = 1 - p$$
$$q = 1 - 0.60$$
$$q = 0.40 = 40\%$$

19.4 Using the Hardy-Weinberg Law in Human Genetics

The Hardy-Weinberg Law is one of the foundations of population genetics and has many applications in human genetics and human evolution. We consider only a few of its uses, primarily those that apply to measuring allele and genotype frequencies.

The Hardy-Weinberg Law measures the frequency of autosomal dominant and recessive alleles.

If a trait is inherited recessively, we can use the Hardy-Weinberg Law to calculate the frequency of the recessive allele in the population. We begin by counting the number of homozygous recessive individuals in the population. For example, cystic fibrosis is an autosomal recessive trait, and homozygous recessive individuals can be identified by their distinctive phenotype. Suppose that 1 in 2,500 members of a population is affected with cystic fibrosis. These individuals have the genotype aa. According to the Hardy-Weinberg equation, the frequency of this genotype in the population is equal to q^2. The frequency of the a allele in this population is therefore equal to the square root of q^2:

$$q^2 = 1/2500 = 0.0004$$
$$q = \sqrt{0.0004}$$
$$q = 0.02 = 1/50$$

Once we know that the frequency of the a allele is 0.02 (2%), we can calculate the frequency of the dominant allele A by subtraction:

$$p + q = 1$$
$$p = 1 - q$$
$$p = 1 - 0.02$$
$$p = 0.98 = 98\%$$

In this population, 98% of the alleles for gene A are dominant (A), and 2% are recessive (a). This method can be used to calculate the allele and genotype frequencies for any dominant or recessive trait.

The Hardy-Weinberg Law estimates the frequency of heterozygotes in a population.

Many human genetic disorders are inherited as recessive traits. In such cases, a recessive disorder appears when each parent is heterozygous. For many reasons, it is important to know the population frequency of heterozygotes carrying a deleterious recessive allele. Calculating the frequency of heterozygotes is an important application of the Hardy-Weinberg Law because most disease-causing alleles are carried by heterozygotes. To calculate the frequency of heterozygous carriers for recessive traits, we begin by counting the number of homozygous recessive individuals (all of whom show the recessive phenotype) in the population. For example, cystic fibrosis, an autosomal recessive disorder, has a frequency of 1 in 2,500 among Americans of European ancestry. (The disease is much rarer among American blacks and Asians.)

The frequency of cystic fibrosis in this population is 1 in 2,500 and frequency of the recessive allele q in the population can be calculated as

$$q = \sqrt{q^2}$$
$$q = \sqrt{0.0004}$$
$$q = 0.02 = 2\%$$

Because $p + q = 1$, we can calculate the frequency of the dominant allele p:

$$p = 1 - q$$
$$p = 1 - 0.02$$
$$p = 0.98 = 98\%$$

Knowing the allele frequencies, we can use the Hardy-Weinberg equation to calculate genotype frequencies. Recall that in the Hardy-Weinberg equation, $2pq$ is the frequency for the heterozygous genotype. Using the values we have calculated for p and q, we can determine the frequency of heterozygotes as follows:

$$2pq = 2(0.98 \times 0.02)$$
$$2pq = 2(0.0196)$$
$$\text{Heterozygote frequency} = 0.039 = 3.9\%$$

This means that 3.9%, or approximately 1 in 25, white Americans carry the gene for cystic fibrosis.

Sickle cell anemia is an autosomal recessive disorder affecting 1 in 500 black Americans. Using the Hardy-Weinberg equation, we can calculate that 8.5% of black Americans, or 1 in 12, are heterozygous carriers for sickle cell anemia. ▶ Table 19.3 lists the frequencies of heterozygous carriers of recessive alleles with a frequency range from 1 in 10 to 1 in 10,000,000. ▶ Table 19.4 lists the heterozygote frequencies for some common human autosomal recessive traits.

Many people are surprised to learn that heterozygotes for recessive traits are so

Table 19.3 Heterozygote Frequencies for Recessive Traits

Frequency of Homozygous Recessives (q^2)	Frequency of Heterozygous Individuals ($2pq$)
1/100	1/5.5
1/500	1/12
1/1,000	1/16
1/2,500	1/25
1/5,000	1/36
1/10,000	1/50
1/20,000	1/71
1/100,000	1/158
1/1,000,000	1/500
1/10,000,000	1/1,582

Table 19.4 Frequency of Heterozygotes for Some Recessive Traits (U.S. Population)

Trait	Heterozygote Frequency
Cystic fibrosis	1/22 whites; much lower in blacks, Asians
Sickle cell anemia	1/12 blacks; much lower in most whites and in Asians
Tay-Sachs disease	1/30 among descendants of Eastern European Jews; 1/350 among others of European descent
Phenylketonuria	1/55 among whites; much lower in blacks and those of Asian descent
Albinism	1/10,000 in Northern Ireland; 1/67,800 in British Columbia

common in the population. If a genetic disorder is relatively rare (say 1 in 10,000 individuals), they generally assume that the number of heterozygotes must also be rather low. In fact, if a disorder is found in 1 in 10,000 members of a population, 1 in 50 (2%) members of the population is a heterozygote, and there are about 200 times as many heterozygotes as there are homozygotes.

What are the chances that two heterozygotes will mate and have an affected child? We can calculate the answer to this question as follows: The chance that two heterozygotes will mate is $1/50 \times 1/50 = 1/2,500$. Because they are heterozygotes, the chance that they will produce an affected child is one in four. So the chance that they will mate and produce an affected child is $1/2,500 \times 1/4 = 1/10,000$. In other words, if the disorder is present in 1 of every 10,000 individuals, 1 in 50 individuals must be a heterozygous carrier of the recessive allele.

Keep in mind

■ Estimating the frequency of heterozygotes in a population is an important part of genetic counseling.

19.5 Measuring Genetic Diversity in Human Populations

Understanding our evolutionary history depends on identifying factors that lead to variations in allele frequencies between populations and the way that these variations are acted upon by natural selection. In the following sections, we explore how genetic variation is produced and the role of natural selection and culture as a force in changing allele frequencies.

Mutation generates new alleles but has little impact on allele frequency.

As discussed earlier, the gene pool is reshuffled each generation to produce the genotypes of the offspring. In the process, genetic variation (new combinations of alleles) is produced by recombination and Mendelian assortment. However, these events do not produce any new alleles. Mutation is the ultimate source of all new alleles and is the origin of all genetic variability.

We discussed mechanisms and rates of mutation in Chapter 11. In this section, we consider the effect of mutation on allele frequencies in a population. If the mutation rate for a gene is known, we can use the Hardy-Weinberg Law to calculate the change in allele frequency resulting from new mutations in that gene in each generation. Let's use the dominant trait achondroplasia as an example. Statues and murals indicate that this form of dwarfism was known in Egypt more than 2,500 years ago, or about 125 generations (allowing 20 years per generation) ago. If mutant achondroplasia alleles were introduced into the gene pool by mutation in each generation beginning 2,500 years ago, how much would the frequency of achondroplasia change over this time period? Should we expect a higher frequency of achondroplasia among residents of an ancient city, such as Cairo, than among residents of a recently established city like Houston?

For this calculation we assume that initially, only homozygous recessive individuals with the genotype dd (normal stature) were present in the population and that mutation has added new mutant (D) alleles to each generation at the rate of 1×10^{-5}. The change in allele frequency over time that results from this rate of mutation is shown in ▶ Figure 19.6. To change the frequency of the recessive allele (d) from 1.0 (100%) to 0.5 (50%) at this rate of mutation will require 70,000 generations, or 1.4 million years. Thus, the frequency of achondroplasia need not be any higher in

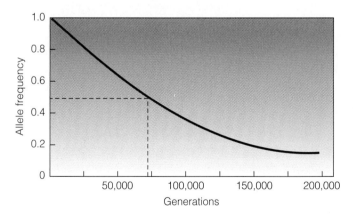

▲ FIGURE 19.6 The rate of replacement of a recessive allele *a* by the dominant allele *A* by mutation alone. Even though the initial rate of replacement is high, it will take about 70,000 generations or 1.4 million years to drive the frequency of the allele *a* from 1.0 to 0.5.

Houston than in Cairo. Our conclusion in this case is that mutation alone has a minimal impact on the genetic variability present in a population.

Keep in mind

- Mutation generates all new alleles, but drift, migration, and selection determine the frequency of alleles in a population.

Genetic drift can change allele frequencies.

Founder effects Allele frequencies established by chance in a population that is started by a small number of individuals (perhaps only a fertilized female).

Genetic drift The random fluctuations of allele frequencies from generation to generation that take place in small populations.

Occasionally populations start with a small number of individuals, or founders. Alleles carried by the founders, whether they are advantageous or detrimental, become established in the new population. These events take place simply by chance and are known as **founder effects.**

Random changes in allele frequency that occur from generation to generation in small populations are examples of **genetic drift.** In addition to founder effects, genetic drift can occur in small populations by drastic reductions in population size. These reductions, called population bottlenecks, are often caused by natural disasters. In extreme cases, drift can lead to the elimination of one allele from all members of the population. Small interbreeding groups on isolated islands often provide examples of genetic drift.

The island of Tristan da Cuhna (▶ Figure 19.7) is located in the southern Atlantic Ocean, 2,900 km (about 1,800 mi) from Capetown, South Africa, and 3,200 km (about 2,000 mi) from Rio de Janeiro, Brazil. British troops were stationed here in 1816 to prevent Napoleon from escaping his exile on the island of St. Helena. After Napoleon's death, Corporal William Glass, his wife, and two daughters received permission to remain after the British army withdrew. Others joined Glass at intervals, and the development of the isolated and highly inbred population that formed here can be traced with great accuracy.

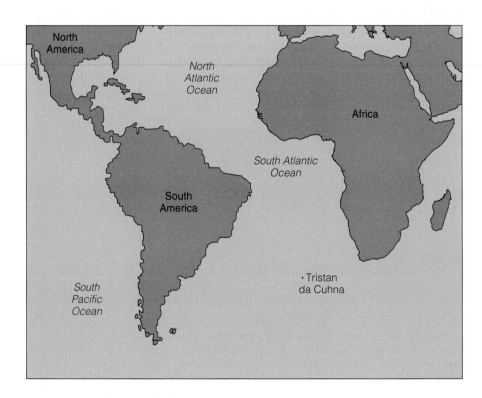

▶ FIGURE 19.7 Location of the island of Tristan de Cuhna, first discovered by a Portuguese admiral in 1506.

The genotypes of the few hundred residents have been studied for over 40 years to see how isolation and inbreeding affect the island's gene pool. As might be expected, one effect of inbreeding is an increase in homozygosity for recessive traits. ▶ Table 19.5 lists some genetic markers for which all the islanders tested are homozygous.

Because of a founder effect, traits carried by one or a small number of early settlers often end up in a large fraction of their descendants. On Tristan, a deformity of the fifth finger, known as **clinodactyly** (OMIM 112700), is present at a high frequency. This autosomal dominant trait is especially prominent in members of the Glass family, the first permanent residents of the island. The high frequency of this trait in the present island population can be explained by its presence in one of the original colonists.

This brief example illustrates how genetic drift can be responsible for changing allele frequencies in populations that are isolated, inbred, and stable for long periods. Most human populations, however, do not live on remote islands and are not subject to prolonged isolation and inbreeding. Yet there are differences in the distribution and frequency of alleles among populations, indicating that founder effects and drift are not the only factors that can change allele frequencies.

Table 19.5 Homozygous Markers among Tristan Residents
Transferrins
Phosphoglucomutase
6-Phosphogluconate dehydrogenase
Adenylate kinase
Hemoglobin A variants
Carbonic anhydrase (2 forms)
Isocitrate dehydrogenase
Glutathione peroxide
Peptidase A, B, C, D

Natural selection acts on variation in populations.

In formulating their theory of evolution, Alfred Russel Wallace and Charles Darwin recognized that some members of a population are better adapted to the environment than others. These better-adapted individuals have an increased chance of leaving more offspring than those with other genotypes. The ability of a given genotype to survive and reproduce is a measure of its **fitness.** Fitter genotypes are better at survival and reproduction and in time, this reproductive difference leads to changes in allele frequencies within the population. The differential reproduction of better-adapted genotypes is **natural selection.**

The relationship between the allele for sickle cell anemia and malaria is an example of how natural selection changes allele frequencies. Sickle cell anemia is an autosomal recessive condition associated with a mutant form of hemoglobin. Affected individuals have a wide range of clinical symptoms (see Chapter 4 for a review). Although many untreated homozygotes die in childhood, the mutant allele is present in very high frequencies in certain populations. In some West African countries, 20% of the population may be heterozygous for this trait, and along rivers such as the Gambia, almost 40% of the population is heterozygous. If homozygotes die before they reproduce, why hasn't this mutant allele been eliminated from the population?

The mutant allele is present at a high frequency in certain West African countries and in certain regions of Europe, the Middle East, and Asia because of malaria, an infectious disease caused by a protozoan parasite, *Plasmodium falciparum*. The parasite is transmitted to humans by infected mosquitoes (▶ Figure 19.8). Once infected, victims suffer recurring episodes of illness throughout life. More than 2 million people die from malaria each year, and more than 300 million individuals worldwide are infected with this disease.

The geographic distributions of malaria and sickle cell are shown in ▶ Figure 19.9. The mutant allele for sickle cell anemia confers resistance to malaria, and experiments on human volunteers have confirmed this. In heterozygotes and in recessive homozygotes, the mutant hemoglobin (Hb S) alters the plasma membrane of red blood cells, making them resistant to infection by the malarial parasite. This resistance makes heterozygotes fitter than those with homozygous normal genotypes. Those with the homozygous recessive genotype are resistant to malaria but also have sickle cell anemia. In this case, selection favors the survival and differential reproduction of heterozygotes.

Because of their resistance to malaria, heterozygotes leave more offspring than the

■ **Clinodactyly** An autosomal dominant trait that produces a bent finger.

■ **Fitness** A measure of the relative survival and reproductive success of a given individual or genotype.

■ **Natural selection** The differential reproduction shown by some members of a population that is the result of differences in fitness.

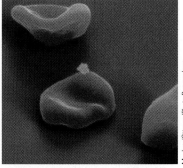

▲ FIGURE 19.8 *Plasmodium* parasites (*yellow*) attacking and infecting red blood cells. Infection by *Plasmodium* causes malaria.

Meckes/Ottawa/Photo Researchers

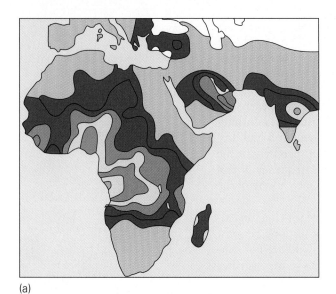

(a)

Allele frequencies of *Hb S* allele

Greater than 0.140	From 0.060 to 0.080
From 0.120 to 0.140	From 0.040 to 0.060
From 0.1100 to 0.120	From 0.020 to 0.040
From 0.080 to 0.100	From 0.000 to 0.020

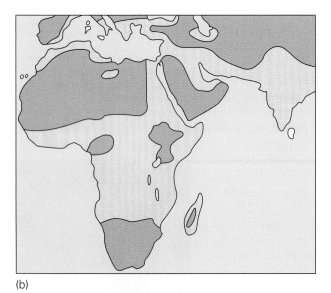

(b)

Regions with malaria

▲ **FIGURE 19.9** (a) The distribution of sickle cell anemia in the Old World. (b) The distribution of malaria in the same region.

other genotypes, and the *Hb S* allele is spread through the population and maintained at a high frequency.

19.6 Natural Selection Affects the Frequency of Genetic Disorders

Many genetic disorders are disabling or fatal, so why are they so common? In other words, what keeps natural selection from eliminating the deleterious alleles responsible for these disorders? In analyzing the frequency and population distribution of human genetic disorders, it is clear that there is no single answer. One conclusion, drawn from the Hardy-Weinberg Law, is that rare lethal or deleterious recessive alleles survive because the vast majority of them are carried in the heterozygous condition and thus are hidden in the gene pool. Other factors, however, can cause the differential distribution of alleles in human populations, and several of these are discussed in the following paragraphs (see Genetics in Society: Lactose Intolerance and Culture).

Almost all affected individuals with Duchenne muscular dystrophy (DMD; OMIM 310200) die without reproducing. If this is the case, eventually the mutant allele should be eliminated from the population. However, because the mutation rate for *DMD* is high (perhaps as high as 1×10^{-4}), mutation replaces the *DMD* alleles lost when affected individuals die before reproducing. Thus the frequency of the *DMD* allele in a population is a balance between alleles introduced by mutation and those removed by the death of affected individuals.

Selection can cause detrimental alleles to have high frequencies in large, well-established populations. Heterozygote advantage in sickle cell anemia is an example. For other genetic disorders with a high frequency of the disease allele, the selective advantage of heterozygotes may be less obvious or perhaps no longer exists.

Tay-Sachs disease (OMIM 272800) is an autosomal recessive disorder that is fatal in early childhood. Although this is a rare disease in most populations, Ashkenazi Jews (those who live in or have ancestors from Eastern Europe) have a tenfold higher incidence of the disease. In these populations, heterozygote frequency can be as high as 11%. There is indirect evidence that Tay-Sachs heterozygotes are more resistant to tuberculosis, a disease endemic to cities and towns, where most of the European Jews lived. As in sickle cell anemia, the death of homozygous Tay-Sachs individuals is the genetic price paid by the population to allow the higher fitness and survival of the more numerous heterozygotes.

These examples illustrate that several factors contribute to disease allele frequency and that each disease must be analyzed individually. In some cases mutant alleles are maintained by mutation. In other cases migration and founder effects increase the

Genetics in Society

Lactose Intolerance and Culture

Lactose is the principal sugar in milk (human milk is 7% lactose) and is a ready energy source. In infants, the enzyme lactase converts lactose into glucose and galactose, sugars that are easily absorbed by the intestine. Lactase production in most humans slows at the time of weaning, and lactase production stops as children grow into adults (review the biochemistry of lactose breakdown in Chapter 10). Adults with low lactase levels cannot metabolize lactose (OMIM 223100), and after eating lactose-containing foods, they develop gas, cramps, diarrhea, and nausea. In these individuals, undigested lactose passes from the small intestine into the colon, where it is metabolized by bacteria, resulting in gas and diarrhea. However, in some human populations, lactase is produced throughout adulthood; these individuals are called lactose absorbers (LA). Genetic evidence indicates that adult lactose utilization (and the adult production of lactase) is inherited as an autosomal dominant trait. Across different populations, the frequency of the lactose absorption allele varies from 0.0% to 100%.

Why does the frequency of this allele vary so widely across populations? The answer is that cultural practices are a selective force. The human species originally was lactose-intolerant as adults, as are all other land mammals. As dairy herding developed in some populations, adult LA had the selective advantage of being able to derive nutrition from milk. This improved their chances of survival and success in leaving offspring. As a result, the cultural practice of maintaining dairy herds was the selective factor that provided a fitness advantage for the LA genotype.

frequency of deleterious alleles. In still other cases, natural selection favors heterozygous carriers, and affected homozygotes bear the burden of conferring advantages on other genotypes.

19.7 Genetic Variation in Human Populations

Evidence indicates that a great deal of genetic variation is present in the human genome. The process of mutation has introduced all this variation. Natural selection and drift are the primary mechanisms by which alleles spread through local population groups.

How can we measure gene flow between populations?

Anthropologists and geneticists have had a long-standing interest in estimating gene flow between populations to reconstruct the origin and history of populations formed when European and non-European populations come into contact (see Genetics in Society: Genghis Khan Lives On). The best-documented case is gene flow into the American black population from Europeans, but other populations have also been studied.

Most of the black population in the United States originated in West Africa, and the majority of the white population arrived from Europe. The Duffy blood group has three alleles: *FY*A*, *FY*B*, and *FY*O* (OMIM 110700). The frequency of the *FY*O* allele is close to 100% in West African populations, whereas in Europeans, this allele has a frequency close to zero. Almost all Europeans carry *FY*A* or *FY*B*, and these alleles are very rare in African populations. By measuring the frequencies of the $FY*A$ and $FY*B$ alleles in the U.S. black population, we can estimate how much genetic mixing has occurred between these populations over the last 300 years. ▶ Figure 19.10 shows the frequency of the $FY*A$ allele among black populations in West Africa and in several locations in the United States. Using this as an average gene, we can calculate that about 20% of the genes in the black population in some northern cities are derived from Europeans.

In another study, using 52 alleles of 15 genes in U.S. blacks from the Pittsburgh region (including 18 unique alleles of African origin), it was found that the proportion of European genes in this black population is approximately 25%. These studies are all consistent with the idea that members of the U.S. black population derive approximately 20% of their genes from Europeans.

However, not all contact between genetically distinct populations leads to a reduction in genetic differences. Using a combination of genetic and historical methods, researchers investigated the origin and extent of European admixture in the Gila River American Indian community of central Arizona. Results show a European admixture of 5.4%, whereas the demographic data indicate an admixture of 5.9%. The first contact between Europeans and the Gila River community occurred in 1694. Using 20 years as the estimate for one generation, approximately 15 generations of contact have occurred. Because gene flow is approximately 0.4% per year, the Gila River American Indian community has retained almost 95% of its native gene pool even after close contact with other gene pools for 300 years.

Genetics in Society

Genghis Khan Lives On

Although he died in 1227, Genghis Khan's legacy to the world includes more than the empire he assembled, and his descendants expanded, stretching from Central Europe to the Pacific Ocean, across the Middle East, and most of Asia, except for northern Siberia, India, and Southeast Asia. A survey of human population variation using markers on the Y chromosome has uncovered evidence that a Y-chromosome lineage with distinctive features is present in 8% of the men in the region from the Caspian Sea to the Pacific. This Y chromosome is carried by about 0.5% of men in the world, all of whom live within the borders of the former Mongol empire.

The pattern of markers on the chromosome and its geographic distribution indicate this lineage originated in Mongolia about 1,000 years ago, and spread rapidly. Although selection is a possible explanation, the survey team ruled this out because of the speed with which this Y chromosome spread.

Their explanation is that Ghengis Khan, his sons, and their male descendants spread this genetic lineage as they conquered and ruled an expanding empire. Many times, males in the conquered populations were slaughtered, and in addition to widespread rapes, the ruling Mongols descended from Ghengis Khan established large harems and had many children. One Persian historian wrote in 1260 that more than 20,000 descendants of Ghengis Khan lived in affluence in the empire.

Although some geneticists agree that the results show a rapid and recent expansion of a Mongolian population, it is hard to link this chromosome directly to Ghengis Khan. Others point out that there are no other historical explanations for this event, and that the culture of Mongolian warfare and society are consistent with the royal Mongol line being the source of this chromosome. Final proof awaits the discovery of the grave of Ghengis Khan or members of his family.

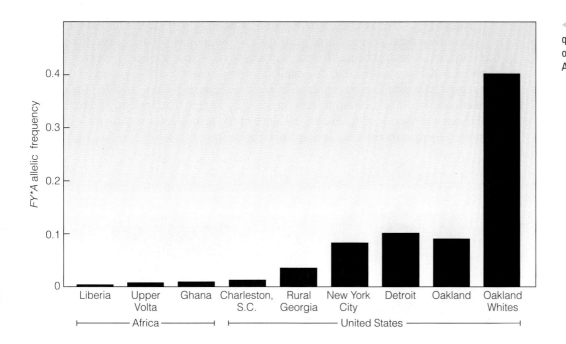

◀ FIGURE 19.10 Frequency of the *FY*A* in various populations in Africa and America.

Are there human races?

If there are significant differences in allele frequencies between populations, is there any justification for classifying human populations into racial groups? In the nineteenth century, biologists used the term *race* to describe groups of individuals within a species that were phenotypically different from other groups in that species. For example, biologists of that time might find two-spotted beetles and four-spotted beetles in the same population and call them separate races, even though some two-spotted and four-spotted beetles might be siblings.

Clearly there are phenotypic differences among humans. Residents of the Kalahari Desert are rarely mistaken for close relatives of Aleutian seal hunters, for example. In spite of these visible phenotypic differences, from a genetic point of view, is there a need or a value in classifying humans into racial groups?

In one sense, it is easy and seemingly logical to assume that the physical differences we see in human populations are evidence for underlying genetic differences. However, from the standpoint of genetics, the decision to divide our species into races depends on showing that there are significant genetic differences between the races, not simply phenotypic differences.

Beginning in the early 1970s, enough information was available to study allele frequencies in various populations and to relate these differences to the concept of human races. In one study, Richard Lewontin used data on protein variants from populations across the globe to analyze allele frequencies at 15 loci. His results show that 85% of the detected variation is present within populations and that less than 7% of the variation is present between populations classified into racial groups.

Beginning in the 1990s, geneticists have used DNA markers to search for genetic variation between and among populations. One study analyzed the distribution of variation within populations and among populations using 109 DNA markers. The differences were analyzed for members of the same population, between and among populations on the same continent, and among four or five different geographic groups. Again, as in the protein studies, most of the genetic variation was within groups (about 85%), and only about 10% of the variation was among groups on different continents (equivalent to racial groups).

Taken together, the work on proteins and DNA indicates that most of the genetic variation in the species is present within human populations and that there is little variation among populations, including those classified as different racial groups

(▶ Figure 19.11). In keeping with the genetic definition of race, if humans are to be divided into racial groups, large-scale genetic differences should occur along sharp boundaries. If such genetic differences exist, they have not yet been observed. On the other hand, is it possible that the small amount of genetic variation observed among groups is enough to warrant the classification of humans into racial groups? Many geneticists would answer no to that question and agree that presently there is no genetic basis for subdividing our species into racial groups, although work with additional genetic markers may turn up such variation.

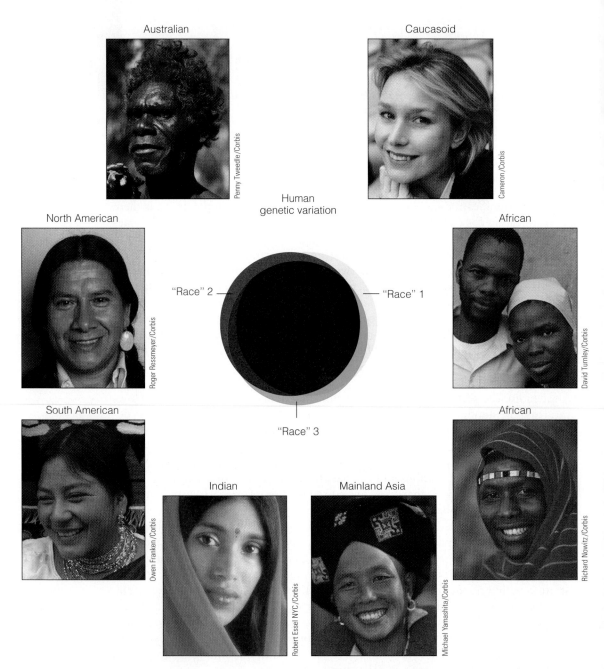

▲ FIGURE 19.11 The amount of genetic variation within populations is far greater than the variation between populations. Each colored circle represents genetic variation within populations classified as racial groups. The variations overlap greatly, with few or no genetic differences belonging to a single racial group. Because most variation is found within groups, many geneticists think there is no basis for classifying humans into racial groups.

What are the implications of human genetic variation?

If genetic diversity is analyzed by the size of populations, ranging from small villages to large nations and continents, a large fraction of the genetic variation in our species is present in small populations. This is consistent with the idea that our species has undergone a recent expansion from a small, common population. This insight plays an important part in our current understanding of how, when, and where our species originated.

19.8 The Appearance and Spread of Our Species (*Homo sapiens*)

Tracing the origins of our species has become a multidisciplinary task, using the tools and methods of anthropology, paleontology, and archaeology; the techniques of genetics and recombinant DNA technology; and more recently, satellite mapping from space. These methods are being used to reconstruct the origins and ancestry of populations of *H. sapiens* and to determine how and when our species originated and became dispersed across the globe.

Two theories differ on how and where *Homo sapiens* originated.

From the evidence provided by fossils and artifacts, it is clear that our ancestral species, *H. erectus,* originated in Africa and began migrating from there 1 to 2 million years ago, spreading through parts of the Middle East and Asia. What currently divides anthropologists and evolutionary biologists is the question of how and where *H. sapiens* originated. In a general sense, there are two opposing views about the origin of modern humans. One idea (often called the out-of-Africa hypothesis) argues that after *H. erectus* moved out of Africa, populations that remained behind continued to evolve and gave rise to *H. sapiens* about 200,000 years ago. From this single source in Africa, modern humans migrated to all parts of the world, displacing and driving into extinction members of our related species, *H. erectus.* According to this model, modern human populations are all derived from a single speciation event that took place in a restricted region within Africa. As a result, the human populations of today should show a high degree of genetic relatedness (which they do).

This model of human evolution is based on evidence that members of African populations have the greatest amount of genetic diversity, measured by differences in mitochondrial DNA nucleotide sequence. Members of non-African populations show much less diversity. The underlying assumption in these studies is that these mutational changes accumulate at a constant rate, providing a "molecular clock" that can be calibrated by studying the fossil record.

Studies of mitochondrial DNA reveal a single ancestral mitochondrial lineage for our species that originated in Africa. Calculations using the molecular clock indicate that our species originated about 200,000 years ago from an African population that might have been made up of about 10,000 individuals.

The second idea about the origin of our species is called the multiregional hypothesis. According to this idea, after populations of *H. erectus* spread from Africa over the Middle East and Asia, *H. sapiens* developed as the result of an interbreeding network descended from the original colonizing populations of *H. erectus.* The evidence to support this model is derived from a combination of genetic and fossil evidence. The fossil record shows a gradual transition from archaic to modern humans that took place at multiple sites outside of Africa. In this model, *H. erectus* became gradually transformed into *H. sapiens,* instead of being replaced by *H. sapiens.*

The two opposing ideas can be summarized as follows: one favors speciation and replacement (the out-of-Africa model), and the other favors evolution and transition within a single species (the multiregional model). These ideas are hotly debated and have received a great deal of attention in the press and other media. These alternative explanations for the appearance of modern humans show that scientists can reach different conclusions about the same problem.

The accuracy of the molecular clock and the method used to construct evolutionary relationships based on mitochondrial sequences has been called into question. However, studies of genetic variation in nuclear genes support a single point of origin and a timescale consistent with the out-of-Africa hypothesis. In addition, the distribution of RFLPs on the Y chromosome (which is passed from father to son) is consistent with the origin of our species at a single site about 200,000 to 270,000 years ago. In sum, the genetic evidence supports the out-of-Africa model across studies of mitochondrial DNA, microsatellite data, autosomal genes, and Y chromosomes. In spite of this, the issue over the origins of *H. sapiens* has not been resolved for several reasons: some genetic evidence is difficult to reconcile with the out-of-Africa model, the fossil record is difficult to interpret, and human population dynamics are very complex. Although we have considered only two possible origins for our species here, many theories abound and further work may produce additional ideas. Each theory will have to be evaluated by using the available information, by acquiring new information, and perhaps by applying new techniques.

Humans have spread across the world.

As summarized previously, there is strong evidence to support the idea that *H. sapiens* originated in Africa and spread from there to other parts of the world (▶ Figure 19.12). Based on the molecular evidence and some fossil evidence, it appears that modern *H. sapiens* originated in Africa some 130,000 to 170,000 years ago and that a small subset of this population emigrated from Africa about 137,000 years ago. There may have been one primary migration or several from a base in northeast Africa. The emigrants carried a subset of the variation present in the African population, consistent with the finding that present-day non-African populations have a small amount of genetic variation.

Modern forms of *H. sapiens* spread through Central Asia some 50,000 to 70,000 years ago and into Southeast Asia and Australia about 40,000 to 60,000 years ago. *H. sapiens* moved into Europe some 40,000 to 50,000 years ago, displacing the Ne-

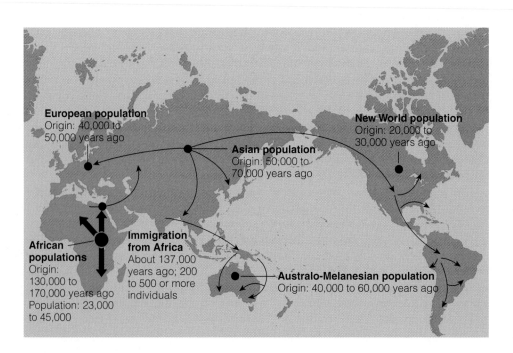

▶ FIGURE 19.12 The origin and spread of modern *H. sapiens* reconstructed from genetic and fossil evidence.

anderthals who lived there from about 100,000 years ago to about 30,000 years ago. Recent analysis of DNA from the bones of Neanderthal skeletons indicates that they were not the ancestors to European populations of *H. sapiens,* further supporting the out-of-Africa hypothesis.

Genetic data and recent archaeological findings indicate that North and South America were populated by three or four waves of migration that occurred 15,000 to 30,000 years ago. Migrations from Asia across the Bering Sea are well supported by archaeological and genetic findings, but Asia may not have been the only source of the first Americans. Some skeletal remains, such as Kennewick man and the Spirit Cave mummy, have features that more closely resemble Europeans than Asians. Evidence from a mitochondrial DNA variant called haplotype X, found only in Europeans, as well as a reinterpretation of stone tool technology, all make it seem likely that Europeans migrated to North America more than 10,000 years ago. Although a model with migrations from two sources explains most of the data available, there are other issues that remain to be resolved. Nonetheless, genetic analysis of present-day populations coupled with anthropology, archaeology, and linguistics can provide a powerful tool for reconstructing the history of our species.

Genetics in Practice

Genetics in Practice case studies are critical thinking exercises that allow you to apply your new knowledge of human genetics to real-life problems. You can find these case studies and links to relevant websites at **http://biology.brookscole.com/cummings7**

CASE 1

Jane, a healthy woman, was referred for genetic counseling because she had two siblings, a brother Matt and a sister Edna, with cystic fibrosis who died at the ages of 32 and 16, respectively. Jane's husband, John, has no family history of cystic fibrosis. Jane wants to know the probability that she and John will have a child with cystic fibrosis. The genetic counselor used the Hardy-Weinberg model to calculate the probability that this couple will have an affected child.

The counselor explained that there is a two-in-three chance that Jane is a carrier for the mutant *CFTR* allele. She used a Punnett square to illustrate this. The probability that John is a carrier is equal to the population carrier frequency ($2pq$). The probability that John and Jane will have a child who has cystic fibrosis equals the probability that Jane is a carrier (2/3) multiplied by the probability that John is a

carrier ($2pq$) multiplied by the probability that they will have an affected child if they are both carriers (1/4).

1. Using the heterozygote frequency for cystic fibrosis among white Americans to estimate the probability that John is a carrier, what is the likelihood that their child would have the disease?

2. If you were their genetic counselor, would you recommend that Jane and John be genetically tested before they attempt to have any children?

3. It is now possible to use preimplantation testing, which involves *in vitro* fertilization plus genetic testing of the embryo before implantation, to ensure that a heterozygous couple has a child free of cystic fibrosis. Do you see any ethical problems or potential future dangers associated with this technology?

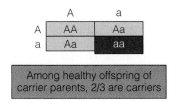

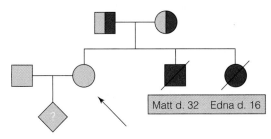

(continued)

CASE 2

Natural selection alters genotypic frequencies by increasing or decreasing fitness (that is, differential fertility or mortality). There are several examples of selection associated with human genetic disorders. Sickle cell anemia and other abnormal hemoglobins are the best examples of selection in humans. Carriers of the sickle and other hemoglobin mutations are more resistant to malaria than either homozygous class. Therefore, in areas where malaria is endemic, carriers are less likely to die of malaria and will have proportionally more offspring than homozygotes, thus passing on more genes. Balancing selection may also have influenced carrier frequencies for more "common" recessive diseases, such as cystic fibrosis in Europeans and Tay-Sachs in the Ashkenazi Jewish population, but the selective agent is not known for certain.

Selection may favor homozygotes over heterozygotes, resulting in an unstable polymorphism. One example is selection against heterozygous fetuses when an Rh^- mother carries an Rh^+ (heterozygous) fetus. This should result in a gradual elimination of the Rh^- allele. However, the high frequency of the Rh^- allele in so many populations suggests that other unknown factors may maintain the Rh^- allele in human populations.

1. If you suspected that heterozygous carriers of a particular disease gene had a selective advantage in resisting a type of infection, how would you go about testing this hypothesis?

2. If allele frequencies in the hemoglobin gene are influenced by sickle cell anemia on one hand, and by resistance to malaria on the other, what factors may cause a change in these allele frequencies over time?

Summary

19.1 The Population as a Genetic Reservoir

■ In the early decades of the twentieth century, genes were recognized as the agents that cause phenotypic variations, giving rise to the field of population genetics. After the mathematical and theoretical basis of this field was established, experimentalists began to study allele frequencies in populations rather than in the offspring of a single mating. This work has produced the basis for our understanding of evolution.

19.2 How Can We Measure Allele Frequencies in Populations?

■ In some cases, including codominant alleles, allele frequency can be measured directly by counting phenotypes, because in these cases phenotypes are equivalent to genotypes. In other cases, the Hardy-Weinberg Law provides a means of measuring allele frequencies within populations.

19.3 The Hardy-Weinberg Law Measures Allele and Genotype Frequencies

■ The Hardy-Weinberg equation assumes that the population is large and randomly interbreeding and that factors such as mutation, migration, and selection are absent. The presence of equilibrium in a population explains why dominant alleles do not replace their recessive alleles. In equilibrium populations, the Hardy-Weinberg equation can be used to measure allele and genotype frequencies from generation to generation.

19.4 Using the Hardy-Weinberg Law in Human Genetics

■ The Hardy-Weinberg Law can also be used to estimate the frequency of autosomal and recessive alleles in a population. It can also be used to detect when allele frequencies are shifting in the population. The conditions that lead to changing allele frequencies in a population are those that produce evolutionary change. One of the law's most common uses is to measure the frequency of heterozygous carriers of deleterious recessive alleles in a population. This information can be used to calculate the risk of having an affected child.

19.5 Measuring Genetic Diversity in Human Populations

■ Studies indicate that human populations carry a large amount of genetic diversity. All genetic variants originate by mutation, but mutation is an insignificant force in bringing about changes in allele frequency. Other forces, including genetic drift, act on the genetic variation in the gene pool and are responsible for changing the frequency of alleles in the population. Drift is a random process that acts in small, isolated populations to change allele frequency from generation to generation. Examples include island populations or those separated from general population by socioreligious practices. Natural selection acts on genetic diversity in populations to drive the process of evolution.

19.6 Natural Selection Affects the Frequency of Genetic Disorders

■ Selection increases the reproductive success of fitter genotypes. As these individuals make a disproportionate contribution to the gene pool of succeeding generations, genotypes change. The differential reproduction of fitter genotypes is known as natural selection. Darwin and Wallace identified selection as the primary force in evolution that leads to evolutionary divergence and the formation of new species. The high frequency of genetic disorders in some populations is the result of selection that often confers increased fitness on the heterozygote.

19.7 Genetic Variation in Human Populations

■ The biological concept of race evolved from emphasis on phenotypic differences to genotypic differences. Information from variations in proteins, microsatellites, and nuclear genes shows that most human genetic variation is present within populations rather than between populations. For this reason, there is no genetic basis for dividing our species into races.

19.8 The Appearance and Spread of Our Species (*Homo sapiens*)

■ A combination of anthropology, paleontology, archaeology, and genetics is being used to reconstruct the dispersal of human populations across the globe. The evidence available suggests that North and South America were populated by waves of migration sometime during the last 15,000 to 30,000 years.

Questions and Problems

Genetics Now™ Preparing for an exam? Assess your understanding of this chapter's topics with a pre-test, a personalized learning plan, and a post-test at **http://now.ilrn.com/cummings7**

The Population as a Genetic Reservoir

1. Explain why a population carries more genetic diversity than an individual carries.

How Can We Measure Allele Frequencies in Populations?

2. Define the following terms:
   a. population
   b. gene pool
   c. allele frequency
   d. genotype frequency
3. Why are codominant alleles ideal for studies of allele frequencies in a population?
4. Explain the connection between changes in population allele frequencies and evolution, and relate this to the observations made by Darwin and Wallace concerning natural selection.
5. Do you think populations can evolve without changes in allele frequencies?
6. Design an experiment to determine if a population is evolving.

The Hardy-Weinberg Law Measures Allele and Genotype Frequencies

7. What are four assumptions of the Hardy-Weinberg Law?
8. Drawing on your newly acquired understanding of the Hardy-Weinberg equilibrium law, point out why the following statement is erroneous: "Because most of the people in Sweden have blond hair and blue eyes, the genes for blond hair and blue eyes must be dominant in that population."
9. In a population where the females have the allelic frequencies $A = 0.35$ and $a = 0.65$ and the frequencies for males are $A = 0.1$ and $a = 0.9$, how many generations will it take to reach Hardy-Weinberg equilibrium for both the allelic and genotypic frequencies? Assume random mating, and show the allelic and genotypic frequencies for each generation.

Using the Hardy-Weinberg Law in Human Genetics

10. Suppose you are monitoring the allelic and genotypic frequencies of the MN blood group locus in a small human population. You find that for 1-year-old children the genotypic frequencies are $MM = 0.25$, $MN = 0.5$, and $NN = 0.25$, whereas the genotypic frequencies for adults are $MM = 0.3$, $MN = 0.4$, and $NN = 0.3$.
    a. Compute the M and N allele frequencies for the 1-year-olds and adults.
    b. Are the allele frequencies in equilibrium in this population?
    c. Are the genotypic frequencies in equilibrium?
11. Using Table 19.3, determine the frequencies of p and q that result in the greatest proportion of heterozygotes in a population.
12. In a given population, the frequencies of the four phenotypic classes of the ABO blood groups are found to be $A = 0.33$, $B = 0.33$, $AB = 0.18$, and $O = 0.16$. What is the frequency of the O allele?
13. If a trait determined by an autosomal recessive allele occurs at a frequency of 0.25 in a population, what are the allelic frequencies? Assume Hardy-Weinberg equilibrium, and use A and a to symbolize the dominant and recessive alleles, respectively.

Measuring Genetic Diversity in Human Populations

14. Why is it that mutation, acting alone, has little effect on gene frequency?

15. Successful adaptation is defined by:
    a. evolving new traits
    b. producing many offspring
    c. increasing fitness
    d. moving to a new location

16. What is the relationship between founder effects and genetic drift?

17. How would a drastic reduction in a population's size affect the population's gene pool?

18. The major factor causing deviations from Hardy-Weinberg equilibrium is
    a. selection
    b. nonrandom mating
    c. mutation
    d. migration
    e. early death

19. A specific mutation in the *BRCA1* gene has been estimated to be present in approximately 1% of Ashkenazi Jewish women of Eastern European descent. This specific alteration, 185delAG, is found about three times more often in this ethnic group than the combined frequency of the other 125 mutations found to date. It is believed that the mutation is the result of a founder effect from many centuries ago. Explain the founder principle.

20. The theory of natural selection has been popularly summarized as "survival of the fittest." Is this an accurate description of natural selection? Why or why not?

Natural Selection Affects the Frequency of Genetic Disorders

21. Will a recessive allele that is lethal in the homozygous condition ever be removed from a large population by natural selection?

22. Do you think that our species is still evolving or are we shielded from natural selection by civilization? Is it possible that misapplications of technology will end up exposing our species to more rather than less natural selection (consider the history of antibiotics)?

Genetic Variation in Human Populations

23. a. Provide a genetic definition of race.
    b. Using this definition, can modern humans be divided into races? Why or why not?

The Appearance and Spread of Our Species (*Homo sapiens*)

24. a. Briefly describe the two major theories discussed in this chapter about the origin of modern humans.
    b. Which of these two theories would predict a closer relationship for the various modern human populations?
    c. Which of the two theories is best supported by the genetic evidence?

Internet Activities

 Internet Activities are critical thinking exercises using the resources of the World Wide Web to enhance the principles and issues covered in this chapter. For a full set of links and questions investigating the topics described below, log on to **http://biology.brookscole.com/cummings7**

1. *Comparing DNA Sequences.* GenBank is the National Institutes of Health's (NIH) database of all known nucleotide and protein sequences, including supporting bibliographic and biological data. Use GenBank's Entrez system to search for a DNA sequence and BLAST to find similar sequences in GenBank.

2. *Exploring the Hardy-Weinberg Equilibrium Equation.* The *Access Excellence Activities Exchange* site includes several Hardy-Weinberg–related exercises. To see how selection can affect a population's allele frequencies, try the *Fishy Frequencies* activity. This exercise can be run alone or as part of a group—and you get to eat fish crackers as you work!

3. *DNA, Archaeology, and Human History.* Read the article "Scientists Rough Out Humanity's 50,000 Year Old History" at the *New York Times Learning Network* site.

How would you vote now?

The Human Genome Diversity Project discussed at the beginning of this chapter is attempting to study the evolution and divergence of our species and to identify genes for disease susceptibility and resistance. To do this, the project is collecting DNA samples from members of isolated, indigenous populations around the world. Although the project has reformulated its methods in reaction to widespread criticism, opponents, many of whom are members of indigenous groups, still feel that the project exploits their genetic heritage and disrupts their social structure. Now that you know more about population genetics and the evolution of our species, what do you think? Do the benefits of the project outweigh the possible ethical, legal, and social complications? Visit the Human Heredity Companion website to find out more on the issue, then cast your vote online at **http://biology.brookscole.com /cummings7**

For further reading and inquiry, log on to **InfoTrac College Edition,** your world-class online library, including articles from nearly 5,000 periodicals, at **http://infotrac.thomsonlearning.com**

Answers to Selected Questions and Problems *Appendix*

Chapter 1

2. Population genetics studies genetic variations found in individuals in a population and the forces that alter the frequency of these variations as they are passed from generation to generation.
5. A genome is the set of genes carried by an individual. Genomics is the study of genomes and their genetic content, organization, function, and evolution.

Chapter 2

3. There are 44 autosomes in a body (somatic) cell and 22 autosomes in a gamete.
5. d
7. Cells undergo a cycle of events involving growth, DNA replication, and division. Daughter cells undergo the same series of events. During S phase, DNA synthesis and chromosome replication occurs. During M phase, mitosis takes place.
8. a, e
10. Meiosis II, the division responsible for the separation of sister chromatids, would no longer be necessary. Meiosis I, wherein homologues segregate, would still be required.
18. Cell cycle genes can turn cell division on and off. If a gene normally promotes cell division, it can mutate to cause too much cell division. If a gene normally turns off cell division, it can mutate so that it can no longer repress cell division.
19.

	Mitosis	Meiosis
Number of daughter cells produced	2	4
Number of chromosomes per daughter cell	$2n$	n
Number of cell divisions	1	2
Do chromosomes pair? (Y/N)	N	Y
Does crossing over occur? (Y/N)	N	Y
Can the daughter cells divide again? (Y/N)	Y	N
Do the chromosomes replicate before division? (Y/N)	Y	Y
Type of cell produced	SOMATIC	GAMETE

20.

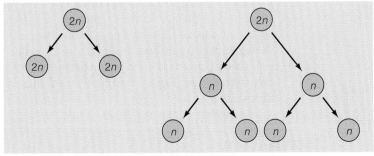

24. **a.** Mitosis; **b.** Meiosis I; **c.** Meiosis II
27. Meiotic anaphase I: no centromere division, chromosomes consisting of two sister chromatids are migrating; Meiotic anaphase II: centromere division, the separating sister chromatids are migrating. Meiotic anaphase II more closely resembles mitotic anaphase by these two criteria.

Chapter 3

1. **a.** A gene is the fundamental unit of heredity. The gene encodes a specific gene product (e.g., a pigment involved in determining eye color). Alleles are alternate forms of a gene that may cause various phenotypic effects. For example, a gene may have blue, brown, and green eye color alleles. Locus is the position of a gene on a chromosome. In a normal situation, all alleles of a gene would have the same locus.

   b. Genotype refers to the genetic constitution of the individual (*AaBb* or *aabb*). Notice that the genotype always includes at least two letters, each representing one allele of a gene pair in a diploid organism. A gamete would contain only one allele of each gene because of its haploid state (*Ab* or *ab*). Phenotype refers to an observable trait. For example, *Aa* (the genotype) will cause a normal pigmentation (the phenotype) in an individual, whereas *aa* will cause albinism.

   c. Dominance refers to a trait that is expressed in the heterozygous condition. Therefore, only one copy of a dominant allele needs to be present to express the phenotype. Recessiveness refers to a trait that is not expressed in the heterozygous condition. It is masked by the dominant allele. To express a recessive trait, two copies of the recessive allele must be present in the individual.

   d. Complete dominance occurs when a dominant allele completely masks the expression of a recessive allele. For example, in pea plants, yellow seed color is dominant to green. In a heterozygous state, the phenotype of the seeds is yellow. This is the same phenotype seen in seeds homozygous for the yellow allele.

      Incomplete dominance occurs when the phenotype of the heterozygote is intermediate between the two homozygotes. For example, in *Mirabilis*, a red flower crossed with a white flower will give a pink flower.

2. Phenotypes: b, d; Genotypes: a, c, e

5. **a.** 1/2 *A*, 1/2 *a*
   b. All *A*
   c. All *a*

11. **a.** All F$_1$ plants will be long-stemmed.
    b. Let *S* = long-stemmed and *s* = short-stemmed. The long-stemmed P$_1$ genotype is *SS*, the short-stemmed P$_1$ genotype is *ss*. The long-stemmed F$_1$ genotype is *Ss*.
    c. Approximately 225 long-stemmed and 75 short-stemmed
    d. The expected genotypic ratio is 1 *SS*:2 *Ss*:1 *ss*.

13. **a.** 1/2 *A_B_*, 1/2 *A_bb*
    b. 1/4 *A_B_*, 1/4 *A_bb*, 1/4 *aaB_*, 1/4 *aabb*
    c. 9/16 *A_B_*, 3/16 *A_bb*, 3/16 *aaB_*, 1/16 *aabb*

15. **a.** Both are 3:1
    b. 9:3:3:1
    c. Swollen is dominant to pinched, yellow is dominant to green.
    d. Let *P* = swollen and *p* = pinched; *C* = yellow and *c* = green. Then: P$_1$ = *PPcc* × *ppCC* or *PPCC* × *ppcc*. F$_1$ = *PpCc*

17. **a.** Let *S* = smooth and *s* = wrinkled and *Y* = yellow and *y* = green. The parents are *SSYY* × *ssyy*. The F$_1$ offspring are *SsYy*.
    b. The smooth, yellow parent is *SsYy*. The genotypes of the offspring are *SsYy*, *Ssyy*, *ssYy*, and *ssyy*.

21. 3/4 for *A* × 1/2 for *b* × 1 for *C* = 3/8 for *A*, *b*, *C*

23. The P$_1$ generation is *FF* × *ff*. The F$_1$ generation is *Ff*. The mode of inheritance is incomplete dominance.

26. Because neither species produces progeny resembling a parent, simple dominance is ruled out. The species producing pink-flowered progeny from red and white (or very pale yellow) suggests incomplete dominance as a mode of inheritance. However, in the second species, the production of orange-colored progeny cannot be explained in this fashion. Orange would result from an equal production of red and yellow; instead in this case, codominance is suggested, with one parent producing bright red flowers and the other producing pale yellow flowers.

31. During meiotic prophase I, the replicated chromosomes synapse or pair with their homologues. These paired chromosomes align themselves at the equator of the cell during metaphase I. During anaphase I it is the homologues (each containing two chromatids) that separate from each other. There is no preordained orientation for this process—it is equally likely that a maternal or a paternal homologue will migrate to a given pole. This provides the basis for the law of random segregation. Independent assortment results from the fact that the polarity of one set of homologues has absolutely no influence on the orientation of a second set of homologues. For example, if the maternal homologue of chromosome 1 migrated to a certain pole, it will have no bearing on whether the maternal or paternal homologue of chromosome 2 migrates to that same pole.

Chapter 4

2. d
3. **a.** Female
   b. Yes
   c. 3 siblings; The proband is the youngest child.
5. Autosomal dominant with incomplete penetrance

7. a.

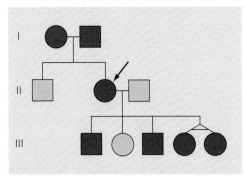

b. The mode of inheritance is consistent with an autosomal dominant trait. Both of the proband's parents are affected. If this trait were recessive, all of their children would have to be affected (*aa* × *aa* can only produce *aa* offspring). As we see in this pedigree, the brother of the proband is not affected, indicating that this is a dominant trait. His genotype is *aa*, the proband's genotype is *AA* or *Aa*, and both parents' genotype is most likely *Aa*.

c. Because the proband's husband is unaffected, he is *aa*.

9. a. This pedigree is consistent with autosomal recessive inheritance.

b. If inheritance is autosomal recessive, the individual in question is heterozygous.

10.

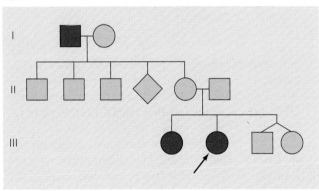

14. Due to the rarity of the disease, we assume the paternal grandfather is heterozygous for the gene responsible for Huntington disease. His son has a 1/2 chance of possessing the deleterious allele. In turn, should the father carry the *HD* allele, his son would have a 1/2 chance of inheriting it. Therefore, at present, the child has a 1/2 × 1/2 = 1/4 chance of having inherited the *HD* allele.

16. a. 50% chance for sons, 25% of all children

b. 50% for daughters, 25% of all children

19. In the autosomal case, the parents are *Aa* and *Aa* where the disease is recessive. They can have a son or daughter that is affected (*aa*) or unaffected (*AA* or *Aa*). In the X-linked situation, the parents are $X^A X^a$ × $X^A Y$ in a recessive case. The children would be $X^A X^a$ and $X^a Y$. Because an unaffected daughter and affected son are possible in each case, this limited in-

formation is not enough to determine the inheritance pattern.

20. X-linked dominant

21. X-linked recessive

23. Mitochondria contain DNA carrying genetic information and are maternally inherited.

27. 20% of 90 = 18

Chapter 5

2. a. Height in pea plants is determined by a single pair of genes with dominant and recessive alleles. Height in humans is determined by polygenes.

b. For traits determined by polygenes, the offspring of matings between extreme phenotypes show a tendency to regress toward the mean phenotype in the population.

5. a. F_1 genotype = *A′AB′B*, phenotype = height of 6 ft

b. *A′AB′B* × *A′AB′B*
↓

Genotypes	Phenotypes
A′A′B′B′	7 ft
A′A′B′B	6 ft 6 in.
A′A′BB	6 ft
A′AB′B′	6 ft 6 in.
A′AB′B	6 ft
A′ABB	5 ft 6 in.
AAB′B′	6 ft
AAB′B	5 ft 6 in.
AABB	5 ft

6. In the case of polygenes, the expression of the trait depends on the interactions of many genes, each of which contributes a small effect to the expression of the trait. Thus, the differences between genotypes often are not clearly distinguishable. In the case of monogenic determination of a trait, the alleles of a single locus have major effects on the expression of the trait, and the differences between genotypes are usually easily discerned.

7. Liability is caused by a number of genes acting in an additive fashion to produce the defect. If exposed to certain environmental conditions, the person above the threshold will most likely develop the disorder. The person below the threshold is not predisposed to the disorder and will most likely remain normal.

11. Relatives are used because the proportion of genes held in common by relatives is known.

13. No. Dizygotic twins arise from two separate fertilized eggs. Only monozygotic twins can be Siamese, since they originate from the same fertilized egg and are genetically identical.

14. b

18. a. The study included only men who were able to pass a physical exam that eliminated markedly obese individuals, and so the conclusions cannot be generalized beyond the group of men inducted into the armed forces.

b. To design a better study, include MZ and DZ twin men and women, maybe even children. Include a cross section of various populations (ethnic groups, socioeconomic groups, weight classifications, etc.). Control the diet so that it remains a constant. Another approach is to study MZ and DZ twins that were reared apart (and presumably in different environments), or adopted and natural children that were raised in the same household (same environment). There are other possible answers.

22. First, intelligence is difficult to measure. Also, for such a complex trait, many genes and a significant environmental component are likely to be involved.

24. The heritability difference observed between the racial groups for this trait cannot be compared because heritability measures variation within one population at the time of the study. Heritability cannot be used to estimate genetic variation between populations.

Chapter 6

1. a. Chemical treatment of chromosomes resulting in unique banding patterns
   b. Q banding and G banding with Giemsa
6. Advanced maternal age, previous aneuploid child, presence of a chromosomal rearrangement, presence of a known genetic disorder in the family history
8. Triploidy
13. The embryo will be tetraploid. Inhibition of centromere division results in nondisjunction of an entire chromosome set. After cytoplasmic division, some cytoplasm is lost in an inviable product lacking genetic material, and the embryo develops from the tetraploid product.
18. Condition 2 is most likely lethal. This condition involves a chromosomal aberration, trisomy. This has the potential for interfering with the action of all genes on the trisomic chromosome. Condition 1 involves an autosomal dominant lesion to a single gene, which is more likely to be tolerated by the organism.
22. Turner syndrome (45,X) is monosomy for the X chromosome. A paternal nondisjunction event could contribute a gamete lacking a sex chromosome to result in Turner syndrome. The complementary gamete would contain both X and Y chromosomes. This gamete would contribute to Kleinfelter syndrome (47,XXY).
23. Loss of a chromosome segment — deletion
    Extra copies of a chromosome segment — duplication
    Reversal in the order of a chromosome segment — inversion
    Movement of a chromosome segment to another, nonhomologous chromosome — translocation
25. In theory, the chances are 1/3.

26. Two or three possibilities should be considered. The child could be monosomic for the relevant chromosome. The child has the paternal copy carrying the allele for albinism (father is heterozygous) and a nondisjunction event resulted in failure to receive a chromosomal copy from the homozygous mother. The second possibility is that the maternal chromosome carries a small deletion, allowing the albinism to be expressed. The third possibility is that the child represents a new mutation, inheriting the albino allele and having the other by mutation. Because monosomy is lethal, either the second or third possibility seems likely.

Chapter 7

3. Meiosis began before the birth of the parent and is completed shortly after fertilization. The time taken is therefore approximate. Shortest time: (from Jan. 1, 1980, to July 1, 2004) 24.5 yrs. Longest time: (from April 1, 1979, to July 1, 2004) 25.25 yrs.
5. Significant economic and social consequences are associated with FAS, including the costs of surgery for facial reconstruction, treatment of learning disorders and mental retardation, and caring for institutionalized individuals. Prevention depends on the education of pregnant women and the early treatment of pregnant women with alcohol dependencies. Other answers are possible.
8. d
9. Female
11. A mutation causing the loss of the SRY gene, testosterone, or testosterone receptor gene function. Also a defect in the conversion from testosterone to DHT can cause the female external phenotype until puberty.
15. Pattern baldness acts as an autosomal dominant trait in males and an autosomal recessive trait in females.
18. Random inactivation occurs in females, so the genes from both X chromosomes are active in the body as a whole.

Chapter 8

2. Proteins are found in the nucleus. Proteins are complex molecules composed of 20 different amino acids; nucleic acids are composed of only 4 different nucleotides. Cells contain hundreds or thousands of different proteins; only 2 main types of nucleic acids.
4. Protease destroyed any small amounts of protein contaminants in the transforming extract. Similarly, treatment with RNAse destroyed any RNA present in the mixture.
6. The process is transformation, discovered by Frederick Griffith. The P bacteria contain genetic information that is still functional even though the cell has been heat-killed. However, it needs a live recipient

host cell to accept its genetic information. When heat-killed P and live D bacteria are injected together, the dead P bacteria can transfer its genetic information into the live D bacteria. The D bacteria are then transformed into P bacteria and can now cause polkadots.

10. Chargaff's Rule: A = T and C = G
If A = 27%, then T must equal 27%
If G = 23%, then C must equal 23%
Base composition:
A = 27%
T = 27%
C = 23%
G = 23%
    100%

12. b and e
13. b
15. c
20.

	DNA	RNA
a. Number of chains:	2	1
b. Bases used:	A, C, G, T	A, C, G, U
c. Sugar used:	deoxyribose	ribose
d. Function:	blueprint of genetic information	transfer of genetic information from nucleus to cytoplasm

23. a

Chapter 9

2. Replication is the process of making DNA from a DNA template. Transcription makes RNA from a DNA template, and translation makes an amino acid chain (a polypeptide) from an mRNA template. Replication and transcription happen in the nucleus, and translation occurs in the cytoplasm.
3. There would be 4^4, or 256, possible amino acids encoded.
8. b
9. 1. Removal of introns: to generate a contiguous coding sequence that can make an amino acid chain
   2. Addition of the 5′ cap: ribosome binding
   3. Addition of the 3′ polyA tail: mRNA stability
12. Codons are triplets of bases on an mRNA molecule. Anticodons are triplets of bases on a tRNA molecule and are complementary in sequence to the nucleotides in codons.
13. Answer: 25%   Total length: 10 kb
                        Coding region: 2.5 kb
16. tRNA: UAC   UCU   CGA   GGC
    mRNA: AUG   AGA   GCU   CCG
    DNA: TAC   TCT   CGA   GGC-sense strand
    protein: met   arg   ala   pro
    Hydrogen bonds present in the DNA: 31
    7 GC pairs × 3 = 21
    5 AT pairs × 2 = 10

24. a. No
    b. Yes

Chapter 10

3. c
4. a. Buildup of substance A, no substance B or C
   b. Buildup of substance B, no substance C
   c. Buildup of substance B, as long as A is not limiting factor
   d. 1/2 the amount of C
5. a. Yes. Each will carry the normal gene for the other enzyme. (Individual 1 will be mutant for enzyme 1 but normal for enzyme 2. This is because two different genes encode enzymes 1 and 2.)
   b. Let D = dominant mutation in enzyme 1, let normal allele = d
   Let A = dominant mutation in enzyme 2, let normal allele = a

   Ddaa   ×   ddAa
   Offspring: DdAa   mutation in enzyme 1 and 2, A buildup, no C
               Ddaa   mutation in enzyme 1, A buildup, no C
               ddAa   mutation in enzyme 2, B buildup, no C
               ddaa   no mutation, normal
   Ratio would be 1:2:1 for substance B buildup, no C: substance A buildup, no C: normal
6. Alleles for enzyme 1: A (dominant, 50% activity); a (recessive, 0% activity). Alleles for enzyme 2: B (dominant, 50% activity); b (recessive, 0% activity).

	Enzyme 1	Enzyme 2	Compound A	B	C
1AABB	100	100	N	N	N
2AaBB	50	100	N	N	N
4AaBb	50	50	N	N	N
2AABb	100	50	N	N	N
1Aabb	100	0	N	B	L
2Aabb	50	0	N	B	L
1aaBB	0	100	B	L	L
2aaBb	0	50	B	L	L
1aabb	0	50	B	L	L

N, normal; B, buildup; L, less.
12. b
17. No, because individuals who are GD/GD show 50% activity. The g allele reduces activity by 50% so heterozygotes appear normal. It is not until the level of activity falls below 50% that the mutant phenotype is observed.
18. Let H = the mutant allele for hypercholesterolemia
    Let h = the normal allele
    Answer: HH Heart attack as early as the age of 2, definite heart disease by age 20, death in most cases by 30. No functional LDL receptors produced.

Hh Heart attack in early 30s. Half the number of functional receptors are present and twice the normal levels of LDL.

hh Normal. Both copies of the gene are normal and can produce functional LDL receptors.

23. It would cause a frameshift mutation very early in the protein. Most likely, the protein would lose all of its functional capacity.

26. Drugs usually act on proteins. Different people have different forms of proteins. Different proteins are inherited as different alleles of a gene.

27. People have different abilities to smell and taste chemical compounds such as phenylthiocarbamide (PTC); some people are unable to smell skunk odors; different reactions to succinylcholine, a muscle relaxant, and to primaquine, an antimalarial drug. Others are sensitive to the pesticide parathion.

Chapter 11

1. Mutation rate measures the occurrence of mutations per individual per generation.

2. 245,000 births represent 490,000 copies of the achondroplasia gene, because each child carries two copies of the gene. The mutation rate is therefore 10/490,000, or 2×10^{-5}, per generation.

7. Muscular dystrophy is an X-linked disorder. A son receives an X chromosome from his mother and a Y chromosome from his father. In this case, the mother was a heterozygous carrier of muscular dystrophy and passed the mutant gene to her son. The father's exposure to chemicals in the workplace is unrelated to his son's condition.

10. Missense, same; Nonsense, shorter; Sense, longer

16. **a.** See pedigree below.

19. Our bodies can correct base-pairing errors made during DNA replication. We can also repair UV-damaged DNA after thymine dimers form.

Chapter 12

3. a

6. d

9. An oncogene is a mutated proto-oncogene that promotes uncontrolled cell division that leads to cancer. A tumor suppressor gene is a normal gene that stops cell division when it is not needed. If mutated, a tumor suppressor gene can lose its function and no longer be able to control cell division. The result is too much cell division.

10. The inheritance of predisposition is dominant because only one mutant allele causes the predisposition to retinoblastoma. However, the second allele must also be mutated in at least one eye cell to produce the disease. Therefore, the expression of retinoblastoma is recessive.

15. Conditions a and d would produce cancer. The loss of function of a tumor suppressor gene would allow cell growth to go unchecked. The overexpression of a proto-oncogene would promote more cell division than normal.

19. Mutations in *APC* form hundreds or thousands of benign tumors. If the right mutations occur in one of these tumors, the benign growths can progress to cancer. The large number of benign tumors makes the chance of acquiring other mutations likely.

21. c-*myc* lies at the breakpoint of a translocation involving chromosome 8 and either chromosome 14, 22, or 2. The translocation places the *myc* gene in an altered chromosomal milieu and thus disrupts its normal expression. Altered expression of c-*myc* is thought to be necessary for the production of Burkitt's lymphoma.

25. Diet is suspected as the cause in both cases. When Japanese move to the United States and adopt an American diet, the rate of breast cancer goes up, but the rate of colon cancer goes down. The reverse is also the case: the Japanese diet in Japan predisposes to colon cancer, but not to breast cancer.

See Chapter 11, answer 16a.

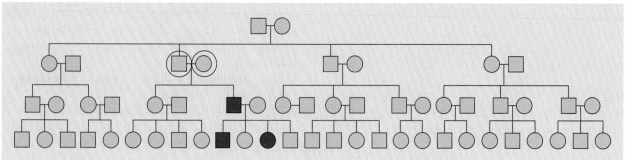

Chapter 13

2. d

4. b

8. They leave compatible ends that can be pasted (ligated) to other DNA cut with the same enzyme.

9. *Eco*RI: 2 kb, 11.5 kb, 10 kb
   *Hin*dIII/*Pst*I: 7 kb, 3 kb, 7.5 kb, 6 kb
   *Eco*RI/*Hin*dIII/*Pst*I: 2 kb, 5 kb, 3 kb, 3.5 kb, 4 kb, 6 kb

13. b

15. a

20. DNA derived from individuals with sickle cell anemia will lack one fragment contained in the DNA from normal individuals. In addition, there will be a large (uncleaved) fragment not seen in normal DNA.

Chapter 14

4. c

8. The first recommendation would be for another alpha-fetoprotein test to confirm the initial result. If the second test is still positive or ambiguous, amniocentesis could be considered in order to examine alpha-fetoprotein levels in the amniotic fluid.

12. d

14. a. $1/100 \times 1/500 = 1/50,000$ individuals with this combination of alleles are present in the population.
    b. This is not very convincing, because in a large city, say with a population of 3 million, there will be approximately 60 individuals with this profile.
    c. The lab should test two or more additional loci to reduce the probability of another individual having this profile to a much lower number, such as 1 in 50 million or more.
    d. The answer is an opinion and a point of discussion.

17. There are several potential problems. For example with gene transfer, are there harmful side effects with human proteins grown in animals or plants? Do genetically engineered crops pose any short- or long-term harmful effects to humans or to the environment? Extensive testing and a rigorous approval process should accompany any recombinant DNA technology before it is put to use.

Chapter 15

1. Genes that are said to show "linkage" are located near each other on the same chromosome. Linked genes tend to be inherited together.

6. The human genome contains about 3.2 billion nucleotides.

10. The first organism to have its genome entirely sequenced was the bacterium *Haemophilus influenzae*. The sequencing was carried out by a consortium, co-ordinated by Craig Venter and The Institute for Genome Research, and used a shotgun cloning strategy. The sequencing was completed in 1995.

13. Genes make up about 5% of the total DNA sequence of the human genome. It is not clear what, if any, function the remaining DNA has. About half the total DNA in the genome is made up of various kinds of repeated sequences.

16. c

19. In addition to the scientific elements of the project, the organizers of the HGP set up a program called ELSI to address the ethical, legal, and social implications of genomics research. This project has used meetings, grants, workshops, and other forums to discuss various issues related to genomics research and to help bring about legislation to protect against the abuse of genetic information.

Chapter 16

2. RU-486 is controversial because it is a medication that terminates, rather than prevents, pregnancy. Unlike birth control methods, which prevent ovulation, sperm transport, fertilization, or implantation, RU-486 blocks the hormones necessary for maintaining a pregnancy.

4. In gamete intrafallopian transfer (GIFT), eggs and sperm are collected and placed in the oviduct for fertilization. In intracytoplasmic sperm injection, a single sperm is selected and injected into an egg, fertilizing it.

15. a. The story of nonpaternity during family genetic counseling is familiar to genetic counselors. When deciding how to approach this type of unexpected findings, counselors need to weigh the benefits and harms of nondisclosure against that of disclosure. The first considerations include the relevance of the information to the patient's situation and the consequences of the findings. The 1983 President's Commission recommends that patients be advised before testing that unexpected information may be revealed.
    b. It is reasonable for the counselor to call the woman beforehand and explain the results and the implications of the findings. Given the sensitivity of this information, the long-term effect on the couple's relationship may be dramatic, and disclosure may do more harm than good.

Chapter 17

1. a. Microorganisms that penetrate the skin infect cells, which then release chemical signals such as histamine. This causes an increased blood flow into the area, resulting in an increase in temperature.
   b. The heat serves to inhibit microorganism growth, to mobilize white blood cells, and to raise the

metabolic rate in nearby cells, thereby promoting healing.

3. Immunoglobulins: IgD, IgM, IgG, IgA, IgE

4. Helper T cells activate B cells to produce antibodies. Suppressor T cells stop the immune response of B and T cells. Cytotoxic (killer) T cells target and destroy infected cells.

10. Express the cloned gene to make the protein product, isolate the protein, and inject it into humans. The immune system should make antibodies to that protein. When the actual live virus is encountered, the immune system will have circulating antibodies and T cells that will recognize the protein (antigen) on the surface of the virus.

16. The antigens of the donor/recipient are more important. The antigen of the donor will be rejected if the recipient does not have the same antigen. The antigen of the recipient determines which antibodies can be produced. For example, a blood type A individual will make B antibodies if exposed to the B.

20. a. The mother is Rh$^-$. She will produce antibodies against the Rh antigen if her fetus is Rh$^+$. This happens when blood from the fetus enters the maternal circulation.

   b. The mother already has circulating antibodies against the Rh protein from her first Rh$^+$ child. She can mount a greater immune response against the second Rh$^+$ child by generating a large number of antibodies.

24. One approach is to clone the human gene that suppresses hyperacute rejection and inject it into pig embryos. The hope is that the pig's cells will express this human protein on the cell surface. The human recipient may then recognize the transplanted organ as "self." In addition, transplants of bone marrow from donor pigs into human recipients may help in preventing rejection mediated by T cells. This dual bone marrow system will recognize the pig's organ as "self" but will retain the normal human immunity.

25. a. They need to test one cell of each eight-cell embryo for an ABO and Rh blood type match and also an HLA complex match. If a match exists, they will implant the embryo(s) into the mother and hope that pregnancy occurs. When the baby is born, bone marrow will be extracted and transplanted into the existing child.

   b. Ethically, it is difficult to imagine having a child for the primary purpose of being a bone marrow donor. The new child may come to feel demeaned or less valued. Also, what happens to the embryos that are not a match to the couple's existing child? These embryos are completely healthy; they simply have the wrong blood type and histocompatibility complex. On the other hand, if the couple will love and provide for this new child, it may be a wonderful experience that the new child has the opportunity to save the life of his or her sibling.

28. Antihistamines block the production or action of histamine. The allergen causes a release of IgE antibodies, which bind to mast cells. These cells release histamine, which causes fluid accumulation, tissue swelling (such as swollen airways or eyes), and mucus secretion (such as a runny nose).

31. The HIV virus infects and kills helper T4 cells, the very cells that normally trigger the antibody-mediated immune response. Therefore, as the infection progresses, the immune system gets weaker and weaker as more T cells are killed. The AIDS sufferer is then susceptible to various infections and certain forms of cancer.

Chapter 18

2. The definition must be precise enough to distinguish the behavior from other, similar behaviors and from the behavior of the control group. The definition of the behavior can significantly affect the results of the genetic analysis, and even the mode of inheritance of the trait.

4. *Drosophila* has many advantages for the study of behavior. Mutagenesis and screening for behavior mutants allows the recovery of mutations that affect many forms of behavior. The ability to perform genetic crosses and recover large numbers of progeny over a short period of time also enhances the genetic analysis of behavior. This organism can serve as a model for human behavior, because cells of the nervous system in both *Drosophila* and humans use similar mechanisms to transmit impulses and store information.

6. Huntington disease is caused by expansion of a CAG trinucleotide repeat within the *HD* gene. Expansion of the repeat causes an increase in the number of glutamines in the encoded protein, causing the protein to become toxic to neurons. In regions of the nervous system expressing the mutant protein, cells fill with clusters of the protein, degenerate, and die.

10. It may be argued that using such a test to identify potentially violent individuals would allow them to be given appropriate therapy (such as drugs to increase MAOA activity) before they harm anyone. On the other hand, the use of such a test would result in individuals being labeled as aggressive or violent not based on their behavior, but on their genotype. This could have significant consequences in their work and personal lives.

13. No, it means that there are probably other genes involved or the environment plays a significant role. The linkage to chromosome 7 is still valid, and the next goal would be to find the gene on chromosome 7 that is linked to manic depression. Also, finding the other genes involved is important. A researcher may find that a subset of manic depressive individuals has a defect in the gene on chromosome 7 and another

subset in a gene on chromosome 12. Both defects can contribute to the same disease.

16. The heritability of Alzheimer disease, a multifactorial disorder, cannot be established because of interactions between genetic and environmental factors. Less than 50% of Alzheimer cases can be attributed to genetic causes, indicating that the environment plays a large role in the development of this disease.

18. Probably not, because correlations must eventually be related to a causal relationship, in this case, between body hair and intelligence. Many factors contribute to intelligence, including environmental factors. Lacking an explanation for the relationship between body hair and intelligence, the assumption is unwarranted. Testing would depend on the definition of intelligence used in the study, and may involve IQ testing by individuals who know nothing about the person's body hair. Alternatively, testing could be done for *g*, a measure of cognitive ability in blind testing, where the presence of the subject's body hair is unknown to the test administrator.

Chapter 19

2. a. Population: local groups of individuals occupying a given space at a given time.
   b. Gene pool: the set of genetic information carried by a population.
   c. Allele frequency: the frequency of occurrence of particular alleles in the gene pool of a population.
   d. Genotype frequency: the frequency of occurrence of particular genotypes among the individuals of a population.

8. The frequency of an allele in a population has no relationship to its mode of inheritance. For example, a dominant allele may exist at a very low frequency in a population and cannot ultimately overtake a recessive allele in frequency.

10. a. Children: $M = 0.5$, $N = 0.5$; Adults: $M = 0.5$, $N = 0.5$

b. Yes. Allelic frequencies are unchanged.
c. No. The genotypic frequencies are changing within each generation.

20. No. It is not a particularly accurate description. Natural selection depends not just on an ability to survive but also to reproduce. It is the differential reproduction of some individuals that is the essence of natural selection.

22. This is an open question. Culture, in the form of society and technology, has shielded humans from many forms of selective forces in the environment but has also created new forms of selection for humans and for infectious agents. The abuse of antibiotics may increase the effect of selection on human populations, as will the expansion of the human population into new geographic areas, increasing exposure to endemic agents of disease.

23. a. Genetically speaking, races are populations with significant differences in allele frequencies compared with other populations.
   b. No. No systematic differences have been identified in allele frequencies within modern human populations that would justify the use of the term *race*. Studies into the level of genetic variation within and between populations have consistently found that there is much more variety within each population than between them.

24. a. The out-of-Africa hypothesis holds that modern humans first appeared in Africa and then left the continent to replace all of the then-existing hominid populations in the world. The multiregional hypothesis proposes that, through a network of interbreeding populations, *Homo erectus* gradually evolved into modern *Homo sapiens* in different regions of the world.
   b. The out-of-Africa hypothesis
   c. Both mitochondrial and nuclear DNA evidence favor the out-of-Africa hypothesis.

Glossary

Acquired immunodeficiency syndrome (AIDS) A collection of disorders that develop as a result of infection with the human immunodeficiency virus (HIV).

Acrocentric Describes a chromosome whose centromere is placed very close to, but not at, one end.

Adenine One of two nitrogen-containing purine bases found in nucleic acids, along with guanine.

Affect A short-term expression of feelings or emotion.

Alkaptonuria An autosomal recessive trait with altered metabolism of homogentisic acid. Affected individuals do not produce the enzyme needed to metabolize this acid, and their urine turns black.

Allele One of the possible alternative forms of a gene, usually distinguished from other alleles by its phenotypic effects.

Allele frequency The frequency with which alleles of a given gene are present in a population.

Allelic expansion Increase in gene size caused by an increase in the number of trinucleotide sequences.

Allergens Antigens that provoke an inappropriate immune response.

Alpha thalassemia Genetic disorder associated with an imbalance in the ratio of alpha and beta globin caused by reduced or absent synthesis of alpha globin.

Alzheimer disease (AD) A heterogeneous condition associated with the development of brain lesions, personality changes, and degeneration of intellect. Genetic forms are associated with loci on chromosomes 14, 19, and 21.

Amino group A chemical group (NH_2) found in amino acids and at one end of a polypeptide chain.

Amniocentesis A method of sampling the fluid surrounding the developing fetus by inserting a hollow needle and withdrawing suspended fetal cells and fluid; used in diagnosing fetal genetic and developmental disorders; usually performed in the sixteenth week of pregnancy.

Anaphase A stage in mitosis during which the centromeres split and the daughter chromosomes begin to separate.

Anaphylaxis A severe allergic response in which histamine is released into the circulatory system.

Androgen insensitivity An X-linked genetic trait that causes XY individuals to develop into phenotypic females.

Aneuploidy A chromosomal number that is not an exact multiple of the haploid set.

Annotation The analysis of genomic nucleotide sequence data to identify the protein-coding genes, the non-protein-coding genes, their regulatory sequences, and their function(s).

Antibody A class of proteins produced by B cells that bind to foreign molecules (antigens) and inactivate them.

Antibody-mediated immunity Immune reaction that protects primarily against invading viruses and bacteria using antibodies produced by plasma cells.

Anticipation Onset of a genetic disorder at earlier ages and with increasing severity in successive generations.

Anticodon A group of three nucleotides in a tRNA molecule that pairs with a complementary sequence (known as a codon) in an mRNA molecule.

Antigens Molecules carried or produced by microorganisms that initiate antibody production.

Assisted reproductive technologies (ART) The collection of techniques used to help infertile couples have children.

Assortment The result of meiosis I that puts random combination of maternal and paternal chromosomes into gametes.

Autosomes Chromosomes other than the sex chromosomes. In humans, chromosomes 1 to 22 are autosomes.

B cells A type of lymphocyte that matures in the bone marrow and mediates antibody-directed immunity.

Background radiation Radiation in the environment that contributes to radiation exposure.

Barr body A densely staining mass in the somatic nuclei of mammalian females. An inactivated X chromosome.

Base analogs A purine or pyrimidine that differs in chemical structure from those normally found in DNA or RNA.

Beta thalassemia Genetic disorder associated with an imbalance in the ratio of alpha and beta globin caused by reduced or absent synthesis of beta globin.

Bioinformatics The use of computers and software to acquire, store, analyze, and visualize the information from genomics.

Biotechnology The use of recombinant DNA technology to produce commercial goods and services.

Bipolar disorder An emotional disorder characterized by mood swings that vary between manic activity and depression.

Blastocyst The developmental stage at which the embryo implants into the uterine wall.

Blastomere A cell produced in the early stages of embryonic development.

Blood type One of the classes into which blood can be separated based on the presence or absence of certain antigens.

Bulbourethral glands Glands that secrete a mucus-like substance that provides lubrication for intercourse.

Camptodactyly A dominant human genetic trait that is expressed as immobile, bent, little fingers.

Cap A modified base (guanine nucleotide) attached to the 5′ end of eukaryotic mRNA molecules.

Carboxyl group A chemical group (COOH) found in amino acids and at one end of a polypeptide chain.

Caretaker genes Genes that help maintain the integrity of the genome, for example, DNA repair genes.

Cell cycle The sequence of events that takes place between successive mitotic divisions.

Cell-mediated immunity Immune reaction mediated by T cells directed against body cells that have been infected by viruses or bacteria.

Centimorgan (cM) A unit of distance between genes on chromosomes. One centimorgan equals a value of 1% crossing-over between two genes.

Centromere A region of a chromosome to which microtubule fibers attach during cell division. The location of a centromere gives a chromosome its characteristic shape.

Cervix The lower neck of the uterus opening into the vagina.

Chorion A two-layered structure formed from the trophoblast.

Chorionic villus sampling (CVS) A method of sampling fetal chorionic cells by inserting a catheter through the vagina or abdominal wall into the uterus. Used in diagnosing biochemical and cytogenetic defects in the embryo. Usually performed in the eighth or ninth week of pregnancy.

Chromatid One of the strands of a duplicated chromosome, joined by a single centromere to its sister chromatid.

Chromatin The complex of DNA and proteins that makes up a chromosome.

Chromosomes The threadlike structures in the nucleus that carry genetic information.

Clinodactyly An autosomal dominant trait that produces a bent finger.

Clones Genetically identical molecules, cells, or organisms all derived from a single ancestor.

Clone-by-clone method A method of genome sequencing that begins with genetic and physical maps and sequences clones after they have been placed in order.

Codominance Full phenotypic expression of both members of a gene pair in the heterozygous condition.

Codon Triplets of nucleotides in mRNA that encode the information for a specific amino acid in a protein.

Color blindness Defective color vision caused by reduction or absence of visual pigments. There are three forms: red, green, and blue blindness.

Comparative genomics Compares genomes of different species in order to look for clues to the evolutionary history of genes or a species.

Complement system A chemical defense system that kills microorganisms directly, supplements the inflammatory response, and works with (complements) the immune system.

Complex traits Traits controlled by multiple genes and the interaction of environmental factors where the contributions of genes and environment are undefined.

Concordance Agreement between traits exhibited by both twins.

Continuous variation A distribution of phenotypic characters that is distributed from one extreme to another in an overlapping, or continuous, fashion.

Correlation coefficients Measures of the degree to which variables vary together.

Covalent bonds Chemical bonds that result from electron sharing between atoms. Covalent bonds are formed and broken during chemical reactions.

Cri du chat syndrome A deletion of the short arm of chromosome 5 associated with an array of congenital malformations, the most characteristic of which is an infant cry that resembles a meowing cat.

C-terminus The end of a polypeptide or protein that has a free carboxyl group.

Cystic fibrosis A fatal recessive genetic disorder associated with abnormal secretions of the exocrine glands.

Cytogenetics The branch of genetics that studies the organization and arrangement of genes and chromosomes using the techniques of microscopy.

Cytokinesis The process of cytoplasmic division that accompanies cell division.

Cytosine One of three nitrogen-containing pyrimidine bases found in nucleic acids along with thymine and uracil.

Deoxyribonucleic acid (DNA) A molecule consisting of antiparallel strands of polynucleotides that is the primary carrier of genetic information.

Deoxyribose One of two pentose sugars found in nucleic acids. Deoxyribose is found in DNA, ribose in RNA.

Dihybrid cross A cross between individuals who differ with respect to two gene pairs.

Diploid (2n) The condition in which each chromosome is represented twice as a member of a homologous pair.

Discontinuous variation Phenotypes that fall into two or more distinct, nonoverlapping classes.

Dizygotic (DZ) Twins derived from two separate and nearly simultaneous fertilizations, each involving one egg and one sperm. Such twins share, on average, 50% of their genes.

DNA A helical molecule consisting of two strands of nucleotides that is the primary carrier of genetic information.

DNA fingerprint Detection of variations in minisatellites used to identify individuals.

DNA microarray A series of short nucleotide sequences placed on a solid support (such as glass) that have several different uses, such as detection of mutant genes or differences in the pattern of gene expression in normal and cancerous cells.

DNA polymerase An enzyme that catalyzes the synthesis of DNA using a template DNA strand and nucleotides.

DNA profile The pattern of STR allele frequencies used to identify individuals.

DNA sequencing A technique for determining the nucleotide sequence of a fragment of DNA.

Dominant trait The trait expressed in the F_1 (or heterozygous) condition.

Dosage compensation A mechanism that regulates the expression of sex-linked gene products.

Ecogenetics A branch of genetics that studies genetic traits related to the response to environmental substances.

Edible vaccine A vaccine produced in fruit or vegetables by gene transfer.

Effector cells Daughter cells of B cells, which synthesize and secrete 2,000 to 20,000 antibody molecules per second into the bloodstream.

Ejaculatory duct A short connector from the vas deferens to the urethra.

Endometrium The inner lining of the uterus that is shed at menstruation if fertilization has not occurred.

Endoplasmic reticulum (ER) A system of cytoplasmic membranes arranged into sheets and channels that function in synthesizing and transporting gene products.

Enhancement gene therapy Gene transfer to enhance traits such as intelligence or athletic ability rather than to treat a genetic disorder.

Environmental variance The phenotypic variance of a trait in a population that is attributed to differences in the environment.

Epidemiology The study of the factors that control the presence, absence, or frequency of a disease.

Epididymis Where sperm are stored.

Essential amino acids Amino acids that cannot be synthesized in the body and must be supplied in the diet.

Essential hypertension Elevated blood pressure, consistently above 140/90 mm Hg.

Eugenics The attempt to improve the human species by selective breeding.

Exons DNA sequences that are transcribed, joined to other exons during mRNA processing, and translated into the amino acid sequence of a protein.

Expressivity The range of phenotypes resulting from a given genotype.

Familial adenomatous polyposis (FAP) An autosomal dominant trait resulting in the development of polyps and benign growths in the colon. Polyps often develop into malignant growths and cause cancer of the colon and/or rectum.

Familial hypercholesteremia Autosomal dominant disorder with defective or absent LDL receptors. Affected individuals are at increased risk for cardiovascular disease.

Fertilization The fusion of two gametes to produce a zygote.

Fetal alcohol syndrome (FAS) A constellation of birth defects caused by maternal alcohol consumption during pregnancy.

Fitness A measure of the relative survival and reproductive success of a given individual or genotype.

Follicles A developing egg surrounded by an outer layer of follicle cells, contained in the ovary.

Founder effects Allele frequencies established by chance in a population that is started by a small number of individuals (perhaps only a fertilized female).

Fragile X An X chromosome that carries a nonstaining gap, or break, at band q27; associated with mental retardation in males.

Frameshift mutations Mutational events in which a number of bases (other than multiples of three) are added to or removed from DNA, causing a shift in the codon reading frame.

Friedreich ataxia A progressive and fatal neurodegenerative disorder inherited as an autosomal recessive trait with symptoms appearing between puberty and the age of 25.

Galactosemia A heritable trait associated with the inability to metabolize the sugar galactose. If it is left untreated, high levels of galactose-1-phosphate accumulate, causing cataracts and mental retardation.

Gamete intrafallopian transfer (GIFT) A procedure in which gametes are collected and placed into a woman's oviduct.

Gametes Unfertilized germ cells.

Gatekeeper genes Genes that regulate cell growth and passage through the cell cycle, for example, tumor suppressor genes.

Gene pool The set of genetic information carried by the members of a sexually reproducing population.

Genes The fundamental units of heredity.

Gene therapy Procedure in which normal genes are transplanted into humans carrying defective copies as a means for treating genetic diseases.

General cognitive ability Characteristics that include verbal and spatial abilities, memory, speed of perception, and reasoning.

Genetic code The sequence of nucleotides that encodes the information for amino acids in a polypeptide chain.

Genetic counseling A process of communication that deals with the occurrence or risk that a genetic disorder will occur in a family.

Genetic drift The random fluctuations of allele frequencies from generation to generation that take place in small populations.

Genetic library In recombinant DNA terminology, a collection of clones that contains all of the genetic information in an individual.

Genetic screening The systematic search for individuals in a population who have certain genotypes.

Genetic testing The use of methods to determine if someone has a genetic disorder, will develop one, or is a carrier.

Genetic variance The phenotypic variance of a trait in a population that is attributed to genotypic differences.

Genetically modified organisms (GMOs) A general term used by the media to refer to transgenic plants or animals.

Genetics The scientific study of heredity.

Genome The set of genes carried by an individual.

Genomic imprinting Phenomenon in which the expression of a gene depends on whether it is inherited from the mother or the father. Also known as genetic or parental imprinting.

Genomics The study of the organization, function, and evolution of genomes.

Genotype The specific genetic constitution of an organism.

Germ-line therapy Gene transfer to gametes or the cells that produce them. Transfers a gene to all cells in the next generation, including germ cells.

Golgi apparatus Membranous organelles composed of a series of flattened sacs. They sort, modify, and package proteins synthesized in the ER.

Gonads Organs where gametes are produced.

Guanine One of two nitrogen-containing purine bases found in nucleic acids, along with adenine.

Haploid (n) The condition in which each chromosome is represented once in an unpaired condition.

Haplotype A cluster of closely linked genes or markers that are inherited together. In the immune system, the *HLA* alleles on chromosome 6 are a haplotype.

Hardy-Weinberg Law The statement that allele and genotype frequencies remain constant from generation to generation when the population meets certain assumptions.

Helper T cell A lymphocyte that stimulates production of antibodies by B cells when an antigen is present.

Hemizygous A gene present on the X chromosome that is expressed in males in both the recessive and dominant condition.

Hemoglobin variants Alpha and beta globins with variant amino acid sequences.

Hemolytic disease of the newborn (HDN) A condition of immunological incompatibility between mother and fetus that occurs when the mother is Rh$^-$ and the fetus is Rh$^+$.

Hereditarianism The idea that human traits are determined solely by genetic inheritance, ignoring the contribution of the environment.

Hereditary nonpolyposis colon cancer (HNPCC) A form of colon cancer associated with genomic instability of microsatellite DNA sequences.

Heritability An expression of how much of the observed variation in a phenotype is due to differences in genotype.

Hermaphrodite An individual possessing both ovaries and testes and their associated duct systems.

Heterozygous Carrying two different alleles for one or more genes.

Histamine A chemical signal produced by mast cells that triggers dilation of blood vessels.

Histones DNA-binding proteins that help compact and fold DNA into chromosomes.

Homologous chromosomes Chromosomes that physically associate (pair) during meiosis. Homologous chromosomes have identical gene loci.

Homozygous Having identical alleles for one or more genes.

Huntington disease An autosomal dominant disorder associated with progressive neural degeneration and dementia. Adult onset is followed by death 10 to 15 years after symptoms appear.

Hydrogen bond A weak chemical bonding force between hydrogen and another atom.

Hypophosphatemia An X-linked dominant disorder. Those affected have low phosphate levels in blood and skeletal deformities.

Immunoglobulins (Ig) The five classes of proteins to which antibodies belong.

In vitro fertilization (IVF) A procedure in which gametes are collected and fertilized in a dish in the laboratory, and the resulting zygote is implanted in the uterus for development.

Inborn error of metabolism The concept advanced by Archibald Garrod that many genetic traits result from alterations in biochemical pathways.

Incomplete dominance Expression of a phenotype that is intermediate between those of the parents.

Independent assortment The random distribution of genes into gametes during meiosis.

Inflammatory response The body's reaction to invading microorganisms, a nonspecific active defense mechanism that the body employs to resist infection.

Initiation complex Formed by the combination of mRNA, tRNA, and the small ribosome subunit. The first step in translation.

Inner cell mass A cluster of cells in the blastocyst that gives rise to the embryonic body.

Intelligence quotient (IQ) A score derived from standardized tests that is calculated by dividing the individual's mental age (determined by the test) by his or her chronological age and multiplying the quotient by 100.

Interphase The period of time in the cell cycle between mitotic divisions.

Intracytoplasmic sperm injection (ICSI) A treatment to overcome defects in sperm count or motility; an egg is fertilized by microinjection of a single sperm.

Introns DNA sequences present in some genes that are transcribed but are removed during processing and therefore are not present in mature mRNA.

Ionizing radiation Radiation that produces ions during interaction with other matter, including molecules in cells.

Karyotype A complete set of chromosomes from a cell that has been photographed during cell division and arranged in a standard sequence.

Killer T cells T cells that destroy body cells infected by viruses or bacteria. These cells can also directly attack viruses, bacteria, cancer cells, and cells of transplanted organs.

Klinefelter syndrome Aneuploidy of the sex chromosomes involving an XXY chromosomal constitution.

Leptin A hormone produced by fat cells that signals the brain and ovary. As fat levels become depleted, secretion of leptin slows and eventually stops.

Linkage A condition in which two or more genes do not show independent assortment. Rather, they tend to be inherited together. Such genes are located on the same chromosome. By measuring the degree of recombination between linked genes, the distance between them can be determined.

Lipoproteins Particles that have protein and phospholipid coats that transport cholesterol and other lipids in the bloodstream.

Locus The position occupied by a gene on a chromosome.

Lymphocytes White blood cells that originate in bone marrow and mediate the immune response.

Lyon hypothesis The proposal that dosage compensation in mammalian females is accomplished by partially and randomly inactivating one of the two X chromosomes.

Lysosomes Membrane-enclosed organelles that contain digestive enzymes.

Mad-cow disease A prion disease of cattle, also know as bovine spongiform encephalopathy, or BSE.

Major histocompatibility complex (MHC) A set of genes on chromosome 6 that encode recognition molecules that prevent the immune system from attacking a body's own organs and tissues.

Marfan syndrome An autosomal dominant genetic disorder that affects the skeletal system, the cardiovascular system, and the eyes.

Meiosis The process of cell division during which one cycle of chromosomal replication is followed by two successive cell divisions to produce four haploid cells.

Membrane-attack complex (MAC) A large, cylindrical multiprotein that embeds itself in the plasma membrane of an invading microorganism and creates a pore through which fluids can flow, eventually bursting the microorganism.

Memory B cell A long-lived B cell produced after exposure to an antigen that plays an important role in secondary immunity.

Messenger RNA (mRNA) A single-stranded complementary copy of the nucleotide sequence in a gene.

Metabolism The sum of all biochemical reactions by which cells convert and utilize energy.

Metacentric Describes a chromosome that has a centrally placed centromere.

Metaphase A stage in mitosis during which the chromosomes move and become arranged near the middle of the cell.

Metastasis A process whereby cells detach from the primary tumor and move to other sites, forming new malignant tumors in the body.

Millirem Each rem is equal to 1,000 millirems.

Minisatellite Nucleotide sequences 14 to 100 base pairs long organized into clusters of varying lengths; used in construction of DNA fingerprints.

Missense mutations Mutations that cause the substitution of one amino acid for another in a protein.

Mitochondria (singular: mitochondrion) Membrane-bound organelles, present in the cytoplasm of all eukaryotic cells, that are the sites of energy production within the cells.

Mitosis Form of cell division that produces two cells, each of which has the same complement of chromosomes as the parent cell.

Molecular genetics The study of genetic events at the biochemical level.

Molecules Structures composed of two or more atoms held together by chemical bonds.

Monohybrid cross A cross between individuals who differ with respect to a single gene pair.

Monosomy A condition in which one member of a chromosomal pair is missing; having one less than the diploid number ($2n - 1$).

Monozygotic (MZ) Twins derived from a single fertilization involving one egg and one sperm; such twins are genetically identical.

Mood A sustained emotion that influences perception of the world.

Mood disorders A group of behavior disorders associated with manic and/or depressive syndromes.

Müllerian inhibiting hormone (MIH) A hormone produced by the developing testis that causes the breakdown of the Müllerian ducts in the embryo.

Multifactorial traits Traits that result from the interaction of one or more environmental factors and two or more genes.

Multiple alleles Genes that have more than two alleles.

Multipotent The restricted ability of a stem cell to form only one or a few different cell types.

Muscular dystrophy A group of genetic diseases associated with progressive degeneration of muscles. Two of these, Duchenne and Becker muscular dystrophy, are inherited as X-linked, allelic, recessive traits.

Mutation rate The number of events that produce mutated alleles per locus per generation.

Natural selection The differential reproduction shown by some members of a population that is the result of differences in fitness.

Nitrogen-containing base A purine or pyrimidine that is a component of nucleotides.

Nondisjunction The failure of homologous chromosomes to separate properly during meiosis or mitosis.

Nonsense mutations Mutations that change an amino acid specifying a codon to one of the three termination codons.

N-terminus The end of a polypeptide or protein that has a free amino group.

Nucleolus (plural: nucleoli) A nuclear region that functions in the synthesis of ribosomes.

Nucleosomes A bead-like structure composed of histones wrapped with DNA.

Nucleotide The basic building block of DNA and RNA. Each nucleotide consists of a base, a phosphate, and a sugar.

Nucleotide substitutions Mutations that involve substitutions of one or more nucleotides in a DNA molecule.

Nucleus The membrane-bounded organelle in eukaryotic cells that contains the chromosomes.

Oncogenes Genes that induce or continue uncontrolled cell proliferation.

Oocyte Female gamete.

Oogenesis The process of oocyte production.

Oogonia Mitotically active cells that produce primary oocytes.

Open reading frame (ORF) The codons in a gene that encode the amino acids of the gene product.

Organelles Cytoplasmic structures that have a specialized function.

Ovaries Female gonads that produce oocytes and female sex hormones.

Oviduct A duct with fingerlike projections partially surrounding the ovary and connecting to the uterus. Also called the fallopian or uterine tube.

Ovulation The release of a secondary oocyte from the follicle; usually occurring monthly during a female's reproductive lifetime.

Palindrome A word, phrase, or sentence that reads identically in both directions. In DNA, a sequence of nucleotides that reads identically on both strands when read in the 5' to 3' direction.

Pathogens Disease-causing agents.

Pattern baldness A sex-influenced trait that acts like an autosomal dominant trait in males and an autosomal recessive trait in females.

Pedigree A diagram listing the members and ancestral relationships in a family; used in the study of human heredity.

Pedigree analysis The construction of family trees and their use to follow the transmission of genetic traits in families. It is the basic method of studying the inheritance of traits in humans.

Pedigree construction Use of family history to determine how a trait is inherited and to determine risk factors for family members.

Penetrance The probability that a disease phenotype will appear when a disease-related genotype is present.

Pentose sugar A five-carbon sugar molecule found in nucleic acids.

Peptide bond A covalent chemical link between the carboxyl group of one amino acid and the amino group of another amino acid.

Pharmacogenetics A branch of genetics concerned with the inheritance of differences in the response to drugs.

Pharmacogenomics Analyzes genes and proteins to identify targets for therapeutic drugs.

Phenotype The observable properties of an organism.

Phenylketonuria (PKU) An autosomal recessive disorder of amino acid metabolism that results in mental retardation if untreated.

Philadelphia chromosome An abnormal chromosome produced by translocation of parts of the long arms of chromosomes 9 and 22.

Phosphate group A compound containing phosphorus chemically bonded to four oxygen molecules.

Pluripotent The ability of a stem cell to form most of the cell types in the body.

Polar body biopsy Removing a polar body by micromanipulation to test whether it or the oocyte carries an allele for an X-linked recessive disorder. This test can be done prior to fertilization.

Poly-A tail A series of A nucleotides added to the 3′ end of mRNA molecules.

Polygenic traits Traits controlled by two or more genes.

Polymerase chain reaction (PCR) A method for amplifying DNA segments using cycles of denaturation, annealing to primers, and DNA-polymerase directed DNA synthesis.

Polypeptide A molecule made of amino acids joined together by peptide bonds.

Polyploidy A chromosomal number that is a multiple of the normal haploid chromosomal set.

Polyps Growths attached to the substrate by small stalks. Commonly found in the nose, rectum, and uterus.

Population genetics The branch of genetics that studies inherited variation in populations of individuals and the forces that alter gene frequency.

Populations Local groups of organisms belonging to a single species, sharing a common gene pool.

Porphyria A genetic disorder inherited as a dominant trait that leads to intermittent attacks of pain and dementia. Symptoms first appear in adulthood.

Positional cloning A recombinant DNA-based method of mapping and cloning genes with no prior information about the gene product or its function.

Precocious puberty An autosomal dominant trait expressed in a sex-limited fashion. Heterozygous males are affected, but heterozygous females are not.

Primary structure The amino acid sequence in a polypeptide chain.

Prion A protein folded into an infectious conformation that is the cause of several disorders, including Creutzfeldt-Jakob disease and mad-cow disease.

Proband First affected family member who seeks medical attention for a genetic disorder.

Probe A labeled nucleic acid used to identify a complementary region in a clone or genome.

Product The specific chemical compound that is the result of enzymatic action. In biochemical pathways, a compound can serve as the product of one reaction and the substrate for the next reaction.

Promoter region A region of a DNA molecule to which RNA polymerase binds and initiates transcription.

Prophase A stage in mitosis during which the chromosomes become visible and split longitudinally except at the centromere.

Prostaglandins Locally acting chemical messengers that stimulate contraction of the female reproductive system to assist in sperm movement.

Prostate gland A gland that secretes a milky, alkaline fluid that neutralizes acidic vaginal secretions and enhances sperm viability.

Proteome The set of proteins present in a particular cell at a given time under a given set of environmental conditions.

Proteomics The study of the expressed proteins present in a cell at a given time under a given set of circumstances.

Proto-oncogenes Genes that initiate or maintain cell division and that may become cancer genes (oncogenes) by mutation.

Pseudogenes Nonfunctional genes that are closely related (by DNA sequence) to functional genes present elsewhere in the genome.

Pseudohermaphroditism An autosomal genetic condition that causes XY individuals to develop the phenotypic sex of females.

Purine A class of double-ringed organic bases found in nucleic acids.

Pyrimidine A class of single-ringed organic bases found in nucleic acids.

Quantitative trait loci (QTLs) Two or more genes that act on a single polygenic trait.

Quaternary structure The structure formed by the interaction of two or more polypeptide chains in a protein.

R group A term used to indicate the position of an unspecified group in a chemical structure. They can be positively or negatively charged or neutral.

Radiation The process by which electromagnetic energy travels through space or a medium such as air.

Recessive trait The trait unexpressed in the F_1 but reexpressed in some members of the F_2 generation.

Recombinant DNA technology A series of techniques in which DNA fragments are linked to self-replicating vectors to create recombinant DNA molecules, which are replicated in a host cell.

Regression to the mean In a polygenic system, the tendency of offspring of parents who have extreme differences in phenotype to exhibit a phenotype that is the average of the two parental phenotypes.

Rem The unit of radiation exposure used to measure radiation damage in humans. It is the amount of ionizing radiation that has the same effect as a standard amount of x-rays.

Restriction enzyme A bacterial enzyme that cuts DNA at specific sites.

Retinoblastoma A malignant tumor of the eye arising in retinoblasts (embryonic retinal cells that disappear at about 2 years of age). Because mature retinal cells do not transform into tumors, this is a tumor that usually only occurs in children. It is associated with a deletion on the long arm of chromosome 13.

Ribonucleic acid (RNA) A nucleic acid molecule that contains the pyrimidine uracil and the sugar ribose. The several forms of RNA function in gene expression.

Ribose One of two pentose sugars found in nucleic acids. Deoxyribose is found in DNA, ribose in RNA.

Ribosomal RNA (rRNA) RNA molecules that form part of the ribosome.

Ribosomes Cytoplasmic particles composed of two subunits that are the site of protein synthesis.

Schizophrenia A behavioral disorder characterized by disordered thought processes and withdrawal from reality. Genetic and environmental factors are involved in this disease.

Scrotum A pouch of skin outside the male body that contains the testes.

Secondary structure The pleated or helical structure in a protein molecule that is brought about by the formation of bonds between amino acids.

Segregation The separation of members of a gene pair from each other during gamete formation.

Semen A mixture of sperm and various glandular secretions containing 5% spermatozoa.

Semiconservative replication A model of DNA replication that provides each daughter molecule with one old strand and one newly synthesized strand. DNA replicates in this fashion.

Seminal vesicles Glands that secrete fructose and prostaglandins into the sperm.

Seminiferous tubules Small, tightly coiled tubes inside the testes where sperm are produced.

Sense mutations Mutations that change a termination codon into one that codes for an amino acid. Such mutations produce elongated proteins.

Severe combined immunodeficiency disease (SCID) A genetic disorder in which affected individuals have no immune response; both the cell-mediated and antibody-mediated responses are missing.

Sex chromosomes In humans, the X and Y chromosomes that are involved in sex determination.

Sex-influenced traits Traits controlled by autosomal genes that are usually dominant in one sex but recessive in the other sex.

Sex-limited genes Loci that produce a phenotype in only one sex.

Sex ratio The proportion of males to females, which changes throughout the life cycle. The ratio is close to 1:1 at fertilization but the female to male ratio increases as a population ages.

Short tandem repeat (STR) Short nucleotide sequences 2 to 9 base pairs long organized into clusters of varying lengths; used in the construction of DNA profiles.

Shotgun cloning A method of genome sequencing that selects clones at random from a genomic library, and after sequencing them, assembles the genome sequence using software analysis.

Sickle cell anemia A recessive genetic disorder associated with an abnormal type of hemoglobin, a blood transport protein.

Sister chromatids Two chromatids joined by a common centromere. Each chromatid carries identical genetic information.

Somatic cell nuclear transfer A cloning technique that transfers a somatic cell nucleus to an enucleated egg, which is stimulated to develop into an embryo. Inner cell mass cells are collected from the embryo and grown to form a population of stem cells. Also called therapeutic cloning.

Somatic gene therapy Gene transfer to somatic target cells to correct a genetic disorder.

Southern blot A method for transferring DNA fragments from a gel to a membrane filter, developed by Edward Southern for use in hybridization experiments.

Sperm Male gamete.

Spermatocytes Diploid cells that undergo meiosis to form haploid spermatids.

Spermatogenesis The process of sperm production.

SRY A gene, called the sex-determining region of the Y, located near the end of the short arm of the Y chromosome, plays a major role in causing the undifferentiated gonad to develop into a testis.

Start codon A codon present in mRNA that signals the location for translation to begin. The codon AUG functions as a start codon.

Stem cells Cells in bone marrow that produce lymphocytes by mitotic division.

Stop codon A codon present in mRNA that signals the end of a growing polypeptide chain. The codons UAG, UGA, and UAA function as stop codons.

Structural genomics Derives three-dimensional structures for proteins.

Submetacentric Describes a chromosome whose centromere is placed closer to one end than the other.

Substrate The specific chemical compound that is acted upon by an enzyme.

Suppressor T cells T cells that slow or stop the immune response of B cells and other T cells.

T-cell receptors (TCRs) Unique proteins on the surface of T-cells that bind to specific proteins on the surface of cells infected with viruses, bacteria, or intracellular parasites.

T-cells A type of lymphocyte that undergoes maturation in the thymus and mediates cellular immunity.

Telophase The last stage of mitosis, during which division of the cytoplasm occurs, the chromosomes of the daughter cells disperse, and the nucleus re-forms.

Template The single-stranded DNA that serves to specify the nucleotide sequence of a newly synthesized polynucleotide strand.

Teratogen Any physical or chemical agent that brings about an increase in congenital malformations.

Terminator region The nucleotide sequence at the end of a gene that signals the end of transcription.

Tertiary structure The three-dimensional structure of a protein molecule brought about by folding on itself.

Testes Male gonads that produce spermatozoa and sex hormones.

Testosterone A steroid hormone produced by the testis; the male sex hormone.

Tetraploidy A chromosomal number that is four times the haploid number, having four copies of all autosomes and four sex chromosomes.

Thalassemias Disorders associated with an imbalance in the production of alpha or beta globin.

Thymine One of three nitrogen-containing pyrimidine bases found in nucleic acids, along with uracil and cytosine.

Thymine dimer A molecular lesion in which chemical bonds form between a pair of adjacent thymine bases in a DNA molecule.

Totipotent The ability of a stem cell to form every cell type in the body; characteristic of embryonic stem cells.

Tourette syndrome (GTS) A behavioral disorder characterized by motor and vocal tics and inappropriate language. Genetic components are suggested by family studies that show increased risk for relatives of affected individuals.

Trait Any observable property of an organism.

Transcription Transfer of genetic information from the base sequence of DNA to the base sequence of RNA, mediated by RNA synthesis.

Transfer RNA (tRNA) A small RNA molecule that contains a binding site for a specific type of amino acid and a three-base segment known as an anticodon that recognizes a specific base sequence in messenger RNA.

Transformation The process of transferring genetic information between cells by DNA molecules.

Transforming factor The molecular agent of transformation; DNA.

Transgenic Refers to the transfer of genes between species by recombinant DNA technology; transgenic organisms have received such a gene.

Translation Conversion of information encoded in the nucleotide sequence of an mRNA molecule into the linear sequence of amino acids in a protein.

Transmission genetics The branch of genetics concerned with the mechanisms by which genes are transferred from parent to offspring.

Trinucleotide repeats A form of mutation associated with the expansion in copy number of a nucleotide triplet in or near a gene.

Triploidy A chromosomal number that is three times the haploid number, having three copies of all autosomes and three sex chromosomes.

Trisomy A condition in which one chromosome is present in three copies, whereas all others are diploid; having one more than the diploid number $(2n + 1)$.

Trisomy 21 Aneuploidy involving the presence of an extra copy of chromosome 21, resulting in Down syndrome.

Trophoblast The outer layer of cells in the blastocyst that gives rise to the membranes surrounding the embryo.

Tubal ligation A contraceptive procedure for women in which the oviducts are cut, preventing eggs from reaching the uterus.

Tumor suppressor genes Genes encoding proteins that suppress cell division.

Turner syndrome A monosomy of the X chromosome (45,X) that results in female sterility.

Uniparental disomy A condition in which both copies of a chromosome are inherited from one parent.

Unipolar disorder An emotional disorder characterized by prolonged periods of deep depression.

Uracil One of three nitrogen-containing pyrimidine bases found in nucleic acids, along with thymine and cytosine.

Urethra A tube that passes from the bladder and opens to the outside. It functions in urine transport and, in males, also carries sperm.

Uterus A hollow, pear-shaped muscular organ where a fertilized egg will develop.

Vaccine A preparation containing dead or weakened pathogens that elicits an immune response when injected into the body.

Vagina The opening that receives the penis during intercourse and also serves as the birth canal.

Vas deferens A duct connected to the epididymis, which sperm travels through.

Vasectomy A contraceptive procedure for men in which each vas deferens is cut and sealed to prevent the transport of sperm.

Vectors Self-replicating DNA molecules that are used to transfer foreign DNA segments between host cells.

X inactivation center (Xic) A region on the X chromosome where inactivation begins.

X-linked The pattern of inheritance that results from genes located on the X chromosome.

X-linked agammaglobulinemia (XLA) A rare, sex-linked, recessive trait characterized by the total absence of immunoglobulins and B cells.

Xenotransplant Organ transplants between species.

XYY karyotype Aneuploidy of the sex chromosomes involving XYY chromosomal constitution.

Y-linked The pattern of inheritance that results from genes located only on the Y chromosome.

Yeast artificial chromosome (YAC) A cloning vector that has telomeres and a centromere that can accommodate large DNA inserts and uses the eukaryote yeast as a host cell.

Zygote The fertilized egg that develops into a new individual.

Credits

Nature Reviews Genetics 2:280–291, Table 1, p. 283, with permission. Copyright 2001 Macmillan Magazines, Ltd.

CHAPTER 7

Opener David M. Phillips/Photo Researchers

Active Figure 7.1 Art by Raychel Ciemma and Lisa Starr from *Biology: Today and Tomorrow* by Cecie Starr, 2005, Brooks/Cole.

Active Figure 7.2 Art by Raychel Ciemma from *Biology: Today and Tomorrow* by Cecie Starr, 2005, Brooks/Cole.

Active Figure 7.3 Art by Raychel Ciemma from *Biology: Today and Tomorrow* by Cecie Starr, 2005, Brooks/Cole.

Active Figure 7.4 Art by Raychel Ciemma from *Biology: Today and Tomorrow* by Cecie Starr, 2005, Brooks/Cole; photo by Lennart Nilsson from *A Child Is Born*, © 1966, 1977 Dell Publishing Company, Inc.

Active Figure 7.5 Art by Raychel Ciemma from *Biology: Today and Tomorrow* by Cecie Starr, 2005, Brooks/Cole; photo by Don W. Fawcett/Photo Researchers

Active Figure 7.6 Art by Raychel Ciemma from *Biology: Today and Tomorrow* by Cecie Starr, 2005, Brooks/Cole.

Figure 7.7 Art by Raychel Ciemma from *Biology: Today and Tomorrow* by Cecie Starr, 2005, Brooks/Cole; photos by Lennart Nilsson from *A Child Is Born*, © 1966, 1977 Dell Publishing Company, Inc.

Active Figure 7.8 Art from *Biology: The Unity and Diversity of Life* by Cecie Starr, 2004, Brooks/Cole.

Figure 7.9 Courtesy of Dr. Marilyn Miller, University of Illinois at Chicago

Active Figure 7.10 Art by Precision Graphics and Gary Head from *Biology: The Unity and Diversity of Life* by Cecie Starr, 2004, Brooks/Cole; photos (left to right) 2001 PhotoDisc; 2001 Eye Wire

Figure 7.11 Sandra Wavick/Photonica

Figure 7.12 Art from *Biology: The Unity and Diversity of Life* by Cecie Starr, 2004, Brooks/Cole; photo by Lennart Nilsson from *A Child Is Born*, © 1966, 1977 Dell Publishing Company, Inc.

Figure 7.14 From P. Zourlas et al. (1965). Clinical histologic and cytogenetic findings in male hermaphroditism. *Obstetrics and Gynecology* 25:768-778, Figure 7.

Figure 7.15a Visuals Unlimited

Figure 7.15b M. Abbey/Photo Researchers

Active Figure 7.16 Myrleen Ferguson Cate/Photo Edit

Figure 7.17 Art by Raychel Ciemma and Gary Head from *Biology: The Unity and Diversity of Life* by Cecie Starr, 2004, Brooks/Cole; photo by Dr. Karen Dyer Montgomery

Figure 7.18 Courtesy of Dr. Neil Brockdorff, Imperial College of Medicine, London. From S. Duthie et al. (1999). Xist RNA exhibits a banded localization of the inactive X chromosome and is excluded from autosomal material in *cis. Human Molecular Genetics* 8:195–204.

Figure 7.20 Leonard Lessin/Peter Arnold, Inc.

CHAPTER 8

Opener Courtesy of Dmitry Pruss, National Institutes of Health

Active Figure 8.1 Art by Raychel Ciemma and Gary Head from *Biology: The Unity and Diversity of Life* by Cecie Starr, 2004, Brooks/Cole.

Figure 8.2b © Lee D. Simon/Science Source/Photo Researchers

Active Figure 8.3 Art by Lisa Starr from *Biology: Today and Tomorrow* by Cecie Starr, 2005, Brooks/Cole.

Figure 8.6 Art by Gary Head from *Biology: Today and Tomorrow* by Cecie Starr, 2005, Brooks/Cole.

Figure 8.8 From R. Franklin and R. G. Gosling (1953). Molecular configuration in sodium thymonucleate. *Nature* 171:740–741.

Active Figure 8.9 Art by Lisa Starr from *Biology: The Unity and Diversity of Life* by Cecie Starr, 2004, Brooks/Cole; Lisa Starr with PDB ID: 1BBB

Figure 8.12 Art by Raychel Ciemma and Gary Head from *Biology: Today and Tomorrow* by Cecie Starr, 2005, Brooks/Cole.

Active Figure 8.13 Art by Lisa Starr and Raychel Ciemma from *Biology: The Unity and Diversity of Life* by Cecie Starr, 2004, Brooks/Cole.

Figure 8.14 Photo by Thomas Cremer. Courtesy of William C. Earnshaw from A. I. Lamond and W. C. Earnshaw, Structure and function in the nucleus. Reprinted with permission from *Science* 280:547–553. © American Association for the Advancement of Science.

Active Figure 8.15 Art by Precision Graphics from *Biology: Today and Tomorrow* by Cecie Starr, 2005, Brooks/Cole.

CHAPTER 9

Opener Science Photo Library/Photo Researchers

Figure 9.2 Dr. Namboori B. Raju, Stanford University

Active Figure 9.5 Art by Lisa Starr from *Biology: The Unity and Diversity of Life* by Cecie Starr, 2004, Brooks/Cole.

Active Figure 9.7 Art by Gary Head from *Biology: The Unity and Diversity of Life* by Cecie Starr, 2004, Brooks/Cole.

Figure 9.9 Art by Lisa Starr from *Biology: Today and Tomorrow* by Cecie Starr, 2005, Brooks/Cole.

Figure 9.10 Art by Lisa Starr from *Biology: Today and Tomorrow* by Cecie Starr, 2005, Brooks/Cole.

Active Figure 9.11 Art by Lisa Starr from *Biology: Today and Tomorrow* by Cecie Starr, 2005, Brooks/Cole.

Active Figure 9.12 Art by Lisa Starr from *Biology: The Unity and Diversity of Life* by Cecie Starr, 2004, Brooks/Cole.

Figure 9.13a Photo by David M. Allen

Figure 9.14 Lisa Starr with PDB ID: 1DGF

CHAPTER 10

Opener Walter Reinhart/Phototake

Figure 10.1 Museo del Prado, Madrid, Spain/Giraudon, Paris/Superstock

Figure 10.5 Photo Researchers

Active Figure 10.12 Art by Preface, Inc. from *Biology: The Unity and Diversity of Life* by Cecie Starr, 2004, Brooks/Cole; photos by Stanley Fleger/Visuals Unlimited

Figure 10.13 B. Carragher, D. Bluemke, and R. Josephs, Electron Microscope and Image Processing Laboratory, University of Chicago

Active Figure 10.14 Art by Lisa Starr and Gary Head from *Biology: The Unity and Diversity of Life* by Cecie Starr, 2004, Brooks/Cole.

Figure 10.17 Grant Heilman Photography

Figure 10.18a Ray Coleman/Photo Researchers

Figure 10.18b Ken Brate/Photo Researchers

CHAPTER 11

Opener Lior Rubin/Peter Arnold, Inc.

Figure 11.3 Giraudon/Art Resource

Figure 11.4 © Corbis-Sygma

Active Figure 11.10 Art by Precision Graphics from *Biology: The Unity and Diversity of Life* by Cecie Starr, 2004, Brooks/Cole.

Active Figure 11.11 Art by Lisa Starr from *Biology: The Unity and Diversity of Life* by Cecie Starr, 2004, Brooks/Cole.

Active Figure 11.14 Art by Lisa Starr from *Biology: The Unity and Diversity of Life* by Cecie Starr, 2004, Brooks/Cole.

Figure 11.16 Kenneth E. Greer/Visuals Unlimited

CHAPTER 12

Opener Nancy Kedersha/Science Photo Library/Photo Researchers

Figure 12.1 Visuals Unlimited

Active Figure 12.2 Art from *Biology: The Unity and Diversity of Life* by Cecie Starr, 2004, Brooks/Cole.

Figure 12.4a Science Photo Library/Photo Researchers

Figure 12.4b (left) Ken Greer/Visuals Unlimited; (middle) Biophoto Associates/Photo Researchers; (right) James Stevenson/SPL/Photo Researchers

Figure 12.5 Art by Raychel Ciemma and Gary Head from *Biology: Today and Tomorrow* by Cecie Starr, 2005, Brooks/Cole.

Figure 12.6 Custom Medical Stock Photo

Figure 12.11 Benjamin/Custom Medical Stock Photo

Figure 12.13 Drs. Michael Spelcher and David Ward, Yale University

CHAPTER 13

Opener David Parker/Science Photo Library/Photo Researchers

Figure 13.2 E. Webber/Visuals Unlimited

Figure 13.4 Le Corre-Ribeiro/Liaison/Getty Images

Figure 13.5b May 27, 1999 by Dave Au. Courtesy University of Hawaii

Figure 13.9 SPL/Custom Medical Stock Photo

Active Figure 13.11 Art by Lisa Starr from *Biology: Today and Tomorrow* by Cecie Starr, 2005, Brooks/Cole.

Figure 13.12 Michael Gabridge/Custom Medical Stock Photo

Active Figure 13.15 Art by Lisa Starr from *Biology: The Unity and Diversity of Life* by Cecie Starr, 2004, Brooks/Cole.

Active Figure 13.16 Art by Gary Head from *Biology: Today and Tomorrow* by Cecie Starr, 2005, Brooks/Cole; photo by David Parker/SPL/Photo Researchers

Figure 13.17 Alfred Pasieka/SPL/Photo Researchers

Figure 13.18 1992 NIR. All rights reserved.

Active Figure 13.20 Art by Lisa Starr from *Biology: Today and Tomorrow* by Cecie Starr, 2005, Brooks/Cole; photo by TEK IMAGE/Photo Researchers

CHAPTER 14

Opener Courtesy CropTech Corp.

Figure 14.2 Courtesy of Drs. Ans Van der Ploeg and Arnold Reuser, ErasmusMC, Rotterdam, The Netherlands and Pharming Group N.V., Leiden, The Netherlands

Active Figure 14.3 (top) Herve Chaumeton/Agence Nature; (bottom) Lowell Georgia/Corbis

Figure 14.4 Agricultural Research Service/USDA

Figure 14.5 Reprinted with permission of the Department of Soil and Crop Sciences, Colorado State University. Available at: http://www.colostate.edu/programs/lifesciences/TransgenicCrops/current.html

Figure 14.6 Professor Ingo Potrykus, Chair of the Golden Rice Humanitarian Board

Figure 14.7 EURELIOUS/Phototake

Figure 14.9 Sergei Evsikov, Ph.D., ViaGene Fertility, LLC

Figure 14.10 Sergei Evsikov, Ph.D., ViaGene Fertility, LLC

Figure 14.11 Courtesy of Affymetrix, Inc., Santa Clara, California

Figure 14.12 Courtesy of Tyson Clark, Department of Molecular, Cellular, and Developmental Biology, University of California, Santa Clara

Figure 14.15 (top) David Parker/SPL/Photo Researchers; (bottom) Cellmark Diagnostics, Abingdon, U.K.

unnumbered photo, page 334 UPI/Corbis

Figure 14.16a Cydney Conger/Corbis

Figure 14.16b Yann Arthus-Bertrand/Corbis

CHAPTER 15

Opener Celera, Inc.

Figure 15.5 Source: Adapted from International Human Genome Sequencing Consortium. 2001. Initial sequencing and analysis of the human genome. *Nature* 409:860–921, Fig. 1, p. 862. Adapted with permission from *Nature*. Copyright 2001, Macmillan Magazines Limited.

Figure 15.6 Courtesy the author

Figure 15.7 Volker Steger/SPL/Photo Researchers

Figure 15.12 From J. C. Venter et al. (2001). The sequence of the human genome. *Science* 291:1304–1351, Fig. 15, p. 1335. Copyright 2001 American Association for the Advancement of Science. Reprinted with permission.

Figure 15.14 Courtesy of Sirano Dhe-Paganon and Steven Shoelson

Figure 15.15 Dr. John N. Weinstein, M.D., Ph.D., Laboratory of Molecular Pharmacology, NCI

Figure 15.16 From F. Collins et al. (2003). A vision for the future of genomics research. *Nature* 442:835–847, Fig. 2, p. 836. Reprinted with permission. www.nature.com.

CHAPTER 16

Opener Lennart Nilsson

Figure 16.3 Sovereign/Phototake

Figure 16.4 Courtesy of the Nash Family

Figure 16.6 Van De Silva, courtesy Dr. W. French Anderson, Director Gene Therapy Laboratory, Keck School of Medicine, University of Southern California

Figure 16.7 From M. Edelstein et al. (2004). Gene therapy clinical trials worldwide 1989–2004—an overview. *Journal of Gene Medicine* 6:597–602. Used by permission of John Wiley & Sons, Ltd.

Figure 16.8b James Thomson Research Laboratory, University of Wisconsin–Madison

Figure 16.9 Martha Cooper/Peter Arnold, Inc.

CHAPTER 17

Opener J. L. Carson/Custom Medical Stock

Figure 17.1 Robert R. Dourmashkin, courtesy of Clinical Research Centre, Harrow, England

Active Figure 17.2 Art by Lisa Starr from *Biology: The Unity and Diversity of Life* by Cecie Starr, 2004, Brooks/Cole.

Active Figure 17.3 Art by Raychel Ciemma from *Biology: The Unity and Diversity of Life* by Cecie Starr, 2004, Brooks/Cole.

Active Figure 17.4 Art by Lisa Starr from *Biology: The Unity and Diversity of Life* by Cecie Starr, 2004, Brooks/Cole.

Active Figure 17.5 Art by Lisa Starr from *Biology: Today and Tomorrow* by Cecie Starr, 2005, Brooks/Cole.

Active Figure 17.6 Art by Lisa Starr from *Biology: Today and Tomorrow* by Cecie Starr, 2005, Brooks/Cole.

Figure 17.7 Courtesy of Dorothea Zucker-Franklin, New York University School of Medicine

Active Figure 17.8 Art by Lisa Starr from *Biology: The Unity and Diversity of Life* by Cecie Starr, 2004, Brooks/Cole.

Active Figure 17.9 Art by Lisa Starr from *Biology: Today and Tomorrow* by Cecie Starr, 2005, Brooks/Cole.

Figure 17.10 Jean Claude Revi/Phototake

Active Figure 17.12 Nadine Sokol after G. J. Tortora and N. P. Anagnostakos, *Principles of Anatomy and Physiology*, 6th ed. © 1990 by Biological Sciences Textbook, Inc., A&P Textbooks, Inc., and Elia-Sparta, Inc. Reprinted with permission of John Wiley & Sons, Inc.

Figure 17.16 Baylor College of Medicine/Peter Arnold, Inc.

Active Figure 17.17 Art by Raychel Ciemma from *Biology: Today and Tomorrow* by Cecie Starr, 2005, Brooks/Cole; photo by NIBSC/Photo Researchers

CHAPTER 18

Opener Tim Beddow/SPL/Photo Researchers

Figure 18.1 Columbus Instruments/Visuals Unlimited

Figure 18.2 Jonathan Flint, Wellcome Trust Centre for Human Genetics, Oxford University

Figure 18.3 Courtesy of Dr. Donald Price, Johns Hopkins University

Figure 18.4 Malcolm S. Kirk/Peter Arnold, Inc.

Figure 18.5 Courtesy of P. Hemachandra Reddy, Neurological Sciences Institute, Oregon Health and Science University

Figure 18.6 From C. Lai et al. (2000). The *SPCH1* region on human 7q31: Genomic characterization of the critical interval and localization of translocations associated with speech and language disorder. *American Journal of Human Genetics* 67:357–368, Figure 1, p. 358. Adapted with permission from University of Chicago Press.

Active Figure 18.8 Art by Lisa Starr from *Biology: The Unity and Diversity of Life* by Cecie Starr, 2004, Brooks/Cole; photo by Dr. Constantino Sotelo.

Figure 18.9 From N. Craddock et al. (1995). Mathematical limits of multilocus models: The genetic transmission of bipolar disorder. *American Journal of Human Genetics*, 57:690–702. Reprinted with permission of University of Chicago Press.

Figure 18.10 From J. B. Potash and J. R. DePaulo (2000). Searching high and low: A review of the genetics of bipolar disorder. *Bipolar Disorders*, 2:8–26, Figure 3, p. 14.

Figure 18.11 Corbis

Figure 18.12 Dr. Monte Buchsbaum, Mt. Sinai Medical Center, New York

Figure 18.14 From Y. Hayak et al. Genome-wide expression analysis reveals dysregulation of myelination-related genes in chronic schizophrenia. *Proceedings of the National Academy of Sciences* 98:4746–4751. Used by permission of the National Academy of Sciences.

Figure 18.15 Dr. Dennis Selkoe, Center for Neurologic Diseases, Harvard Medical School, Boston, MA

CHAPTER 19

Opener Tim Davis/Stone/Getty Images

Figure 19.1 C. Mayhew and R. Simmon/(NASA/GSFC) NOAA/NGDC, DMSP Digital Archive

Active Figure 19.2 Art from *Biology: Today and Tomorrow* by Cecie Starr, 2005, Brooks/Cole.

Figure 19.3a Michael Newman/Photo Edit

Figure 19.3b Fred E. Cohen, M.D., Ph.D., and Cedric Govaerts, Ph.D., Department of Cellular and Molecular Pharmacology, University of California, San Francisco

Figure 19.8 Meckes/Ottawa/Photo Researchers

Figure 19.11 Clockwise from top right: Cameron/Corbis; David Turnley/Corbis; Richard Nowitz/Corbis; Michael Yamashita/Corbis; Robert Essel NYC/Corbis; Owen Franken/Corbis; Roger Ressmeyer/Corbis; Penny Tweedle/Corbis

Index

Italic page numbers indicate material in tables or figures.

Garrod, Archibald, 199, 221, 223
gatekeeper genes, *279, 279–280*
Gaucher disease, 18–19, 22
G-banding, 126, *126*
Gelsinger, Jesse, 373, 375
gene expansion and anticipation, *258, 258–259*
gene expression
    antibiotics and protein synthesis, 207
    anticipation and, *258, 258–259*
    cases, 214
    differential. *See* genomic imprinting
    genetic code, *201–202, 202*
    metabolism and genes, *199–200, 200*
    polypeptide folding, 210
    prion diseases, 198–199
    protein structure and function
        functions of proteins, 212, *213*
        levels of structure, 210, *211*
        protein folding and disease, *211–212, 212*
    relationship between genes and enzymes, 200
    storage of genetic information, 200–201
    transcription
        introns and exons, 204, *204*
        pre-mRNAs, *204–205, 205*
        stages, 203, *203*
    transfer of genetic information, 202, *203*
    translation
        interactions, *205–209, 205–208*
        steps, 208–209
    variations shown in pedigree analysis
        changes with age, 85–87
        penetrance and expressivity, 87–88, *88*
        of X chromosome genes
            dosage compensation, *171, 171–172*
            mosaic pattern of expression, *172–173, 173*
            X chromosome inactivation, *173–174, 174*
gene pool, 435–436
general cognitive ability, 115
genes and genetics, 24
    approaches to study, 5–6
    artistic depictions of disorders, 2
    cancer and. *See* cancer
    definition, 1–2
    described, 3
    expression of genes. *See* gene expression
    impact of, 11–13
    patenting of genes, 194
    social policy and law. *See* public policy
    trait determination and, *46–47, 47*
    transmission from parents to offspring, 4, *5*
    transmission through generations. *See* heredity studies; pedigree analysis
    use in research, 6–7
    variations in genes. *See* mutations
gene switching to treat hemoglobin disorders, *234–235, 235*
gene therapy
    early successes, 373
    ethical concerns, 374–376
    gene transfer strategies, 372, *373*
    setbacks and restarts, *373–374, 374*
    techniques, 374, *375*
    testing of newborns, *12, 12–13*, 219
genetically modified foods
    concerns about, 322–323
    transfer of genes into crops, *319, 319–320*
    transgenic crops as herbicides, *320, 320–321*
    transgenic crops to enhance nutritional value, *322, 322*
genetically modified organisms (GMOs), 320
genetic code, 3, *201–202, 202*. *See also* DNA
genetic counseling, 376–378
genetic disorders. *See also specific disorders*
    associated with accelerated aging, *29, 29*

of the blood, 42–43, 86
    cell membranes and, 20
    in culture and art, 2
    DNA repair systems and, *260, 261*
    embryo testing, *326–327, 327*
    genomics used to study, *355, 355–356*
    lysosomes and, 189
    ownership of a test, 328
    prenatal testing, 325–326
    prenatal testing risks, 327
    presymptomatic testing, 329
    protein folding and disease, 211
    screening of newborns, *12, 12–13*, 43, 219
genetic drift, *444–445, 445*
genetic engineering, 397
genetic library, *301–302, 303, 303–304, 304*, 349–350
genetic mapping
    linkage, *343–344, 344*
    positional cloning, *345–346, 346, 347*
    recombination frequencies, *344–345, 345*
genetic modification of food, 13
genetic testing and screening. *See* testing and screening for genetic disorders
genetic variance, 103
Genghis Khan, 448
genital herpes, 163
genome, 11. *See also* Human Genome Project
genomic imprinting, 262, 368
genomics
    annotation to find genes in a sequence, *351–352, 352*
    bioinformatics and, 350–351
    cases, 359–360
    challenges for, *357, 357–359*
    comparison of selected genomes, *354*
    ethical concerns, *356–357, 357*
    gene function identification, 352, *353*
    genetic disorders study, *355, 355–356*
    genetic mapping
        linkage, *343–344, 344*
        positional cloning, *345–346, 346, 347*
        recombination frequencies, *344–345, 345*
    genome ownership issues, 358
    goals of, *348, 348–349*
    human genome features, *352–354, 354*
    Human Genome Project. *See* Human Genome Project
    new findings, 354
    proteomics, *356, 356*
    sequencing strategies, *349–350, 351*
genotype, 47
George III of England, *87, 87*
germ cells, 31
germ-line therapy, 375
Gerstmann-Straussler disease, 211
GIFT. *See* gamete intrafallopian transfer
Gila River American Indians, 448
Glass, William, *444, 445*
Gleevec, 282
globins, 229
glucose-6-phosphate dehydrogenase (G6PD), 238
glucosylceramides, 18
glycogen storage disease type II, 239–240
GMOs. *See* genetically modified organisms
goitrous cretinism, 225
Golde, Dr., 358
golgi apparatus, 21, *23*
gonads, 151
Griffith, Frederick, 180
GTS. *See* Tourette syndrome
guanine, 185, 187

H
Haldane, J. B. S., 343
haploid chromosomes, 31
haplotype, *396, 397*

HapMap project, 360
Hardy, Godfrey, 439
Hardy-Weinberg Law
    allele and genotype frequency calculation, *439, 439–440, 440*
    assumptions, 439
    autosomal dominant and recessive allele frequency measurement, 440–441
    frequency of heterozygotes in a population, *441–442, 442*
Hayflick limit, 28
hCG. *See* human chorionic gonadotropin
HDL. *See* high-density lipoprotein
HDN. *See* hemolytic disease of the newborn
health care
    genetic testing and screening. *See* testing and screening for genetic disorders
    targeted cancer treatments, 282–283
    treatment of hemoglobin disorders, *234–235, 235*
    treatment of PKU, 222, 224–225
height, inheritance of, *95–96, 96, 97*
helper T cells, 387, 390
heme, 229
hemizygous, 79
hemoglobin
    genetically engineered, 397
    genetic disorders categories, 231
    mutations by nucleotide substitutions, *254–255, 255*
    role in the body, *229–231, 229–231*
    sickle cell anemia, *231, 232, 233*
    thalassemias, *232–234, 233, 234*
    treatment of disorders, *234–235, 235*
    variants, 231
hemolytic disease of the newborn (HDN), 395
hemophilia, 84, 86, 247, 316
hereditarianism, 8
hereditary nonpolyposis colon cancer (HNPCC), *278–279, 286*
heredity studies. *See also* pedigree analysis
    codominant alleles and, *53, 54, 54*
    DNA. *See* DNA
    incomplete dominance, *54, 55*
    inheritable susceptibility to cancer, *268–269, 270–271, 271, 286*
    meiosis and, *52, 52, 53*
    Mendelian inheritance in humans
        independent assortment, *55, 56*
        pedigree construction, *57–59, 58, 59, 60*
        segregation, 55–56
    Mendel's experimental design, *44, 45, 46*
    Mendel's use of pea plants, 43–44
    multiple alleles and, *52, 53, 54, 54*
    principle of segregation, *47–49, 48*
    single-trait cross study, *45, 46–47, 46, 47*
    two-trait cross study
        experimental design and result, *49, 49–50, 50*
        independent assortment, *50, 51*
heritability and phenotype variation, *103–104, 104*
hermaphrodites, 169
Hershey, Alfred, 182
heterochromatin, 350
heterozygous, 49
HGP. *See* Human Genome Project
high-density lipoprotein (HDL), *110, 111*
Hirschsprung disease, 354
histamine, 384, 400
histidine, 221
histones, *191, 191*
Hitler, Adolf, 10
HIV/AIDS, 84, 163, *402, 403*
HLA complex, 396
HLA complex and organ transplants, 43–44
HNPCC. *See* hereditary nonpolyposis colon cancer